Aquatic Resources Management of the Colorado River Ecosystem

Edited by

V. Dean Adams
Vincent A. Lamarra

ANN ARBOR SCIENCE
THE BUTTERWORTH GROUP

This book is to be returned on
or before the date stamped below

UNIVERSITY OF PLYMOUTH

PLYMOUTH LIBRARY

Tel: (01752) 232323
This book is subject to recall if required by another reader
Books may be renewed by phone
CHARGES WILL BE MADE FOR OVERDUE BOOKS

Aquatic Resources Management of the Colorado River Ecosystem

Proceedings of the 1981 Symposium
on the Aquatic Resources Management
of the Colorado River Ecosystem,
November 16–18, 1981,
Las Vegas, Nevada

Sponsored by
Office of Water Research and Technology
(U.S. Department of Interior),
Utah Water Research Laboratory and
Utah State University

Copyright © 1983 by Ann Arbor Science Publishers
230 Collingwood, P.O. Box 1425, Ann Arbor, Michigan 48106

Library of Congress Catalog Card Number 82-72349
ISBN 0-250-40594-6

Manufactured in the United States of America
All Rights Reserved

Butterworths, Ltd., Borough Green, Sevenoaks
Kent TN15 8PH, England

The Colorado River system has often been referred to as
"the most regulated river system in the world." The Colorado
River Basin serves millions of people through agricultural,
energy, municipal and industrial uses, fish and wildlife
activities, and recreation. The symposium was conceived and
organized to allow researchers, private industry, consultants,
water users, regulatory agencies, and concerned citizens the
opportunity to express needs, desires, and concerns about the
vast resources of the Colorado River.

We found that there were a diverse number of problems
confronting the individuals who are involved in the management
of this important ecosystem. A variety of broad topics have
been presented which include: water policy and major
diversions; energy impacts; oil shale development--resources
and impacts; Lake Mead and the other major reservoirs in the
system; the ecology and management of the watershed and the
riparian habitat in the system; fisheries; salinity problems;
sedimentation; eutrophication; flow depletion; and water
augmentation.

This timely symposium brought together many individuals,
representing a variety of disciplines, to discuss and transfer
information appropriate to the needs of the Colorado River
Basin. The results of this symposium, which have been compiled
herein, are an attempt to examine current and projected
effects of water and land management within the Colorado River
Basin and to provide a basis for determining what can be done
to better manage the resources within the total context of
activities affecting the Colorado River Ecosystem.

V. Dean Adams

A project such as this cannot be accomplished without the outstanding help, cooperation, and assistance of many individuals. We want to thank Ms. Mary Ann Nielsen for the tremendous effort she put forth in organizing the initial phases of the symposium and proceedings. Ms. Kathy Bayn also provided invaluable assistance, contributing to the peer review evaluations, editing, typing and analyzing the proceedings, and overseeing the compilation of all the manuscripts. We owe a tremendous debt of gratitude to both of these women. Ms. Donna Lake also contributed significantly to preparing the final manuscript for publication. We wish to recognize her efforts and also the efforts of the many typists, assistants, and graduate students at the Utah Water Research Laboratory who helped with the technical aspects of presenting the symposium, as well as the proceedings.

The Office of Water Research and Technology and the Utah Water Research Laboratory were sponsors of this project and have our sincere appreciation, for without them, a project such as this would not have been possible. Special recognition is given to Dr. L. D. James, Director of the Utah Water Research Laboratory for his support and contribution to the symposium and proceedings.

Much of the success of the symposium we owe to our session chairpersons whom we thank for their leadership, research efforts, and time spent in preparation for the sessions. These session chairpersons were: Dr. J. G. Carter, Dr. A. B. Davis, Dr. J. G. Dickson, Dr. E. R. Harris, Dr. L. D. James, Dr. A. J. Medine, Mr. J. B. Miller, Dr. L. J. Paulson, and Ms. M. E. Pitts.

We also want to thank all of the peer reviewers for the time and effort they gave in their attempt to evaluate and suggest improvements for the submitted papers. Their efforts have definitely enhanced the quality of this publication.

Finally, we express our gratitude to the invited speakers, the panelists, and the many contributing authors, who, by their research efforts, presentations, and written subject matter have provided us with excellent material regarding the Colorado River System, its resources and management.

V. Dean Adams is Associate Professor and Head of the Division
of Environmental Engineering, Utah Water Research Laboratory,
Utah State University. He received his BS in Chemistry from
Idaho State University and his PhD in Organic Chemistry from
Utah State University. Dr. Adams has been involved in numerous research projects, including effects of organic ligand
complexation on heavy metals, evaluation of commercially
available home water purifiers, and blue-green algae control
in the protection of reservoir water quality against toxic
organics. He is the author of several publications, technical
reports and papers, and co-author of Analytical Procedures
for Selected Water Quality Parameters. Dr. Adams is a member
of many professional societies, including the American Chemical Society, Water Pollution Control Federation and Utah
Water Pollution Control Association.

Vincent A. Lamarra is Adjunct Professor, Department of Civil
and Environmental Engineering, Utah State University, and
Co-director of the Ecosystems Research Institute. He received
his BS in Natural Science from Fresno Pacific and his PhD in
Limnology from the University of Minnesota. Dr. Lamarra is
currently involved in several research projects, including
the use of power plant effluent waters in fish aquaculture,
and biological and chemical effects of hatchery effluent on
stream ecosystems. He is the author or co-author of several
publications and papers. Dr. Lamarra is a member of the
American Society of Limnology and Oceanography, American
Fisheries Society, International Association of Theoretical
and Applied Limnology, and Lake Management Society.

CONTENTS

ix

Part 3: Main Streams

Part 4: Watersheds

Part 5: Oil Shale Development

Part 8: Fisheries

Part 9: Water: Economics

Part 10: Panel Discussion

PART 1

KEYNOTE SPEAKERS

CHAPTER 1

THE COLORADO, A RIVER FOR MANY PEOPLE

Bill Plummer
    Bureau of Reclamation
    Boulder City, NV

The Colorado River is a life-sustaining water resource that winds more than 1,400 miles through seven stages and two countries. Because this ribbon of water descending from the snowcapped Rockies to Mexico's Gulf of California is the primary source of water for much of the basin it drains, the wellbeing of many communities in the basin is directly related to management of the river.

The Colorado's drainage basin encompasses 242,000 miles^2 in the United States, or one-twelfth of the country's land area, and 2,000 miles^2 in Mexico. Within the basin, the River's waters are used for irrigation, municipal and industrial purposes, hydroelectric power generation, fish and wildlife enhancement, and recreation. In addition, some of its waters are exported outside the basin to densely populated metropolitan areas.

Many water resource people have labeled the Colorado "the World's most regulated river." This need for control is the result of: 1) the scarcity of water in the areas served by the River, and 2) a long history of competition and struggle for this resource.

Use of the River's waters is regulated by various legislative and other legal acts -- known collectively as the "law of the river"-- that have been implemented through the years. The Secretary of the Interior operates the Colorado, in consultation with the seven basin states, according to the mandates of these documents.

The first of these major documents was the Colorado River Compact, dated November 24, 1922. This compact divided

use of the River's water between the upper and lower basins
at a point about a mile below historic Lee's Ferry, near the
Paria River, in northern Arizona. In essence, it apportioned
7.5 million acre-feet of water annually to both the Upper and
Lower Basins and paved the way for construction of works to
control, regulate, and utilize the stream.

In 1928, the Boulder Canyon Project Act, which authorized
Hoover Dam, apportioned the Lower Basin's 7.5 million acre-
foot entitlement in a manner which annually provides California with 4.4 maf; Arizona with 2.8 maf, and Nevada with 0.3
maf. This apportionment was reaffirmed by a 1964 United
States Supreme Court decision.

In 1944, the United States entered into an international
treaty with Mexico which assured that country 1.5 maf of
Colorado River water annually. The Upper Colorado River Compact, signed in 1948, permitted Arizona to use 50,000 acre-
feet of water annually from the Upper Basin, and apportioned
the remaining water among the Upper Basin states. By percentage, that distribution was: Colorado, 51.75 percent; New
Mexico, 11.25 percent; Utah, 23 percent; and Wyoming, 14 percent. This compact also provided that the Upper Basin states
could divert more than their entitlement if return flows were
sufficient to make up the delivery requirement to the Lower
Basin states and Mexico.

After all these allocations had been made, the availability of water began to be questioned. The 1922 apportionment
between the basins had been based upon river data collected
between 1906 and 1921 -- 15 years which now appear to have
provided the system with more water than might be expected
for the long-term average runoff.

The implications of this revised riverflow information
soon became obvious. After delivering the guaranteed 7.5
maf average annual release to the Lower Basin, the Upper
Basin may or may not have 7.5 maf available for use.

As it became evident that less water was available than
earlier supposed, as agricultural and municipal water supply
projects became a reality, and as population and use rates
in the southwest soared, prudent management of the Colorado
became an absolute necessity. Its waters, after all, are
used for many purposes, and serve many diverse interests,
throughout the Basin.

For instance, more water is delivered for agriculture in
the basin than for any other need. But because of a shorter
growing season, the agriculture of the Upper Basin is generally less intensive. In the Lower Basin, agriculture is

4

almost entirely dependent on Colorado River water. The availability of this water, coupled with a year-round growing season, has resulted in some of the world's most productive farmland.

Municipal and industrial water is also a need in both basins. However, the demands for such water in the Lower Basin are currently about 10 times greater than the demands of the Upper Basin. When the Central Arizona Project begins delivering water to Phoenix and Tucson in a few years, and the Southern Nevada Water Project delivers additional water to Las Vegas, that figure may go higher.

Flood control is also a need of both basins. Here again, mainstem flooding of the Colorado has been a far more serious problem on the Lower Basin, particularly at the lower end of the River.

Hydroelectric power generation is another use for the River's waters in both basins, although this is considered an important by-product of the storage and delivery of water for other purposes. During 1980, the Bureau's hydroelectric powerplants on the Colorado River and its tributaries generated 13 billion kilowatt-hours of energy -- enough to supply the needs of 4.3 million people for one year. Much of this power was generated in response to peak demands for electricity. Hydroelectric plants are extremely valuable sources of electricity because of their ability to provide immediate peaking power without costly warmups.

Recreation is an important fringe benefit of our water resource projects in both basins. Two of the most significant recreation areas are Lake Powell, in the Upper Basin, and Lake Mead, in the Lower Basin. Together, these areas attracted approximately seven and a half-million visitors in 1980.

The Colorado River and its adjacent riparian areas continue to provide valuable habitat for fish and wildlife. Trout, largemouth and striped bass, and channel catfish are the dominant gamefish population in the river basin. The Colorado River flyway has long been recognized as a major migration and wintering area for many game and nongame species of birdlife. Working with the Fish and Wildlife Service and state and local agencies, Reclamation has helped improve fish and wildlife habitat along selected sections of the river. Beal Slough, a filled backwater renovated to enhance fish and wildlife values on the Lower Colorado River near Needles, California, is an example of this type of work. Modification of the powerplant intakes at Flaming Gorge Dam in Utah is another example. This work was performed to help restore the blue-ribbon trout fishery on the Green River below the dam.

It is obvious from these very brief user summaries that
"managing" the Colorado River means different things to
different people. To some, it means a life-sustaining supply
of water, to others, flood control for protecting their property, and to still others, it means creation or enhancement
of significant recreational resources.

All Colorado River water users would probably agree that
management of this river for many people has changed it from
a natural menace to a national resource. What some of them
overlook is that the benefit they derive from the River is
just one of many provided by our multiple-use management programs. The use problems stem from the fact that all of these
benefits cannot be fully satisfied without some conflict.

Solving these conflicts is a difficult task, but not an
insurmountable one. Reclamation does have defined responsibilities for managing the River. And we perform the task,
without owning a drop of the administered water, for the
benefit of the people comprising the communities and states
of the Basin. In performing this task, we coordinate and
consult a great deal with other Federal agencies, state agencies, water users, and other interested parties.

Consider Reclamation's responsibilities and priorities
for managing the Colorado River. Current operation of the
Colorado River by the Bureau of Reclamation is based largely
on forecast of runoff, available storage, and requirements or
demand for water -- all according to applicable laws.

As required by the Colorado River Basin Project Act,
operation of Reclamation reservoirs in the Basin is coordinated under long-range criteria issued in June, 1970. These
criteria state that the objective shall be to maintain a
minimum release of 8,230,000 acre-feet of water from Lake
Powell annually, and also state that a reservoir operating
plan must be developed annually for the Colorado River.

Under these criteria, the Secretary of the Interior
determines how much water must be retained in Upper Basin
reservoirs each year in order to meet obligations to the
Lower Basin without impairing the Upper Basin's consumptive
uses. When Upper Basin storage is greater than the amount
needed, releases above the minimum are made to maintain, as
near as possible, active storage in Lake Mead equal to active
storage in Lake Powell.

A third facet of the criteria is that they provide that
all reasonable consumptive use requirements of all mainstem

users in the Lower Basin will be met without cutback until such time as deliveries commence from the Central Arizona Project.

Releases in excess of downstream water requirements were made in 21 of the 27 years of operation between completion of Hoover Dam and completion of Glen Canyon Dam. With closure of Glen Canyon Dam in March, 1963, the storage capability of the Colorado River reservoir system was essentially doubled. While Lake Powell was filling, essentially all excess water in the Colorado was put into storage -- an annual average of two million acre-feet. However, a combination of three successive years of above average flow, coupled with the June 1980 filling of Lake Powell, resulted in nearly five million acre-feet of water in excess of downstream requirements being released from lower Colorado River dams from May, 1979 to January, 1981.

These excess releases were made in accordance with provisions of the Boulder Canyon Project Act. This legislation tends to alleviate one of the Lower Basin's most pressing management conflicts: when water should be stored for future use, and when water should be released to provide flood storage space in the reservoir.

Basically, the Boulder Canyon Project Act states that flood control will be the number one priority in operating Hoover Dam. Water storage and delivery and hydroelectric power generation have lesser priority.

The criteria for operating Hoover Dam under flood control conditions have been developed jointly by Reclamation and the U.S. Army Corps of Engineers. These criteria are reviewed and modified from time to time as conditions warrant. A public involvement program was conducted in 1979 to obtain updated input from the many people affected by the Dam's operation. A report citing the findings of this program should be published within the next few months.

The report stresses a plan for controlling flood flows to nondamaging levels while simultaneously making optimum use of these flows for hydroelectric generation. It also integrates the Upper Basin reservoirs into the overall flood control capability of Hoover Dam and Lake Mead.

Incidentally, all these excess flows were released through our hydroplants on the lower river. Although we release water only when it has been requested, or when it is dictated by flood control requirements, we do put the water to work as it flows through the system.

7

Mexico also used these excess flows for leaching, double cropping, and irrigating additional lands. To the extent we were able, we scheduled these excess flows to try to accommodate Mexico's needs and use capabilities.

In January of this year, with a below average runoff forecast, we cut river flows back to the routine condition of water being released only in sufficient amounts to meet downstream requirements. And, although we are temporarily relieved from the threat of high flood control releases, we still foresee a fairly high probability of encountering a similar situation during the next few years.

Encroachment upon the river floodway, particularly in the Lower Basin, has become a serious problem in recent years. Much of this land is in private ownership and not federal control. In the absence of routine flood control releases, development has occurred in and near the floodway that was designated to accommodate such releases. When the Central Arizona Project begins operation in 1985, the additional water used will significantly reduce the likelihood of having to operate the reservoirs under flood control regulations.

Legislation also defines the position of fish and wildlife interests in the operation of the River. The Fish and Wildlife Coordination Act requires that planning for any federally funded water project must include consideration of the project's impact on fish and wildlife. We also operate under direction of the National Environmental Policy Act of 1969 and the Endangered Species Act, and consult regularly with state and federal fish and wildlife authorities.

As an example of our commitment to fish and wildlife interests, consider a study being conducted at Lake Mead. Each spring the water orders from downstream irrigation districts increase. Unfortunately, this coincides with the annual bass spawning period. These increased releases generally lead to a decrease of the lake level, a condition which may affect the bass spawn. Although the reservoir must operate according to the established priorities, a five-year study of the Lake Mead bass population has been initiated in cooperation with the states of Nevada and Arizona which will attempt to identify the role of fluctuating lake levels on the bass population.

For many years we were concerned primarily about the quantity of water available in the River. More recently, we have also become concerned about the quality of this water--specifically, the salinity of Colorado River water. The push for salinity control was given emphasis when Mexico complained

8

in the early 1960s about the increase in the salinity of water
being delivered to them under terms of the 1944 treaty. After
several years of negotiation between the two countries, and
adoption of interim control measures, we entered into an
international agreement for a permanent and definitive solution relative to the salinity of Colorado River water delivered to Mexico.

In order to meet the terms of the agreement, the Colorado
River Basin Salinity Control Act was signed into law in 1974.
The Act had two parts, Title I and Title II. The Title I
portion was concerned with salinity control measures upstream
of Imperial Dam. Although the act contained no provisions for
fish and wildlife mitigation measures, Title I has since been
amended to include this provision; to date, no mitigation
measures have been included for Title II.

The heart of the Title I measures is the Yuma Desalting
Plant, which will remove enough salt from irrigation return
flows to make the water acceptable for delivery to Mexico.
Preparation of the plant site, four miles west of Yuma, is
nearly complete. Contracts have been awarded for the production of the reverse osmosis membrane units, and one of the
two manufacturers has been notified to begin production.

Other water salvage operations of the Title I work are
also nearing completion. Lining the first 49 miles of the
Coachella has now been completed, and our protective and
regulatory well field near the U.S.- Sonoran border is partially operative. When completed, Title I features are expected to make over 300,000 acre-feet of additional water
available for use in the arid Southwest. Title II measures
are designed to reduce salt inflows into the Colorado from
particularly saline areas upstream of Imperial Dam.

Four projects -- two in Colorado, and one each in Utah
and Nevada -- were originally authorized for construction.
Two of these, the Grand Valley and Paradox Valley units in
Colorado, are under construction and advance planning is
underway on the Las Vegas Wash Unit in Nevada. There is no
activity on the Crystal Geyser, Utah, Unit. Planning studies
are also underway on twelve additional areas -- four in
Colorado, five in Utah, and one each in Wyoming, Nevada and
California.

About one-half of the dissolved salts in the River today
can be attributed to man's development and utilization of
this resource. Increased salinity lessens the quality of the
water for both agricultural and municipal use. For every
milligram per liter, we can reduce the salinity of water
arriving at Imperial Dam, a benefit of about $472,000 may be

realized by water users.

What about future management of this highly complex river
system? Despite the fact that we have more than the equivalent of three years of average runoff stored in Colorado
River reservoirs, and despite our concerns over potentially
high flows in the future, there is no overlooking the fact
that eventually we must deal with water shortages in the
basin. While we cannot absolutely predict when, how long,
or how severe these shortages may be, our studies indicate
that there is a strong possibility of significant shortage
in the Colorado's water supply within the next twenty to
twenty-five years.

Because of the importance of the Colorado River for the
many millions of people and the wildlife it serves, we must
plan for and implement measures that will enable us to minimize the effects of drought periods and make maximum use of
the water available during high runoff years.

There are several avenues available to stretch present
uses of water to help meet the dry times. Principal among
potential water saving methods are better onfarm irrigation
efficiency, lining water conveyance facilities, perfecting
water transport schedules, recycling return flows, and managing high water-consuming vegetation, to mention only a few.
Water supplementing techniques, such as cloud seeding or
upper watershed management have also been proposed.

Over the past several years, reclamation has been
actively developing "Irrigation Management Services" (IMS).
This is a method of providing the farmer with solid recommendations for managing his irrigation practices to assure
effective use of the land and water resources. The program
basically determines when and how much crops should be irrigated for maximum production and maximum water use. Ultimately, we foresee when an irrigation district's water will
be based more precisely on crop need and water holding capability of individual fields rather than on convenience and
historic practice.

Future trends in water use have been developing for some
time. Present use must frequently be reexamined to ascertain
that these trends will preserve water quality and at the
same time meet people's needs. Thanks to the existence of
reservoirs like Lake Mead and Lake Powell, Colorado River
users have both a reliable and a sufficient water supply for
some years to come.

But many questions are being asked about future uses of

10

Colorado River water. To cite a few: Should we use the
water to irrigate more lands? Expand cities and industries?
Cool thermal electric plants? Develop shale deposits?
Improve fish and wildlife habitats? Should we stretch the
water supply by encouraging a shift from crops that use a lot
of water to those that use less water? The answer to these
questions must come from the basin states. They must decide
on priorities, within the "Law of the River," and thus direct
future water use. We are looking forward to long-term coordination with water, power, wildlife, and land interests to
manage the Colorado River to meet the water supply needs of
the basin states.

CHAPTER 2

COLORADO RIVER MANAGEMENT
TO ENHANCE AQUATIC RESOURCES

Bob Jacobsen
U.S. Fish and Wildlife Service
Salt Lake City, Utah

River management problems as they relate to wildlife and
fish will be dealt with first. Problem identification is
really quite simple. It is man's uses of water versus fish
and wildlife uses of water. The solution to these problems
is similar to placing man on Mars: it is going to require a
great deal of scientific exploration to achieve a balanced
management of water and fish and wildlife resources.

Man's uses of the Colorado River are well documented
and they will be further documented in this symposium. Traditional uses, such as dams for irrigation purposes, municipal and industrial purposes are well known, but all too
often these projects and uses of waters have continued to
result in losses of fish and wildlife. It is an insidious,
ever growing loss. Currently, we are beginning to see rapidly
expanding losses of fish and wildlife habitat. The expanded,
unimpeded coal leasing program, oil and gas leasing, the oil
shale program, uranium development, and power production are
just a few of those uses by man that we are all too aware of.
Losses of riparian habitat are expanding as development occurs.
Riparian habitat losses can be lost through transportation
and sand and gravel operations which provide for increased
populations. The projections for population growth in the
Colorado River system point to potentially three and a half-
million people living in the upper basin. Salinity control
is another problem which may very well result in additional
losses of fish and wildlife resources.

Man's needs are readily understood in the Colorado River
system. Institutionally, they are well known. There are
local organizations and support. There is industrial support
for uses of water. There are state water laws, compacts and
agreements, all very well understood. However, fish and
wildlife needs and uses remain poorly understood to this day.

Traditionally, most everyone in the Colorado River drainage area thought that wildlife was a vast, expendable resource. Tremendous losses have taken place and now listings of species as endangered or threatened are appearing. There has been a poor understanding of the biological requirements of these species. To this day we still have a poor understanding of the numbers of fish and wildlife: deer, for example.

All too often there is never enough time to address a project adequately in terms of what its true impact will be to fish and wildlife. And so the Fish and Wildlife Service and state game and fish agencies are viewed as organizations that are in opposition to the developer when we are merely saying we need to study, study, study. We are forced to try and baffle people with rhetoric. Our posture quite often has been to oppose projects which have less than satisfactory data with less than satisfactory data. There is a lack of true grass roots public support. Thorough public understanding is lacking on fish and wildlife values. For example, the Colorado River Squaw fish, Humpback Chub, Bonytail Chub are considered trash fish by people in Montrose, Colorado. The snail darter has become a symbol to the developer because it stood in the way of progress, and then, when all of a sudden progress took place, we began finding snail darters everywhere. Everybody points to the snail darter. Environmental groups are totally supportive of fish and wildlife values, but quite often for the wrong purpose, using the Endangered Species Act to stop projects when there is no other way to do so. However, I should point out that there is national support for fish and wildlife. Polls recently conducted show that fish and wildlife values are of utmost concern to the United States public. Therefore, I would propose that in working with the Colorado River system, we must look at fish and wildlife as a use of water.

There are a number of laws protecting fish and wildlife resources: state and federal laws. Most of these are poorly understood except for hunting laws. Everybody knows full well that you are not to hunt out of season. However, what acts protect the wildlife when you are not hunting is poorly understood. The Endangered Species Act, Fish and Wildlife Coordination Act, Migratory Bird Treaty Acts, Bald Eagle Act, and the Clean Water Act, Section 404, all protect fish and wildlife. Defining just what species we are actually concerned with in the Colorado River system becomes a problem: trout versus squaw fish, consumptive versus nonconsumptive species, deer versus dicky birds. Consider aquatic resource management. In the biological field we tend to separate aquatic and terrestrial resources, but they are very interdependent. Although this symposium is dealing with aquatics, the same issues apply when dealing with terrestrial resources.

Instream flow is really the bottom line in dealing with
aquatic resources. There are legal problems. Few states
recognize the value of instream flows for fish. However,
Colorado is one state in the Colorado River Basin that actually
does recognize instream flow and has the legal mandate to
protect instream flow for fishes. Instream flows are poorly
understood by most developers and generally the public. When
looking at aquatic resources and instream flows, often the
developer's standpoint is that if the lowest flow occurred in
1922, that is all the water you need to protect fish. The
dynamics of instream flow issues need to be recognized by
everyone. These issues include: water quality; watershed
inputs in terms of sediments; particulate organic matter
and nutrients; flow regime; physical habitat structure, such
as channel form; substrate distribution; and riparian vegetation.

Certain specific action has been taken to deal with
instream flow issues. In 1980, due in large part to Bill
Plummer's efforts, we entered into an agreement which was
signed by the Department of the Interior, the Governor of
Utah, and the Central Utah Water Conservancy District to
recognize instream flows for fish. We are working on a number of Bureau of Reclamation projects, state projects such as
the White River Dam, private projects such as Rawley Fisher's
Juniper Springs, Cross Mountain Project, and other private
projects to determine how the projects can proceed and still
provide fish and wildlife habitat. We are looking specifically at the endangered species problem in the Colorado
River system.

The Bureau of Reclamation, Bureau of Land Management,
National Park Service and Fish and Wildlife Service have
funded a long term study (over two years) of the Colorado
River fishes in terms of trying to find out what their life
history is and what the flow requirements are, not only for
those that are listed but also for those that are in danger
of becoming extinct. We can no longer afford the time to go
out and look and study in the field for the appropriate amount
of time precisely what a project is going to do in terms of
regional impacts on fish and wildlife. Thus, we are employing
the latest in computer technology and considering rapid
assessment methodology. We've developed a map indexing system
for the states of Colorado and Utah. This system, which is
also being used in other basin states, allows the user to
look through the computer at what maps might be available for
fish and wildlife resources in a given area. We are considering a water-for-energy computer model, which will give the
user an opportunity to look at a project and translate the
flow all the way down the Colorado River system.

15

Perhaps more important than these uses of computer technologies is an early input into planning. We are getting
involved in a Bureau of Land Management planning for billions
of acres in an effort to get fish and wildlife values included
before final decisions are made. One question relating to
future planning is, "Are we in time to make a difference?"
Quite often, we feel that we are losing our resources by bits
and pieces when in fact, it is a slow, insidious loss that is
hardly recognizable. However, right now we are faced with
development in massive proportions and so we need to deal with
that. We are looking at regional environmental impact statements for overthrust oil and gas leasing. We are looking at
regional environmental impact statements for coal development.
We are presently participating in a regional environmental
impact statement for synfuels development in the Uintah Basin.
We are also looking at pipelines for synfuels delivery. Communication problems between fish and wildlife resource managers
and the developers are a vital concern. There are institutional barriers between universities and federal agencies and
among federal agencies, and, in some cases, state/federal
relations are not the best. However, there are some positive
actions which will be completed within the near future.

We will soon be completing our field studies on the
Colorado River endemic species. Field work will be completed
by January, 1982, and reports issued. We will be preparing
a conservation plan for the endangered species of the Colorado
River system, which will largely deal with how we can protect
and preserve those species that are so near extinction, in
some cases, and still allow water development to take place.
We are about to complete our data entry into various computer
systems which should be useful to many state, federal and
private agencies.

Managing fish and wildlife resources in the Colorado
River has a long way to go. Many communication problems and
problems in getting public recognition of fish and wildlife
resource needs still exist and hopefully, this symposium will
at least add information which can be useful to all of us.

PART 2

RESERVOIRS

CHAPTER 3

SALINITY AND PHOSPHORUS ROUTING
THROUGH THE COLORADO RIVER/RESERVOIR
SYSTEM

Jerry B. Miller
David L. Wegner
Donald R. Bruemmer
      Upper Colorado Region
      Bureau of Reclamation
      Salt Lake City, Utah

INTRODUCTION

The development of storage reservoirs on the Colorado
River system, accompanied by the water and land use development
during the past 30 to 40 years, has brought about significant
changes to the physical, chemical, and biological balances of
the Colorado River Basin.

Although sediment transport, temperature, biological
productivity, and light penetration are mentioned, we have
chosen to focus on salinity and phosphorus relationships in
the reservoir sequence of Fontenelle, Flaming Gorge, Lake
Powell, and Lake Mead (Figure 1).

A river/reservoir system is never in a state of static
equilibrium, but is dynamic in its response to changing
hydrological and chemical conditions. Each of the four
reservoirs that we are examining has, to varying degrees
altered this dynamic equilibrium, requiring the system to
establish a new balance.

This paper is an overview of the complex physical and
biological interactions which presently define the Colorado
River Basin.

In order to develop a perspective of the influence of
the four-reservoir sequence, it is necessary to take into
consideration the climate, geology, and hydrologic character
of the river basin.

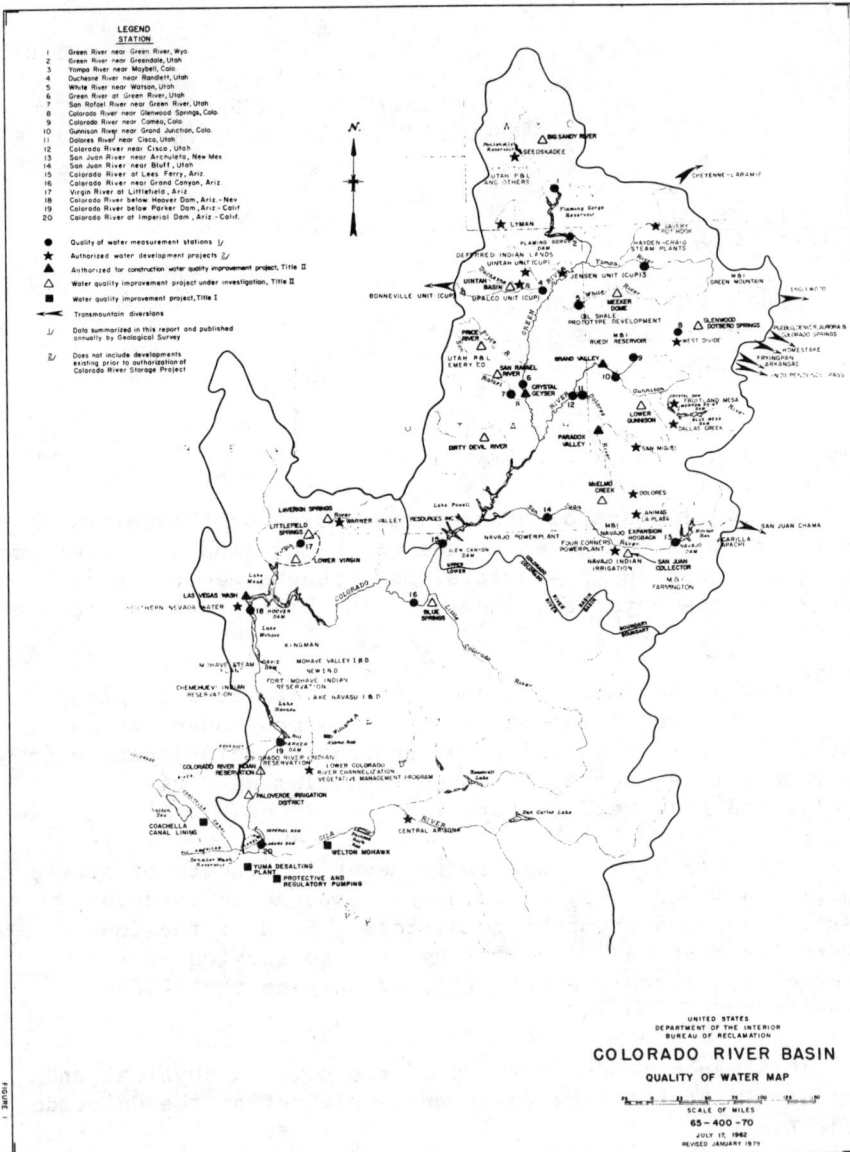

Figure 1. Colorado River Basin: Quality of Water Map.

Climate

The Colorado River Basin ranges in elevation from sea
level to over 1450 m (14,000 ft). The component tributaries
flow through complex mountain systems, elevated plateaus,
and deserts with all but the mountains being primarily
arid. While the greatest climate contrast is between the
desert province in the south and the mountain province to
the north, significant local variances are also present.
Temperatures may vary from -40 to 120° F seasonally, with
precipitation ranging from 6 to 60 in. per year. In the
majority of the river basin's surface area, the evaporation
potential far exceeds local precipitation. This high
evaporation rate concentrates the dissolved minerals in the
remaining water and is a factor in determining the high
salinity values in the Colorado River.

Geology

The geology of the Colorado River Basin is as varied as
the climate. The igneous and metamorphic rock forming the
headwaters region produces cold, crystal clear streams often
lacking in sufficient dissolved minerals to support a diverse
aquatic community. However, as the river flows downstream
it contacts marine deposits containing salts and fine-grain
sediments which can over a few miles change the pristine
streams into torrents of mud, salts, and nutrients.

Large quantities of salts, sediment, fossil fuels, and
evaporite minerals are available, particularly in numerous
marine deposits of the geosynclinal basins. The flow regime
of the basin has mobilized many of these salt and phosphate
deposits. In those areas where the salt has become most
mobile, either naturally or due to man's influence, salinity
control projects have been designated.

In addition to the readily available salts, deep
geosynclinal basins and impermeable aquicludes temporarily
isolate highly saline, static, ground water from the hydrologic
forces. Control of the major mobilized salt sources in the
14 USBR designated Colorado River salinity control project
areas [1] is very important to the future development
of the remaining water resources. It is equally important
that with the development of the mineral and energy resources
within the basin, care be taken not to mobilize presently
static saline systems.

Hydrology

Wet and dry cycles have played a significant role in

bringing about the development of the Colorado River/Reservoir complex. In the past, the annual flow of the river has varied from less than 6 million acre-feet to over 20 million acre-feet per year [1]. The reservoir system allows storage of sufficient water to maintain the flows of the river to meet downstream needs during dry periods.

The construction and filling of the mainstem reservoirs of the Colorado River Basin have brought about significant changes in the hydrologic cycle. In addition to the major reservoirs, numerous smaller reservoirs are found throughout most of the tributaries. Since major storage began with Lake Mead in 1935, to the conclusion of the initial filling of Lake Powell in 1980, the reservoirs in the Colorado River Basin have developed a storage capacity equal to approximately four times the total average annual flow of the entire Colorado River (Figure 2).

Reservoir Limnology/Downstream Hydrology

During initial reservoir filling, the rising waters inundate soils rich in nutrients, organic matter, and salts. Initially, the rising water will leach out the soils, particularly during the shoreline wave action phase. Quantities of water also go into bank storage, varying with the permeability of the underlying geology.

The typical reservoir of this system is usually highly productive during initial filling. As filling continues, the inflowing sediment is distributed over the reservoir area and temporarily locks in the underlying nutrients and salts. Leaching of those materials below the wind mixed epilimnion is substantially decreased due to the lack of mechanical action. Eventually a chemical balance will develop between the water column, sediments, and bank storage. Again, a fluctuating reservoir is not a static environment, but is in a state of dynamic equilibrium and responds to changes in the hydrologic, climatic, and chemical conditions.

The most readily noted impact from the reservoir is the change in the suspended sediment, water temperature, and flow patterns downstream. Figures 3 and 4 illustrate how the flow and salinity patterns have changed below Flaming Gorge and Glen Canyon Dams.

It is the transition of the inflow conditions through the internal reservoir circulation which determines downstream conditions. The two most significant factors determining

22

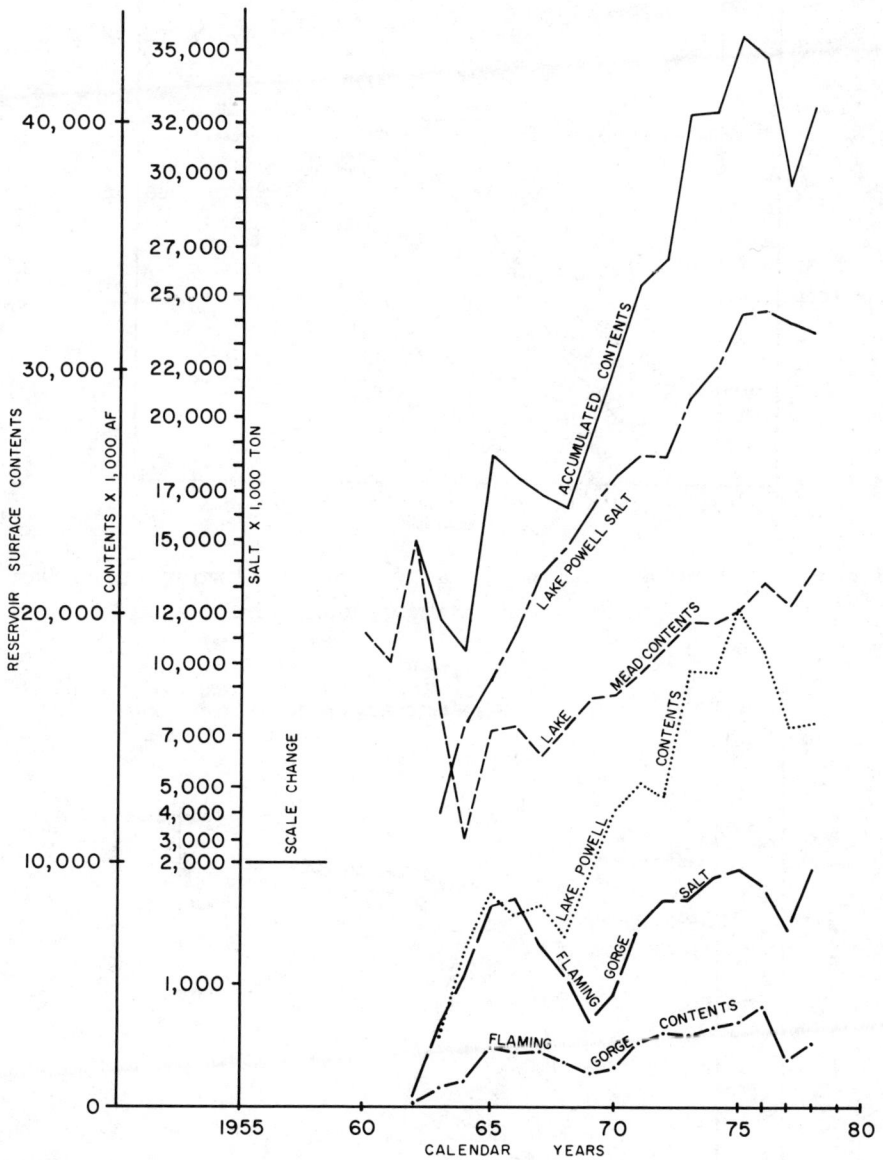

Figure 2. Storage capacity for Flaming Gorge, Lake Powell and Lake Mead

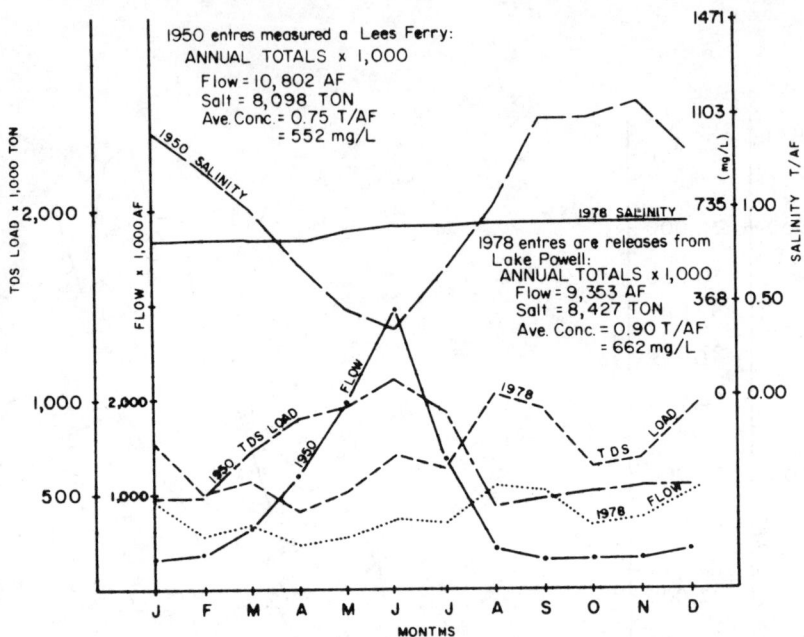

Figure 3.   FLow and salinity pattern below Flaming Gorge Dam

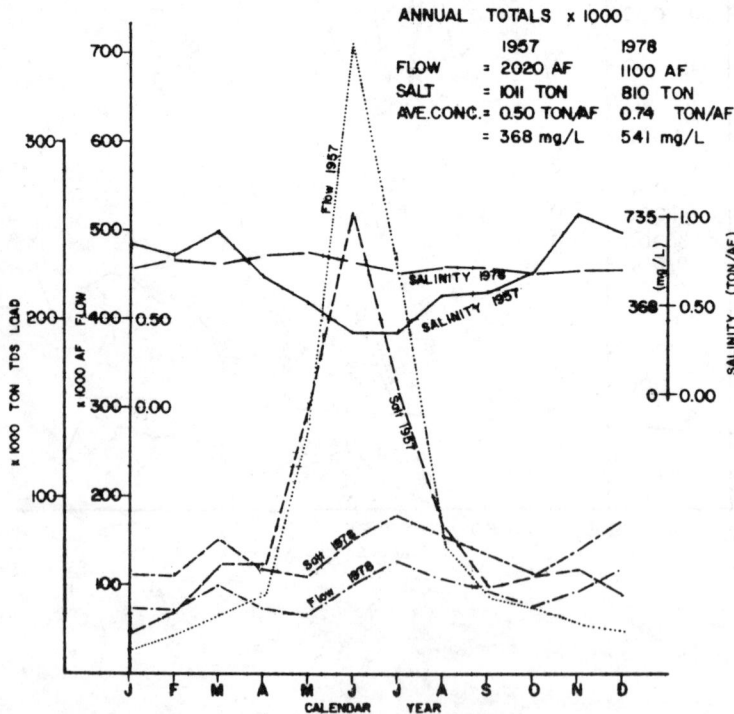

Figure 4.   Flow and salinity patterns below Glen Canyon Dam

24

Figure 6. Salinity trends at Imperial Dam

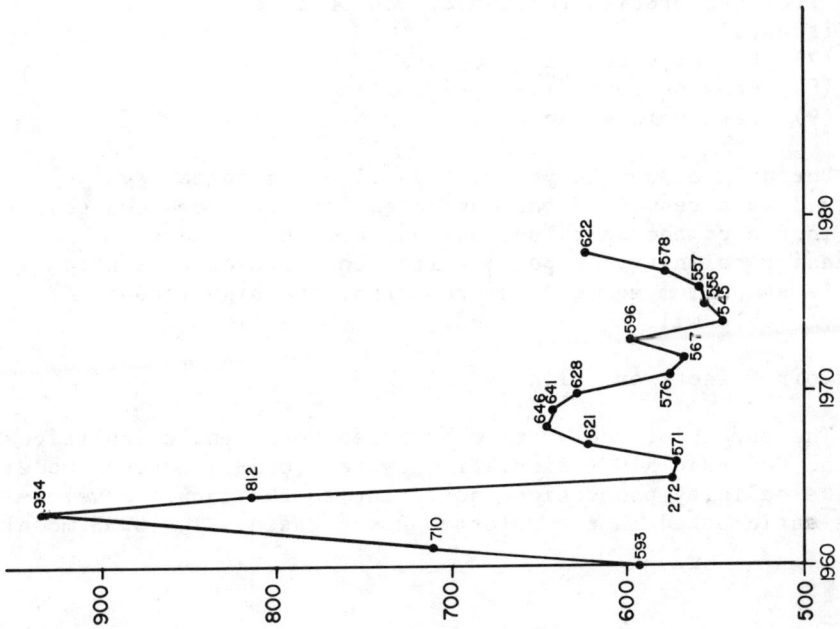

Figure 5. Salinity trends at Lees Ferry

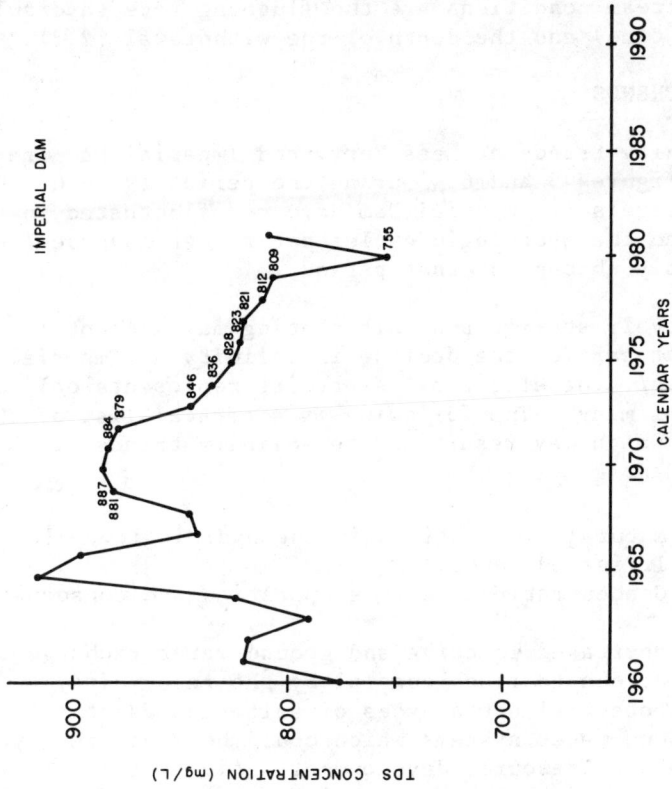

the downstream conditions are the flushing rate (hydraulic detention time) and the depth of the withdrawal [2,3].

SALINITY TRENDS

Salinity trends at Lees Ferry and Imperial Dam are shown in Figures 5 and 6. During the period 1970-80, the salinity levels at Imperial Dam have not fluctuated in relation to the hydrologic cycle, but rather have declined continuously throughout that period.

Reservoir storage and salt routing may account for a significant part of the decline in salinity at Imperial Dam; however, the effect of reservoirs represents only one variable of many. The following is a partial list of the variables which may result in the salinity trends at Imperial Dam:

(1)   Natural fluctuations in the hydrologic cycle.
(2)   Irrigated lands.
(3)   Concentration due to evaporation and consumptive use.
(4)   Decreased leaching and ground water recharge of the flood plain, due to flood control by the reservoirs.
(5)   Potential new sources of salt such as static saline ground water systems which could be mobilized by various natural resource development activities.
(6)   Switching the reporting of total dissolved solids (TDS) from evaporation residue at 180° C to sum of the constituents.
(7)   Salinity control projects.
(8)   Erosion control.
(9)   Reservoir effects.

During the past 20 years, significant changes have occurred as a result of the variables listed. Some changes have increased the salt load while others have led to decreasing salinity. Hopefully from the studies presented at this symposium and ongoing research, the significance of each variable will be addressed.

Reservoir Effects on Salinity

The Bureau of Reclamation has used both hand calculations and the Colorado River Simulation System (CRSS) computer model to make salinity projections according to the future develop-ments anticipated in the Colorado River Basin. The CRSS model

26

is currently based on the assumption of a once-a-month complete reservoir mix and does not allow for analysis of in-
reservoir salinity reactions and movements. Consequently,
the predictions may not be accurate in respect to the actual
salinity processes occurring. Studies now indicate that
reservoir processes which affect salinity include leaching,
precipitation, selective storage and routing, concentration
due to evaporation, bank storage, and the flow weighted
averaging over a period of several years. The Bureau of
Reclamation is currently studying these problems to improve
the salinity modeling capabilities in the near future.

With the closure of Glen Canyon Dam in 1963, the actual
surface storage in Flaming Gorge, Lake Powell, and Lake Mead
increased from about 20 million acre-feet (MAF) to over 43
MAF by 1975. Figure 2 indicates that the majority of this
storage occurred during the relatively wet period of 1968 to
1975. As total storage increased, the annual salinity
fluctuations downstream were dampened due to a 2- to 4-year
hydraulic detention time being developed within Lake Powell
and Lake Mead.

The flow weighted annual salinity trends at Lees Ferry
and Imperial Dam (Figures 5 and 6) show several interesting
variations. From 1960 to 1970 the salinity at Imperial Dam
generally fluctuated at a 1-to-2 year time lag, and with
the same directional trends as Lees Ferry. Contrarily,
for the time period from 1970 to 1980, the salinity levels
at Imperial Dam continuously declined and did not reflect
the increases at Lees Ferry in 1973 or 1977-78. The sharp
decrease in salinity in 1980 was primarily due to increased
(anticipating flood control) releases past Imperial Dam.

The salinity reduction at Imperial Dam may also be
correlated with the fact that Lake Powell selectively
retained the most saline inflows during the dry periods
of 1967 and 1977. Figure 7 shows that at the Wahweap site
(near the dam) the TDS increased at elevations 975 m
(3200 ft.) and elevations 1036 m (3400 ft.) by 260 mg/1
and 350 mg/1, respectively. The differences in TDS between
the reservoir surface and bottom at the Wahweap site (1067 m
to 975 m) during 1967 average about 300 mg/1. This difference
declined gradually to about 140 mg/1 in 1977. This sequence
was repeated as the 1977-78 fall and winter inflows arrived
as an underflow, density current at the Wahweap site in
December through April of 1978. This salinity trend then
reversed and again increased to a difference of over 300 mg/1.
This suggests that Lake Powell can temporarily retain
higher density waters with greater salinity, particularly

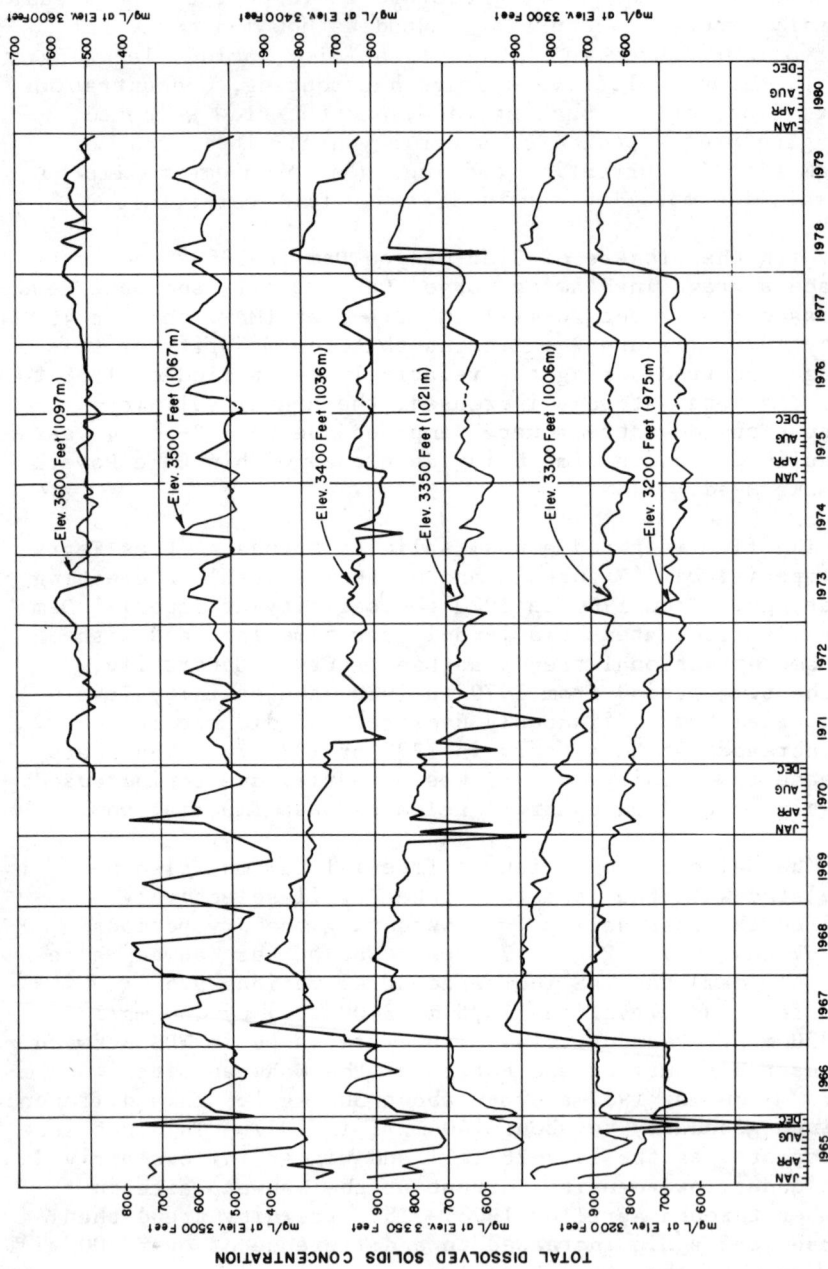

Figure 7. Salinity changes in Lake Powell at the Wahweap Site.

28

during low inflow periods. The depth of the outlet, flushing rate of the hypolimnion, and density (produced by temperature and salinity) of the inflow all contribute to this process. Once the water is stored, dilution, leaching, precipitation, mixing, and bank storage all affect the saline water retained in the reservoir.

A similar salinity sequence has been observed at Flaming Gorge Reservoir. A pronounced salinity profile and chemocline have been documented by USGS [4,5] and Bureau of Reclamation [6] limnological surveys at Flaming Gorge Reservoir. Figures 8 and 9 show that the surface-to-bottom salinity variance near the dam was 300 micromhos in 1971 and 170 micromhos in 1981. This is very similar to the TDS trends observed at the Wahweap site in Lake Powell.

In 1978 a major operational change to a selective withdrawal system was made at Flaming Gorge Reservoir. The summer releases in 1978 were changed from the hypolimnion to the epilimnion for downstream temperature control. Limnological surveys and analysis are still ongoing, and a final determination on the impact of selective withdrawal on the salt routing, water quality, and aquatic ecology of Flaming Gorge Reservoir has not been made. However, the salinity profile shown in Figure 9 indicates that the chemocline in Flaming Gorge Reservoir is decreasing and the hypolimnion may mix during the fall turnover in 1981 and spring turnover in 1982. The addition of the selective withdrawal and the continuation in the changing salinity levels of the chemocline indicate that Flaming Gorge Reservoir is still undergoing minor chemical adjustments towards a dynamic equilibrium.

Bolke and Waddell [4] reported that Flaming Gorge Reservoir increased the load of sulfate and decreased the load of bicarbonate in the Green River from 1963 to 1972. They predicted that the rate of sulfate leaching would decrease after initial filling in 1972.

Fontenelle, the headwater reservoir in the sequence, has a high flushing rate and a deep hypolimnion outlet. No significant variations in the TDS profiles have been observed with depth in this reservoir.

Under present conditions, Lake Mead does not exhibit significant variation in salinity with depth. The maximum conductivity variations in the lower basin of Lake Mead were

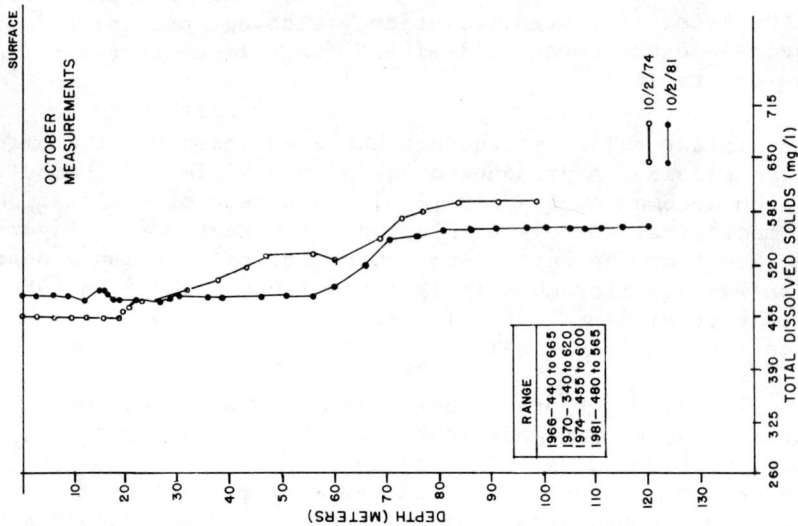

Figure 9.   October salinity profiles at Flaming Gorge Dam.

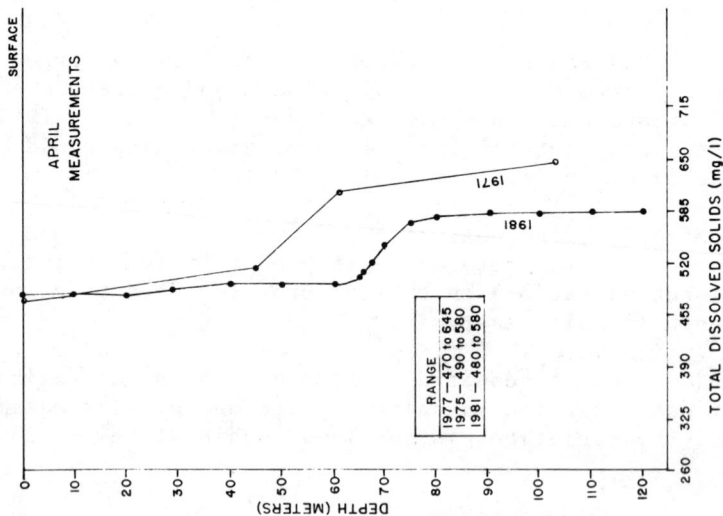

Figure 8.   April salinity profiles at Flaming Gorge Reservoir

30

30 to 65 mg/l in 1978 [7]. This is excluding the inflow density currents shown only in the shallow inflow areas. Lake Mead has a deep hypolimnion withdrawal and low seasonal TDS variance in the Colorado River inflow due to the attenuating effects of Lake Powell.

The reservoirs have a significant impact on the seasonal salinity variation downstream, and also have a cumulative effect on the long-term salinity trends at Imperial Dam. There is evidence that the reservoirs trap bicarbonate due to calcium carbonate precipitation, but also leach sulfate (gypsum) [4]. The long term impacts on salinity cannot yet be precisely predicted because the period of record represents the initial filling for over 50 percent of the storage capacity.

Hydrologically, the reservoir pool levels and operation pattern observed since 1975 are probably typical of the future expected conditions. However, Lake Powell did not complete initial filling until 1980, and Flaming Gorge Reservoir is still undergoing minor chemical adjustments. Therefore, minor modifications to predicted trends may be observed.

The data presented also suggest that Lake Powell and Flaming Gorge can selectively trap the most saline inflows and retain these waters for several years. However, there is no assurance that this process will continue as the river/reservoir system approaches a more steady-state condition.

The Bureau of Reclamation has several investigations ongoing to determine the long term effects the reservoirs will have on salinity. These include:

(1) A two-dimensional thermodynamic/salinity reservoir model of Lake Powell and Mead.

(2) An ion constituent study to determine changes in the chemical characteristics of the water, the causes of these changes, and their longevity.

(3) Limnology surveys of Lake Powell, Lake Mead, and Flaming Gorge Reservoirs.

(4) The continued development and improvement of the Colorado River Simulation System.

(5) A study to improve evaporation estimates for the Colorado River Reservoirs.

31

(6) A study of selective withdrawal from Lake Mead.

As the reservoirs approach a dynamic equilibrium
and our understanding of long term effects on salinity
improves, the accuracy of the Bureau's salinity predictive
tools will also improve.

PHOSPHORUS

Eutrophication refers to the enrichment of nutrients
and increases in primary productivity in water [8]. The
trophic status, as it is reflected by the types and quantities
of algae, largely determines the fishing and recreational
potential, dissolved oxygen, aesthetics, and general water
quality for potable uses. In fresh water it is most generally
considered that phosphorus is the key "limiting nutrient"
which regulates primary productivity and determines trophic
status. [2].

Eutrophication and its relationship to the available
phosphorus supply have been the subject of keen interest in
international research. Numerous empirical models have been
developed to predict the trophic status of a lake based on
the phosphorus budget [9]. It is important to note, however,
that the lake must be phosphorus limited.

These empirical phosphorus models are beneficial,
but their basic assumptions are conditional to well mixed
lakes, not to stratified run-of-the-river reservoirs with
deep outlets.

In the Upper Colorado River Basin phosphorus has
been mined and exported as fertilizer. In addition, vast
formations of oil shale were deposited in ancient eutrophic
lakes in Utah, Wyoming, and Colorado. The oil itself is the
product of the tremendous algal biomass accumulations in
these eutrophic lakes. These eutrophic lake deposits
indicate that phosphorus is possibly more abundant in the
geochemistry of the Upper Colorado River Basin than is
typically found in other river basins.

The seasonal thermal stratifications, deep outlets,
and high flushing rates typical of many reservoirs present
obstacles to applying the empirical phosphorus models. In
addition, the climatic effects particularly on reservoirs
over 4000 ft. in elevation cause seasonal light and temperature
variables which become physically limiting factors to
primary productivity.

32

The factors which determine a lake's primary productivity
have been classified into three groups [10]:   (1) variables
related to solar energy input (temperature and light), (2)
variables in nutrient supply, and (3) variables in lake
morphometry. Flushing rates for hydraulic detention time
have also been recognized as being important in determining
the residence time and availability of phosphorus [2]. In
addition, the depth of the outlet, internal mixing, and
density currents are key parameters which influence the
residence time and physical/biological availability of
phosphorus in reservoirs.

An often overlooked but important factor which physically
induces light limitation is the relationship between the
euphotic zone (sufficient light for photosynthesis) and the
zone of wind-driven turbulent mixing known as the epilimnion.
The deeper the mix zone (epilimnion) relative to the euphotic
zone, the less time the algae spend in the light, the lower
the average amount of light available to the algae, and thus
the lower their net rates of photosynthesis and growth
(Figure 10)[11]. The algae are physically displaced into
the dark portion of the epilimnion.

In reservoirs with deep outlets, the thermocline
tends to migrate downward as the cooler hypolimnion water is
withdrawn. This deepened thermocline should be considered,
as it may induce physical light limitation. The magnitude
of this downward thermocline migration is a function of the
hypolimnion flushing rate and depth of withdrawal. Deep
outlets and high flushing rates may also reduce phosphorus
retention.

A summary of these variables which influence a reservoir's
primary productivity is illustrated in Figures 10 and 11.
It is estimated that on a world wide basis variables related
to solar energy input have the greatest influence on primary
productivity [12].

The depth of the outlet and the vertical placement of
the inflow based on its density (primarily a function of
temperature) may have a major influence on the availability
and retention of phosphorus in a reservoir. Not only must
the algae remain in the lighted portion of the water column,
but the inflowing nutrient supply must also be physically
and chemically available in the euphotic zone as well.

In addition to physical and chemical restrictions
to primary productivity, limitations may also be the result
of biological actions between groups of algae. Generally, a
succession of phytoplankton species occurs throughout the

33

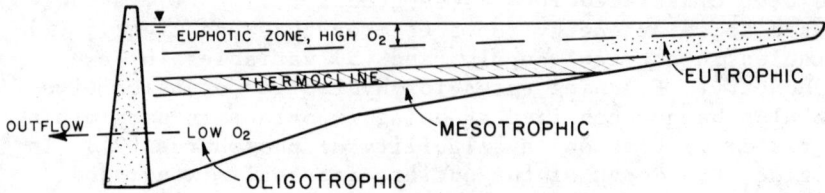

Summer/Fall conditions: Light limitation is determined by the depth of the epilimnion (wind driven mixing) in relation to the depth of the euphotic zone (sufficient light for photosynthesis). The algae are physically mixed down into the dark; thus, the lower the average amount of light available to the algae, the lower the net rate of photosynthesis and growth. The algae themselves may cause most of the turbidity. A hypolimnion outlet draws the epilimnion deeper.

Figure 10.   Reservoir Limnology/hypolimnion outlet

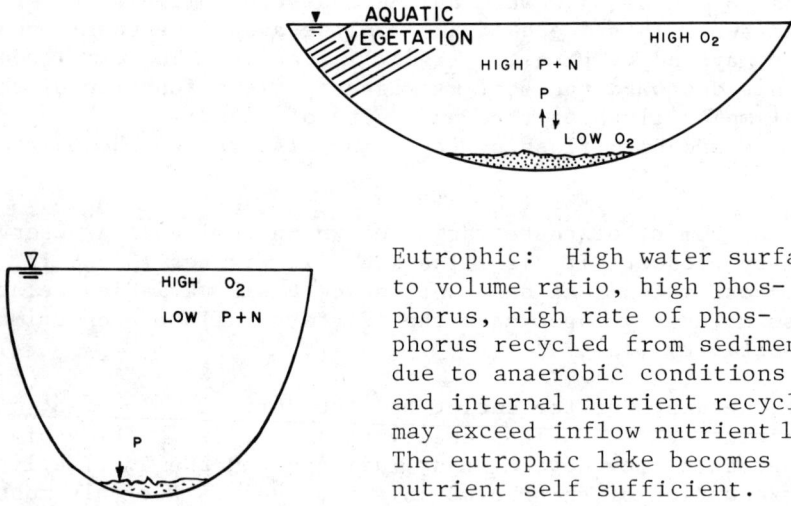

Eutrophic: High water surface to volume ratio, high phosphorus, high rate of phosphorus recycled from sediments due to anaerobic conditions and internal nutrient recycling may exceed inflow nutrient load. The eutrophic lake becomes nutrient self sufficient.

Oligotrophic: Low water surface to volume ratio, low productivity, high phosphorus sedimenting out, HIGH O_2.

Figure 11.   Limnology/morphometry variables that influence primary productivity.

34

year. This sequence of algae succession may trend from desirable species to those which are potentially harmful, such as certain species of blue-green.

Blue-green algae have several distinct characteristics, such as buoyancy and nitrogen fixation, which give them the capability to outcompete more desirable species [13]. Certain species of blue-green algae have the following harmful effect: (1) undesirable as food to grazing zooplankton species; (2) cause reduced light penetration and aesthetics; (3) reduced dissolved oxygen which may mobilize iron and manganese thereby causing additional potable water use problems; (4) the production of organic toxins at death which affect both aquatic and terrestrial life; (5) taste and odor problems in municipal water diversions; (6) production of complex organic compounds which may contribute to the formation of trihalomethanes after chorination; and (7) in excesses they may increase domestic water treatment costs.

The following section is a review of the phosphorus dynamics in the example four-reservoir sequence.

Phosphorus Dynamics

The reservoirs in the Upper Colorado River Basin, which contain natural phosphate deposits in their drainages, are generally eutrophic even though much of the phosphorus is sediment bound and usually biologically unavailable. This could account for the high phosphorus retention rates observed in the reservoirs. Based on the EPA National Eutrophication Surveys [14,15] and U.S. Geological Survey Water Resources data for Water Year 1975 [16], it has been estimated that the four Colorado River Basin Reservoirs can retain 70 to 96 percent of their inflowing phosphorus loads.

Fontenelle Reservoir has natural phosphate deposits in its drainage basin and was calculated to retain approximately 87 percent of the inflowing total phosphorus for Water Year 1975. Reservoirs with bottom releases and high flushing rates tend to retain less phosphorus. Fontenelle's high phosphorus retention rate is apparently due to the mineralized form of the phosphorus which is bound to the sediment and remains predominantly biologically unavailable.

Much of the May/June phosphorus budget is associated with the sediment laden spring runoff and is physically and chemically unavailable in the hypolimnion of Fontenelle Reservoir. Subsequent chemical reductions, organic decomposi-

35

tions, in-reservoir movement, and release of phosphorus from
Fontenelle in July through September may contribute to the
substantial blue-green algae blooms downstream in the Green
River Arm of Flaming Gorge Reservoir.

Flaming Gorge Reservoir is over 90 miles long and
has a low surface area to volume ratio. It is a very
efficient phosphorus trap and was calculated to have retained
84 percent of the calculated phosphorus load for Water
Year 1975. This estimate may be conservative since the
single outflow from the dam is probably a better estimate of
phosphorus releases than can be made from the multiple
inflows. The phosphorus measured at the point sources and
from Fontenelle Reservoir exceeded the measured loads in the
Green River Arm above Flaming Gorge Reservoir. Flaming Gorge
can be classified as eutrophic in both the Green River and
Black's Fork inflow areas, mesotrophic through the middle
section, and oligotrophic in the downstream canyon portion
[17].

Blue-green Algae Relationships

Blue-green algae population levels represent a seasonally
significant impact on the Colorado River Basin reservoirs.
The occurrence of excessive blooms of blue-green algae in
the headwater reservoirs appears to be related to phosphorus
dynamics, reservoir dynamics, and other water quality interactions. Blue-green algae exist throughout the basin and the
effect of the blue-green algae in the basin reservoirs
varies seasonally and annually.

In September 1981, Fontenelle Reservoir experienced
fall overturn and mixed the entire water column. The
blue-green algae were dispersed throughout the water column
with primary productivity being physically limited by light
availability. The extent of blue-green population expansion
is limited by the elevation, temperature levels, and light
intensity.

Blue-green algae blooms are a substantial problem
in both the Green River and Black's Fork arm of Flaming
Gorge Reservoir. Depending on the magnitude of the blue-green
blooms and the climatic conditions of the fall, the cold
water fishery may not be continuously maintained in this
area. Primarily, this is due to the low dissolved oxygen and
high water temperatures. The determining or limiting
factor to the blue-green algae blooms in the inflow area of
Flaming Gorge Reservoir may be phosphorus controlled, but it

is more likely a combination of the length of summer stagnation period, the fall meteorological conditions and the phosphorus supply. In addition, the blue-green algae can often outcompete many of the green algae and diatom species based on their ability to fix nitrogen and control their position in the water column.

A cool wet spring and/or a cool wet fall can greatly reduce the length of summer stagnation due to a reduction in the period that the reservoir is stratified.

The location and stability of the fall thermocline appear to be key factors in determining the timing and strength of the blue-green algae bloom and the amount of reservoir which it affects.

A more thorough investigation of the variables controlling primary productivity in the inflow of Flaming Gorge Reservoir is necessary to determine if phosphate can be reduced sufficiently in the fall to become a limiting factor.

With the shift in reservoir releases from a deep hypolimnion release to a multiple level withdrawal scheme, the availability and concentration of available phosphorus may have been increased. Wright [18] has hypothesized that deep discharge reservoirs may progressively decline in fertility due to withdrawal of nutrient rich hypolimnion water. Conversely, shallow discharge reservoirs may experience an increase in fertiility. This hypothesis and its application to the Colorado River system have been supported by Paulsen [19].

Flaming Gorge Reservoir is also above 6000 ft. in elevation and the fall climate can vary considerably. The reservoir begins to turn over in the inflow areas in early September. There appears to be a relationship between this turnover and the extent of blue-green algae blooms in the fall. Considerable work needs to be done on high elevation reservoirs (such as Fontenelle and Flaming Gorge) to determine the relationship between the fall blue-green algae blooms and available phosphorus supply in the inflow area as a function of the internal phosphorus recycling and meteorological variations.

Lake Powell is the next major downstream reservoir on the Colorado River system. We have estimated that for Water Year 1975, Lake Powell retained approximately 97 percent of the total phosphorus that flowed into it. The reason for this high retention can be directly related to the morphometry of the reservoir basin.

Lake Powell is 170 miles long and is at an elevation of 3650 ft. It was developed within incised sandstone canyons and consequently it is very deep and narrow. This type of morphology and geology is not conducive to high physical or chemical availability of phosphorus, particularly in the horizontal movement down reservoir. The natural phosphorus loads that would have passed Lake Powell have also been reduced by upstream storage in tributary reservoirs, including Fontenelle and Flaming Gorge. However, many unregulated tributaries contribute additional nutrients to the Colorado River above Lake Powell. The high turbidity of the inflow and the short sun day due to the shading effect in the bottom of Cataract Canyon appear to have an influence on primary productivity in the upper end of Lake Powell. No significant blue-green algae blooms have been documented that could impact aquatic or terrestrial life.

Lake Mead, the lowest reservoir in our analysis, has shown significant shifts in its trophic status due to changes in phosphorus availability since 1970. An analysis of the 1975 Water Year data indicates a 69 percent phosphorus retention rate. This lower phosphorus retention rate has been hypothesized to have resulted from several factors, including upstream storage and reduced phosphorus availability due to physical changes in the inflow (primarily temperature) and the deep hypolimnion outlet [7]. The aquatic ecology and nutrient chemistry of Lake Powell and Lake Mead will be further discussed in other symposium papers.

CONCLUSIONS

1. Hydrologically, the reservoir pools and operation did not stabilize until about 1975. Lake Powell completed initial filling in 1980 and the operation of Flaming Gorge Reservoir was changed by the addition of a selective epilimnion withdrawal in 1978. The major chemical and biological adjustments due to reservoir effects are progressing towards an equilibrium. A dynamic equilibrium responsive to hydrologic and climatic conditions is anticipated.

2. The reservoirs have caused major changes in salinity and phosphorus routing in the Colorado River System.

3. The observed salinity trends at Imperial Dam during the 1970 to 1980 period may not be totally understood without an additional period of record.

4.   Ongoing studies would provide the needed information
     to improve predictive capabilities, particularly
     regarding future salinity and nutrient conditions in
     the Colorado River System.

5.   As water and energy resources are developed in the
     Upper Colorado River Basin, and the reservoirs become a
     more important source of water supply.

6.   The relationship between blue-green algae population
     levels, reservoir limnology, and phosphorus dynamics
     must be defined.

7.   Phosphorus retention in the reservoirs above Lake
     Mead has caused a significant reduction in its nutrient
     inflow.

LITERATURE CITED

1.   U.S. Bureau of Reclamation. 1981. Quality of Water,
     Colorado River Basin, Progress Report No. 10,
     U.S. Department of the Interior, Salt Lake City,
     Utah. 190 p.

2.   Vollenweider, R. A. 1968. Scientific fundamentals
     of the eutrophication of lakes and flowing waters,
     with particular reference to phosphorus and nitrogen
     as factors in eutrophication. OECD Technical Report
     DAS/CSI/68.27. 159 p.

3.   Martin, R. G. and Stroud, R. H. 1973. Influence
     of reservoir discharge location on water quality,
     biology, and sport fisheries of reservoirs and
     tailwaters, 1968-1971. U.S. Corps of Engineers,
     Waterway: Experiment Station. Contract
     No. DACW31-67-C-0083. 128 p.

4.   Bolke, E. L. and Waddell, K. M. 1975. Chemical
     quality and temperature of water in Flaming Gorge
     Reservoir, Wyoming and Utah, and the effect of the
     Reservoir on the Green River. U.S. Geological Survey
     Water-Supply Paper 2039-A. 27 p.

5.   Bolke, E. L. 1979. Dissolved-oxygen depletion and
     other effects of storing water in Flaming Gorge
     Reservoir, Wyoming and Utah, U.S. Geological Survey
     Water Supply Paper 2058. 41 p.

6.   U.S. Bureau of Reclamtion. 1981. Limnological
     surveys of Flaming Gorge Reservoir. U.S. Department
     of the Interior. Upper Colorado Region, Salt Lake
     City, Utah.

7.   Paulson, L. J., Baker, J. R., and Deacon, J. E. 1980.
     The limnological status of Lake Mead and Lake Mohave
     under present and future powerplant operations of
     Hoover Dam. Lake Mead Limnological Research Center,
     University of Nevada, Las Vegas, Nevada. Technical
     Report. No. 1. 229p.

8.   Lee, G.F., Rast, W., and Jones, R. A. 1978. Eutrophi-
     cation of water bodies. Insights for an age-old
     problem. Environmental Science and Technology,
     Volume 12, No. 8: 900-908.

9.   Reckhow, K. H. 1979. Quantitative techniques for
     the assessment of lake quality. Department of
     Natural Resource, Michigan State University.
     EPA-440/5-79-015.

10. Rawson, D. S. 1955. Morphometry as a dominant
     factor in the productivity of large lakes. Verh.
     International Verein. Limnology 12: 164-175.

11. Cullen, P. and Rusich, R. 1978. A phosphorus budget
     for Lake Burley Griffin and management implications
     for urban lakes. Australian Water Resources Council,
     Technical Paper No. 31, Australian Government
     publishing service, Canberra.

12. Brylinsky, M. and Mann, K. H. 1973. An analysis of
     factors governing productivity in lakes and reservoirs.
     Limnology and Oceanography. 18: 1-14.

13. Lorenzen, M. W., Procella, D. B., and Grieb, T. M.
     1981. Phytoplankton-environmental interactions in
     reservoirs. Environmental and Water Quality Operational
     Studies. Technical Report E-81-13. 98 p.

14. U.S. EPA National Eutrophication Survey, 1977. Report
     on Lake Powell, Conconino County, Arizona; Garfield,
     Kane, and San Juan Counties, Utah. EPA Regions VIII
     and IX. Working Paper No. 733. 77 p.

15. U.S. EPA, National Eutrophication Survey. 1976.
     Preliminary Report on Flaming Gorge Reservoir,
     Sweetwater County, Wyoming, and Daggett County, Utah.
     EPA Region VIII. 58 p.

16. U.S. Geological Survey 1976. Water Resources
    Data for Wyoming, Water Year 1975. Report No.
    USGS/WRD/40-76/038. 646 p.

17. Nielson, Bryce. 1976. Physicochemical limnology
    of Flaming Gorge Reservoir 1963-1971. Utah
    Division of Wildlife Resources, Salt Lake City, Utah.

18. Wright, J.C. 1967. Effect of impoundments on
    productivity, water, chemistry, and heat budgets
    of rivers, p. 188-199. In Lane, C. E. (ed.),
    Reservoir Fisheries Resources. Symp. Amer. Fish.
    Soc. Spec. Publ. No. 6. 569 p.

19. Paulson, L. J. 1981. Nutrient management with hydro-
    electric dams on the Colorado River System. Technical
    Report No. 8, Lake Mead Limnological Research Center,
    University of Nevada. Las Vegas, Nevada. 39 p.

20. Reckhow, K. H. and Simpson, J. T. 1980. An empirical
    study of factors affecting blue-green versus non-
    blue-green algal dominance in lakes. Institute of
    Water Research, Michigan State University. OWRT
    Project No. A-102-MICHIGAN. 99 p.

21. Renk, R. R., V. D. Adams, and D. B. Porcella. 1978.
    Naturally occurring organic compounds in eutrophic
    Hyrum Reservoir, Utah. Water Quality Series Report
    Q-78-001. Utah Water Research Laboratory.

CHAPTER 4

THE EFFECTS OF MAINSTREAM DAMS ON PHYSICOCHEMISTRY
OF THE GUNNISON RIVER, COLORADO

J. A. Stanford
    University of Montana Biological Station
    Bigfork, Montana

J. V. Ward
    Colorado State University
    Fort Collins, Colorado

INTRODUCTION

The ways by which dams and diversions impact ecological
processes in rivers have received increasing scientific inquiry
in recent years [1]. However, almost all knowledge of effects
of hydrologic regulation on riverine physicochemistry is based
on measurements made at one or a few locations immediately
downstream from the point of regulation. While it is obvious
that dams alter downstream physicochemical regima profoundly,
such impacts have not usually been placed in the context of an
entire river system (see [2] for a notable exception).

As a part of a holistic approach to assess the ecology of
stream regulation in the Gunnison River, Colorado, we report
herein the physicochemical impacts of four mainstream dams on
the river system from headwaters to mouth. The changes manifested by this intense regulation greatly influenced patterns
and processes within the biotic communities extant in the
various river segments [3]. Results reported here are limited
to a physicochemical description of this major tributary of
the Colorado River before and after regulation.

STUDY AREA

The Gunnison River flows westerly from the Continental
Divide in central Colorado to its confluence with the Colorado
River near Grand Junction, Colorado. The 20,533 km^2 drainage
basin may be divided into two parts, based on basin geology.
The upstream portion, above the confluence of the Cimarron
River (Figure 1), lies primarily in mountainous terrain and
drains granitic soils and relatively insoluble crystalline

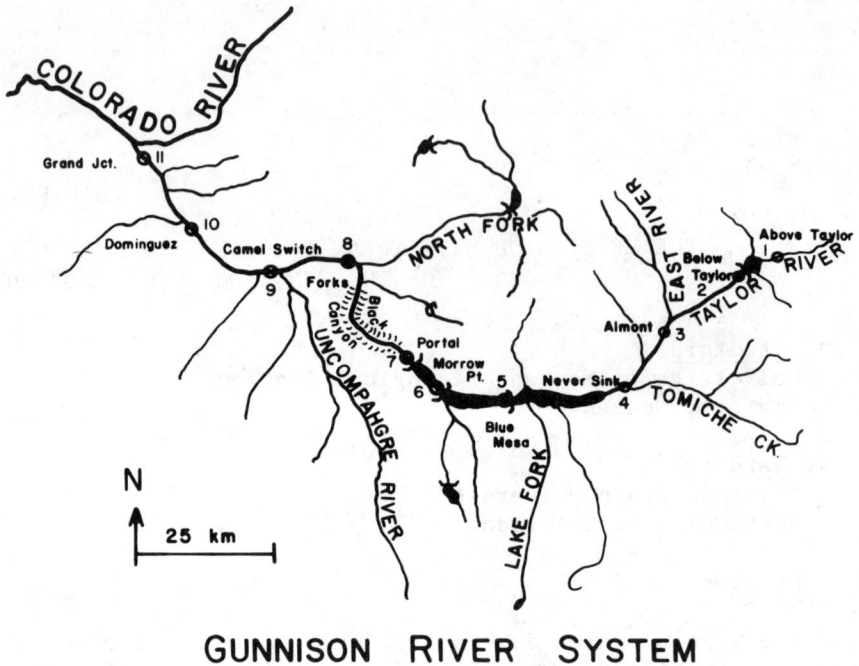

GUNNISON RIVER SYSTEM

Figure 1.   Location of the mainstream dams and eleven sampling
            sites on the Gunnison River, Colorado.

bedrock.   Downstream from the Cimarron, the river drains a
variety of mineral-rich sedimentary formations (especially
gypsum shales), which characterize the semi-arid, high plateau
of western Colorado.

     The average monthly extremes in discharge of the Gunnison
River at Grand Junction in the last 25 years have varied
between a low of <1 m^3/sec to a high of >230 m^3/sec.   However,
the annual hydrograph is intensely regulated by hydropower and
irrigation demands.   Four mainstream reservoirs, Taylor Park,
Blue Mesa, Morrow Point and Crystal (Figure 1), are impounded
behind high dams and severely influence riverine hydrology.
All four dams are deep-release (i.e., hypolimnial drain)
systems.   Taylor Park is an irrigation storage reservoir built
in 1936, while the other three comprise the Aspinal Unit of
the Colorado River Storage Project.   Blue Mesa Dam was
finished in 1965; Morrow Point Dam was closed in 1969.
Crystal Reservoir began operation in 1975 as a re-regulation
dam to dampen the extreme flow fluctuations below Morrow Point
Reservoir.   Considerable irrigation return flow occurs in the

downstream river segment, especially via the lower Uncompahgre
River and adjacent areas.

Few data are available concerning the limnology of these
mainstream reservoirs. They are impounded within deep, granite
walled canyons at ca. 2200 m elevation, where winter
temperatures prevail from October – April. Consequently, these
impoundments have a low heat budget. All, except Crystal,
apparently stratify seasonally; surface temperatures may exceed
20°C for short periods during summer, but the majority of the
stored water volume remains below 8°C year around (see Methods).

METHODS

We established eleven sampling sites along the Gunnison
River from a headwater location above Taylor Park Reservoir to
a point just upstream from the confluence with the Colorado
River (Figure 1). Sampling was conducted on eleven occasions
during the period September 1979 to October 1980.

Water samples for analyses of ion concentrations were
collected in high-density polyethylene bottles, while grab
samples for analyses of carbon fractions were collected in
acid-washed teflon or glass bottles. All samples were stored
on ice and air-freighted to the University of Montana
Biological Station for analysis in the Freshwater Research
Laboratory. Conductivity (YSI meter) and pH (Corning meter)
measurements and alkalinity titrations (as $CaCO_3$) were made in
the field. We installed Ryan© thermographs at two locations
to augment records provided by Colorado Division of Wildlife.

Ions (Ca^{++}, Mg^{++}, K^+, Na^+, $NO_3^=$ and $SO_4^=$) were quantified
by raw water injection into a model 16 Dionex© Ion Chroma-
tograph with output integrated and digitized on a Hewlitt-
Packard Model 3388 terminal.

Organic carbon present in water samples was separated into
two fractions, particulate organic carbon (POC) and dissolved
organic carbon (DOC), with glass-fiber filters (Gelman© 0.2 μm
pore size). Organic carbon in filtrates was considered to be
in the DOC fraction. POC and DOC were converted to CO_2 by hot
persulfate digestion in sealed ampules and concentrations subsequently determined by quantification of the liberated CO_2
using an Oceanography International© infrared detector.

Every fifth analysis (ions or carbon) was replicated
(i.e., multiple determinations, usually three, of the same
parameter on the same sample) and samples were duplicated
(i.e., two samples from the same location and time) to permit
calculation of analytical precision and natural variation

45

within sample locations. Reagent spikes were utilized (again
every fifth analysis) to check accuracy of analytical technique. Standard deviations of replicates and duplicates were
consistently less than one percent of the mean (i.e., high
precision) and 90-110 percent of the sample spikes were
recovered in analyses leading to the data reported herein.

     Some chemical data were available in the STORET file of
the U.S. Environmental Protection Agency for comparison to
those generated during the present study. Discharge data
were provided by the U. S. Geological Survey for various river
sites and the U. S. Bureau of Reclamation for the dam sites.
Available time-series flow data enabled us to compare discharge regima during the study period, with pre- and post-
impoundment regimes (i.e., 1900-64 and 1965-present) on the
mainstream river. Time-series temperature data were derived
from unpublished literature, such as theses and various agency
reports. Thermograph records for Sites 8 and 9 were provided
by the Colorado Division of Wildlife, while data for Site 11
were provided by the U. S. Bureau of Reclamation.
Relationships between discharge and thermal regima were established with the use of polynomial regression analyses and
simple plots of annual degree days (a sum of mean daily temperatures over an annual period, [4]) along the river profile.

RESULTS AND DISCUSSION

     The pre-regulation discharge regime of the Gunnison River
varied from minimum flows during autumn and winter to spring
maxima as a result of melting snowpack in the headwaters
(Figure 2). The post-regulation flow has been considerably
higher in winter and lower during spring (Figure 2), as runoff
is stored in the reservoirs and discharged primarily from
November to March. Greater than 90 percent of the average
annual discharge is derived from precipitation in the headwaters; downstream side flows (i.e., below the North Fork
River) in the lowland sedimentary formations contribute significant amounts of water only during short spates in spring
and after heavy summer thunderstorms.

     Historically, the upstream segment carried substantial
sediment and bed loads during spring runoff which were
deposited in the lower gradient downstream segment. Thus, for
much of the year, the upstream segment flowed low and clear
over a cobble and boulder bottom that was annually scoured
and re-distributed by the spring freshet. The downstream
segment was also fairly clear at base flow, but the bottom was
predominantly silt. Occasional rubble riffles occurred in
areas where side flows carried large materials into the river
channel (Dolan et al. [5] describe this process of riffle or

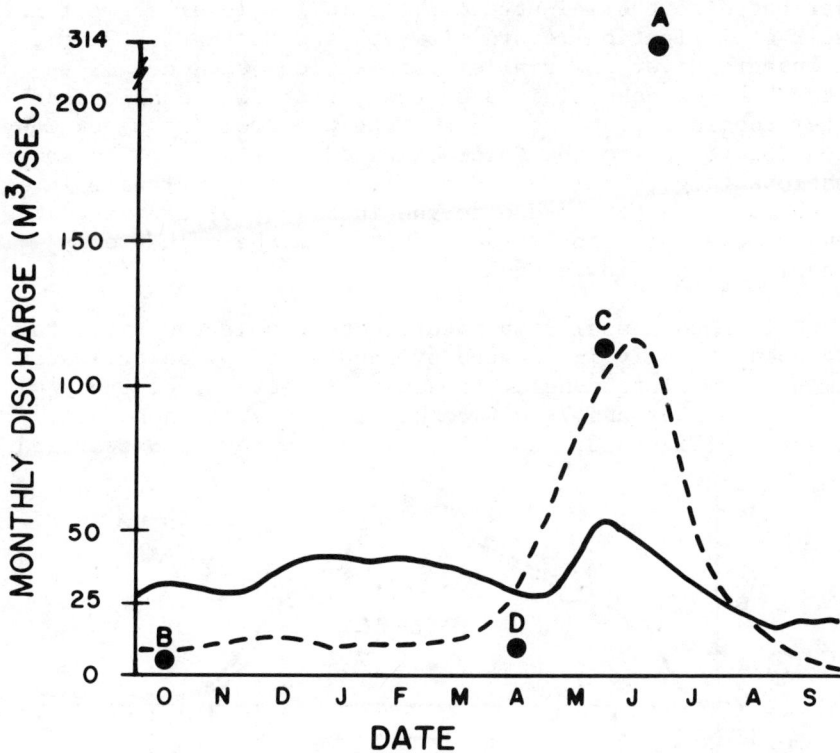

Figure 2. Discharge measured at Site 7, the U.S.G.S. gauging
station below Crystal Dam, before regulation
(broken line: monthly means 1948-1964) and after
construction of mainstream dams (solid line:
monthly means 1965-1980). Points A and B identify
maximum and minimum pre-regulation discharge
(monthly means 1934 and 1957); points C and D represent the maximum (1974) and minimum (1977)
monthly flows since regulation (based on U. S.
Geological Survey data).

rapids building by side flows on the mainstream Colorado River).
Since regulation, silt loads accompanying runoff have been
retained in the reservoirs. Thus, discharge below the dams is
continually without significant amounts of suspended solids;
the river from Taylor Park Reservoir to the East River and
from Crystal Dam to the Colorado River is being continually
sluiced by clear-water discharges that are of a higher mean
volume July to March than prior to impoundment. The result is
considerable armoring of the river bottom, the substrata being
composed of firmly imbedded large rocks [6]. This situation
presently characterizes the Taylor River and Black Canyon
segments downstream from the dams to the East River and North
Fork River, respectively. Although considerable sediment is

47

contributed to the mainstream river in its lower segment as a result of irrigation return flow via the North Fork River, Uncompahgre River and smaller tributaries, the once silty bottom has now been sluiced to the extent that cobbles and larger rubble predominate in the thalweg from the Black Canyon reach downstream to the Colorado River confluence. In several locations (e.g., Dominguez Canyon), rapids are growing in length and wave height due to the inability of the regulated flow to move large boulders deposited in the mainstream channel by side flow spates.

Daily and annual temperature patterns in the river have also been strongly influenced by regulation. The tailwater segments immediately below the dams are several degrees warmer in winter and $7-20°C$ colder in summer than before regulation (Figure 3; Table I), because water is discharged

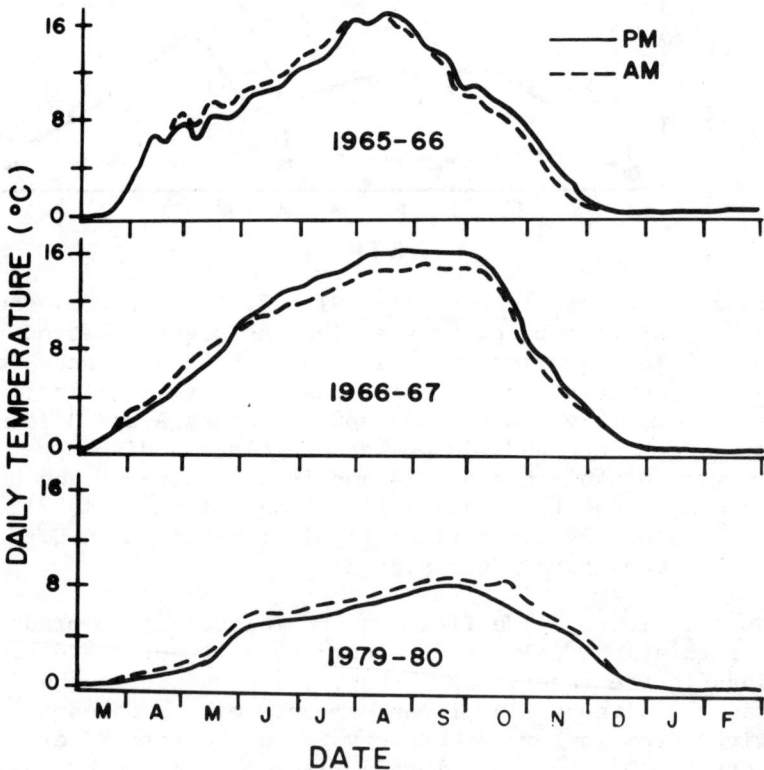

Figure 3.   Pre- (1965-66, 1966-67 [from 7]) and post-regulation (1979-80) temperature patterns measured at Site 7, three km downstream from Crystal Dam.

Table I. Comparison of temperature patterns along the Gunnison River continuum before and after construction of the mainstream dams (modified from [3]).

Station No.	Km from Headwaters	Before Regulation		After Regulation	
		Annual Degree Days (Annual Thermal Range)	Daily[a] ΔT	Annual Degree Days (Annual Thermal Range)	Daily[a] ΔT
1	18	1950 (0–15.0)	–	1950 (0–15.0)	–
2*	24	2000 (0–15.0)	+0.1	1000 (2.5–7.2)	–2.6
3	54	2250 (0–16.5)	+0.7	2150 (0–15.5)	+3.2
4	81	2550 (0–18.8)	+0.8	2250 (0–18.8)	+1.1
5*	115	2650 (0–19.0)	+0.3	2323 (3.3–11.1)	–0.6
6*	130	–	–	–	–
7*	144	2895 (0–20.0)	+0.7	1361 (0–9.4)	–3.8
8	195	–	–	–	–
9	228	3606 (0–24.0)	+2.0	3694 (2.8–21.7)	+6.5
10	271	–	–	–	–
11	290	4132 (0–26.6)	+1.5	3432 (0–23.3)	–0.7

[a]calculated mean daily thermal gain or loss from upstream site (see text).
*tailwater area.

from near the bottom of the reservoirs. Prior to regulation, the annual mean temperature of the river progressively increased downstream (Table I). The daily thermal gain (averaged over 12 months) between the headwater site and the Colorado River was about 6°C. In the Black Canyon National Monument the granite walls and shading greatly influenced the daily thermal regime. Kinnear [7] observed that vernal temperatures in the Black Canyon were actually warmer during the night, than during daytime (Figure 3), due to differential heating and cooling of the canyon walls. Since regulation, this daily cycle has been eliminated by the high-volume, cold discharge from Crystal Reservoir. The post-regulation river thermal regime is summarized in Table I. The major conclusion from these data is that the Taylor River and lower mainstream segments are much colder than before regulation and the thermal gain in Black Canyon is more dramatic (simply because the water is so cold at the head of the canyon during the warmest time of the year). Rhithron conditions [8] now extend well into the lower river segment.

The negative thermal gain of –0.7°C observed between Sites 9 and 11 remains largely inexplicable and may be an artifact of limited time-series data (the post-regulation thermal regime at this site were based only on data for one year, 1978), or a response to groundwater input. Several

warm springbrooks (e.g., Tongue and Buttermilk Creeks) flow
into the river between Sites 8 and 9. The lower Uncompahgre
River is also apparently fed by considerable flow from surface
aquifers. These side flows may warm the Gunnison River
slightly; a subsequent thermal loss could then eventuate in
downstream areas not influenced by groundwaters. Thus, the
thermal gain estimate at Site 9 could be slightly high.

A strong correlation (r = .87) between the flow rate from
Crystal Dam and river temperatures below the Black Canyon
(Sites 8 and 9) was observed. At minimum flows (ca. 16 m^3/sec),
which occurred only during spring and summer during the period
for which thermograph records exist (1978-81), thermal gain in
the canyon was 10-12°C; whereas, high flows (ca. 30 m^3/sec and
greater) limited thermal gain to 2-3°C. Thus, a very predictable relationship exists between discharge temperature,
discharge volume and temperature of the river at any point
downstream, given some knowledge of seasonal trends in air
temperature. However, heat storage in the granite walls of
the Black Canyon undoubtedly limits variance in this
relationship; river channels in more open, low-gradient terrain
probably exhibit greater diurnal fluctuations.

The observed significant difference between pre- and post-
impoundment temperature minima at Site 9 (0° vs. 2.8°, Table I)
may be related to the flow-thermal gain relationship within the
Black Canyon. Even though the midwinter thermal gain is
generally low, high volume discharge limits heat loss. The
canyon walls apparently absorb enough heat to ameliorate heat
loss. Prior to regulation, low flows coincided with cold,
midwinter air temperatures. Thus, the river froze over for
periods of a few days to several weeks until air temperatures
moderated to the extent that a thermal gain occurred relative
to flow rate.

Concentrations of major ions in solution were highest at
the downstream sites, indicating substantial salt loading in
the lower river segment. Ion concentration was inversely
related to seasonal trends in flow at the least regulated
sites during 1979-80 (Figure 4). Dissolved solids in tailwater segments were consistently lower than at upstream sites
and concentrations were much less variable (i.e., influenced
by flow volume) over all sampling dates (Figures 5 and 6).

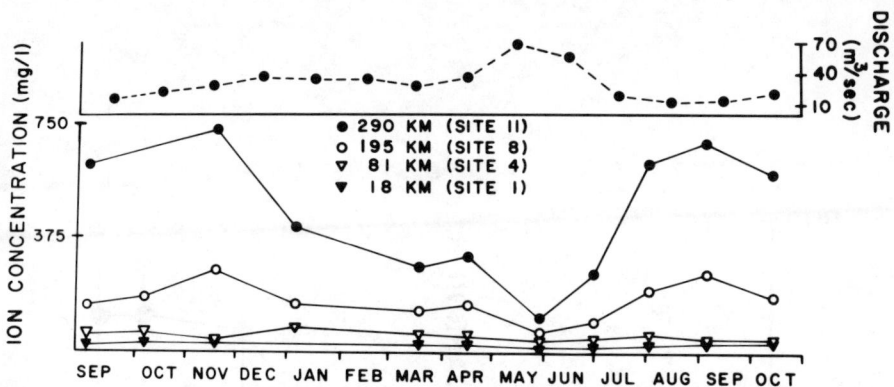

Figure 4. Concentration of the major ions (i.e., sum of Ca++, Mg++, Na+ and $SO_4^=$ concentrations) in water samples taken in time-series during 1979-80 at four locations along the Gunnison River profile. Flow rates (monthly means) are plotted only for Site 11 located near the confluence with the Colorado River.

Calcium was the dominant ion by percentage composition in the upper river (above the Black Canyon), while sulfate loading from side flows draining gypsum formations characterized the lower river segment (Figure 5). Sulfate-containing salts were observed in high concentrations (e.g., > 3000 mg/l) in the side flows (especially springbrooks and irrigation return flows) between Sites 7 and 9. The propensity of the reservoirs to sediment or precipitate dissolved solids was evinced in our data, but this loss was countered by loading rates nearly two orders of magnitude greater in the lower river segment (Figure 5).

Nitrate concentrations also increased in a downstream direction over the river continuum, but values were consistently elevated in tailwaters in comparison to sites above the reservoirs (Figure 6). The mobilization of nitrate is attributed to mineralization of organic matter (i.e., nitrification) and perhaps nitrogen fixation within the water column of the reservoirs. Nitrates were apparently utilized by autotrophic processes in riverine segments downstream from the dams (Figure 6). This was particularly evident in the Black Canyon, which is the segment least influenced by side flows. Benthic algae, particularly Cladophora spp., grow in profusion in all tailwater segments and are a dominant feature of the river bottom from Crystal Dam to Site 8. Tributary effects and turbid irrigation return flows apparently limited excessive growths of filamentous algae below Site 8, even though nutrient loading was apparent (Figure 6). However, thick accumulations of aufwuchs were present at the Dominguez Canyon Site (10) where we measured 3-5 cm accumulations of algae, fine silts, clays and organic detritus firmly attached to cobbles in riffle areas.

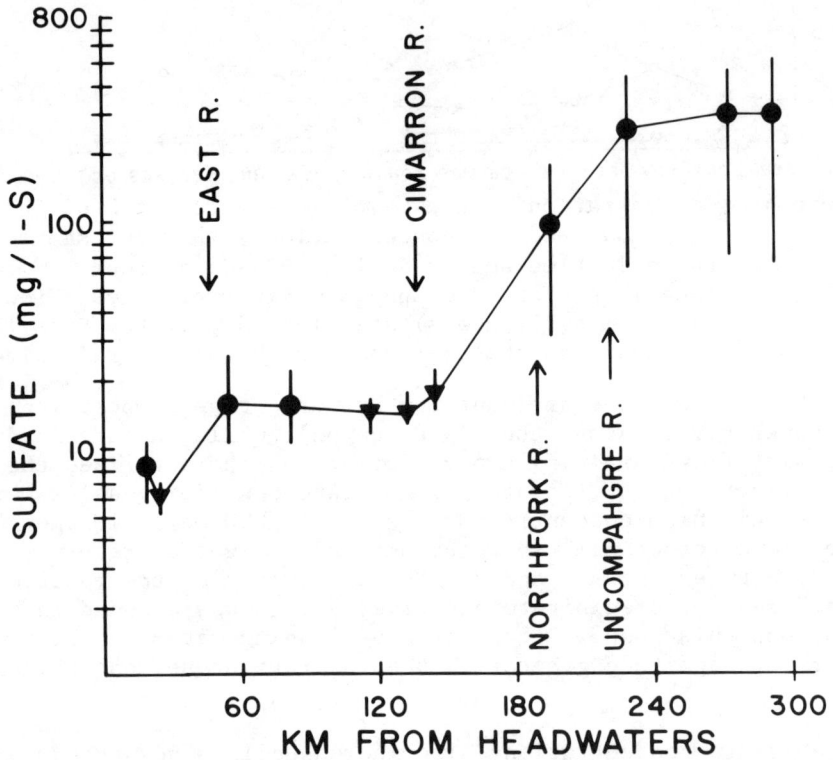

Figure 5. Mean annual sulfate concentrations (mg/l as S)
measured at 11 sites on the Gunnison River.
Inverted triangles indicate tailwater sites below
mainstream dams; bars indicate ranges of values for
11 sampling periods during 1979-80. Location of
major side flows are indicated by arrows.

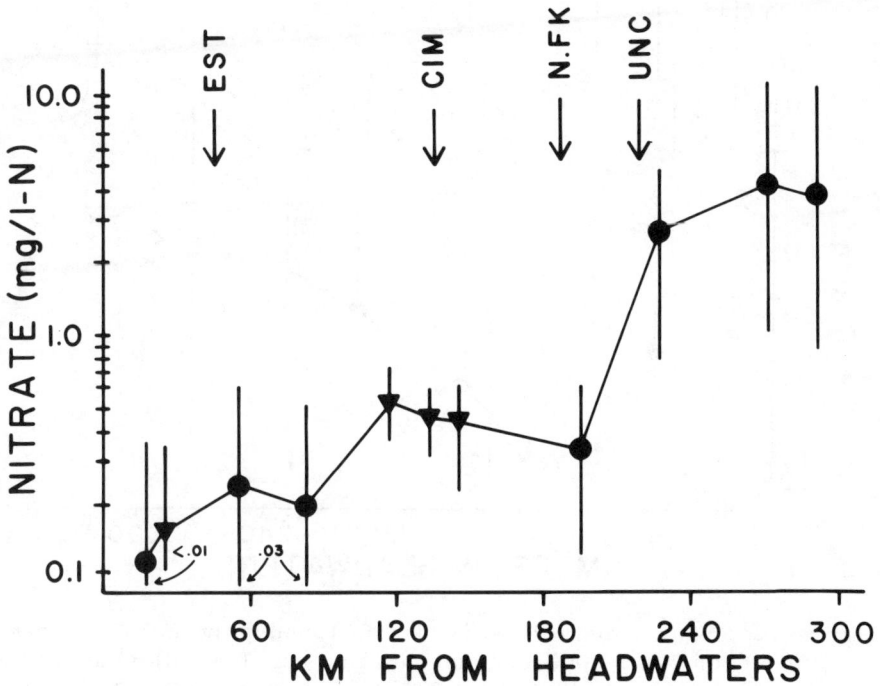

Figure 6.   Mean annual nitrate concentrations (mg/1 as N)
            measured at 11 sites on the Gunnison River.
            Inverted triangles indicate tailwater sites below
            mainstream dams; bars indicate ranges of values
            for 11 sampling periods during 1979-80. Location
            of major side flows are indicated by arrows.

     The mineralization effect of the reservoirs was very
evident in time-series measurements of particulate and dissolved
organic carbon. Despite exports of plankton from the reservoirs,
POC levels below the dams were consistently lower than in river
segments immediately upstream from the impoundments and vice
versa for DOC values. The total organic carbon pool in the
river increased from ca. 1.0 to 10.0 mg/1, on the average from
headwaters to the mouth (Figure 7). Agglutination processes
(i.e., demobilization of dissolved solids by conversion to
particulate carbon forms) were responsible for progressively
increasing POC values downstream from Taylor Park and Crystal
Dams. Much of the seston drift in these segments was due to

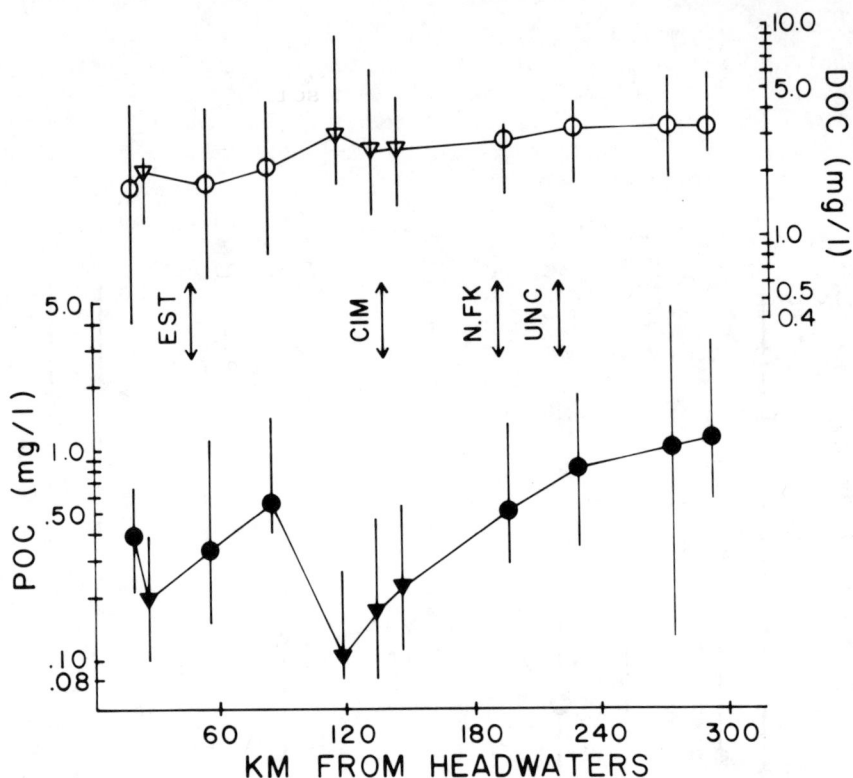

Figure 7. Mean annual dissolved (DOC) and particulate organic carbon (POC) concentrations (mg/l as C) measured at 11 sites on the Gunnison River. Inverted triangles indicate tailwater sites below mainstream dams; bars indicate range of values for 11 sampling periods during 1979-80. Location of major side flows are indicated by arrows.

sloughed filaments of *Cladophora* and other benthic algae. In lower river segments, side flows contributed significant amounts of allochthonous particulates; however, agglutination by autotrophic and micro-heterotrophic activity undoubtedly played a major role in size fractions and POC concentrations in this river segment, except during the spring freshet.

Thus, during 1979-80 the dissolved solids and organic carbon pool increased dramatically in a downstream direction; but, concentrations in the intensely regulated segments were greatly influenced by mineralization and precipitation within the reservoirs and, by agglutination as materials moved downstream in riverine segments. Time-series chemical data for periods previous to our study were limited to Site 4, upstream from Blue Mesa Reservoir. Our data were remarkably similar to these measurements (Table II) indicating that the trends reported here have been the norm since the Gunnison River was regulated. Dissolved and particulate solids loading undoubtedly occurred prior to regulation, but concentrations exported to the Colorado River were likely much lower and more erratic before irrigation return-flows were a significant feature of the lower river.

Table II. Comparison of data in the U. S. Environmental Protection Agency's STORET file to those obtained in the present study. Both data sets were generated from samples collected in time-series at the same location on the Gunnison River 5 km west of Gunnison, Colorado.

	STORET File 1968–1980		This Study 1979–1980	
	Mean (Range)	N	Mean (Range)	N
Magnesium	8.7 (4.0–18.0)	68	7.8 (5.0–12.9)	11
Sodium	5.4 (1.0–15.0)	63	4.0 (2.3–7.4)	11
Sulfate	19.0 (3.0–31.0)	69	15.6 (10.8–22.3)	11
Nitrate	0.19 (*–1.60)	59	0.19 (0.04–0.50)	11

*less than detection limit.

CONCLUSIONS

Hypolimnial-release impoundments on the Gunnison River have altered the physicochemistry of the riverine environment, mainly by reducing seasonal variability. Summer-cold, winter-warm conditions prevail in the river downstream from the dams. Dissolved solids (except $NO_3^=$) and particulate organic matter (POM) are reduced in concentration within reservoir tailwaters in comparison to concentrations in river segments above the reservoirs. Mobilization of $NO_3^=$ and other nutrients in reservoir effluents has stimulated thick growths of periphyton thalweg substrata, which has stabilized (armored) in response to elimination of spring flood flows. Inherent biophysical processes (e.g., communition and agglutination of POM; thermal gain via insolation) and side flows ameliorate or reset the consequences of regulation, as distance downstream from impoundments increases. Although the dissolved solids pool increases down the river profile, conditions 30–40 km downstream from the last dam (i.e., at Site 8) mimic the rhithron environment 115 km upstream (i.e., at Site 4). Physicochemistry of the Gunnison River near its confluence with the Colorado River is similar to pre-regulation, except that annual variance in discharge has decreased and dissolved solids increased.

REFERENCES

[1] Ward, J. V. and J. A. Stanford (Eds.) 1979. The Ecology of Regulated Streams. Plenum Press, New York. 398 pp.

[2] Penáz, M., F. Kubícek, P. Marvan and M. Zelinka. 1968. Influence of the Vír River Valley Reservoir on the hydrobiological and ichthyological conditions in the River Svratka. Acta Academiae Scientiarum Hungaricae 2(1): 1-60.

[3] Stanford, J. A. and J. V. Ward. 1981. Preliminary interpretations of the distribution of Hydropsychidae in a regulated river. pp. 323-328. *IN*: Moretti, G.P. (Ed.) Proceedings of the Third International Symposium on Trichoptera. Series Entomologica, Volume 20. Dr. W. Junk. The Hague.

[4] Baskerville, G. L. and P. Emin. 1969. Rapid estimation of heat accumulation from maximum and minimum temperatures. Ecology 50(3): 514-517.

[5] Dolan, R., A. Howard and A. Gallenson. 1974. Man's impact on the Colorado River in the Grand Canyon. American Scientist 62: 392-401.

[6] Simons, D. B. 1979. Effects of stream regulation on channel morphology. pp. 95-112. *IN*: Ward, J. V. and J. A. Stanford (Eds.) The Ecology of Regulated Streams. Plenum Press. New York. 398 pp.

[7] Kinnear, B. 1967. Fishes of Black Canyon. M. S. Thesis, Colorado State University. 45 pp.

[8] Hynes, H. B. N. 1970. The Ecology of Running Waters. University of Toronto Press. Toronto. 555 pp.

CHAPTER 5

THE INFLUENCE OF LAKE POWELL ON THE
SUSPENDED SEDIMENT-PHOSPHORUS DYNAMICS
OF THE COLORADO RIVER INFLOW TO LAKE MEAD

T.D. Evans
L.J. Paulson
  Lake Mead Limnological Research Center
  University of Nevada, Las Vegas

INTRODUCTION

The Colorado River has been successively modified by
the construction of several reservoirs, beginning in 1935
with the formation of Lake Mead by Hoover Dam. These reservoirs are located in a chain, and each one has an influence
on the nutrient dynamics and productivity of the river and
downstream reservoir [1]. Lake Mead derives 98% of its annual inflow from the Colorado River [2]. Historically, the
Colorado River inflow was unregulated into Lake Mead. Regulation occurred in 1963, when Lake Powell was impounded by
the construction of Glen Canyon Dam, approximately 450 km
upstream. The formation of Lake Powell drastically altered
the physical characteristics of the Colorado River inflow to
Lake Mead [1]. Regulated releases from Glen Canyon Dam have
eliminated the spring discharge peaks that historically resulted from spring flooding in the Upper Colorado River
drainage basin. Temperatures in the Colorado River below
Lake Powell have been reduced 5-10°C during the spring and
summer, due to cold hypolimnetic releases from Glen Canyon
Dam. There were also marked reductions in the suspended
sediment loads due to decreases in spring and summer discharge peaks. The turbid overflows that once extended across
the Upper Basin of Lake Mead [3] during spring were not
evident in 1977-78 [2]. The Upper Basin of Lake Mead is now
severely phosphorus deficient, and this appears to have been
caused by reductions in suspended sediment loading [1].
Phosphorus has been reported by many investigators as
the most common nutrient limiting phytoplankton productivity
[4]. Phosphorus loading models are generally based on total
phosphorus (total-P), but this fraction may not accurately
reflect the amount of phosphorus available for biological
uptake in turbid river systems [5]. Total-P loading models
greatly overestimate the trophic states in Lake Powell and

Lake Mead [2,6].

Little emphasis has been placed on the interaction between suspended sediments and dissolved inorganic phosphorus in rivers [7,8]. The removal of inorganic phosphorus by suspended sediment, however, does appear to be a sorption rather than a precipitation process [9]. Loosely bound phosphorus on suspended sediments is more readily available than precipitated phosphorus [10]. Wang and Brabec [11], in their work on the Illinois River at Peoria Lake, found that dissolved inorganic phosphorus was actively adsorbed by suspended sediments. Other workers have also observed this process occurring in oxygenated rivers and lakes [12,13]. Mayer and Gloss [14] have shown that phosphorus buffering by suspended sediments in the turbid Colorado River is an important mechanism for sustaining the dissolved inorganic phosphorus pool in Lake Powell. It appears that this same mechanism occurred in Lake Mead when it received turbid inflows from the Colorado River.

The intent of this paper is to discuss the possible effects that the formation of Lake Powell has had on the suspended sediment-phosphorus dynamics of the Colorado River inflow to Lake Mead. This is based on results from recent investigations and on preliminary results of research conducted in the late-summer and early-fall of 1981.

STUDY AREA

This study focuses on a 1000 km stretch of the Colorado River which includes two of the largest reservoirs in the Western Hemisphere, Lake Powell and Lake Mead (Figure 1). Comparative morphometric characteristics for the reservoirs are presented in Table I.

Lake Powell was formed by the construction of Glen Canyon Dam in 1963. The reservoir covers a 300 km stretch of Glen Canyon, and is morphometrically complex with over 3000 km of shoreline. The Colorado and San Juan Rivers provide 96% of the total annual inflow to this reservoir. Approximately 60% occurs in late-spring and early-summer (May-July), as a result of snowmelt [15] from the Upper Colorado River Basin (Figure 2). Lake Mead is the second of four major reservoirs on the main stem Colorado River. It is a large deep-storage reservoir 180 km in length, extending from the mouth of the Grand Canyon at Pierce Ferry to Hoover Dam in Black Canyon (Figure 1). The dominant hydrologic input to this reservoir is from the Colorado River which provides approximately 98% of the total annual inflow. The Virgin and Muddy Rivers discharge approximately 1% into the Overton Arm of Lake Mead. The remainder is derived from Las Vegas Wash, a secondarily-treated sewage and industrial effluent stream from metropolitan Las Vegas, which discharges into Las Vegas Bay [2].

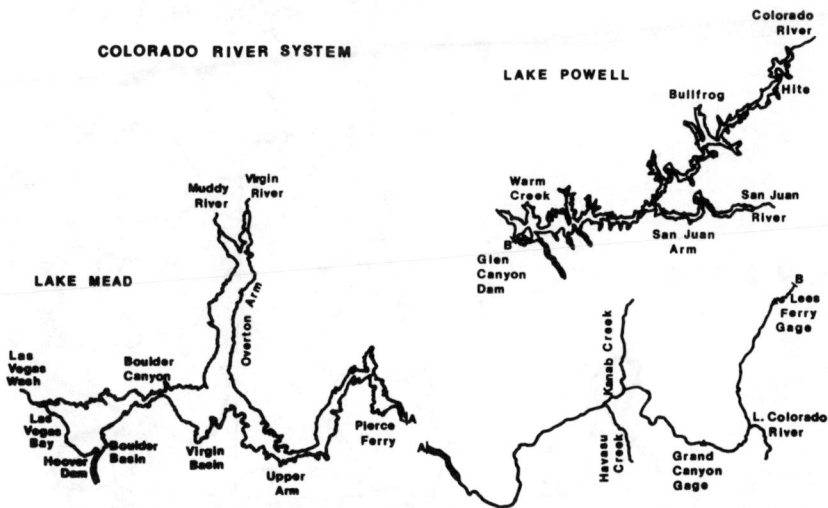

Figure 1. Map of the Colorado River System from Lake Powell
to Lake Mead.

Table I. Morphometric Characteristics of Lake Powell and
Lake Mead.

Parameter	Lake Powell	Lake Mead
Maximum operating level (m)	1128	374
Maximum depth (m)	171	180
Mean depth (m)	51	55
Surface area (km^2)	653	660
Volume ($m^3 \times 10^9$)	33	36
Maximum length (km)	300	183
Maximum width (km)	25	28
Shoreline development	26	10
Discharge depth (m)	70	100
Approximate storage ratio (years)	2	4

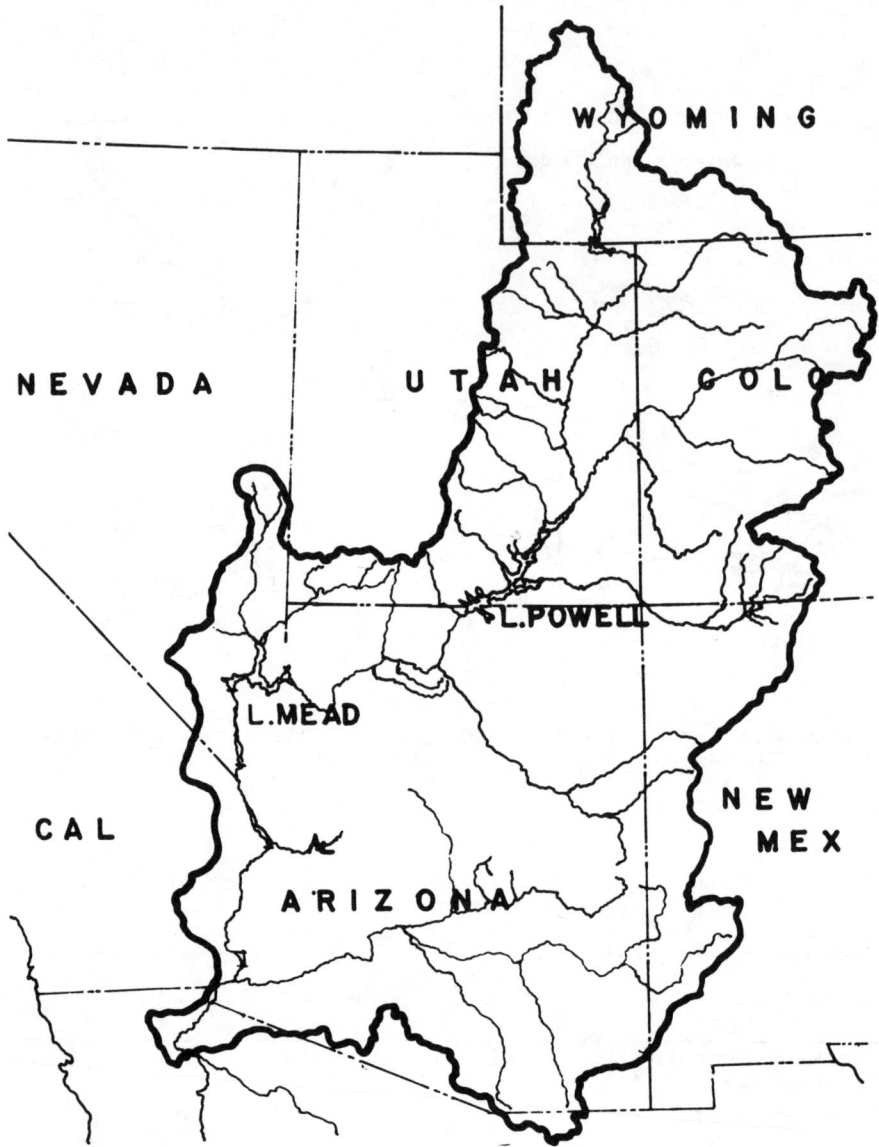

Figure 2. Map of the Colorado River Drainage Basin.

METHODS

The primary inflows to Lake Powell and Lake Mead were sampled monthly from August through October, 1981. A composite sample, consisting of several tows with a 3-liter Van Dorn bottle, was collected from the surface at each station. River water was analyzed for total-P, total particulate phosphorus (part-P), and ortho-phosphorus (ortho-P). Ortho-P was determined by methods described in Kellar, Paulson, and Paulson [16] on samples that were filtered immediately upon collection through 0.45 μm membrane filters. Some clay-sized sediment particles may be as small as 0.06 μm in diameter. However, turbidity measurements using a spectrophotometer showed no difference between 0.45 μm filtered river water and a sediment-free distilled water blank. Total-P was determined by persulfate digestion on unfiltered 50 ml samples. Total part-P was determined on suspended sediments collected on 0.4 μm Nucleopore filters. These sediment-filters were dried, weighed to determine sediment concentration, and digested in a 50 ml solution of distilled water and ammonium persulfate. Available sediment-P was also determined on 0.4 μm Nucleopore filtered samples. The NaOH extraction technique described by Sagher [17], and Williams, Shear and Thomas [18] was used to estimate biologically available sediment-P. NaOH extractable-P gives an approximate estimate of the amount of inorganic phosphorus that is biologically available through sorption reactions with suspended sediments. This fraction includes non-occluded inorganic phosphorus that is loosely bound to iron and aluminum in sediments. Much of the work that has been done on suspended sediment-P dynamics uses only the total-P fraction in high sediment:water ratios. High sediment:water ratios are indicative of soils and sediments rather than suspended riverine sediments [14]. The estimates of total part-P and available sediment-P are based on using natural river water, which has a low sediment:water ratio. The sediment:water ratio appears to be an important factor influencing sorption reactions by suspended riverine sediments.

DATA SOURCES

Suspended sediment data for the Grand Canyon gaging station were derived from "Quality of Surface Waters for the United States," and discharge data were obtained from "Surface Water of the United States," U.S. Geological Survey Water-Supply Papers Part 9. Colorado River Basin (1940-1970). After 1970, these data were taken from "Water Resources Data for Arizona or Nevada" prepared jointly by the U.S. Geological Survey and state agencies.

RESULTS AND DISCUSSION

Suspended Sediment Loads

        Suspended sediment loads in the Colorado River at Grand
Canyon were extremely high prior to the formation of Lake
Powell (Figure 3). In years of high runoff, up to 140 million tons per year of suspended sediments flowed into Lake
Mead. The majority of this occurred during the spring runoff
periods (Figure 3). Impoundment of Lake Powell in 1963 resulted in a 70-80% reduction in suspended sediment loads in
the Grand Canyon (Figure 3). The direct drainage area to
Lake Mead was reduced to a few tributary inputs in Grand
Canyon (Figure 2). The Little Colorado River, which enters
the main stem Colorado River 40 km above the Grand Canyon
gaging station, is now the only appreciable source of sediments to the river [19]. Suspended sediment inputs from the
Little Colorado River can be quite high when floods occur,
but annual loading to Lake Mead is still far below that
which occurred in pre-Lake Powell periods (Figure 3).

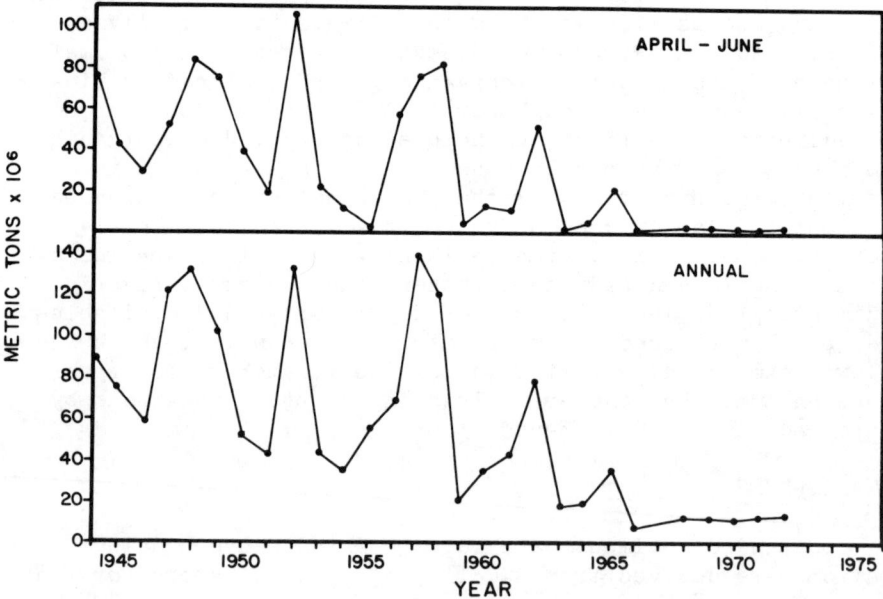

Figure 3. Historical Annual and Spring Suspended Sediment
          Loads at Grand Canyon Gaging Station. (USGS Data)

Effects on Total-P

Gloss, Mayer, and Kidd [20] demonstrated that total-P
concentrations were closely associated with suspended clays
in river water. A similar relationship has been observed by
other researchers [21]. Preliminary measurements made on the
Colorado River above Lake Powell and Lake Mead, and on the
San Juan River inflow to Lake Powell, also show a close correlation between total-P and suspended sediment concentrations (Figure 4). The relationship appears to be linear in
the range of suspended sediment concentrations that occurred
in the Colorado River from August to October, 1981. This research is continuing to determine if the relationship also
holds for other seasons.

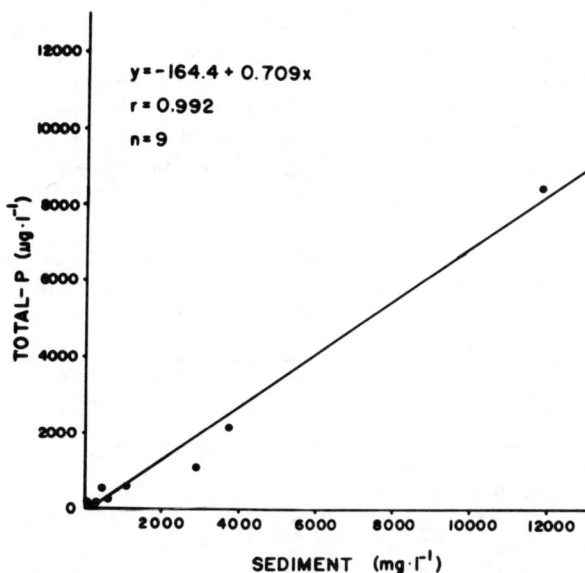

Figure 4. Total-P Concentrations as Related to Suspended
Sediment Concentrations in the Colorado River
System.

Phosphorus budgets were recently determined for Lake
Powell [6] and Lake Mead [22] (Table II). It is readily
apparent that Lake Powell is serving not only as a sediment
trap, but also as a phosphorus sink. Gloss et al. [6] reported that over 95% of the phosphorus loads entering Lake
Powell were in the particulate form. They further concluded

that the phosphorus retention coefficients determined for
Lake Powell were among the highest reported to date. This
probably reflects the strong relationship between phosphorus
and suspended sediments in the Colorado River. The phosphorus retention coefficients determined for Lake Mead were not
as high as Lake Powell. This was caused by high inputs of
ortho-P from the Las Vegas Wash inflow. Las Vegas Wash forms
a density current in Lake Mead [22], resulting in a large
percentage of the phosphorus input being loaded into the
hypolimnion. Hoover Dam is operated from a hypolimnion discharge, which rapidly strips phosphorus from the reservoir
[23]. The combination of these two processes greatly reduces
retention of ortho-P and total-P in Lake Mead. However,
Prentki et al. [24] found that total-P in Lake Mead sediments was high (300-1000 mg/l). Inorganic-P averaged 86% of
total-P. Measurements made on various phosphorus fractions
in the major river inflows to Lake Powell and Lake Mead also
indicate that the majority of total-P is inorganic-P, bound
to suspended sediments (Table III). This trend was consistent in Lake Mead sediment layers for pre- and post-Lake
Powell periods. There was, however, a 93.5% decrease in the
phosphorus sedimentation rates in Lake Mead after Lake
Powell was formed [24]. This agrees well with recent work on
Lake Powell [6], where it was estimated that 96.3% of the
total-P was retained in the reservoir.

Table II.   Phosphorus Budgets for Lake Powell and Lake Mead
            from Gloss et al. [6] and Baker and Paulson [22],
            Expressed as Flow Weighted Estimates in Metric
            Tons Per Year.

Location	Total Phosphorus	Phosphate	Dissolved Phosphorus
Lake Powell			
Colorado River	5224	-	267
San Juan River	785	-	83
Other tributaries	250	-	15
Precipitation	1	-	1
Glen Canyon Dam	229	-	100
R =	.963		.727
Lake Mead			
Colorado River	199	56.8	-
Las Vegas Wash	263	136.6	-
Hoover Dam	123	110.6	-
R =	.734	.428	

R   = Experimentally determined retention coefficient

Table III. Concentrations of Total-P, Part-P and Part-P
           Expressed as a Percentage of Total-P for the
           San Juan and Lower Colorado Rivers for August
           and September, 1981, in µg/l (±95% CL).

River	Month	Total-P	Part-P	% of Total-P
San Juan	Aug	1149 (±53.6)	1022 (±10.1)	89
	Sep	124 (± 3.8)	100 (±44.2)	81
L. Colorado	Aug	239 (±17.4)	268 (±15.0)	112
	Sep	77 (± 4.2)	70 (± 1.9)	90

Availability of Phosphorus from Suspended Sediment

It has been shown [14] that the suspended sediments in
the Colorado River inflow to Lake Powell have the capability
of desorbing approximately 20-30 µg/l of dissolved inorganic-P. We are currently investigating the suspended sediment-
P dynamics in the Colorado River system above and below Lake
Powell. Our work is in the preliminary stages, and must be
considered on that basis. However, our data thus far agree
with findings of other workers. In general, these data indicate that a small percentage (10-30%) of the total-P is biologically available. Lee, Jones, and Rast [25], in their review of availability of part-P to phytoplankton, have established an equation to estimate total available-P. Available-
P = SRP + 0.2 PP_T, where SRP = soluble reactive phosphorus,
and PP_T = total part-P. Prentki et al. [24] found that an
average of 9% of the total sediment-P was available. Our
estimates for August and September range from 7.1-19.2% with
a mean value of 11.3% (Table IV). We also estimated total
available-P on a volumetric basis by combining sediment
available-P with ortho-P values. On a volumetric basis total
available-P represented 7.3% of total-P, with a range of
1.7-14.1%.

Table IV. Available-P and Total Part-P for the Upper and
          Lower Colorado and San Juan Rivers During August
          and September, 1981, in µg/l (±95% CL).

River	Month	Part-P	Available-P	% of Part-P
U. Colorado	Aug	294 (±13.5)	21.0 (± 5.6)	7.1
	Sep	-	2.4 (± 0.4)	-
San Juan	Aug	72 (± 1.1)	6.9 (± 0.4)	9.5
	Sep	1850 (±15.2)	355 (±43.2)	19.2
L. Colorado	Aug	240 (±18.6)	29 (±28.0)	12.1
	Sep	685 (±99.6)	58.2 (±33.4)	8.5

Effects on Productivity

The formation of Lake Powell in 1963 resulted in marked reductions in suspended sediment loading to Lake Mead. Total-P was reduced accordingly, and the Upper Basin of Lake Mead has since become severely phosphorus deficient [1]. Phytoplankton productivity in the Upper Basin averaged 4612 mg C/m^2·day during the 1955-62 period [24]. Productivity decreased to an average of 503 mg C/m^2·day after Lake Powell was formed in 1963. Although only a small percentage of the total-P in the river inflows is biologically available, the historic sediment loads (up to 140 million tons per year) were apparently sufficient to sustain the dissolved inorganic phosphorus pool, and higher productivity.

ACKNOWLEDGEMENTS

We wish to express our appreciation to several people for assistance with this paper. We would like to thank Laurie Vincent and Thom Hardy for typing the manuscript, and Sherrell Paulson and Jim Williams for the drawing and photographing of figures. Also conversations with Richard Prentki, Penelope Naegle, and John Baker were helpful.

REFERENCES

1. Paulson, L.J. and J.R. Baker. 1981. Nutrient interactions among reservoirs on the Colorado River. Pages 1647-1656 in H.G. Stefan, ed. Symposium on surface water impoundments. June 2-5, 1980. Minneapolis, MN.

2. Paulson, L.J., J.R. Baker and J.E. Deacon. 1980. The limnological status of Lake Mead and Lake Mohave under present and future powerplant operations of Hoover Dam. Lake Mead Limnological Res. Ctr. Tech. Rept. No. 1. Univ. Nev., Las Vegas. 229 pp.

3. Anderson, E.R. and D.W. Pritchard. 1951. Physical limnology of Lake Mead. Lake Mead sedimentation survey. U.S. Navy Electronic Lab. San Diego, California Rept. No. 258. 153 pp.

4. Wetzel, R.G. 1975. Limnology. W.B. Saunders Co., Philadelphia, PA. 743 pp.

5. Ryden, J.C., R.F. Harris and J.K. Syers. 1973. Phosphorus in runoffs and streams. Advan. Agron. 25:1-45.

6. Gloss, S.P., R.C. Reynolds, Jr., L.M. Mayer and D.E.

Kidd. 1980. Reservoir influences on salinity and nutrient fluxes in the arid Colorado River Basin. Pages 1618-1630 in H.G. Stefan, ed. Symposium on surface water impoundments. June 2-5, 1980. Minneapolis, MN.

7. Taylor, A.W. 1967. Phosphorus and water pollution. J. Soil Water Conserv. 5:228-231.

8. Shukla, S.S., J.K. Syers, J.D.H. Williams, D.E. Armstrong, and R.F. Harris. 1971. Sorption of inorganic phosphate by lake sediments. Soil Sci. Soc. Amer. Proc. 35:244-249.

9. Patrick, W.H. 1974. Phosphate release and sorption by soils and sediments: Effect of aerobic and anaerobic conditions. Science 186:53-55.

10. Harter, R.D. 1968. Adsorption of phosphorus by lake sediment. Soil Sci. Soc. Amer. Proc. 32:514-518.

11. Wang, W.C. and D.J. Brabec. 1969. Nature of turbidity in the Illinois River. J. Amer. Water Work Assoc. 3:42-48.

12. Latterell, J.J., R.F. Holt and D.R. Timmons. 1971. Phosphate availability in lake sediments. J. Soi. and Water Cons. 26:21-24.

13. Kunishi, H.M. and A.W. Taylor. 1972. Immobilization of radiostrontium in soil by phosphate addition. Soil Sci. 113(1):1-6.

14. Mayer, L.M. and S.P. Gloss. 1980. Buffering of silica and phosphate in a turbid river. Limnol. Oceanogr. 25:(1)12-22.

15. Iorns, W.V., C.H. Hombree and G.L. Oakland. 1965. Water resources of the Upper Colorado River Basin. Tech. Rept. U.S. Geol. Surv. Prof. Pap. 441. 370 p.

16. Kellar, P.E., S.A. Paulson and L.J. Paulson. 1980. Methods for biological, chemical and physical analysis in reservoirs. Lake Mead Limnological Res. Ctr., Tech. Rept. No. 5. Univ. Nev., Las Vegas. 234 pp.

17. Sagher, A. 1976. Availability of soil runoff phosphorus to algae. Ph.D. thesis, Univ. Wisconsin-Madison. 192 pp.

18. Williams, J.D.H., H. Shear and R.L. Thomas. 1980.

Availability to Scenedesmus quadricauda of different
forms of phosphorus in sedimentary materials from the
Great Lakes. Limnol. Oceanogr. 25(1):1-11.

19.   Cole, G. and D.M. Kubly. 1976. Limnological studies on
       the Colorado River from Lees Ferry to Diamond Creek.
       Colorado River Research Program. Tech. Rept. No. 8.
       Grand Canyon National Park Report Series. 88 pp.

20.   Gloss, S.P., L.M. Mayer and D.E. Kidd. 1980. Advective
       control of nutrient dynamics in the epilimnion of a
       large reservoir. Limnol. Oceanogr. 25(1):219-229.

21.   Schreiber, J.D., D.L. Rauch and A. Olness. 1980. Phos-
       phorus concentrations and yields in agricultural runoff
       as influenced by a small flood retention reservoir.
       Pages 303-313 in H.G. Stefan, ed. Symposium on surface
       water impoundments. June 2-5, 1980. Minneapolis, MN.

22.   Baker, J.R. and L.J. Paulson. 1980. Influence of Las
       Vegas Wash density current on nutrient availability and
       phytoplankton growth in Lake Mead. Pages 1638-1647 in
       H.G. Stefan, ed. Symposium on surface water impound-
       ments. June 2-5, 1980. Minneapolis, MN.

23.   Paulson, L.J. 1981. Nutrient management with hydro-
       electric dams on the Colorado River system. Lake Mead
       Limnological Res. Ctr. Tech. Rept. No. 8. Univ. Nev.,
       Las Vegas. 39 pp.

24.   Prentki, R.T., L.J. Paulson and J.R. Baker. 1980.
       Chemical and biological structure of Lake Mead
       sediments. Lake Mead Limnological Res. Ctr., Tech.
       Rept. No. 6. Univ. Nev., Las Vegas. 89 p.

25.   Lee, G.F., R.A. Jones and W. Rast. 1980. Availability
       of phosphorus to phytoplankton and its implications for
       phosphorus management strategies. Pages 259-307 in R.C.
       Loehr, C.S. Martin, W. Rast, eds. Phosphorus Management
       Strategies for Lakes. Ann Arbor Science Publishers,
       Inc.

T R I B U T E

T O

T E R R Y D , E V A N S

January 6, 1955 - December 2, 1981

He rode a horse, he drove a boat
He helped us all, he gave us hope
He entered science, he did it well
He had great goals, he did excel
He helped us laugh, he made us cry
He's in our hearts, he'll never die.

- L. J. Paulson

Terry D. Evans was killed in a boating accident in Grand
Canyon, a short time after he presented his paper at the
symposium from which this book is compiled. Terry and two
of his associates were attempting to find a suitable site to
set up a water sampler when a large wave swamped the research
boat, forcing the occupants to abandon it. The boat driver
and the research assistant made it to shore. Terry's body was
recovered after a long search by park personnel on December 29.

Terry's thesis research consisted of a study of the
suspended sediment-phosphorus dynamics in the principal
inflows to Lake Powell and Lake Mead. He was employed as a
research associate in the Lake Mead Limnological Research
Center and supervised the field sampling programs on Lake
Powell, Lake Mead, Lake Mohave, and Lake Havasu. The preliminary results of his research were presented at the symposium.

Terry was awarded a posthumous Master of Science in
Biology and the David Bruce Dill Award in Environmental
Biology at his memorial service.

Terry would have made an outstanding limnologist. He had
already made a significant scientific contribution to our
efforts to better manage the Colorado River resources. He
was a special individual, devoted to his family, friends and
profession. He will never be forgotten by those of us who
continue to work on the Colorado River.

CHAPTER 6

MASS BALANCE MODEL ESTIMATION
OF PHOSPHORUS CONCENTRATION IN
RESERVOIRS

David K. Mueller
  Bureau of Reclamation
  Denver, Colorado

INTRODUCTION

The significance of phosphorus in reservoirs and lakes
stems from its association with the process of eutrophication,
or fertilization of the water body. When eutrophication
becomes advanced, severe water quality problems can develop.
These include blooms of nuisance algae and reduction of dissolved oxygen concentration. Such conditions impact fisheries, domestic water supply, and recreational use, both in
the water body and downstream. Phosphorus, in the form of
phosphate, is a necessary nutrient and has long been considered a major limiting factor to algal growth. This theory
is supported by studies demonstrating that introduction of
phosphorus tends to stimulate algal growth [1] and that
control of phosphorus loading has the opposite effect [2].

Consequently, the need arose for predictive techniques
for the evaluation of phosphorus reduction as a means to
control eutrophication. Since 1969, a variety of models has
been proposed using the input-output or mass balance approach.
Though most model verification has been conducted using data
from natural lakes in Europe and eastern North America, there
has been a tendency to extrapolate validity to lakes and
reservoirs throughout the northern temperate zone. The purpose of the present work is to test that assumption using the
extensive data base developed by the U.S. Environmental
Protection Agency's NES (National Eutrophication Survey).
Specifically, several model formulations are compared as to
their accuracy in predicting phosphorus concentrations in
reservoirs in the western United States.

This paper has been previously published in the Water
Resources Bulletin. It is reprinted here with permission
of the American Water Resources Association.

Model and Data Selection

Several models have been developed from the steady-state solution to a phosphorus mass balance equation proposed by Vollenweider [3]. Five of these, commonly used in lake quality assessment, were chosen for evaluation and comparison.

As listed in Table I, these are:

1. The Vollenweider-1975 model [4], which assumes a constant settling velocity;
2. The Jones-Bachmann model [5], which assumes a constant sedimentation coefficient;
3. The Dillon-Rigler model [6], which uses phosphorus retention calculated from observed data;
4. The Dillon-Kirchner model [7], which estimates phosphorus retention as a function of hydraulic load;
5. The Vollenweider-1976 model [8], which estimates phosphorus flushing from the inverse of hydraulic detention.

Variables used in these models are defined as follows:

P = phosphorus concentration (mg/L)
L = areal phosphorus loading rate (g/m^2/yr)
z = mean lake depth (m)
τ = hydraulic detention time (yr)
q_s = areal hydraulic loading rate, z/τ (m/yr)
R = fraction phosphorus retention
R_p = empirical estimate of R

Data were compiled from NES [9]. Selection was based on the following criteria:

1. The water body was a manmade reservoir located in the western continental United States;
2. All data were available for solution of the phosphorus loading models listed in Table I;
3. The phosphorus retention calculated from inflow and outflow data was greater than zero.

The resultant data set included 68 reservoirs, distributed by state as shown in Figure 1. A statistical summary is given in Table II. Using criteria in Table III, 5 reservoirs were classified oligotrophic, 16 mesotrophic, and 47 eutrophic. This data set then represents wide geographic, hydrologic, morphologic, and trophic ranges.

Table I. Forms of Phosphorus Mass Balance Models

1. Vollenweider-1975

$$P = \frac{L}{10 + z/\tau}$$

2. Jones-Bachmann

$$P = \frac{0.84\ L}{z(0.65 + 1/\tau)}$$

3. Dillon-Rigler

$$P = \frac{L\tau}{z}\ (1-R)$$

4. Dillon-Kirchner

$$R_p = 0.426\ \exp\ (-0.271\ q_s) + 0.574\ \exp\ (-0.00949\ q_s)$$

$$P = \frac{L\tau}{z}\ (1-R_p)$$

5. Vollenweider-1976

$$P = \frac{L/q_s}{(1 + \sqrt{\tau})}$$

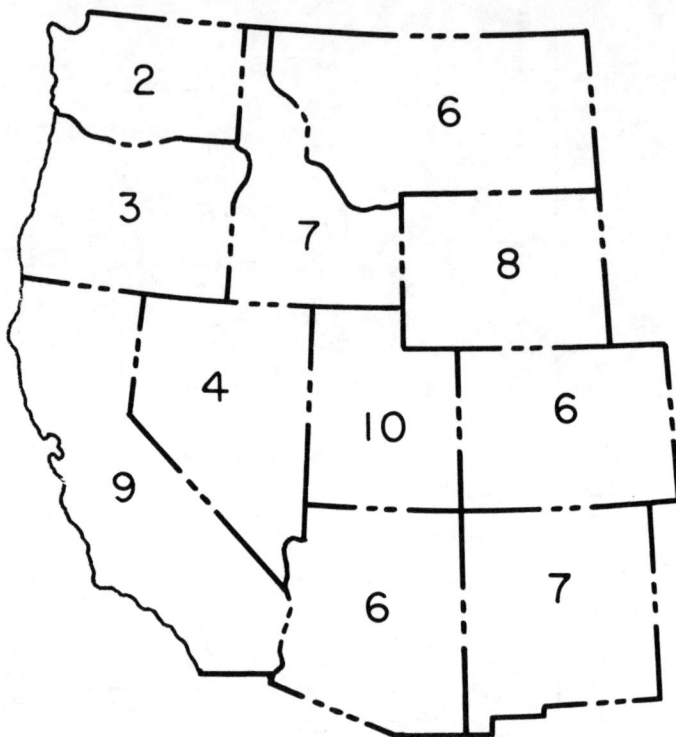

Figure 1. Distribution by state of reservoirs included in the data set.

Table II. Data Statistics

	P (mg/L)	L (g/m²yr)	z (m)	τ (yr)	R −
Arithmetic Mean	0.055	7.44	16.2	1.37	0.51
Geometric Mean	0.037	1.82	11.8	0.67	0.39
Standard Deviation	0.061	25.78	12.7	2.14	0.30
Maximum	0.371	183.51	59.2	15.20	0.99
Minimum	0.007	0.08	0.6	0.003	0.03

Table III. Trophic State Classification Criteria

Comparison	Classification		
	Oligotrophic	Mesotrophic	Eutrophic
Total phosphorus (mg/L)	<0.010	0.010-0.020	>0.020
Chlorophyll a (μg/L)	<4	4-10	>10
Secchi depth (m)	>3.7	2.0-3.7	<2.0
Hypolimnetic D O (% sat.)	>80	10-80	<10

(after Allum et. al.[12])

75

Each model form was fit to the data set using a Gaussian nonlinear fitting algorithm available on the level 8.0 version of the Statistical Package for the Social Sciences [10, 11]. Fitted forms and resulting coefficient values are listed in Table IV. An interesting comparison can be made between the fitted versions of the Vollenweider-1976 and the Jones-Bachmann models. The latter is:

$$P = \frac{0.882\ L}{z(1.61 + 1/\tau)} \qquad (1)$$

Solving to eliminate the coefficient yields:

$$P = \frac{L}{z(1.83 + 1.13/\tau)} \qquad (2)$$

which may be approximated:

$$P = \frac{L}{z(2 + 1/\tau)} \qquad (3)$$

The Vollenweider-1976 best fit version is:

$$P = \frac{L/q_s}{(1 + 2.09\tau^{0.832})} \qquad (4)$$

Multiplication by τ/τ leaves, on rearrangement:

$$P = \frac{L}{z(2.09\tau^{-0.168} + 1/\tau)} \qquad (5)$$

for which equation 3 is, again, a reasonable approximation. For this reason, equation 3 was named the Combination best fit model and was included in the analysis.

ANALYSIS

Two criteria were used to judge model accuracy. The first of these was the root-mean-square error of logarithmically transformed estimations, for which the computational form is:

$$S_m = \left\{ \frac{\sum\limits_{i\ =\ 1}^{68} \left[\log_{10}(P_o)_i - \log_{10}(P_e)_i\right]^2}{d_f} \right\}^{1/2} \qquad (6)$$

76

Table IV. Best Fit Models

Type	Form	Coefficients
Vollenweider-1975	$P = \dfrac{L}{\alpha + z/\tau}$	$\alpha = 16.4$
Jones-Bachmann	$P = \dfrac{\alpha L}{z(\beta + 1/\tau)}$	$\alpha = 0.882$ $\beta = 1.61$
Dillon-Kirchner 1/	$R_p = \alpha \exp(\beta q_s) + (1-\alpha) \exp(\gamma q_s)$ $P = \dfrac{L\tau}{z}(1-R_p)$	$\alpha = 0.290$ $\beta = -0.556$ $\gamma = -0.00483$
Vollenweider-1976	$P = \dfrac{L/q_s}{1 + \alpha \tau^\beta}$	$\alpha = 2.09$ $\beta = 0.832$

1/ Due to numerical problems in the computational algorithm only 64 reservoirs were included in fitting this equation.

where S_m = standard error of estimation for the model
     P^o = observed phosphorus concentration (mg/L)
     P^e = estimated phosphorus concentration (mg/L)
     d_f = model degrees of freedom

The variable d_f was calculated for each model as the difference between sample size and the number of fitted parameters in the model.

Confidence intervals for the estimation at the meso-eutrophic boundary phosphorus concentration can then be calculated from the model error values:

$$CL = 0.020 * 10^{\left[\pm t_{d_f}^{\alpha/2} (S_m) \right]} \qquad (7)$$

where CL is the upper or lower confidence limit in mg/L.

The second accuracy criterion was the correlation between observed and estimated phosphorus concentrations. This was calculated as the Pearson product moment coefficient. A comparison of standard errors, 90 percent confidence intervals, and correlation coefficients is given in Table V. The Dillon-Rigler model is the only one which achieves a tolerable fit with the observed data (R = 0.86). Graphical results of estimated vs. observed phosphorus concentration for the Dillon-Rigler and Combination models are shown in Figures 2 and 3.

Using the two judgment criteria, the models were grouped by relative accuracy as shown in Table VI. Significance of differences between groups was then tested by comparing mean standard errors and correlation coefficients. Squared error values were compared with an F test of variance. Correlation coefficients were tested using the Fisher transformation method, as described by Bryant [13, p. 140]. Test statistic values and significance levels are given in Table VII. These results leave little doubt that the Dillon-Rigler model is indeed more accurate than any other. The differences among the remaining groups are also significant, though to a lesser degree between groups 3 and 4 and between groups 4 and 5.

RESULTS

The Dillon-Rigler and Combination models were used to develop standard Dillon graphs (Figures 4 and 5) of computed areal phosphorus load vs. reservoir depth. These graphs show the effect of model uncertainty on the prediction of trophic state. The dashed line indicating a phosphorus

Table V. Model Statistics for 68 Reservoirs

	Standard 1/ error	Correlation 1/ coefficient	90% Confidence 2/ interval
Original Formulations:			
Vollenweider-1975	0.417	0.52	0.004-0.097
Jones-Bachmann	0.367	0.65	0.005-0.080
Dillon-Rigler	0.200	0.86	0.009-0.043
Dillon-Kirchner	0.387	0.56	0.005-0.086
Vollenweider-1976	0.387	0.64	0.005-0.087
Best Fit Formulations:			
Vollenweider-1975	0.407	0.48	0.004-0.097
Jones-Bachmann	0.327	0.67	0.006-0.069
Dillon-Kirchner	0.371	0.60	0.005-0.081
Vollenweider-1976	0.324	0.68	0.006-0.068
Combination	0.325	0.68	0.006-0.068

1/ Based on $\log_{10}$ transformed values.
2/ For P_e = 0.020 mg/L.

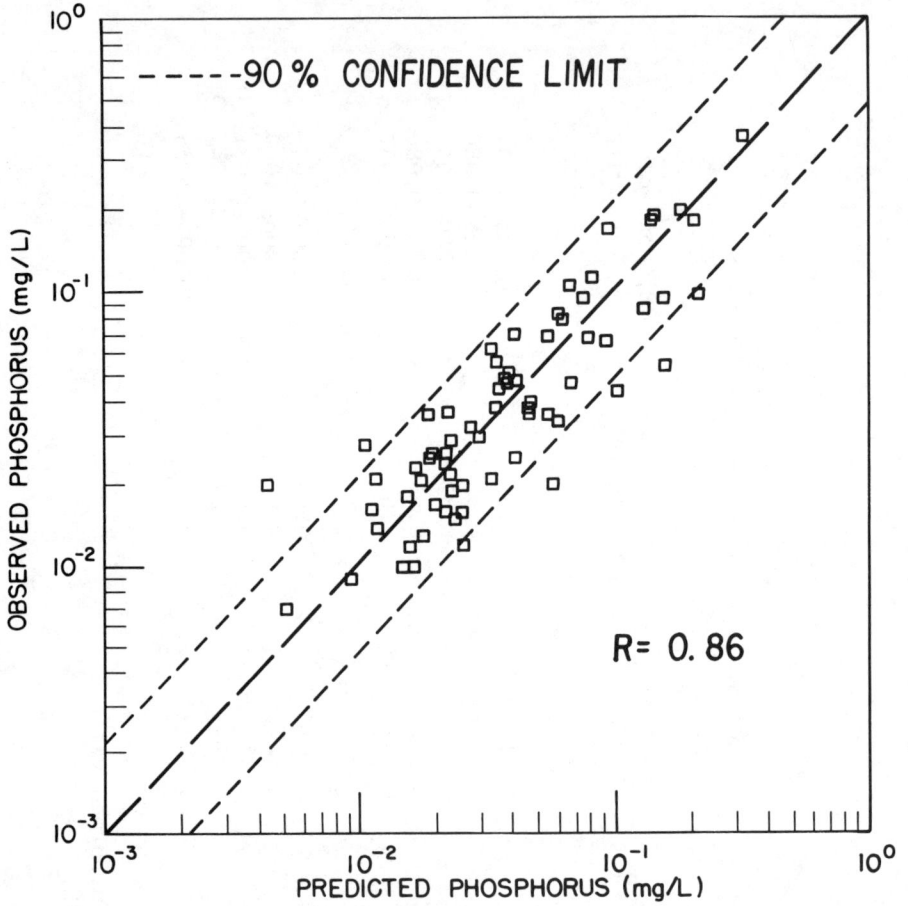

Figure 2. Comparison of observed phosphorus concentrations
and Dillon-Rigler estimations showing the
90 percent confidence interval.

COMBINATION MODEL

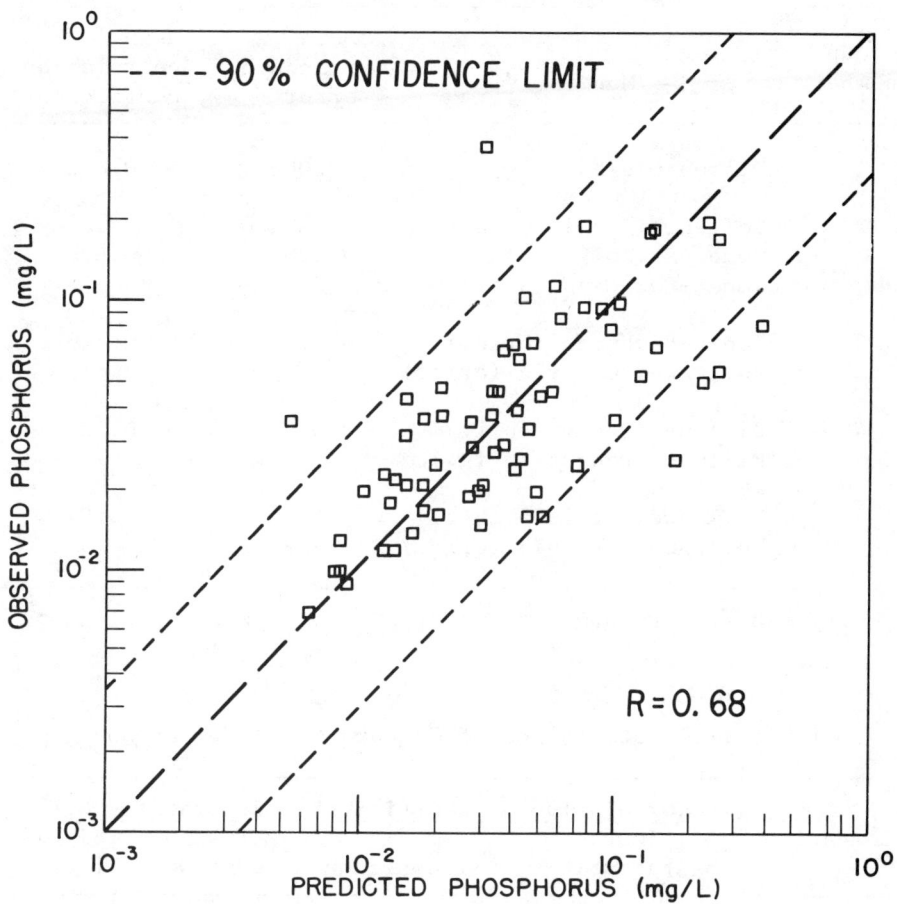

Figure 3. Comparison of observed phosphorus concentrations and Combination model estimations showing the 90 percent confidence interval.

Table VI. Models Grouped by Relative Accuracy

Group number	Models	Mean standard error	Mean correlation coefficient
1	Dillon-Rigler	0.200	0.862
2	Vollenweider-1976 (b.f.) Combination Jones-Bachmann (b.f.)	0.325	0.677
3	Jones-Bachmann (orig.) Vollenweider-1976 (orig.)	0.377	0.643
4	Dillon-Kirchner (b.f.) Dillon-Kirchner (orig.)	0.379	0.581
5	Vollenweider-1975 (b.f.) Vollenweider-1975 (orig.)	0.412	0.501

b.f. = best fit formulation; orig. = original formulation

Table VII. Statistics of Grouped Model Comparison

Comparison	Correlation Coefficient		Standard Error	
	Test statistic	Significance	Test statistic	Significance
1 vs. 2	2.76	<0.01	2.64	<0.01
2 vs. 3	0.35	0.36	1.35	0.11
2 vs. 4	0.92	0.18	1.36	0.11
2 vs. 5	1.57	0.06	1.61	0.03
3 vs. 4	0.58	0.28	1.01	0.48
3 vs. 5	1.24	0.11	1.19	0.24
4 vs. 5	0.66	0.26	1.18	0.24

DILLON PLOT- DILLON-RIGLER MODEL

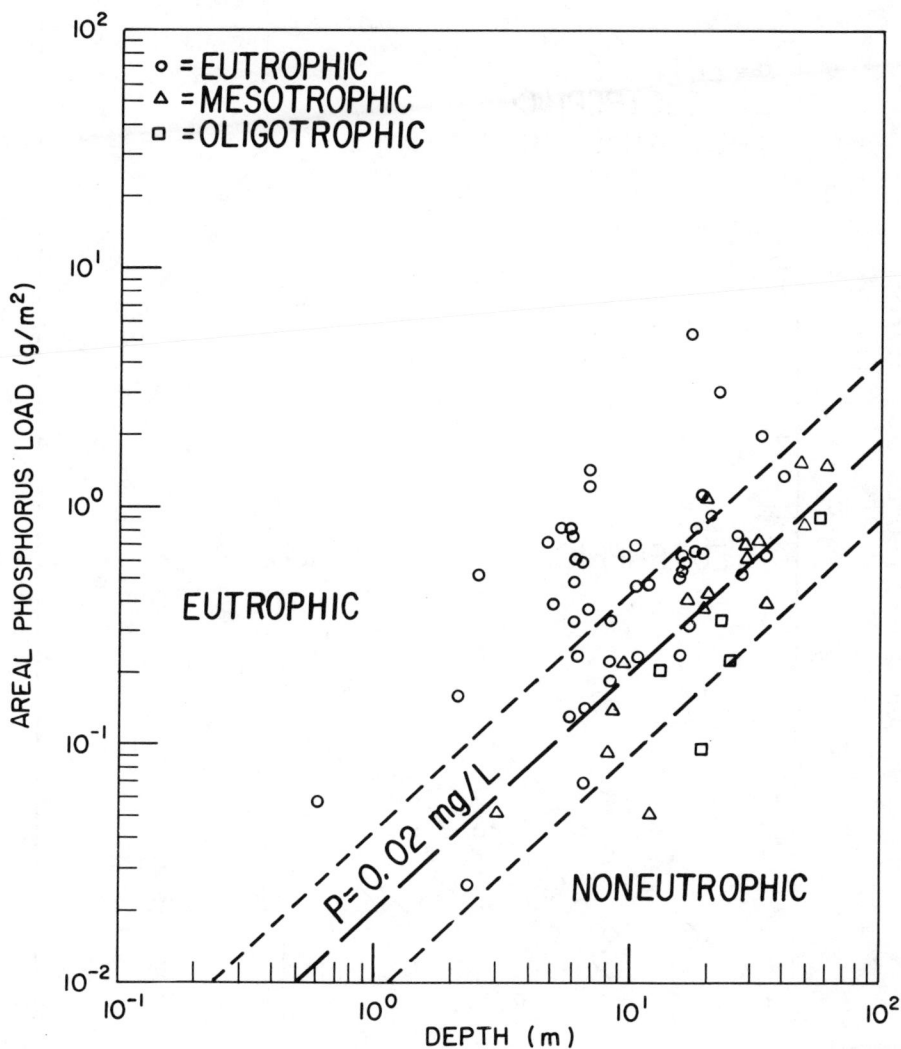

Figure 4. Dillon graph of Dillon-Rigler model results
showing the eutrophic-noneutrophic boundary
phosphorus concentration with a 90 percent
confidence interval.

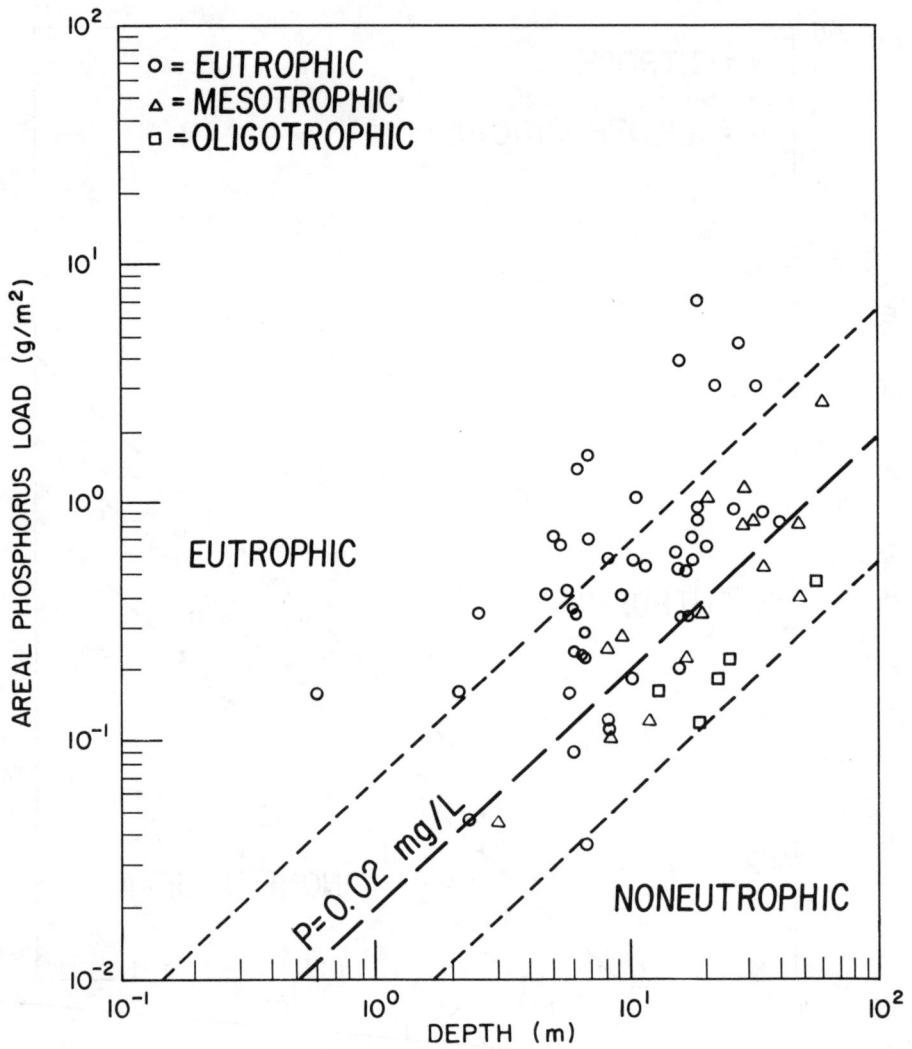

Figure 5. Dillon graph of Combination model results showing
the eutrophic–noneutrophic boundary phosphorus
concentration with a 90 percent confidence
interval.

concentration of 0.020 mg/L, which is usually considered to
be a boundary separating eutrophic and noneutrophic classifi-
cations, is seen here as the center of a range in which classification cannot be made with much confidence. Reservoirs
which plot outside this range can be classified with less
than 5 percent chance of error. The Dillon-Rigler model,
which has the smaller standard error, and, therefore, less
uncertainty, allows confident classification of 28 out of
68 reservoirs, compared to only 18 for the Combination model.

The probability of trophic state classification can
also be plotted as a function of estimated phosphorus concentration (Figure 6), in the manner proposed by Reckhow [14].
Curves in Figure 6 were developed using the standard variate
(z) normalized about a critical concentration of 0.020 mg/L:

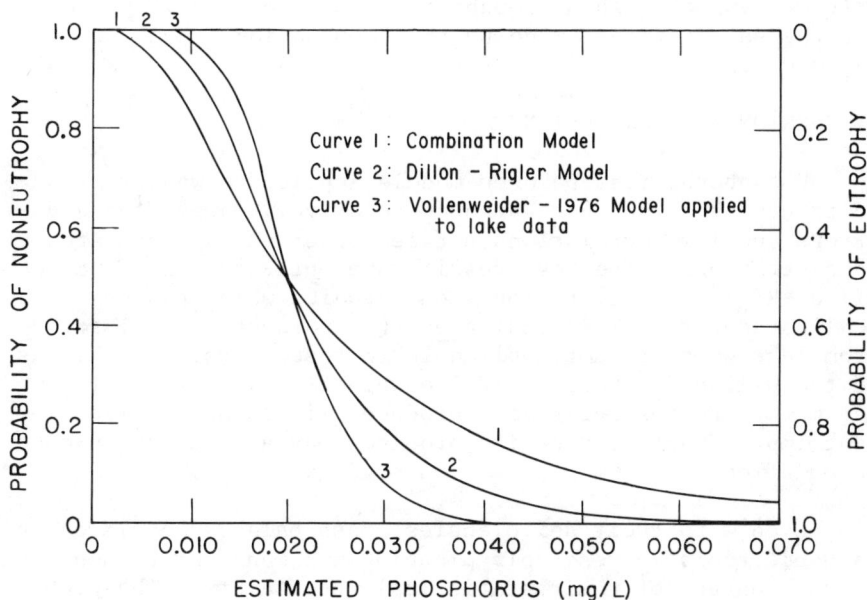

Figure 6. Probabilities of eutrophy and noneutrophy
associated with estimated phosphorus concentration. Data for Curve 3 from Chapra and
Reckhow [15].

$$z = \frac{\log_{10}(0.020) - \log_{10}(P_e)}{S_m} \qquad (8)$$

The three curves represent: (1) the Combination model, (2) the Dillon-Rigler model, and (3) the Vollenweider-1976 model applied to 117 North American natural lakes with a standard error (S_m) of approximately 0.17 [15]. Again, the advantage of a smaller standard error can be seen. As S_m increases, the uncertainty also increases at all values of estimated phosphorus except the critical concentration. For example, a reservoir with a phosphorus concentration of 0.030 mg/L estimated by the Dillon-Rigler model (S_m = 0.200), has an 81 percent probability of being accurately classified eutrophic (having a phosphorus concentration greater than the critical value). This probability is reduced to 71 percent for the same estimate made by the Combination model (S_m = 0.325).

DISCUSSION AND CONCLUSIONS

Phosphorus mass balance models applied to western reservoirs have been shown to produce relatively large standard errors and low correlations between observed and estimated concentrations. The best results are obtained using the Dillon-Rigler model, the only one studied which has no empirical parameters. This indicates that parameters calibrated from lake data are not applicable to reservoirs. In the case of the Dillon-Kirchner and Vollenweider-1975 models, fitting parameters to reservoir data produced little or no significant improvement, indicating problems may also exist in model formulation.

Even with their deficiencies, mass balance models can be valuable tools in reservoir planning and management when used in the context of their statistical uncertainty. The probability of favorable or unfavorable results can be evaluated for the operation of existing reservoirs or the design of new ones. Results of this study indicate that the Dillon-Rigler model is the best choice for application to existing reservoirs. In the case of planned impoundments, for which phosphorus retention data would obviously be unavailable, the best fit Vollenweider-1976 model can be used with least uncertainty. However, the best fit Jones-Bachmann and Combination models would provide statistically similar results.

While it is clear that these methods are not ideally suited for application to western reservoirs, they provide a basis for judgment of improved techniques. New or modified models should be accepted on the basis of their ability to reduce the uncertainty inherent in the currently available ones.

REFERENCES

1. Schindler, D. W., F.A.J. Armstrong, S. K. Holmgren, and G. J. Brunskill. 1971. Eutrophication of Lake 277, Experimental Lakes Area, Northwestern Ontario, by Addition of Phosphate and Nitrate. J. Fish. Res. Board Can. 28: 1763-1782.

2. Edmondson, W. T. 1972. The Present Condition of Lake Washington, Int. Ver. Theor. Agnew. Limnol. Verh. 18: 284-291.

3. Vollenweider, R. A. 1969. Possibilities and Limits of Elementary Models Concerning the Budget of Substances in Lakes. Arch Hydrobiol. 66(1): 1-36.

4. Vollenweider, R. A. 1975. Input-Output Models with Special Reference to the Phosphorus Loading Concept in Limnology. Schweiz. Z. Hydrol. 37(1): 53-84.

5. Jones, J. R., and R. W. Bachmann. 1976. Prediction of Phosphorus and Chlorophyll Levels in Lakes. J. Wat. Pol. Control Fed. 48(9): 2176-2182.

6. Dillon, P. J., and F. H. Rigler. 1974. A Test of a Simple Nutrient Budget Model Predicting the Phosphorus Concentration in Lake Water. J. Fish. Res. Board Can. 31(11): 1771-1778.

7. Kirchner, W. B., and P. J. Dillon. 1975. An Empirical Method of Estimating the Retention of Phosphorus in Lakes. Wat. Res. Res. 11(1): 182-183.

8. Vollenweider, R. A. 1976. Advances in Defining Critical Loading Levels for Phosphorus in Lake Eutrophication. Mem. Ist. Ital. Idrobiol, 33: 53-83.

9. U.S. EPA. 1978. A Compendium of Lake and Reservoir Data Collected by the National Eutrophication Survey in the Weatern United States. Working Paper No. 477, U.S. EPA Corvallis Environ. Res. Lab., Corvallis, Oregon, and Environ. Mon. and Supp. Lab., Las Vegas, Nevada, 168 pp.

10. Robinson, B. 1979. SPSS Subprogram NONLINEAR - Non-linear Regression. Manual No. 433, Vogelback Computing Center, Northwestern University, 27 pp.

11. Nie, N. H., C. H. Hull, J. G. Jenkins, K. Steinbrenner, and D. H. Brent, 1975. SPSS, 2nd edition. McGraw-Hill, New York, New York.

12. Allum, M. O., R. E. Glessner, and J. H. Gakstatter. 1977. An Evaluation of the National Eutrophication Survey Data. Working Paper No. 900, U.S. EPA Corvallis Environ. Res. Lab., Corvallis, Oregon, 79 pp.

13. Bryant, E. C. 1966. Statistical Analysis. McGraw-Hill, New York, New York.

14. Reckhow, K. H. 1976. Uncertainty Analysis Applied to Vollenweiders Phosphorus Loading Criterion. J. Water Pol. Control Fed. 50(8): 2123-2128.

15. Chapra, S. C., and K. H. Reckhow. 1979. Expressing the Phosphorus Loading Concept on Probabilistic Terms. J. Fish. Research Board Can. 36(2): 225-229.

CHAPTER 7

ANALYSIS OF POTENTIAL SEDIMENT TRANSPORT IMPACTS
BELOW THE WINDY GAP RESERVOIR, COLORADO RIVER

Timothy J. Ward
   New Mexico State University,
      Las Cruces
   Research Institute of Colorado,
      Ft. Collins

John Eckhardt
   Northern Colorado Water Conservancy
      District
   Loveland, Colorado

INTRODUCTION

Background

Development of water resources in the upper Colorado
River Basin is a difficult task due to internal and external
demands of users. A significant use of water is by trans-
mountain diversion to the agricultural lands and population
centers of the Colorado Front Range. In order to meet demands of Front Range water users, increased diversions have
become necessary. These diversions will be met in part by
construction of a small forebay reservoir and a pumping plant
capable of up to 16.3 cms withdrawal (at this time) from the
Colorado River. This reservoir, with normal maximum volume
of 0.0005 km^3, will be located near Granby, Colorado
(Figure 1) and is referred to by the name of a nearby geologic feature, Windy Gap. Water pumped from the Windy Gap
Reservoir will be piped back into Lake Granby and then conveyed through the Colorado-Big Thompson Project to Eastern
Colorado.

Owners of the reservoir and pumping plant, the Northern
Colorado Water Conservancy District, saw a need to study the
effects of withdrawing water from the river system. An
aquatic ecologist, Dr. Robert Erickson, was hired and subsequently recommended an indepth investigation of the hydrology
and sediment transport of the river. The investigation
focused on post-pumping effects of potential aggradation in
the stream channel below the reservoir and downstream

91

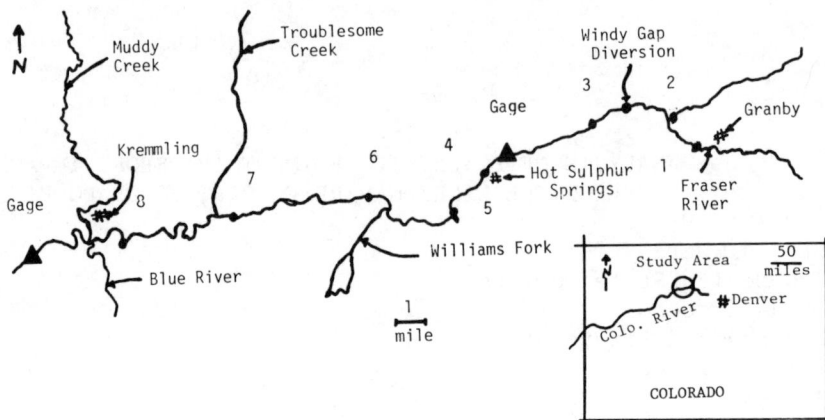

Figure 1. Sketch map of study area showing key locations and sampling sites (one mile equals 1.61 km).

for approximately 48 km (Figure 1). Excessive aggradation created by reductions of the sediment transporting capacity of the river could create situations where habitat conditions of food, protection, and spawning beds would no longer support the current trout population. Excessive aggradation would have a significant impact because this segment of the Colorado River supports several private and public fishing reaches.

What was required were estimates of potential aggradation (or degradation) for several intermediate river reaches using 20 years as a base. Water years (WY) 1958 through 1977 were selected to determine pre- and post-pumping flow conditions. These flow conditions along with collected field data were then used to calculate hydraulic and sediment transport characteristics at selected sites. A mass balance approach for distributing discharges between points was employed. For aggradation computations, another mass balance between end sections of the selected reach was used. Results using this approach indicated that flow conditions were sufficient to prevent excessive aggradation at the selected sites for the assumed conditions. Details of the study are presented by Ward [1].

HYDROLOGY AND CHANNEL MORPHOLOGY

The Colorado River between its headwaters on the Continental Divide and the confluence with the Blue River near

92

Kremmling is controlled by Lake Granby, Grand Lake, and Shadow Mountain Reservoir, all above Windy Gap (WG) Reservoir. Tributaries above WG Reservoir include the Fraser River (710 km^2), and Willow Creek (347 km^2) which is controlled by Willow Creek Reservoir. Contributing area to the WG Reservoir is 2023 km^2, of which 837 are controlled by Lake Granby. The major controlled tributary is the Fraser River which is estimated to produce about 60% of the inflow.

Between the WG Reservoir and the confluence with the Blue River, the river is influenced by two major tributaries, 15 minor tributaries, and numerous diversions for agricultural and domestic needs. The two major tributaries are the Williams Fork River (598 km^2) which is controlled by the Williams Fork Reservoir and Troublesome Creek (440 km^2) which is not regulated to any extent. The minor tributaries include approximately 388 km^2. The major and minor tributaries account for over 90% of the contributing area between the reservoir site and the Blue River confluence. However, inflows from the minor tributaries are relatively insignificant in comparison to the major one. Over 20 diversions have been identified with the largest water right being 1.8 cms (one cms equals 35.3 cfs) [2].

Although complete, long-term records for existing and abandoned U.S. Geological Survey (USGS) gaging sites are not available, enough record is available for simulation of daily flows at key locations or tributaries. Because of the long-term records, the Hot Sulphur Springs flows were selected as a point of discussion for the entire segment.

The 20-year base from WY 1958 through WY 1977 represents the flow conditions during operation of Lake Granby. Flow statistics for this period are shown in Table I for Hot Sulphur Springs. Even with the upstream controls there was significant variation in the flow.

Table I. Flow Statistics for Hot Sulphur Springs
          Gage, Water Years 1958 through 1977.

Statistic	Arithmetic Average	Range
Peak, cms	40	81-9.8
Minimum daily, cms	1.5	2.0-1.2
Yield, km^3	0.20	0.43-0.10

Although the flow has been quite variable and effects of
upstream regulation appear in the long-term record, significant historic changes in river form were not detectable.
Aerial photographs from 1938, 1950, 1967, and 1974 were
obtained and analyzed. During the period only two noticeable
changes occurred, both the Fraser River and Troublesome Creek
straightened naturally or were straightened near their confluences with the mainstem Colorado.

In general, the river flow during the 20-year period was
quite variable. Unfortunately, corresponding discharge measurements were not taken on all the major or minor tributaries
during that period, necessitating a mass balance approach for
distributing river flows from measured and simulated data.
Even with the variability, physical conditions of the river
bed were such that few significant changes occurred.

SEDIMENT TRANSPORT

Sediment available for transport in this segment of the
Colorado River is derived from upstream inflows, tributary
inflows, or the channel bed and banks. In order to ascertain
the type and magnitude of sediments in transport and available for transport, an intensive sampling and measurement
program was conducted during the spring, summer and fall of
1980. Eight sites were selected for sampling of suspended
material and bed load. Measurements for five different flows
at each site were collected for a total of 40 site-samples.
Samples on the rising and falling limbs of the 1980 runoff
hydrograph were fortuitously chosen (Figure 2). In addition

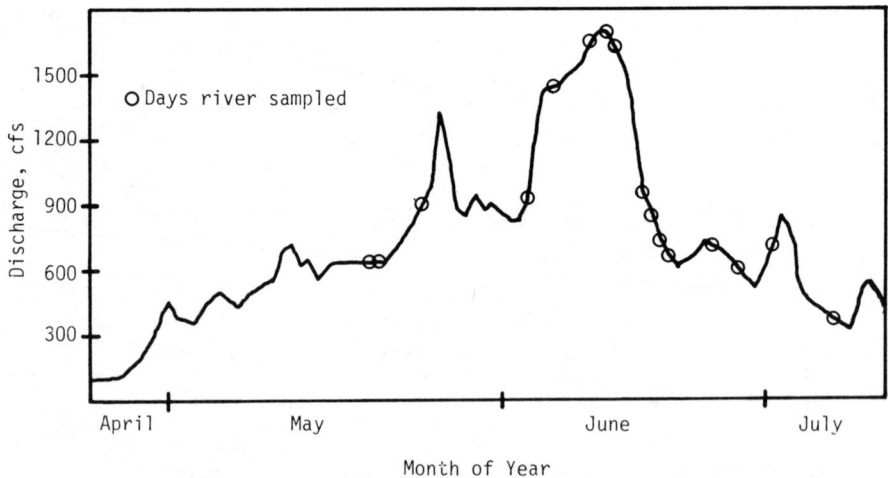

Figure 2.   Daily discharges at Hot Sulphur Spring Gage for
            WY 1980 showing days river sampled at various
            sites. (One cfs equals 0.028 cms).

94

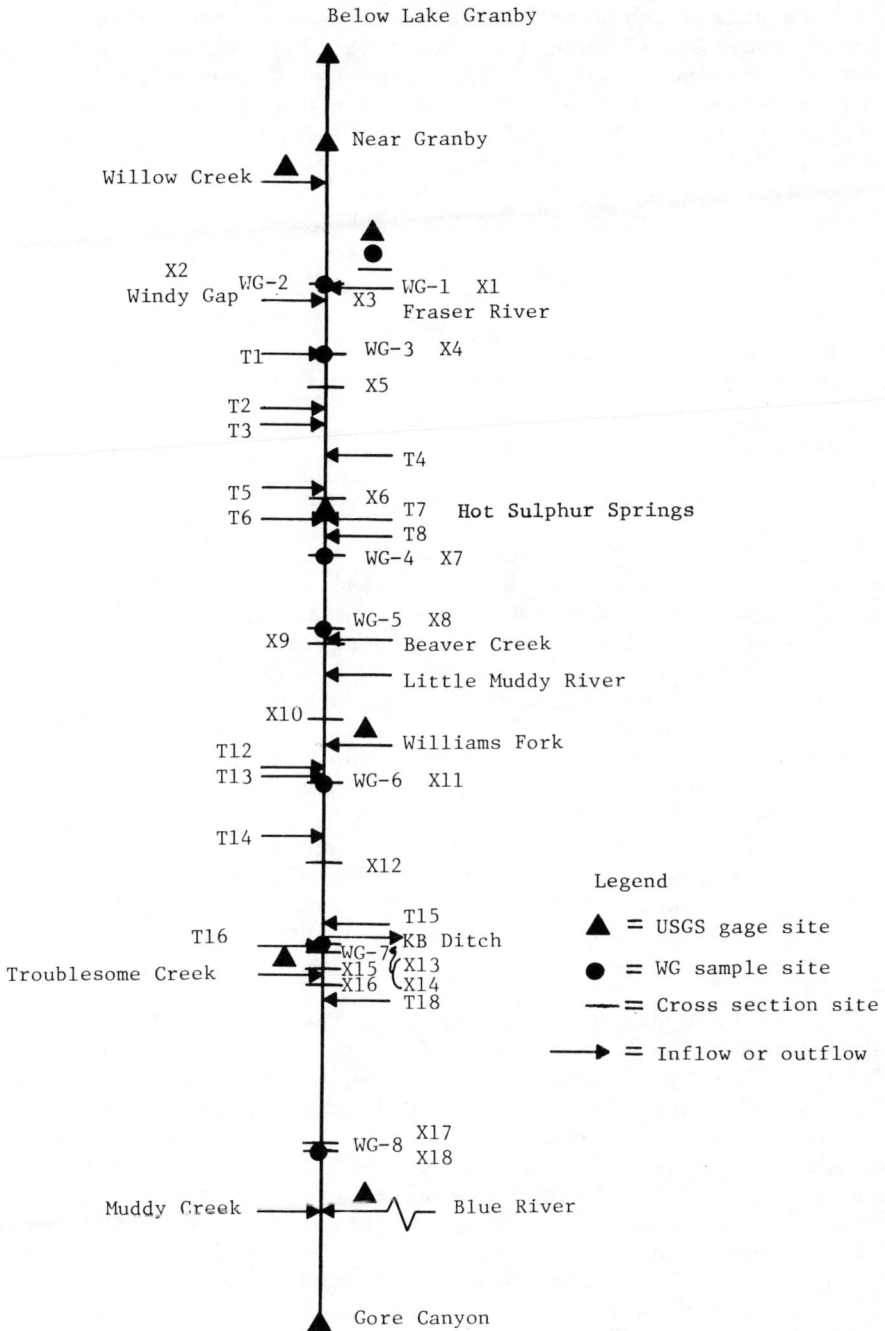

Figure 3. System schematic of the study area. Important tributaries and locations, minor tributaries (T), sampling sites (WG-1) and cross sections (X) are shown. Flow is from top of page. Drawn to scale.

to these sites, ten other cross sections were chosen for
further sampling (Figure 3). These 18 cross sections were
surveyed and samples of surface armor and subsurface material
were collected from the active stream bed and the near-bank
bed. Statistics for the sampled material are presented in
Table II.

Table II. Sediment Size Statistics for Cross
Sections Based on Sieve Analyses.

Statistic	Surface		Subsurface	
	Average	Range	Average	Range
Median size, mm	87	100-62	26	70-5
Gradation [3]	1.26	1.44-1.13	4.88	9.28-2.71

As expected the surface layer was much coarser and better sorted than the subsurface material it protected. There
was not significant difference between the near-bank and
active stream samples. Although a weak relationship between
size and downstream distance could be inferred for the surface layer, none was apparent for the subsurface material.
This supports the previous finding that the channel hasn't
changed its position to any extent thus indicating little, if
any, disturbance and reworking of the subsurface material.

The observation of the intact armour layer also explains
the relatively low transport rates of the 40 samples (Figures
4 and 5). Suspended material (silts with some clay) is not
derived from the channel, suggesting upstream or tributary
inflows as the source.

Field inspections indicated that the suspended material
was being derived from overland flow and fine bank materials
of the tributary watersheds, including the Fraser River. For
the measured conditions, suspended load was significantly
less than transport capacity. Similarly, bed load transport
rates were found to be only 0.1 to 0.01 of the potential
transport rates based on the Meyer-Peter, Müller tractive
force formula [3]. Of the bed load transport, an average of
78% was less than 2mm in size. Again, it was determined that
the minor amounts of bed load were derived from upstream
and lateral inflows. The major lateral inflow of sediment
is Troublesome Creek. The other tributaries are relatively
less active and have a minor impact.

Preliminary transport computations indicated two things.
First, suspended load transport capacity of post-pumping

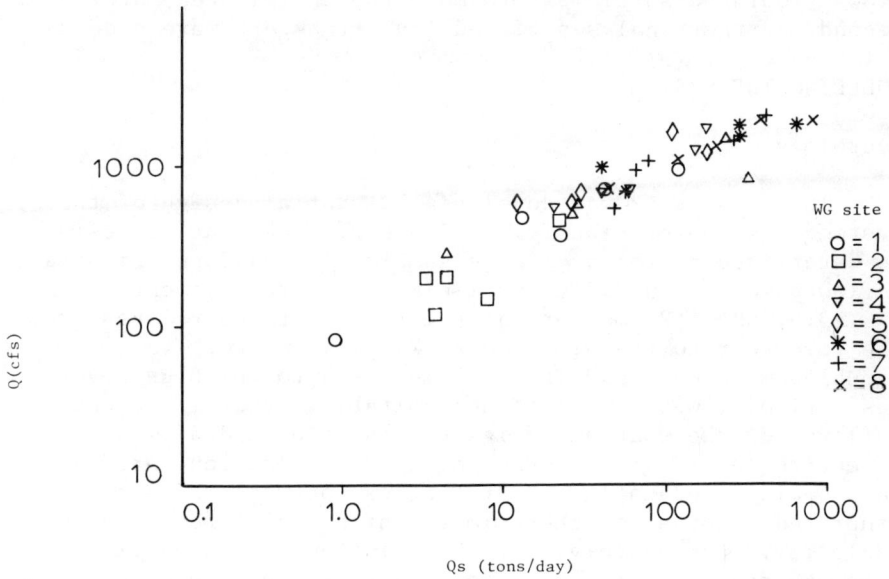

Figure 4. Suspended load transport rates at sample sites. Q is discharge and Qs is sediment transport (one cfs equals 0.028 cms and one ton/day equals 0.91 tonnes/day)

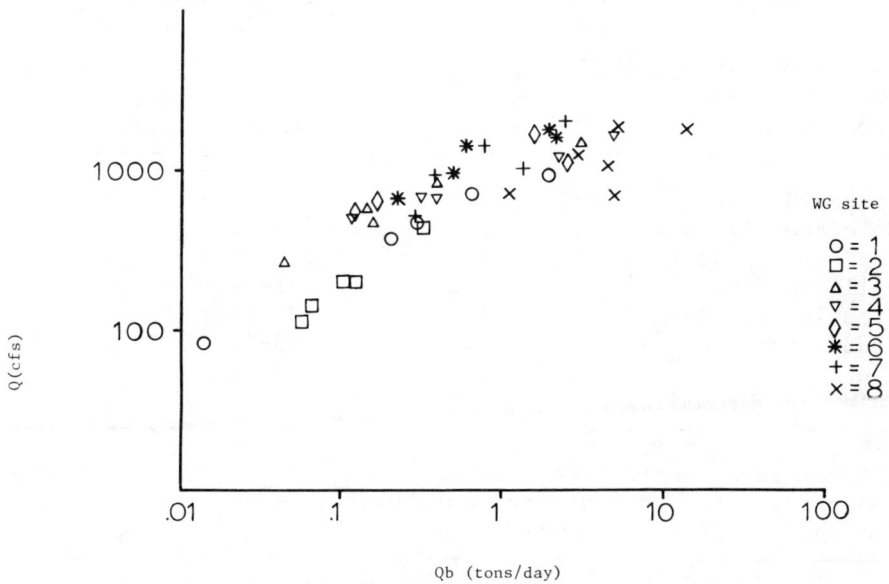

Figure 5. Bed Load transport rates at sample sites. Q is discharge and Qb is sediment transport (one cfs equals 0.028 cms and one ton/day equals 0.91 tonnes/day).

flows would be sufficient to move the anticipated materials. Second, further analyses of bed load transport were needed.

MODELING THE SYSTEM

Hydrology

The first task was developing a spatial design of the system as shown previously in Figure 3. The spatial design is determined by the available sample sites and the important inflow points. In addition, minimum flow requirements of 2.5, 3.8, and 4.2 cms (or natural flow) for the reaches from the reservoir to the Williams Fork, Williams Fork to Troublesome Creek, and Troublesome Creek to the Blue River, respectively, were imposed, necessitating other locations. Inflows for the spatial design points were needed. Partial or entire records for the 20-year period were generated for major points and inflows. Statistics for measured and generated records for the primary inflows are presented in Table III. Fortunately, for those inflows where partial or complete records were generated, historic records existed which were related to other long-term measurements in the same watershed or at nearby stations. Flows were then generated from information at the nearby stations.

Table III. Statistics of Measured and Generated Discharge for Key Inflows (daily flows in cms).

Name	Average	Range	Remarks
Fraser River, WG-1	3.2	0.9-48.5	G
Colorado River, WG-2	2.1	0.3-46.1	C
Hot Sulphur Springs	6.4	1.2-76.7	C
Williams Fork	2.8	0.1-34.1	P
Troublesome Creek	1.2	.03-16.7	G
Muddy Creek	4.7	.03-51.7	G
Blue River	11.9	1.2-53.5	C
Colo. nr. Kremmling	27.3	7.00-185.3	P

C = complete record
P = partial record generated (one year or less)
G = entire record generated

Flow distribution was conducted on a mass balance approach. Two long reaches, the reservoir to Hot Sulphur Springs and Hot Sulphur Springs to the Gore Canyon gage near

98

Kremmling, were utilized. Daily flows were considered.
Gains or losses in the reach were computed as known outflow
minus known (or generated) inflows. Gains were distributed
to the minor tributaries and non-point sources based on
draingae area and elevation. Losses were removed according
to irrigated area, potential diversion, and near-river, non-
irrigated area. When losses occurred, tributary inflows
were assumed to be zero. Once the major, minor, and non-
point inflows and outflows were determined, the appropriate
river flows were distributed to the cross-sections and sample
sites. This provided the initial or base period of pre-
pumping discharges.

The post-pumping discharges were found by imposing the
previously discussed flow constraints at the appropriate
cross-sections, finding the minimum difference between pre-
pumping and constraint values, then using that difference as
the maximum pumping rate if it was not greater than 16.8 cms.
Other constraints on pumping include maximum yearly withdrawals, 10-year average withdrawals, and senior water rights
"calls" on the river. Only the last constraint was considered in addition to the 16.8 cms pumping right and minimum
flows because the other two would permit increased flow and
higher sediment transport, a beneficial result. Generally
"calls" run for about eight months and pumping would be
permitted in the period between about April and August, an
average of 135 days per year. Meeting all these constraints
resulted in the post-pumping flows in the river and the
potential pumping for the diversion.

Sediment Transport

Sediment transport in the river is controlled by upstream and inflow supply because of the heavy bed armor.
Supply is currently (and historically) low so that transport
capacity exceeds supply for the base period. A sediment mass
balance between the WG sampling sites was conducted. Assuming the measured loads were indicative of supply, the
current gains and losses for the 20-year period using daily
flows were computed from empirical relationships. These
loads are presented in Table IV. Except for WG-4, the loads
are very consistent as expected from supply control, i.e.
everything is transported. The load at WG-8 shows the
effects of Troublesome Creek. Gains between sites were
interpreted as lateral inputs that would exist for post-
pumping flows. These loads along with the post-pumping
flows set the conditions for potential aggradation.

Potential sediment transport was computed using a form
of the Meyer-Peter, Müller equation modified to account for

Table IV. Computed Bed Load Passing WG Sampling
Sites for the Base Period.

Site	Load Passing Tonnes	Gain or Loss in Reach Tonnes
WG-1 + WG-2	784	--
WG-3	901	117
WG-4	1992	1091
WG-5	916	-1076
WG-6	913	-3
WG-7	1214	301
WG-8	4935	3721

shear stress against the grain created by the flow velocity.
This theoretical transport was compared with measured values
to confirm that grain movement did occur for the various
sediment sizes collected. Transport rates for individual
grain-size fractions were computed and comparison with
measured data indicated that potential transport was 10 to
100 times greater. This is reasonable as steep channel
(average bed slope was 0.006 in the mainstem) experiments
indicate that bed loads can easily be greater than 500 ppm,
a level never approached in any sample. All of these findings and computations led to the following results.

RESULTS AND CONCLUSIONS

Pumping Rates

Application of three of the five constraints indicated
that pumping could occur 2232 days out of a possible 2692
over the 20-year period. The average rate would be 6.2 cms
or .06 km^3 per year. Fifty-four percent of the time pump
rates of 4.2 cms or less would be permitted. Seventeen percent of the time the maximum rate of 16.8 cms could be
attained.

River Flows

The effect on the river flows varied from year to year.
As comparisons, a "wet", high runoff year (WY 1962) and a
"dry", low runoff year (WY 1977) are shown in Figures 6
and 7. Note the scale differences. The effects on transport
capacity at the WG sites for the two years are presented in
Table V.

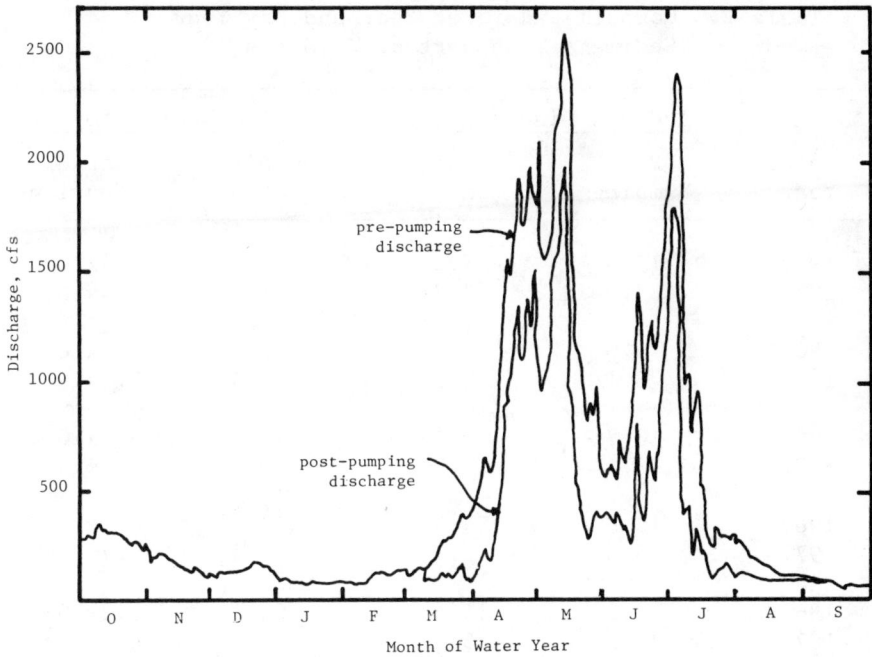

Figure 6. Pre- and post-pumping flows at Hot Sulphur Springs
Gage WY 1962 ("Wet" Year) (one cfs equals 0.028
cms).

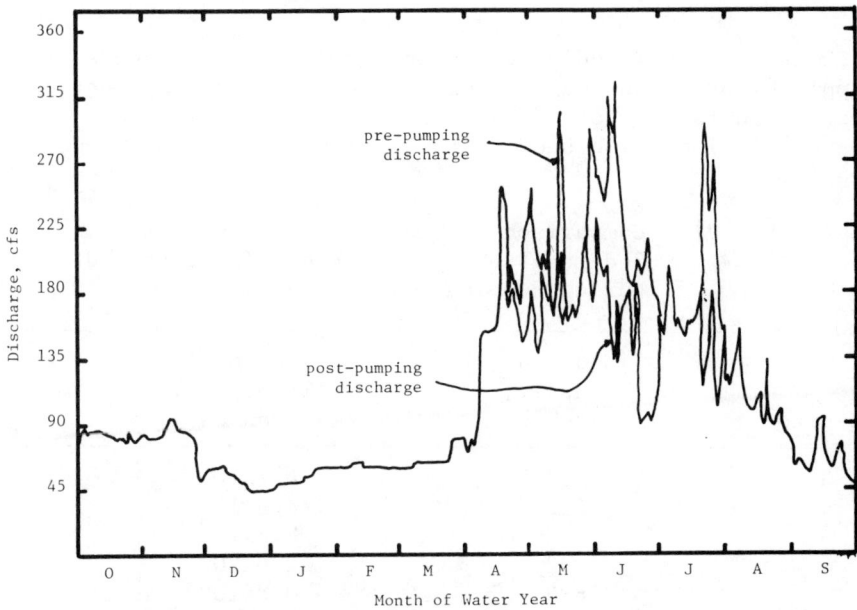

Figure 7. Pre- and post-pumping flows at Hot Sulphur Springs
Gage WY 1977 ("Dry" Year) (one cfs equals 0.028
cms)

101

Table V. Comparison of Wet Year and Dry Year
Sediment Transport at WG Sites.

		Average Discharge, cms		Transport, tonnes	
WG	Water Year	Pre-Pumping	Post-Pumping	Supply equations	Capacity equations
3	1962	12.0	7.2	55	1.4×10^5
	1977	3.0	2.7	1	1.4×10^3
4	1962	14.0	9.1	130	1.5×10^5
	1977	3.1	2.8	0.45	7.3×10^3
5	1962	13.7	9.1	45	6.5×10^7
	1977	3.1	2.8	0.36	3.5×10^6
6	1962	19.8	14.9	79	6.6×10^5
	1977	6.0	5.7	6.4	1.7×10^4
7	1962	19.2	14.4	104	5.2×10^5
	1977	5.5	5.2	19	8.8×10^4
8	1962	21.0	16.1	458	1.2×10^5
	1977	6.0	5.7	43	1.2×10^4

As Table VI. indicates, transport capacity exceeds
supply by orders of magnitude for the entire 20-year period.
These results, composited and averaged for the entire
20-year period, also show that capacity exceeds the assumed
supply.

Table VI. Comparison of Theoretical Transport and
Sediment Supply Over the 20-Year Period
During Pumping.

WG-Site	Sediment Supply, Tonnes	Sediment Transport, Tonnes
3	138	6.7×10^5
4	1232	8.3×10^5
5	1232	8.6×10^7
6	1232	2.6×20^6
7	1683	3.8×10^6
8	5388	$7.5.10^5$

CONCLUSIONS

A field and computer model study was conducted in order to determine potential downstream aggradation from a river diversion. Field data indicate, and process models confirm, that if the observed conditions represent the past and future system, no significant aggradation should occur. However, conditions leading to increased sediment loading to the stream, such as wildfire or flash flooding, may create temporary situations where aggradation can become a problem. Under present conditions and current operating constraints, stream aggradation should not adversely effect the present trout fisheries. Field work, laboratory analyses, and computer simulation of the controlling physical processes all indicate the same conclusion.

REFERENCES

1. Ward, T. J. 1981. Analysis of Aggradation and Degradation Below Proposed Windy Gap Reservoir, Colorado, prepared for Municipal Subdistrict Northern Colorado Water Conservancy District, Loveland, Colorado.

2. Hydro-triad, Ltd. 1980. Colorado River Low Flow Study Windy Gap to Kremmling, Colorado, prepared for Northern Colorado Water Conservancy District, Loveland, Colorado.

3. Simons, D. B., and F. Senturk 1977. Sediment Transport Technology, Water Resources Publications, Fort Collins, Colorado.

CHAPTER 8

HISTORICAL PATTERNS OF PHYTOPLANKTON
PRODUCTIVITY IN LAKE MEAD

R.T. Prentki
L.J. Paulson
    Lake Mead Limnological Research Center
    University of Nevada, Las Vegas

INTRODUCTION

    Lake Mead was impounded in 1935 by the construction of
Hoover Dam. The Colorado River was unregulated prior to then
and therefore was subjected to extreme variations in flows
and suspended sediment loads. Hoover Dam stabilized flows
and reduced suspended sediment loads downstream [1], but
Lake Mead still received silt-laden inflows from the upper
Colorado River Basin. The Colorado River contributed 97% of
the suspended sediment inputs to Lake Mead, and up to 140 x
10^6 metric tons (t) entered the reservoir in years of high
runoff [2]. Most of the sediments were deposited in the
river channel and formed an extensive delta in upper Lake
Mead [3,4]. However, sediments were also transported into
the Virgin Basin and Overton Arm by the overflow that occurred during spring runoff [5]. The limnology of Lake Mead
is thought to have been strongly influenced by this turbid
overflow until Glen Canyon Dam was constructed 450 km upstream in 1963.
    The construction of Glen Canyon Dam and formation of
Lake Powell drastically altered the characteristics of the
Colorado River inflow to Lake Mead [2]. The operation of
Glen Canyon Dam stabilized flows, reduced river temperatures
and cut the suspended sediment loads by 70-80% [2]. Nitrate
loads decreased initially during 1963 and 1964, then increased through 1970, but have since decreased again to a
lower steady state [6]. Phosphorus loads were decreased due
to reductions in suspended sediment inputs [2]. Lake Powell
now retains 70% of the dissolved phosphorus [1] and 96% of
the total phosphorus [7] inputs that once flowed into Lake
Mead. The Colorado River still provides 85% of the inorganic
nitrogen to Lake Mead, but Las Vegas Wash now contributes
60% of the phosphorus inputs [2].
    Wastewater discharges from Las Vegas Wash into Las
Vegas Bay increased steadily during the post-Lake Powell

105

period. The morphometry and hydrodynamics of Lake Mead are
such that the Las Vegas Wash inflow is confined to the Lower
Basin where historically it has elevated phytoplankton productivity. However, high phosphorus loading and productivity
have resulted in decreases in nitrate concentrations, and
the Las Vegas Bay and parts of Boulder Basin have become nitrogen limited since 1972 [6]. A unique situation has therefore developed in Lake Mead in that the Upper Basin has become more phosphorus limited and the Lower Basin more nitrogen limited since the formation of Lake Powell. Paulson and
Baker [2] theorized that these changes in nutrient loading
and limitation must also have been accompanied by decreases
in reservoir-wide productivity.

There is some evidence for this hypothesis in apparent
improvements in water quality of Las Vegas Bay since 1968
[6]. Chlorophyll-a concentrations in the inner Las Vegas Bay
have decreased considerably since the first measurements
were made in 1968 [8] and during the period of the Lake Mead
Monitoring Program [9-12]. Improvements in water quality of
the bay have confounded efforts to establish water quality
standards on effluent discharges and are contrary to predictions made in the early 1970s that water quality would continue to degrade with increased phosphorus loading [13]. The
decline in the largemouth bass fishery documented by the
Nevada Department of Wildlife [14] could also be a symptom
of lower productivity in Lake Mead.

In this paper, the hypothesis that algal productivity
has declined in Lake Mead as a result of impoundment of Lake
Powell is evaluated. The chemical status of six stations in
the Upper and Lower Basins of Lake Mead is analyzed and current and past rates of organic carbon and phosphorus sedimentation are calculated. The relationship between algal
productivity and accretion of organic carbon in sediment is
determined, and this is used to construct a historical record of algal productivity for Lake Mead.

METHODS

Sampling Locations

The productivity and siltation patterns in Lake Mead
are extremely heterogeneous due to the irregular reservoir
morphometry and variable influence of nutrient loading from
Las Vegas Wash and the Colorado River [15]. In order to insure that this heterogeneity was adequately represented in
the survey, multiple sediment cores were collected from
several locations in the reservoir. The location of drilling
sites are shown in Figure 1, and site characteristics are
listed in Table I. Station locations were surveyed with an
echo-sounder and the final sites were selected to provide a

reasonably flat, undisturbed sediment surface. The stations were purposely placed outside the old river channel to avoid possible sediment disturbances from the Colorado River density current. Station 1 was a shallow-water site in a small embayment of the inner Las Vegas Bay, near the point of the sewage inflow from Las Vegas Wash. Stations 2 and 3 were placed in the Lower Basin; one of these in Boulder Basin (Station 3). Two stations were also placed off the old river channel in the Upper Basin: the Virgin Basin (4) and Bonelli Bay (5) stations. The sixth station was located in the Overton Arm, near Echo Bay.

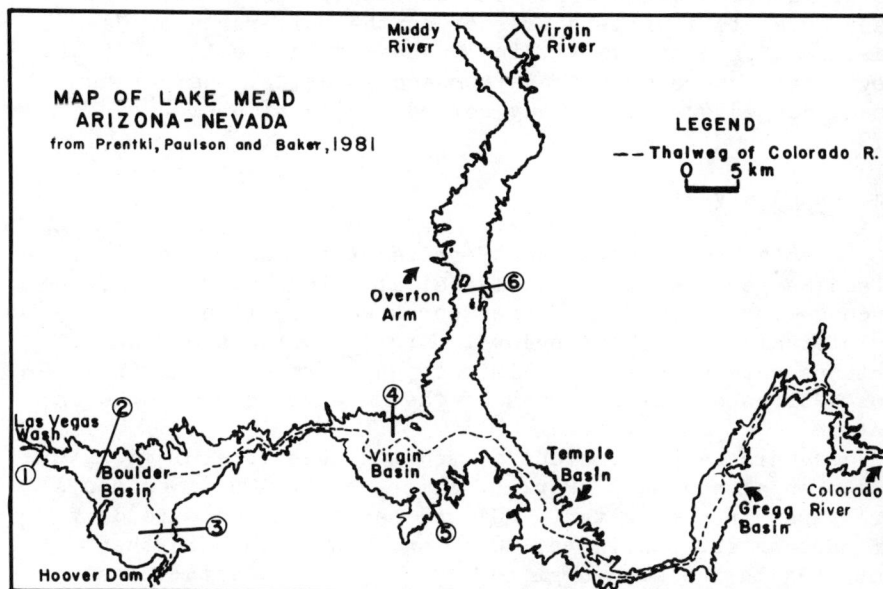

Figure 1.   Map of Lake Mead Sediment Coring Stations.

Table I.   Physical Characteristics at Sediment Coring Stations in Lake Mead.

Station	Water Depth (meters)	Number of cores	Date of submersion (month-year)	Relict material
1*	14	8	6-38	gravel
2	60	11	7-35	gravel
3	90	6	7-35	gravel
4	80-95	10	7-35	soil
5	102	11	7-35	sand
6**	75	8	7-35	sand

* This station was dry in low water years
** Fine sediment was not deposited above sand until 3-40

107

Sediment Coring

The sediment coring was conducted by an oceanographic drilling company (Ocean/Seismic/Survey Inc., Norwood, NJ). A hydraulically-operated vibra-corer was used to obtain undisturbed sediment cores of 8.6 cm effective diameter. Coring rates were monitored with a penetration recorder. Coring was terminated when coring rates indicated that contact had been made with the old reservoir floor. The corer was retrieved and the core was immediately inspected through the Lexan liner for signs of marbling or other disturbance. Undisturbed cores were capped and stored upright. They were transferred to a walk-in freezer on the University of Nevada, Las Vegas campus within 10 hours of collection. Six to eleven cores were collected from each station. The coring was conducted over a ten-day period during mid-October, 1979.

Sediment Analyses

A detailed description of procedures used for analysis of sediments is given in Kellar et al. [16] and will only be discussed briefly here. Frozen cores were sectioned in 1.3-cm intervals from the top down. Outside surfaces of the core sections were scraped to eliminate any surface contamination. Corresponding sections of the several cores from each station were pooled.

Organic carbon content of sediment was determined with an elemental analyzer (Perkin Elmer Model 240B). Sediments were first treated with 1N HCl and heated at 105°C to drive off carbonates. Duplicate, 20-60 mg subsamples were then combusted in the elemental analyzer at 950°C. Total phosphorus was analyzed by the phosphomolybdate method following ignition of 0.5 g samples at 550°C and subsequent extraction of phosphorus from the residue into 1N H_2SO_4.

Sediment bulk density and calcium carbonate content measurements were necessary in order to calculate the organic carbon sedimentation rates but are not reported here. These data and description of their analytical methodology are described by Prentki et al. [17].

The Cesium-137 counting of 500-1000 g samples was performed by Controls for Environmental Pollution Inc. (CEP), a commercial laboratory in Santa Fe, NM. The required sample size necessitated pooling two to three adjoining 1.3-cm sediment sections. A few samples were also counted by the U.S. Environmental Protection Agency, Office of Radiation Programs, Las Vegas, NV, and by the Southern Plains Watershed and Water Quality Laboratory, Durant, OK, for quality assurance purposes.

RESULTS AND DISCUSSION

Sediment Core Dating

Cesium-137 radioactivity from atmospheric bomb fallout
has been widely used to date reservoir sediments [18]. Ce-
sium-137 is strongly adsorbed by fine soil particles and, if
eroded from the watershed, will be deposited in reservoir
sediments. The first occurrence of Cs-137 activity in the
bottom of a sediment profile indicates that the layer was
deposited after the first testing in 1954. The most intensive period of fallout was caused by Russian testing during
1962-64; fallout has decreased steadily since 1963. Peak
fallout, therefore, occurred during the period when Lake
Powell was formed, providing an excellent sediment marker in
Lake Mead.

The Cs-137 concentrations in Lake Mead sediments were
generally low and differed somewhat between the Upper and
Lower Basins (Figure 2). The slightly higher activity in
Upper Basin sediments apparently reflects greater inputs and
deposition of suspended sediments from the Colorado River.
The bottom sediment layers where Cs-137 activity first appeared were evident in all cores from deep stations and were
assigned the 1955 marker. The Cs-137 profiles in middle Las
Vegas Bay, Boulder Basin, Virgin Basin, and the Overton Arm
generally followed the classic pattern that has been found
in other reservoirs. Cs-137 activity increased after 1955,
reached a peak, and then decreased again in recent sediments. The peak activity layer in these cores was assigned
the 1963 marker.

Data collected in Bonelli Bay and the inner Las Vegas
Bay were, however, more difficult to interpret. In Bonelli
Bay, peak Cs-137 activity occurred at 17-19 cm sediment
depth, far below that found at the other Upper Basin Stations. In Virgin Basin, the peak activity occurred at 8-9
cm, and in the Overton Arm, it occurred at 3-4 cm sediment
depth. In order to resolve the obvious discrepancies with
other Upper Basin cores, we assigned the 1963 marker to the
secondary Cs-137 maximum that occurred 3-4 cm from the
sediment surface in Bonelli Bay. This is consistent with
changes in other chemical parameters of this layer [17] and
reasonable in terms of known reductions in suspended sediment loading and siltation in the Upper Basin after 1963.

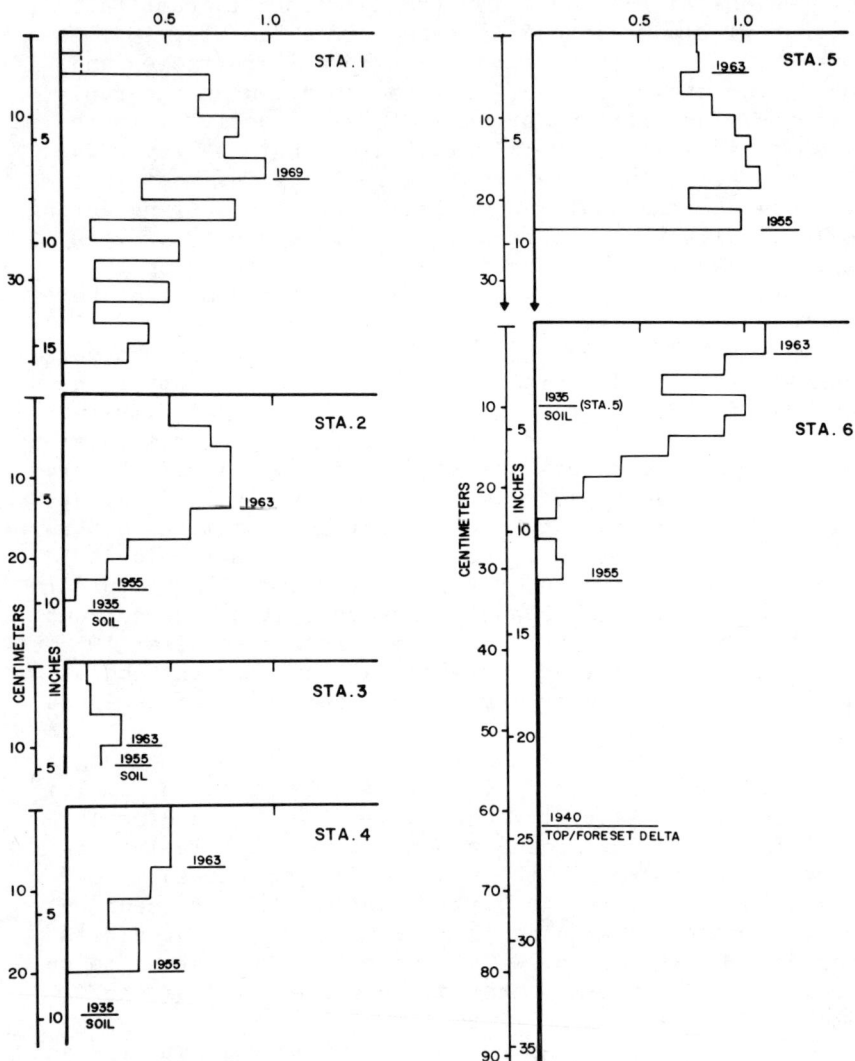

^{137}Cs (pCi g^{-1})

Figure 2.   Cesium-137 Profiles of Lake Mead Sediments and
Dates of Various Sediment Layers.

The Cs-137 profile in the inner Las Vegas Bay was also
difficult to interpret because activity was found in gravel
layers deep in the core. This station was shallow and in the
past has been subject to water level fluctuations and peri-

odic desiccation. This area was dry until 1938, very shallow
(1-2 m) during 1947 and from 1951-57, and then dry again
from 1963-69. Because of possible reworking of sediment during dry or low water years, we were unable to use the disappearance of Cs-137 activity to indicate the 1955 marker.
Moreover, the peak in Cs-137 activity must reflect 1969
rather than 1963, because this area was dry over the period
from 1963-69.

Apart from some difficulties in interpreting Cs-137
profiles in Bonelli Bay and the inner Las Vegas Bay, the
Cs-137 data provide reliable markers of the 1955 and 1963
sediment layers. It is also possible to establish a third
marker, the old reservoir floor of 1935, by obvious dis-
continuities between pre-reservoir soils and reservoir sediments. Sediments were underlain by gravel in the middle Las
Vegas Bay, gravel and soft rock in Boulder Basin, unconsolidated desert soils in Virgin Basin, and sand in Bonelli Bay.
A similar discontinuity existed in Overton Arm, but the
sediment depth here was also influenced by delta deposits
from the Virgin River as the reservoir was filling. Gould
[19] reported that in 1935 and 1936 the mouth of the Virgin
River was located at Bitter Wash, a few kilometers upstream
from our station. He was, therefore, unable to distinguish
between sand deposited by the river and that in the pre-
reservoir deposits. Clay sediments were deposited once lake
levels increased and caused the point of river inflow to
recede up the Overton Arm. This occurred in 1940. Layers
below that represent siltation from the Virgin River inflows
during 1935-40.

Sediment Chemical Structure

Organic carbon in Lake Mead sediments was very low.
Values ranged from 0.3% of sediment dry weight in early
sediments to 1.7% in recent sediments (Figure 3).

Total phosphorus concentrations of Lake Mead sediments
were appreciable and ranged from 300 ppm of dry weight in
old reservoir sediments to 1000 ppm in recent sediments
(Figure 4). In the inner and middle Las Vegas Bay, phosphorus increased steadily in sediments deposited after 1963,
but elsewhere phosphorus concentrations decreased or remained stable. The organic carbon:phosphorus ratios were very
low, ranging from 10 to 20. These ratios are tenfold lower
than found in plankton and considerably lower than those
reported in other lake sediments [17]. The low C/P ratios
were caused by the presence of large amounts of biologically
unavailable particulate phosphorus which entered Lake Mead
from the Colorado River [20].

111

% ORGANIC CARBON

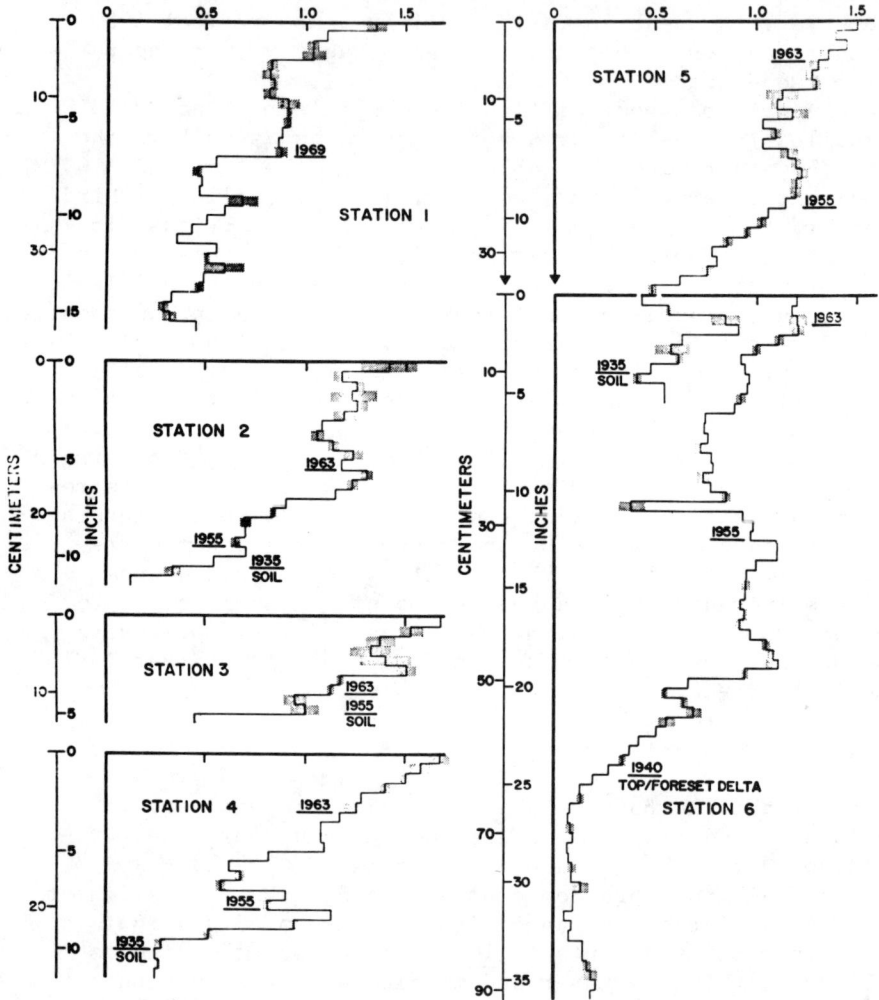

Figure 3. Organic Carbon Content of Lake Mead Sediments
(Range in Replicate Analyses Shown by Shading).

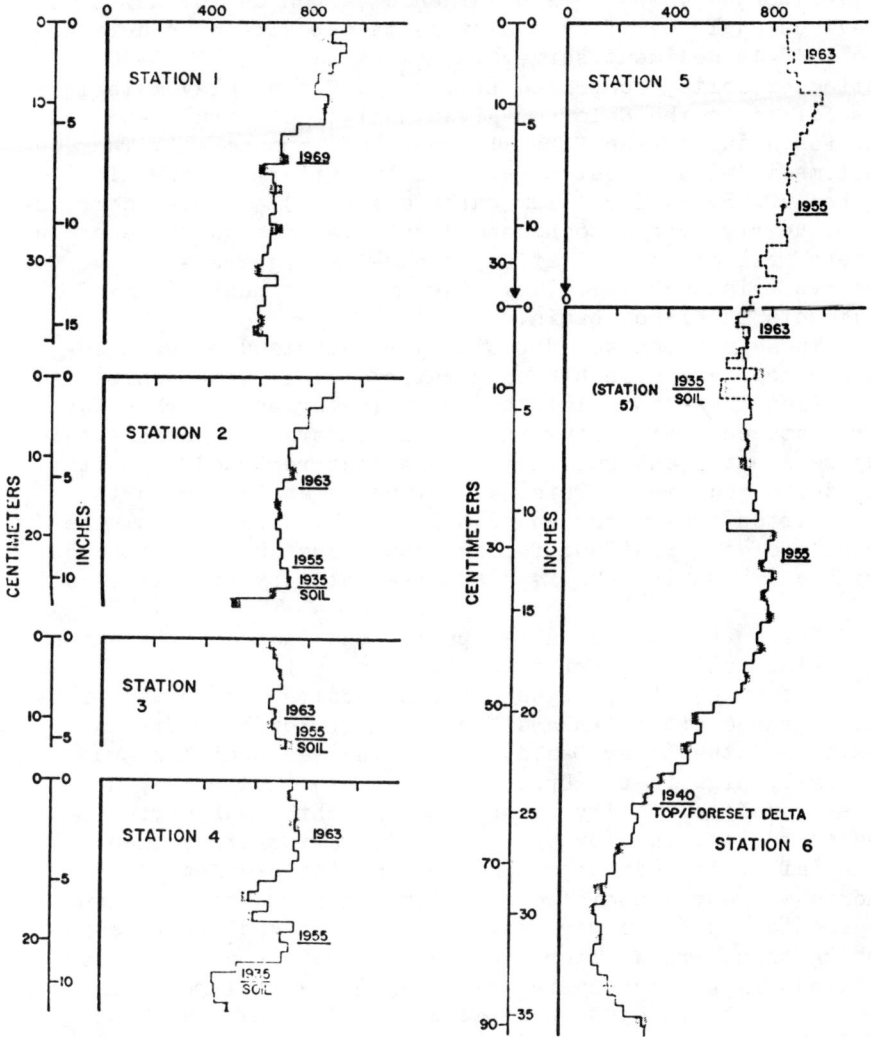

TOTAL PHOSPHORUS (PPM)

Figure 4. Total Phosphorus Content of Lake Mead Sediments
(Range in Replicate Analyses Shown by Shading).

Spatial and Temporal Patterns in Sedimentation

The Cs-137 data and chemical analyses enabled us to
estimate annual sedimentation rates for organic carbon and

phosphorus during three periods of reservoir history (1935-54, 1955-62 and 1963-79). In addition, it was possible to partition autochthonous (in-reservoir) and allochthonous (river-borne) components of organic carbon on the basis of previous analyses of bottomset delta deposits made during the 1948-49 sediment survey in Lake Mead [19]. A 15-30 m bottomset delta, comprised primarily of fine clay materials, was formed in the Colorado River thalweg of Virgin and Boulder Basin during the first 13 years of impoundment. The bottomset delta deposits were fairly uniform in organic carbon (0.55%) and calcium carbonate (16%) and were comprised of nearly pure allochthonous material due to the enormous rate of siltation. Siltation in non-thalweg areas of the reservoir is much lower, since we found at most 46 cm of sediments in either basin.

These non-thalweg deposits are comprised of both autochthonous and allochthonous materials. It was possible to partition these materials by measuring organic carbon and carbonate concentrations in various layers of the non-thalweg sediments, and subtracting out that reported in bottomset delta sediments. This separation is analogous to that for tripton from resuspended sediment by Gasith [21]. The details of calculations for autochthonous and allochthonous organic carbon in Lake Mead are presented by Prentki et al. [17].

There was considerable spatial and temporal variation in sedimentation patterns in Lake Mead (Figure 5). In the period from 1935-54, organic carbon sedimentation was highest in the Overton Arm and Bonelli Bay, lower in Virgin Basin, and the Lower Basin. Phosphorus sedimentation was extremely high in the Upper Basin (up to 17 $g/m^2 \cdot yr$) and closely related to dry weight and allochthonous carbon sedimentation. The low C/P (ca.12:1) ratios of sedimented material again indicated that most of the sediment phosphorus was not associated with limnetic plankton remains. Sedimentation rates were extremely low in the Lower Basin during this period. There was no measureable accumulation of sediment in Boulder Basin prior to 1955. Similarly, in Las Vegas Bay, sedimentation rates were extremely low in the early history of Lake Mead.

Sedimentation rates increased in the Upper Basin during the period from 1955-62. This was especially evident in Bonelli Bay and Virgin Basin where autochthonous carbon sedimentation increased twofold over the preceding period. Phosphorus sedimentation also increased in the Upper Basin but not as drastically as what was observed for carbon. It is somewhat surprising that these sedimentation rates increased during this period because average suspended sediment loading decreased by 34%. The suspended load in the Colorado River averaged 110 x 10^6 t/yr prior to 1955 but then decreased to 73 x 10^6 t/yr during the 1955-62 water years [22]. Allochthonous organic carbon sedimentation rates, however, increased by 20% in the Overton Arm and 400% in Virgin Basin indicating that there must have been a significant change in the distribution of suspended sediment inputs across the Upper Basin.

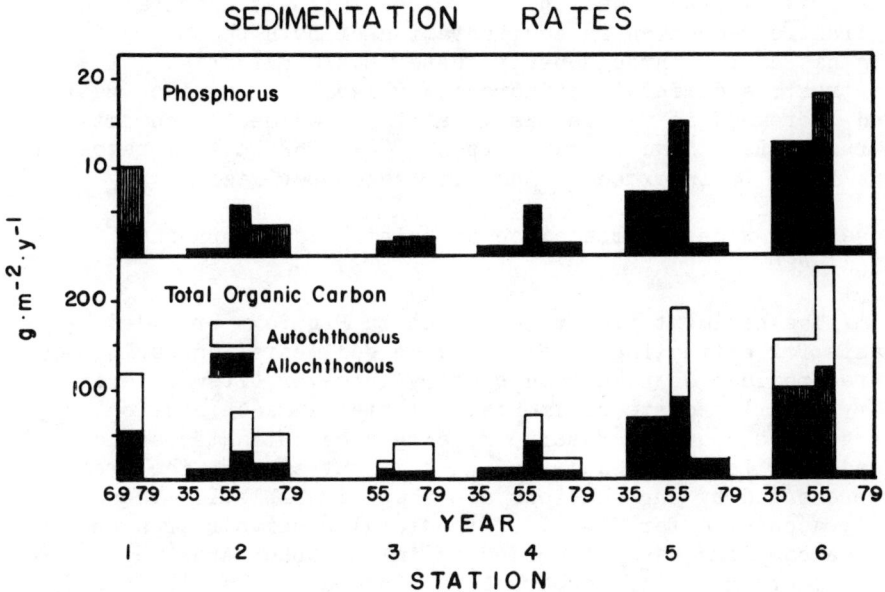

Figure 5.   Sedimentation Rates for Organic Carbon and Phosphorus During Three Periods (1935-54, 1955-62, 1963-79) of Lake Mead History.

The Colorado River has historically formed an overflow during spring and a shallow interflow during summer in the Upper Basin [5]. During spring runoff, this resulted in dispersal of fine suspended sediments across the Upper Arm of Lake Mead (Gregg Basin, Temple Basin). High spring runoff and flooding occurred in the Colorado River during 1956-58 and in 1962 (USGS data), and this apparently caused greater dispersal of suspended sediments into non-delta areas of the Virgin Basin, Bonelli Bay and the Overton Arm. The magnitude of spring runoff and seasonal frequency of flooding appear to be more important factors than is average, annual suspended sediment loading in determining sedimentation in non-delta areas of the reservoir. However, even during years of extreme spring runoff, it does not appear that much Colorado River suspended sediment is transported into the Lower

115

Basin. There was only a slight increase in sedimentation
rates of allochthonous organic carbon in Boulder Basin
during the period 1955-62 (Figure 5). There was a greater
increase in sedimentation in the middle Las Vegas Bay, but
this was probably due to increased discharge of sewage
effluents into the Lower Basin.

Suspended sediment loading in the Colorado River decreased to an average of 16 x 10^6 t/yr in the period after
Lake Powell was formed in 1963 [22]. This was accompanied by
a drastic reduction in sedimentation of both phosphorus and
organic carbon throughout the Upper Basin (Figure 5). In
contrast, sedimentation increased slightly in Boulder Basin
and decreased in middle Las Vegas Bay. Sedimentation patterns in Lake Mead were reversed after 1962 in that rates in
the Lower Basin exceeded those in the Upper Basin.

Reservoir-wide Sedimentation as Related to Phosphorus
Loading

The sedimentation rates given in Figure 5 provided a
basis for estimating reservoir-wide sedimentation during
three periods of Lake Mead history. However, it was necessary to extrapolate sedimentation rates at each station to
larger areas of the reservoir using area estimates of Lake
Mead from Lara and Sander's [4] sediment survey. The areas
represented by our stations are shown in Table II. These
only accounted for 77-78% of the total reservoir area because sampling was not conducted in the Upper Arm (Temple
Basin, Gregg Basin, Iceberg Canyon and Grand Wash). In order
to obtain an estimate of reservoir-wide sedimentation, we
used data from station 5 to characterize the Upper Arm of
Lake Mead.

The formation of Lake Powell markedly reduced phosphorus sedimentation in the Upper Basin of Lake Mead. Phosphorus sedimentation in the Upper Basin was extremely high during the early history of Lake Mead but decreased by 93.5%
after formation of Lake Powell (Table III). Phosphorus sedimentation in the Lower Basin decreased by only 2% in the
post-Lake Powell period. Reservoir-wide phosphorus sedimentation, however, decreased from an average of 5200 t/yr
during 1955-62 to 623 t/yr after 1962.

There are no long-term loading data for phosphorus, but
it must have been high, particularly during 1955-62, to account for the high rates of phosphorus sedimentation during
the pre-Lake Powell years. Phosphorus loading was probably
on the order of that recently measured for Lake Powell by
Gloss et al. [7]. They estimated that the Colorado River
currently provides 5224 t/yr of total phosphorus to Lake
Powell. However, only 229 t/yr of phosphorus is currently
discharged from Glen Canyon Dam [7], and about the same

amount, 198 t/yr enters Lake Mead from the Colorado River
[23]. These numbers represent a 96% reduction in total
phosphorus loading into Lake Mead which accounts for the
abrupt decrease in phosphorus sedimentation in the Upper
Basin.

Table II.   Reservoir Mean Surface Areas (km^2) Characterized
            by Sediment Coring Stations.

Time Interval	Mean* Lake Level (m)	Total Lake Area (km^2)	Station					
			1	2	3	4	5	6
≤ 1954	350	447	**	21.7	101.7	35.8	108.7	80.0
1955–62	352	465	**	22.1	103.6	37.0	112.5	85.1
≥ 1963	353	475	0.8	21.4	104.2	37.7	114.3	87.3

*Lake level from [22] and USGS (unpublished).
**Combined with station 2

Table III. Average Reservoir-Wide and Individual Basin
            Sedimentation of Phosphorus in Lake Mead
            (t/yr).

Time Interval	Whole Reservoir	Lower Basin	Upper Basin	Lower and Upper Basin
≤ 1954	2470	15	1780	1795
1955–62	5200	273	3390	3663
≥ 1963	623	268	220	488

Sewage effluent discharges and nutrient loading from
Las Vegas Wash, however, rose steadily in the post-Lake
Powell period. Las Vegas Wash now contributes 60% of the
phosphorus input to Lake Mead [23]. The morphometry and
hydrodynamics of Lake Mead [15] are such that the phosphorus-rich Las Vegas Wash inflow is confined to the Lower
Basin. Phosphorus sedimentation in the Lower Basin has been
maintained, therefore, at levels equal to that in the 1955-
62 period (Table III).
    The historical patterns of phosphorus sedimentation in
each basin of Lake Mead generally agree with historical
changes in loading. However, there is a considerable difference in sedimentation estimated from nutrient budgets (apparent sedimentation) [23] and absolute sedimentation
measured in this study.
    Phosphorus loading to Lake Mead was 198 t/yr from the
Colorado River and 263 t/yr from Las Vegas Wash in 1977-78
[23]. Total phosphorus loading to Lake Mead was about 460
t/yr because the Virgin and Muddy Rivers contribute minimal
phosphorus to the reservoir [24]. Phosphorus loss from
Hoover Dam was 123 t/yr in 1977-78 [23]. The fish harvest

also resulted in an annual loss of 25 t of phosphorus from the reservoir [25]. The combined phosphorus losses from Lake Mead would therefore be 148 t/yr. Apparent phosphorus sedimentation would be 312 t/yr. Absolute phosphorus sedimentation, as measured in this study, was 268 t/yr in the Lower Basin, 220 t/yr in the Upper Basin and 623 t/yr in the whole reservoir during the post-Lake Powell period (Table III). Absolute sedimentation thus exceeded 1977-78 apparent sedimentation by 311 t/yr. It is unknown whether loading for 1977-78 reflects average annual loading in recent years. However, the discrepancy between the two retention numbers is most likely caused by a higher nutrient output from Lake Powell during the first years of impoundment than is now occurring [2,20].

Organic Carbon Sedimentation and Phytoplankton Productivity

The historical changes in nutrient loading to Lake Mead have also been accompanied by marked changes in organic carbon sedimentation and, as will be shown, phytoplankton productivity. Reservoir-wide autochthonous carbon sedimentation was low prior to 1955 but increased sharply during the period from 1955-62, followed by an abrupt decrease in the post-Lake Powell period (Table IV). The same trends were also evident for allochthonous organic carbon sedimentation. Organic carbon sedimentation was consistently higher in the Upper Basin during the pre-Lake Powell period and accounted for over 90% of reservoir-wide organic carbon sedimentation. This pattern was reversed after 1962, and the Lower Basin now contributes over 50% of organic carbon sedimentation in Lake Mead. However, reservoir-wide sedimentation has still been reduced by 76.8% of that which occurred in the 1955-62 period.

Table IV. Reservoir-Wide and Individual Basin Sedimentation of Autochthonous and Allochthonous Organic Carbon in Lake Mead (t C/yr).

Interval	Whole Reservoir	Lower Basin	Upper Basin	Lower and Upper Basin
		Autochthonous		
≤ 1954	7710	48	6150	6198
1955-62	33400	2290	20300	22590
≥ 1963	7720	3830	2450	6280
		Allochthonous		
≤ 1954	18900	85	13800	13885
1955-62	32500	1710	21700	23410
≥ 1963	3300	1200	1320	2520

For the post-Lake Powell period, autochthonous organic carbon sedimentation in various locations of Lake Mead (Figure 6) was closely related to recent phytoplankton productivity measurements made at these locations by Paulson et al. [15]. There was a good correlation (r=0.979, N=6) between annual autochthonous organic carbon sedimentation and annual phytoplankton productivity (1977-78) at the six sediment sampling stations (Figure 6). Linear regression of organic carbon sedimentation against phytoplankton productivity (Equation 1) provided a means of predicting historical productivity in the reservoir.

$$PPR = -7 + 19.7 \ (AOC) \qquad (1)$$

where PPR = rate of phytoplankton productivity
$(g \ C/m^2 \cdot yr)$
   AOC = autochthonous organic carbon sedimentation
$(g \ C/m^2 \cdot yr)$

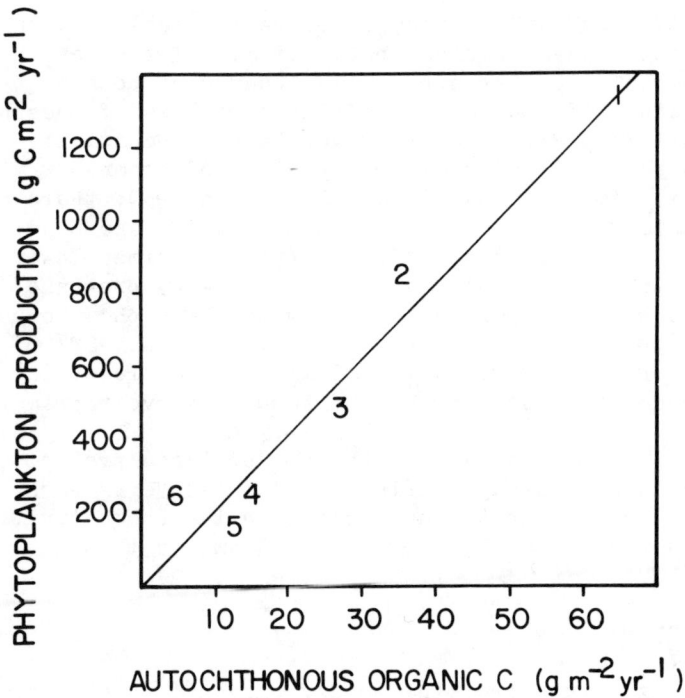

Figure 6.   Relationship of Recent Estimates of Phytoplankton Productivity in Lake Mead to Autochthonous Organic Carbon Sedimentation in the Post-Lake Powell Period.

Rates of phytoplankton productivity estimated for each station with Equation 1 were extrapolated over larger areas of the reservoir to estimate reservoir-wide and individual basin total annual production (Table V). The spatial and historical trends in total production (Table V) necessarily follow those for autochthonous organic carbon sedimentation (Table IV) and thus do not provide different information. However, historical rates in units of productivity enable us to better reconstruct the trophic history of Lake Mead.

Table V. Reservoir-Wide and Individual Basin Estimates of Historical Rates of Phytoplankton Production ($t\ C/yr \times 10^3$).

Interval	Whole Lake	Lower Basin	Upper Basin	Lower and Upper Basin
≤ 1954	146	0.6	117	118
1955-62	651	43	395	438
≥ 1963	144	73	44	117

In the early decades of Lake Mead, only 600 of 146,000 t/yr production occurred in the Lower Basin (Table V). In the subsequent 1955-62 period, productivity of the reservoir increased to 651,000 t/yr apparently because of both high nitrate loading [17] and strong spring overflows of phosphorus-rich, Colorado River water. Lower Basin productivity then accounted for 7% of whole reservoir production.

Since the impoundment of Lake Powell in 1963, there has been a drastic reversal of the productivity of Lake Mead. Productivity has dropped to 144,000 t/yr, 4.5 times lower than in 1955-62 and 49% of the entire, 1935-62, pre-Lake Powell average. The Upper Basin is now severely phosphorus limited and productivity of this basin is now only 22% of the pre-Lake Powell average, 11% of the 1955-62 rate. The Lower Basin now accounts for 51% of total reservoir primary production.

We attribute almost all of this Lower Basin production to fertilization by sewage effluents from Las Vegas Wash. Without this latter input, the decline in the productivity of Lake Mead would have been even more dramatic than documented here.

ACKNOWLEDGEMENTS

We are extremely grateful to Gary Bryant, Bureau of Reclamation, and to John Baker for assistance with all aspects of the study. Darrel Thome, Michael O'Connel, Earl Whittaker and J. Roger McHenry aided in resolving problems with the Cs-137 analyses. We greatly appreciate the technical, graphic and editorial assistance provided by Penelope

Kellar and Sherrell Paulson. Terry Evans, Jim Williams, David Hetzel, Alan Gaddy, Gene Wilde, Theron Miller, Brian Klenk and Mick Reese aided in collection and processing of the sediment cores. Dolf Cardenas assisted with chemical and biological analyses of the sediments. We also wish to thank Laurie Vincent and Thomas Hardy for typing the report.

REFERENCES

1. Dill, W.A. 1944. The Fishery of the Lower Colorado. Calif. Dept. Fish and Game 30:109-211.

2. Paulson, L.J. and J.R. Baker. 1980. Nutrient interactions among reservoirs on the Colorado River. Pages 1647-1656 in H.G. Stefan, ed. Symposium on surface water impoundments. June 2-5, 1980. Minneapolis, MN.

3. Gould, H.R. 1960. Amount of sediment. Pages 195-200 in W.O. Smith, C.P. Vetter, G.B. Cummings, eds. Comprehensive survey of sedimentation in Lake Mead, 1948-49. U.S. Geol. Surv. Prof. Paper 295.

4. Lara, J.M. and J.I. Sanders. 1970. The 1963-64 Lake Mead survey. U.S. Bur. Rec. Rept. No. REC-OCE-70-21. 169 pp.

5. Anderson, E.R. and D.W. Pritchard. 1951. Physical limnology of Lake Mead, Lake Mead sedimentation survey. U.S. Navy Electronics Lab, San Diego, Calif. Rept. No. 258. 153 pp.

6. Paulson, L.J. 1981. Nutrient management with hydroelectric dams on the Colorado River system. Lake Mead Limnological Res. Ctr. Tech. Rept. No. 8. Univ. Nev., Las Vegas. 39 pp.

7. Gloss, S.P., R.C. Reynolds, Jr., L.M. Mayer and D.E. Kidd. 1980. Reservoir influences on salinity and nutrient fluxes in the arid Colorado River Basin. Pages 1618-1629 in H.G. Stefan, ed. Symposium on surface water impoundments. June 2-5, 1980. Minneapolis, MN.

8. Hoffman, D.A., P.R. Tramutt and F.C Heller. 1971. The effect of Las Vegas Wash effluent upon the water quality in Lake Mead. U.S. Bur. Rec. Rept. No. REC-ERC-71-11. 25 pp.

9. Deacon, J.E. and R.W. Tew. 1973. Interrelationship between chemical, physical and biological conditions of the waters of Las Vegas Bay of Lake Mead. Final report

to Las Vegas Valley Water District. 186 pp.

10. Deacon, J.E. 1975. Lake Mead Monitoring Program. Final
     report to Clark County Wastewater Management Agency.
     297 pp.

11. Deacon, J.E. 1976. Lake Mead Monitoring Program. Final
     report to Clark County Sanitation District No. 1, Waste
     Treatment Physical Development Section. 182 pp.

12. Deacon, J.E. 1977. Lake Mead Monitoring Program. Final
     report to Clark County Sanitation District No. 1, Waste
     Treatment Physical Development Section. 55 pp.

13. U.S. Environmental Protection Agency. 1971. Pollution
     affecting Las Vegas Wash, Lake Mead and the Lower
     Colorado River. Division of Field Investigations-Denver
     Center, Denver, CO, and Region IX San Francisco, CA. 52
     pp.

14. Nevada Department of Wildlife. 1980. Job progress
     report for Lake Mead. Proj. No. F-20-17. 209 pp.

15. Paulson, L.J., J.R. Baker and J.E. Deacon. 1980. The
     limnological status of Lake Mead and Lake Mohave under
     present and future powerplant operations of Hoover Dam.
     Lake Mead Limnological Res. Ctr. Tech. Rept. No. 1.
     Univ. Nev., Las Vegas. 229 pp.

16. Kellar, P.E., S.A. Paulson and L.J. Paulson. 1980.
     Methods for biological, chemical, and physical analyses
     in reservoirs. Lake Mead Limnological Res. Ctr. Tech.
     Rept. No. 5. Univ. Nev., Las Vegas. 234 pp.

17. Prentki, R.T., L.J. Paulson, and J.R. Baker. 1981.
     Chemical and biological structure of Lake Mead
     sediments. Lake Mead Limnological Res. Ctr. Tech. Rept.
     No. 6. Univ. Nev., Las Vegas. 89 pp.

18. McHenry, J.R., J.C. Ritchie and J. Verdon. 1976.
     Sedimentation rates in the upper Mississippi River.
     Pages 1339-1349 in Symposium on inland waterways for
     navigation, flood control and water diversions. Colo.
     State Univ., Fort Collins, CO.

19. Gould, H.R. 1960. Character of the accumulated sedi-
     ment. Pages 149-186 in W.O. Smith, C.P. Vetter, G.B.
     Cummings, eds. Comprehensive survey of sedimentation in
     Lake Mead, 1948-49. U.S. Geol. Survey Professional
     Paper 295.

20. Evans, T.D. and L.J. Paulson. (this volume).

21. Gasith, A. 1976. Seston dynamics and tripton sedimentation in the pelagic zone of a shallow eutrophic lake. Hydrobiologia 51:225-231.

22. U.S. Dept. of Interior (USDI), Bureau of Reclamation. 1976. River control work and investigations, Lower Colorado River Basin, 1974-75. 34 pp.

23. Baker, J.R. and L.J. Paulson. 1980. Influence of Las Vegas Wash density current on nutrient availability and phytoplankton growth in Lake Mead. Pages 1638-1646 in H.G. Stefan, ed. Symposium on surface water impoundments. June 2-5, 1980. Minneapolis, MN.

24. U.S. Environmental Protection Agency. 1977. Report on Lake Mead Clark County, Nevada, Mohave County, Arizona, Region IX. Working Paper No. 808. 28 pp.

25. Culp, G.W. 1981. Preliminary nutrient budget. Las Vegas Valley Water quality program. 83 pp.

CHAPTER 9

WATER QUALITY TRENDS IN THE LAS VEGAS WASH WETLANDS

F.A. Morris
L.J. Paulson
Lake Mead Limnological Research Center
University of Nevada, Las Vegas

INTRODUCTION

The Las Vegas Wash is a wetlands ecosystem that acts to buffer the effects of wastewater discharges on the receiving waters of Lake Mead. The wash is the terminus for the 4,144 km^2 Las Vegas Valley drainage basin, emptying into Las Vegas Bay of Lake Mead (Colorado River). It is in the northern Mojave desert, which receives an average of only 10 cm of rainfall annually. The Las Vegas Wash is technically an artificial wetland supported almost entirely by the perennial flows from sewage treatment plants. These flows contribute an average of 3.7 t of nutrients (nitrogen and phosphorus) and 4 t of oxygen consuming organic material (BOD_5) to Lake Mead per day. High nitrate and total dissolved solid loads (2.7 and 603 t/day respectively) are derived primarily from groundwater inputs in the lower wash [1,2,3]. The contaminated groundwater originates from large underground salt mounds that were formed from discharges of industrial effluents into unlined evaporation ponds until 1978.

Conflicting interests among municipal, recreational, and down-river users make the Las Vegas Wash a focal point in current legal disputes regarding the need for advanced wastewater treatment (AWT). In light of rapidly escalating costs, especially for energy and chemicals needed for AWT, many municipalities nation-wide are investigating alternative treatment techniques. Public Law 92-500, Section 210 (parts d and f) specifically encourages the reclamation and recycling of wastewaters. Operation of treatment facilities to produce revenue through the production of agriculture, silviculture, or aquaculture products is encouraged. Combinations of open space and recreational uses with waste treatment management techniques are also emphasized in PL 92-500.

The Las Vegas Wash ecosystem has been identified as a

potential wastewater treatment system. Previous investigations [4,5] indicate that the ecosystem could be removing
substantial amounts of nutrients from wastewaters. Goldman
and Deacon [5] recommended "that a specifically designed
nutrient removal management program be developed and implemented with the flow distribution and erosion control
program necessary to maintain wetland wildlife habitat."

The purpose of this paper is to describe historical and
current water quality and to quantify the degree of nutrient
removal presently occurring in the Las Vegas Wash.

DESCRIPTION OF THE STUDY AREA

Las Vegas Wash is located in Clark County, Nevada,
between the City of Las Vegas and Lake Mead (Figure 1). The
boundaries of the Wash are defined by a large drainage system that was once part of the pluvial Las Vegas River [6].
Our research was focused in the 18 km stretch downstream of
the City of Las Vegas sewage treatment plant (STP) to Las
Vegas Bay of Lake Mead.

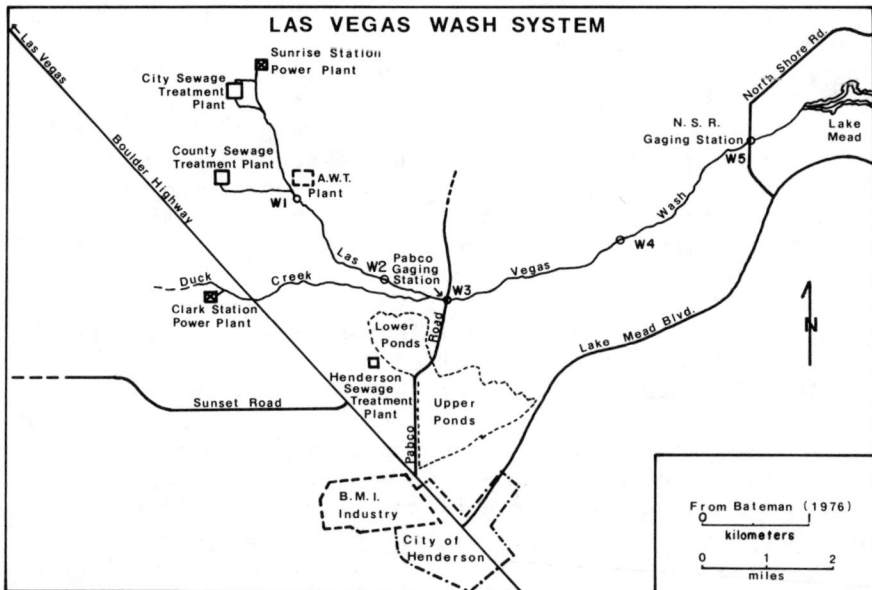

Figure 1.   Sampling Site Location Map.

The water in Las Vegas Wash is comprised of 90%
secondarily-treated wastewater from the City of Las Vegas
and Clark County STPs. Vegas Creek, the last natural creek
in the Las Vegas Valley, dried up in the late 1940's [7].
Flows in the present riparian and marsh wetland have increased with the population of Las Vegas Valley. The Valley

126

is home to nearly one-half million permanent residents and host to approximately nine million tourists annually. Las Vegas has become one of the fastest growing urban areas in the United States. Concomitant with this growth has been an increase in wastewater discharges to Lake Mead. Total discharges currently average 2.8 m^3/sec (100 cfs). This amount is twice the flow rate measured at Northshore Road in 1970 (Figure 2).

MEAN ANNUAL DISCHARGE-LAS VEGAS WASH
1970-1980 (USGS DATA)

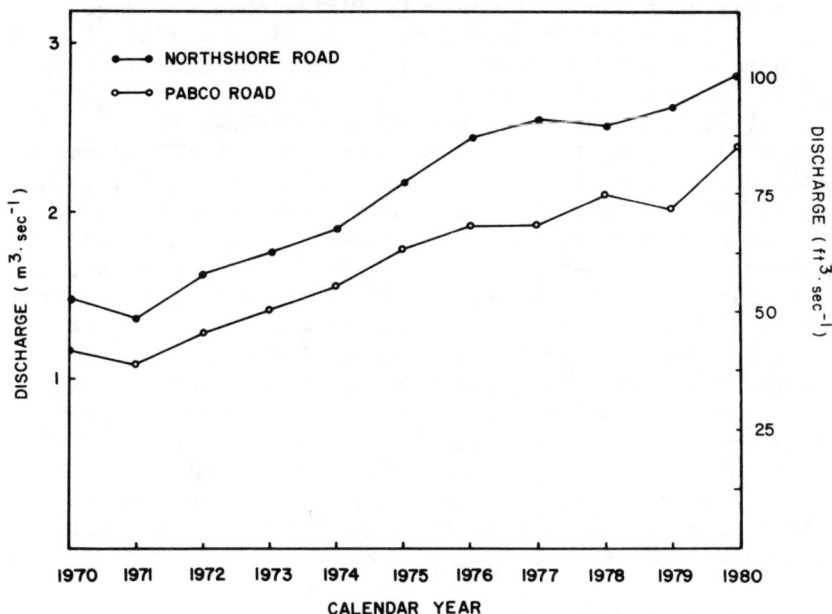

Figure 2.   Historical Discharges From Las Vegas Wash.

Increasing volumes of perennial surface water as well as stormwater discharges have transformed sparse desert shrub and mesquite woodland habitats into dense growths of hydrophytic wetland vegetation dominated by Typha domengensis (cattail) and Phragmites communis (common reed). Extensive growths of the introduced phreatophyte Tamarix petandra (salt cedar) border the wetland and riparian zones.

In 1975, a channelization program was initiated in the upper reach of the wash from the City and County STPs to 1.6 km downstream. This man-made channelization has steadily decreased the extent of wetland from the 1969 to 1975 maximum of approximately 730 ha to 120 ha in 1979. Increased flow velocities and unstable soils in the lower portion of the wetland have also facilitated increased erosion rates at areas known as "headcut" regions for the past 5 yr. Sediment transport in 1979 and 1980 was particularly large. Erosion

127

and headcutting in the lower reach of the Las Vegas Wash is especially prominent during flash flooding and accelerated erosion occurred during this study. The principal headcut region advanced approximately 1.5 km upstream during a single storm event of February 1980. Upstream progression of erosion has resulted in the draining of another 50 ha of wetland creating a riparian habitat with channel depths often exceeding 6 m. Present areal extent of wetland vegetation is 65 ha with 6 ha of shallow (1 m deep) ponds.

Routine collections were taken from five sample stations; W1: confluence of the existing secondary sewage treatment plant effluents (14 km above Las Vegas Bay (LVB)); W2: marsh above Pabco Road (10.7 km above LVB); W3: Pabco Road at culverts (10 km above LVB); W4: headcut area (6.3 km above LVB); W5: Northshore Road (State Highway 41, 1.6 km above LVB) (Figure 1). U.S. Geological Survey (USGS) gaging stations are located at or in close proximity to Stations W1, W3, and W5, facilitating loading rate computations.

Morphologically, the wash can be divided into two components, the Upper (above Pabco Road) and Lower (below Pabco Road) Wash. The stream gradient between Stations W1 and W3 is gradual, dropping 31.7 m in 4 km. The largest extent of Typha occurs in this reach of stream. After crossing Pabco Road via culverts, waters collect in the previously mentioned shallow ponds. Culverts drain these irregularly shaped ponds, emptying into the lower, smaller expanse of wetland vegetation. The gradient between stations W3 and W5 is steeper, dropping 80.5 m in 8.4 km. The major inflows of salt and nitrate laden groundwater occur between Stations W3 and W5 [2].

MATERIALS AND METHODS

These five sampling stations were monitored biweekly from July 1979 to December 1980. Special studies were also conducted to determine diurnal variations in nutrients and flow regimes. Over 40 sampling rounds were conducted during the 18 month study.

Field measurements and sample collections were performed under contractual agreement between the Lake Mead Limnological Research Center, University of Nevada, Las Vegas, and Brown and Caldwell Consulting Engineers of Sacramento, California. Water samples were collected with a large plastic bucket, and subsamples were collected in plastic bottles and preserved on ice. Samples for soluble nutrient analyses were filtered through GF/C filters upon return to the laboratory. The samples were iced and shipped to the Brown and Caldwell laboratory in Emeryville, California for analysis. All analyses were performed as prescribed by U.S. EPA [8].

In addition to physical and chemical measurements, rhodamine WT dye was introduced into segments of the wash during a special study to determine hydraulic retention time. The dye was tracked using a Turner Designs Model 10 fluorometer, and Instrumentation Specialty Corporation (ISCO) automatic water samplers.

RESULTS AND DISCUSSION

Hydraulic retention studies performed jointly by us and Brown and Caldwell Consulting Engineers, during mid-November, 1980, indicate that the wash has a short time of travel from the STPs to Lake Mead (Figure 3). Channelized flows above and below the wetland act to increase flows, and travel time is less than 20 h to Lake Mead. As might be anticipated, the greatest residence time was in the wetlands and ponds (15 h).

LAS VEGAS WASH RETENTION TIME STUDY

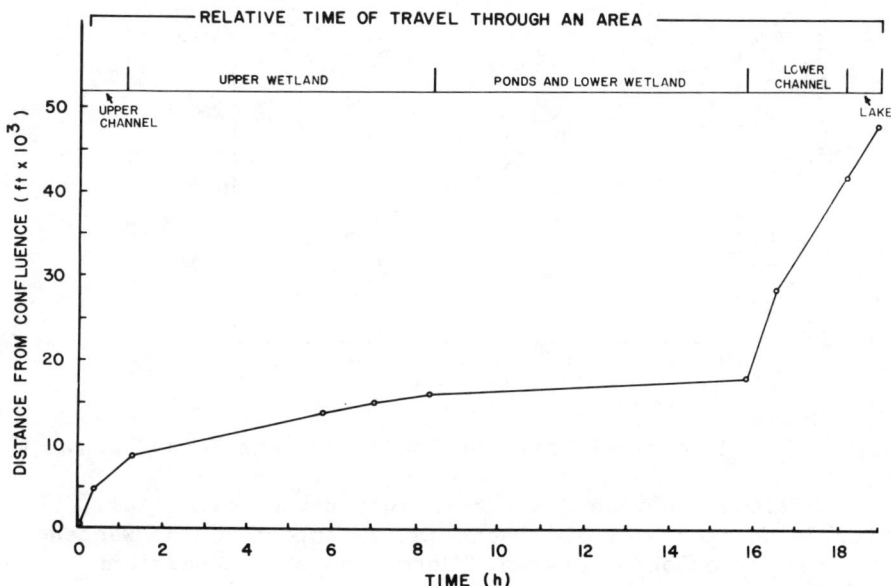

Figure 3. Hydraulic Retention Time Within Las Vegas Wash.

Relatively complete historical nutrient data are available in the USGS records for Las Vegas Wash. Summaries of past data indicate that some dramatic changes have occurred during the last 6 yr of monitoring. Nitrogen loads at Northshore Road have steadily increased since 1977, a drought year (Figure 4). Discharges of industrial wastes into unlined ponds, constructed in the 1940's, was discontinued in the mid-1970's. This has led to a gradual decline of nitrate loads contributed by shallow groundwater aquifers. Ammonia, however, has steadily increased. Ammonia loads at Northshore Road were less than 10% of the total nitrogen load prior to 1977, while current levels exceed 60%.

NITROGEN LOADS AT NORTHSHORE ROAD 1974-1980

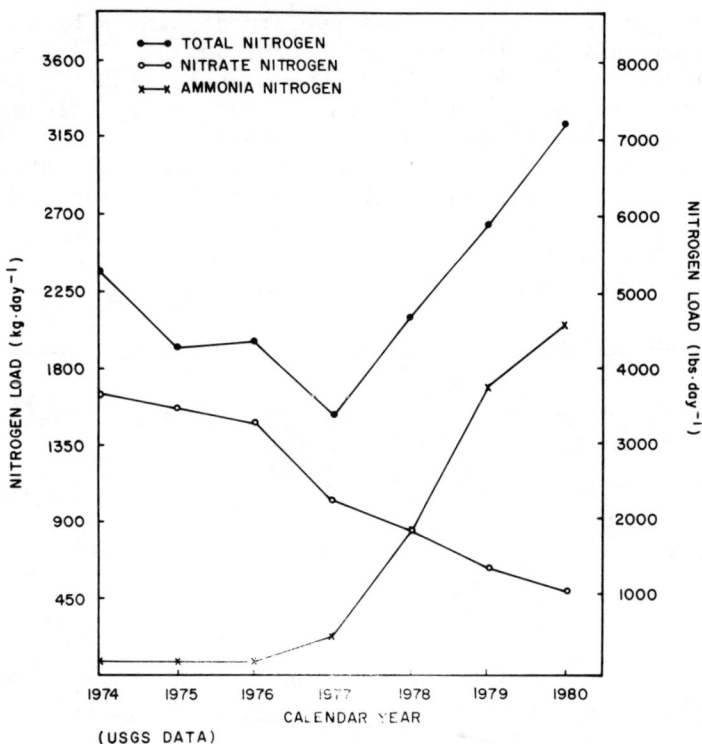

Figure 4. Historical Nitrogen Loads from Las Vegas Wash.

As flows progress downstream into stands of cattail, water velocities slow and bacterial decomposition of wastes causes a depletion of oxygen. These anaerobic conditions favor denitrifying bacteria that effectively convert nitrate to nitrogen gas. No rate measurements are currently available on this, but it appears to result in decreased nitrate concentrations in the upper marsh throughout the year (Figure 5). Denitrification has been generally cited as the major reason that wetlands are nitrogen traps or sinks. Ammonia concentrations increased slightly in this area, apparently due to bacterial decomposition of organic nitrogen compounds.

Overall, total nitrogen concentrations (and loads)

decreased within the wash system. Total nitrogen was reduced by 27% between Stations W1 and W5 during the summer of 1980. This was as high as 47% removal on some occasions and averaged 15.4% during the entire study. Nitrate concentrations at Northshore Road increased to 12 mg/l on three occasions. One event on March 4, 1980 was traced to a leaking pipe which transports industrial wastes to lined evaporation ponds in upland areas of the wash. This resulted in six to seven-fold increases over normal nitrate loads to Lake Mead. Because of these perturbations, the mean removals of nitrogen are conservative.

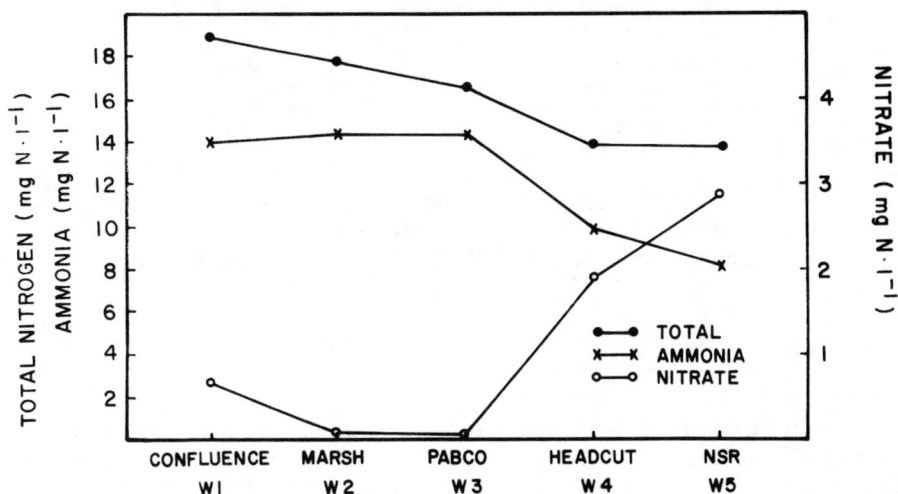

Figure 5.   Mean Concentration of Nutrients Within Las Vegas Wash.

Average nitrogen loads for various seasons are depicted in Figure 6. Loads were calculated from average flows recorded by USGS for the day water samples were taken. There were net removals of total nitrogen and ammonia in the wash. However, there was a net contribution of nitrate, primarily as a result of groundwater inputs in the lower wash. The effects of perturbations discussed earlier can be seen in peaks of nitrate loads during fall of 1979 and spring of 1980.

SEASONAL NITROGEN LOADS
TO AND FROM LAS VEGAS WASH

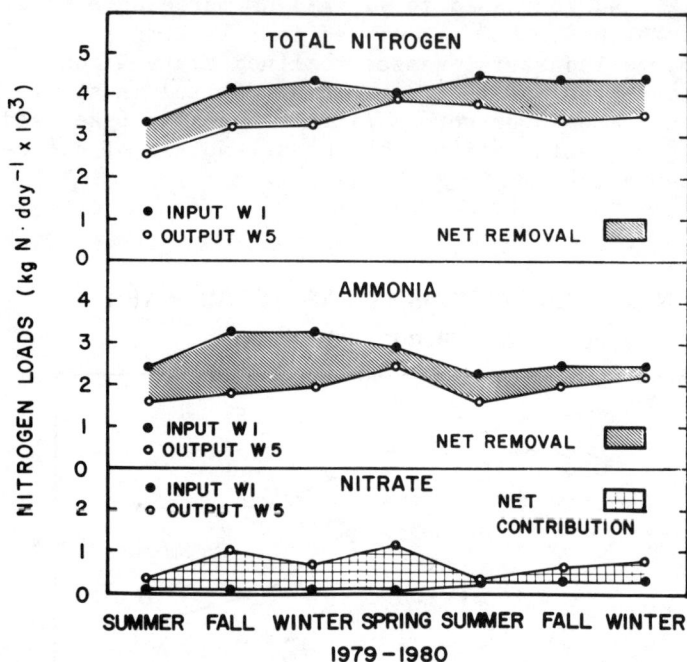

Figure 6. Seasonal Nitrogen Loads to and from
Las Vegas Wash.

A decreasing trend in ammonia removals within the wash
occurred during this study. There are many possible expla-
nations for this. First, loadings of ammonia from the STPs
were lower during later portions of the study. Second, storm
events of winter 1979-1980 caused a major upstream advance-
ment of erosion. Decreased removal efficiencies during lower
loadings were also observed by Morris et al. [9] in related
wetland studies in the Lake Tahoe Basin. They found the most
dramatic nutrient and sediment removals occurred when loads
were greatest. Another relevant conclusion was that one of
the most important factors in determining the effectiveness
of wetland treatment was the degree of sheet flows across
the wetland. In the case of the Las Vegas Wash, channelization limits spatial and temporal contact of waters with
wetland vegetation and, therefore, limits nutrient reducing
capabilities.

Goldman and Deacon [5] suggested that one mechanism
that may be responsible for ammonia removals within the Las
Vegas Wash is the adsorption to clay particles. These in-

132

vestigators indicated 90% reductions in ammonia loading as measured at Northshore Road in comparison to lower (31%) reductions seen during this study. It is possible that active headcutting zones may be eroding strata of less clay content than historical headcut zones. Based on elevational differences between past and present headcut zones, it seems that this may be true. However, a more detailed analysis of the system is required to give a definitive answer as to what mechanism plays a dominant role.

Historical phosphorus data measured at Northshore Road (Figure 7) indicate that recent upgrading of sewage treatment facilities is reducing total phosphorus loads in comparison to previous years. Seasonal analysis of phosphorus (Figure 8) shows a net contribution of total phosphorus in the spring of 1980. This is attributable to high sediment discharges during active headcutting that resulted from the February floods. The apparent removal of dissolved phosphorus during periods of high sediment discharge was probably due to adsorption of inorganic phosphorus to sediments eroded from the headcutting areas.

The results of our study can be summarized with data presented in Table I. Total nitrogen was reduced by an average of 1000 kg/day, and most of this was ammonia. However, there was a net contribution of 697 kg/day of nitrate. Total phosphorus loads were reduced by approximately one-third, and there was a slight decrease in soluble phosphorus.

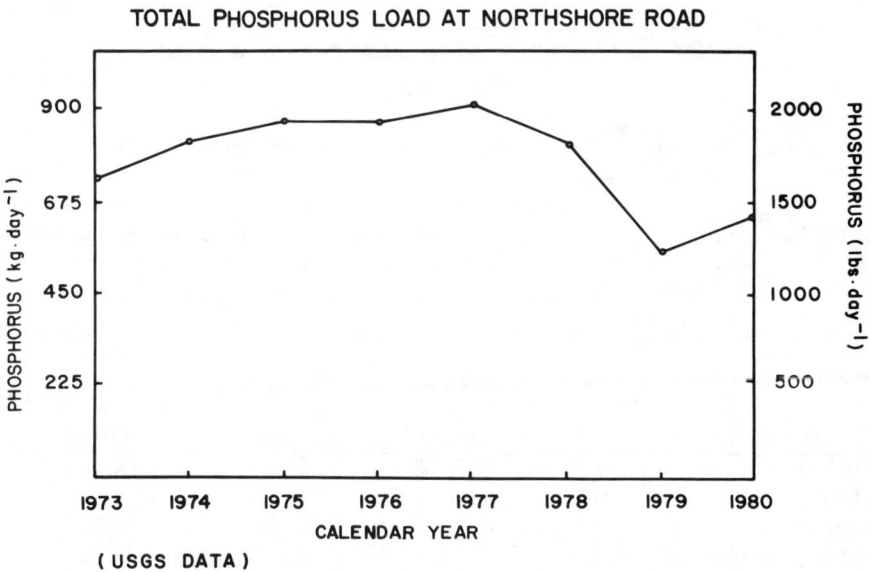

Figure 7. Historical Phosphorus Loads from Las Vegas Wash.

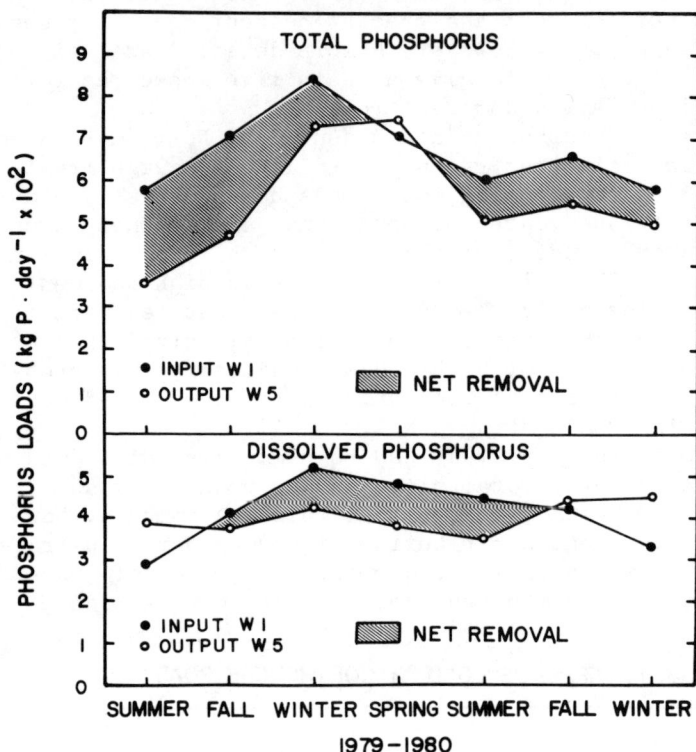

SEASONAL PHOSPHORUS LOADS
TO AND FROM LAS VEGAS WASH

Figure 8.   Seasonal Phosphorus Loads to and from
Las Vegas Wash.

Table I.   Mean Loading and Removal Rates of Nutrients in Las
Vegas Wash (as Measured at Northshore Road,
1.6 km from Lake Mead, July 23, 1979 to December
18, 1980).

Nutrient*	N	Mean Loading Rate (kg/day)	Standard Error (kg/day)	Mean Removal Rate (kg/day)
Total nitrogen	41	3172.0	164.4	1001.1
Nitrate	41	696.9	107.8	–
Ammonia	41	1874.4	82.1	992.0
Total phosphorus	40	539.6	45.0	156.9
Soluble phosphorus	41	380.4	16.7	49.4

*Expressed as elemental form

Although the wastewaters are retained within the Las
Vegas Wash wetlands for a relatively short period of time,
this ecosystem is behaving seasonally as a nitrogen and
phosphorus trap. This results in an improvement of the
quality of water discharged to Lake Mead. Efficiency of
nitrogen and phosphorus removal is a function of the loads
entering the system and the degree of contact of waters with
the wetland. Increasing velocities and volumes of flows have
decreased retention time resulting in less contact time with
the wetland. Rates of nutrient removal in the Las Vegas Wash
as described by URS and Clark County Department of Comprehensive Planning in 1978 [10] appear to be declining as a
result of changes in flow regimes. Improving the efficiency
of nutrient removal by proper management of this wetland
appears to be feasible. Further studies should be conducted
to elucidate specific mechanisms of nutrient removals.

ACKNOWLEDGEMENTS

The authors wish to thank the numerous agencies and
individuals who have supported this work by providing re-
sources, information, and criticisms. We wish to specific-
ally mention:
Brown and Caldwell Consulting Engineers, Sacramento, CA.
Culp/Wesner/Culp, Sacramento, CA.
U.S. EPA Environmental Monitoring Systems Laboratory,
     Las Vegas
U.S. Geological Survey, Carson City, NV.
U.S. Bureau of Reclamation, Boulder City, NV.

REFERENCES

1.   Kaufmann, R.E. 1971. Effects of Basic Management
      Incorporated effluent disposal on the hydrogeology and
      water quality of the Lower Las Vegas Wash area, Las
      Vegas, NV. An interim progress report to the Environ-
      mental Protection Agency on WQA-EPA Project No.
      13030EOB., Desert Research Institute, University of
      Nevada System. 176 pp.

2.   Bateman, R.L. 1976. Analysis of effects of modified
      wastewater disposal practices on Lower Las Vegas Wash.
      Project Report No. 39. Water Resources Center, Desert
      Research Institute, University of Nevada System. 60 pp.

3.   Kaufman, R.F. 1976. Land and water use effects on
      groundwater quality in Las Vegas Valley. Desert
      Research Institute, for the Environmental Protection
      Agency. 215 pp.

4. Hoffman, D.A., P.R. Tramutt, and F.C. Heller. 1971.
    The effect of Las Vegas Wash effluent upon the water
    quality in Lake Mead. U.S. Bur. of Rec. Rept. No.
    REC-ERC-71-11. 25 pp.

5. Goldman, C.R. and J.E. Deacon. 1978. Recommended water
    quality standards for Las Vegas Bay and Las Vegas Wash.
    Submitted to Nevada Environmental Commission. 41 pp.

6. Hubbs, C.L. and R.R. Miller. 1948. The Great Basin with
    emphasis on glacial and postglacial times, Part II. The
    Zoological Evidence. Bull. Univ. Utah Press, 38(20):
    18-144.

7. Bradley, W.G. and J.E. Deacon. 1967. The biotic com-
    munities of Southern Nevada. Nevada State Museum
    Anthropological Papers No. 13, Part 4:201-295.

8. U.S. Environmental Protection Agency. 1979. Methods for
    chemical analysis of water and wastes. EPA Report No.
    600/4-70-020, Environmental Monitoring and Support
    Laboratory, Office of Research and Development,
    Cincinnati, Ohio. 460 pp.

9. Morris, F.A., M.K. Morris, T.S. Michaud and L.R.
    Williams. 1981. Meadowland natural treatment processes
    in the Lake Tahoe Basin: A Field Investigation. U.S.
    Environmental Protection Agency. Environmental Monitor-
    ing Systems Laboratory, Las Vegas, Nevada. 181 pp.

10. URS Co. and Clark County Dept. of Comprehensive Plan-
    ning, Environmental Planning Division. 1978. Clark
    County 208, water quality management plan. Clark
    County, Nevada. 203 pp.

PART 3

MAIN STREAMS

CHAPTER 10

STATUS OF THE COLORADO RIVER
ECOSYSTEM IN GRAND CANYON
NATIONAL PARK AND GLEN CANYON
NATIONAL RECREATION AREA

Steven W. Carothers
  Museum of Northern Arizona and
  National Park Service

R. Roy Johnson
  National Park Service and
    University of Arizona

INTRODUCTION

When the melting snows of the high country on the west
side of the Continental Divide in Rocky Mountain National Park
begin their annual spring journey, the western headwaters of
the Colorado River are a tiny mountain stream. From its simple
snowpack origins, the West's most celebrated river winds southwesterly for almost 1,450 miles (2,330 km), draining 242,000
square miles (626,780 km^2) in Arizona and portions of six
other states, finally crossing the international boundary into
Mexico where 3,000 additional square miles (7,770 km^2) are
drained before the Colorado enters the Gulf of California.

The region drained by the Colorado River contains some of
the most outstanding scenery in the world. Hunt [1] calls
the Grand Canyon and other areas of the Colorado Plateau Province "easily the most colorful part of the United States."
Within the area of the Plateau drained by the Colorado River
the scenery, natural resources, and vast holdings of public
land are sufficiently spectacular and bountiful that six
national parks, four national recreation areas, and more than
a dozen national monuments have been set aside for administration by the National Park Service.

One of these national parks, Grand Canyon, and a national
recreation area, Glen Canyon, have within their boundaries
approximately 255 miles (410 km) or 17.5 percent of the 1,450
mile (2,330 km) Colorado River channel. The 15 miles (24 km)
of free flowing river in Glen Canyon (Glen Canyon Dam to Lees
Ferry) and 240 miles (386 km) in Grand Canyon (Lees Ferry to

Separation Rapid on Lake Mead) represent the largest continuous portion of the river that is presently uninterrupted by structural control (dams, irrigation diversion systems, etc.). The Colorado River is one of the most important natural resources to the several million inhabitants of much of the Rocky Mountains, Colorado Plateau and Basin and Range Province. The water, hydroelectric energy, and recreation potential of the river system are in ever increasing demand [2, 3, 4, 5].

Although the river in Glen and Grand canyons is still in a relatively wild state, the consumptive water and energy demands of both the upper (Colorado, New Mexico, Utah, Wyoming) and lower (Arizona, California, Nevada) basin states have indirectly caused sweeping changes in the natural ecology of this 255 miles (410 km) of free flowing river.

Hoover Dam was completed in 1935, but it was 1941 before Lake Mead was filled to capacity, inundating some 30 miles (48 km) of the Lower Grand Canyon upstream of the Grand Wash Cliffs. But it was the construction and operation of Glen Canyon Dam (finished in 1963, Lake Powell filled in 1980), 15 miles (24 km) upstream of the present boundary of Grand Canyon National Park, that caused the most dramatic changes in the downstream river ecology of Glen Canyon National Recreation Area and Grand Canyon National Park. Superimposed on the ecological changes caused by man's structural control of the river is the direct influence of the annually increasing numbers of river recreationists - the outdoor enthusiasts who look to the National Park Service for the recreational experiences found in pristine natural areas. As the numbers of these visitors increase through time, it is obvious that their actions erode, albeit slowly, some qualities of the total park resource.

This report summarizes the known ecological changes that have occurred within the aquatic and riparian ecosystems of the Colorado River in Glen and Grand canyons that result from structural control of the river and the more recent recreational use taking place. In addition, we discuss the issues and concerns that influence the capability of the National Park Service to carry out its mandate to maintain these administrative areas for the good of present and future generations.

THE INFLUENCE OF DAMS IN GLEN AND GRAND CANYONS

Hoover Dam

Although the scope of this paper will be directed primarily toward the relationship of the pre- and post Glen Canyon Dam river ecology with recreational activities, it is important to discuss the impact Hoover Dam and Lake Mead have had on certain aspects of the aquatic and riparian environments of the

Colorado River in the Grand Canyon area. The dramatic changes that took place in the section of the Lower Grand Canyon actually inundated by Lake Mead, when the river was converted into a reservoir, have been previously reviewed by several authors [see 6 for sedimentation and physical chemistry; 7 for fisheries; 8 for vegetation; and Paulsen et al., this symposium, for chemistry].

The influence the lake system presently has on the aquatic ecology of the upstream portion of the river in Grand Canyon involves exotic fish species and other aquatic organisms invading the upper portions of the canyon from their source in Lake Mead. These exotic species can potentially compete with the remaining native fishes of the Grand Canyon system. However, the extent of the competition is presently unknown [9]. In addition, the river terraces (beach deposits) from the Grand Wash Cliffs (River Mile 275) as far upstream as Diamond Creek (River Mile 225) were apparently deposited by the interaction of the river with Lake Mead when the latter was at its highest stage in 1941 [6]. These deposits have resulted in vegetational communities composed primarily of exotic species which typically invade disturbed areas, e.g., Saltcedar (Tamarix chinensis) and Russian thistle (Salsola kali). Although such exotic species are considered undesirable in national parks, the influence the introduced vegetational assemblages have on native park organisms has been only preliminarily investigated. Some of the highest densities of native park organisms along the river, especially arthropods, reptiles, and small mammals, have been found within Saltcedar thickets [10]. A recent treatise on the birds of Lake Mead [11] mentions Saltcedar but does not evaluate it as avian habitat. Our findings, as well as those of others [12], indicate that Saltcedar communities support fewer riparian birds than most native riparian vegetation, with Whitewing Doves (Zenaida asiatica) - not present in Grand Canyon National Park or Glen Canyon National Recreation Area, and Mourning Doves (Z. macroura) being notable exceptions.

Recreation on Lake Mead, primarily motorboat traffic, commonly extends upstream to the Diamond Creek area (National Park Service regulations prohibit upstream river traffic beyond Diamond Creek). The impacts associated with the recreationists involve the degradation of beach campsites through the accumulation of litter, campfire remains, and fecal wastes.

Glen Canyon Dam

Prior to the 1970's the 240 miles (386 km) of the Colorado River and its tributaries between Lees Ferry and Separation Rapids were virtually unexplored by terrestrial and aquatic ecologists. Due to the harsh environmental conditions and general difficulty of access, basic ecological investigations have been initiated only in the last decade [13]. Recent investigations include surveys of fish [14, 15, 16, 17], aquatic invertebrates [18, 19, 20], aquatic plants [21, 22], physico-chemical research [19, 22], geomorphologic studies [23], and riparian investigations [24]. More recently, Carothers and Minckley and co-workers [25] completed an extensive survey of fishes, aquatic plants and invertebrates of the main stem river and portions of its tributaries from Lees Ferry to Separation Rapids. Much of the material herein is drawn from that study.

Historically, the primitive Colorado represented a unique aquatic habitat, ranging from a swift-flowing, turbid river to a system characterized by long periods of low flows during droughts. Water temperatures fluctuated seasonally with the water being warmer in spring and summer, and cooler during fall and winter. Physico-chemical regimes varied with the flow regimes, i.e., spring run-off, summer flooding, or conversely, summer drought. It was within this system that one of the unique North American fish faunas developed [26], a faunal assemblage which had one of the highest rates of endemism of any river basin in North America [27, 28]. This fauna included the humpback chub (Gila cypha) and razorback sucker (Xyrauchen texanus) as well as the roundtail chub (Gila robusta), speckled dace (Rhinichthys osculus), flannelmouth sucker (Catostomus latipinnis), and bluehead sucker (Catostomus discobolus).

Currently, the Colorado River is drastically changed from its original state, primarily because of the construction and operation of Glen Canyon Dam and the stabilizing effect it has had on the riverine environment. The dam has affected three very basic features of the river: 1) sediment discharge and turbidity, 2) seasonal flow patterns and maximum/minimum flows, and 3) water temperature variation. Prior to the dam, sediment discharge through the Grand Canyon averaged 140 million tons per year with extremes ranging from 50 to 500 million tons. In the post-dam regimen, with most of the sediment now trapped within the 180 mile (290 km) long Lake Powell, an average of 20 million tons is transported through the canyon, presumably to Lake Mead [29]. Thus, the water released through the power-producing turbines of Glen Canyon Dam no longer has the reddish-brown color from which the river derived its Spanish name, Colorado. Now the general color of the river is green, a color apparently caused, in part, by the attached and floating mats of algae (Cladophora spp.) [21] that proliferate in the non-turbid water. Today the waters of the Colorado in Grand Canyon are reddish-brown only when muddied by a major tributary in flood. Parameters of nutrient cycling and transport, currently under investigation further downstream (see Paulsen et al., this symposium) are unknown for this segment of the Colorado.

142

Pre-dam river flow periodicity and discharge patterns were characterized by exceptionally high spring and summer flows, and low flows during fall and winter. The lowest flow recorded in the river canyon by the US Geological Survey between 1895 and the present was 700 cubic feet per second (cfs) (20 cubic meters per second = cms) in December of 1924, and the highest ever recorded was 127,000 cfs (3,556 cms) in July of 1927. Prior to the establishment of the gaging stations, however, peak flows in excess of 300,000 cfs (8,400 cms) were known to have existed within the Grand Canyon; average high flows were usually 20 times greater than the average low flows [30].

Although the effects of the changes in the flow and sediment concentrations of the Colorado are dramatic, the new temperature regimen has probably had the most significant effect on the aquatic ecology of Grand Canyon. Before the dam was built, there was a wide range in water temperature. Winter lows ranged from just above freezing to 40°F (4°C); temperatures gradually warmed to 60° or 70°F (16° or 21°C) during early spring and finally reached 75° to 85°F (24° to 30°C) as the annual floods subsided during July and August. The water that now flows into Grand Canyon is drawn from an area 200 feet (60 m) below the surface of Lake Powell and is released through power-producing turbines located 500 feet (150 m) below the surface, in the perpetually cold hypolimnetic water where the warming wavelengths of sunlight never penetrate. Summer and winter, the temperature of water released from the dam is 45°F (7°C). The river temperature in the dam-influenced environment increased by only 1 to 2 degrees F (1°C) in winter and by 5 to 6 degrees F (3°C) in summer, so there is now an annual temperature range of 6°F (3°C) compared with as much as 50°F (28°C) before construction of the dam [31].

The dramatic impacts to native ecosystems that have apparently resulted from Glen Canyon Dam include 1) extirpation or near extirpation of five of the original eight native fishes, and subsequent domination of the ichthyofauna by exotic species [25], 2) drastic changes in the algal and, possibly, invertebrate regimes, and 3) establishment of riparian communities and associated ecosystems along the lower terraces. Preliminary studies suggest that the species composition of certain plant communities on higher terraces (e.g., Prosopis-Acacia) may be changing, possibly due to the lack of seasonal flooding. These effects are especially well documented by a series of Colorado River technical reports published by Grand Canyon National Park from 1976 through 1977. Dam-related changes are also discussed by Dolan et al. [30], Johnson et al. [32], Dolan et al. [33], Carothers and Minckley [25], Howard and Dolan [23], and Carothers and Dolan [31], as well as in countless other publications, many of which are listed in the "references" section of this paper. A brief summary of the dam-related changes in the aquatic and riparian Colorado River system in Glen and Grand canyons follows.

The Aquatic System

Ichthyofauna

The status, history, and distribution of the ichthyofauna in the main stem Colorado River in Glen and Grand canyons has been more thoroughly studied (National Park Service and Bureau of Reclamation research contracts) than any of the other wild-life systems. A summary of the fisheries data is presented in Table I. At present, 27 fish species, 70 percent of which are introduced, non-native species, are known to occur or to have occurred within Glen and Grand canyons.

Of the eight original native species, the bonytail chub, Colorado (roundtail) chub and Colorado squawfish have apparent-ly been extirpated in Grand Canyon. The disappearance of these species is not restricted to the Grand Canyon area alone, since all three have been officially designated as "endangered" spe-cies [34] and are therefore protected by Federal Law in those portions of their former ranges where populations persist. Another endangered species native to the Colorado River system is the humpback chub. Although still extant in Grand Canyon waters, its population has evidently become markedly reduced by major habitat changes resulting from the construction and oper-ation of Glen Canyon Dam. The humpback chub was previously thought to breed throughout the river system. Recent data in-dicate that it now only reproduces near the confluence of the Little Colorado and Colorado rivers in the canyon. Preliminary indications from ongoing FWS studies on the humpback chub in the Little Colorado River are that the species appears to be main-taining a stable population in this area. The razorback sucker is thought to be in danger of extirpation throughout its former range within the Colorado River Basin [35], and was considered extirpated in Grand Canyon until recently [9].

At present, three native species, the bluehead and flannel-mouth suckers and the speckled dace, are still represented by what appear to be healthy, reproducing populations through most of their former range. The juvenile suckers remain in the pe-rennial tributaries for 2 to 3 years after hatching before mov-ing into the mainstream. Speckled dace are regularly found in the tributaries while their densities in the mainstream are highly variable.

Thus, in Grand Canyon three of the eight native fish spe-cies continue to thrive; one is restricted to a tiny portion of

Table I. The status, history, and distribution of Glen and Grand canyon fishes. After Carothers and Minckley [25].

Species	Status	History[1]	Distribution
NATIVE SPECIES:			
Bonytail chub (Gila elegans)[2]	Ext	Arch 1942 1963	
Humpback chub (Gila cypha)[2]	LC	Arch 1914 1979	Adults and young concentrate in the vicinity of Little Colorado R.
Colorado River (roundtail) chub (Gila robusta)[2]	Ext	1961 1966	
Colorado squawfish (Ptychocheilus lucius)[2]	Ext	Arch 1914 1972	
Speckled dace (Rhinichthys osculus)	C	1932? 1979	Ubiquitous, mostly in tributaries, some adults in mainstream
Razorback sucker (Xyrauchen texanus)	R	1944 1979	Only recent (1978) record from mouth of Paria River
Flannelmouth sucker (Catostomus latipinnis)	C	Arch 1979	Ubiquitous, young in tributaries
Bluehead sucker (Pantosteus (Catostomus) discobolus)	C	Arch 1937 1979	Ubiquitous, young in tributaries
INTRODUCED SPECIES:			
Coho salmon (Oncorhynchus kisutch)	Acc, Ext	1971 1976	

145

Table I. (cont.)

Species	Status	History[1]	Distribution
Rainbow trout (Salmo gairdneri)	Abun	1922 1979	Ubiquitous, adult spawning runs in tributaries
Cutthroat trout (S. clarki)	Abun		Ubiquitous
Brown trout (S. trutta)	FC	1934 1979	Ubiquitous
Brook trout (Salvelinus fontinalis)	FC	1920 1979	Ubiquitous
Carp (Cyprinus carpio)	Abun	late 1800's 1979	Ubiquitous
Golden Shiner (Notemigonus crysoleucus)	Acc	1976 1979	
Virgin River spinedace (Lepidomeda mollispinis)	Acc, Ext	1972	
Woundfin (Plagopterus argentissimus)	Acc, Ext	1972	
Red Shiner (Notropis lutrensis)	Acc, Ext	1968 1976	
Fathead minnow (Pimephales promelas)	LC	1952 1979	Highest densities in tributaries
Channel catfish (Ictalurus punctatus)	LC	1909 1979	Throughout mainstream in low densities, locally common in Little Colorado River and Kanab Ck
Black bullhead (Ictalurus melas)	Acc	1968 1976	
Rio Grande killifish (Fundulus zebrinus)	LC	1938 1979	Unkar, Royal Arch and Kanab creeks

Species		Dates	Comments
Striped bass (Morone saxatilis)	LC	1970 1979	Only near Separation Rapids on Lake Mead
Largemouth bass (Micropterus salmoides)	Acc	1979	
Green sunfish (Chaenobryttus cyanellus)	Acc	1968 1979	
Bluegill sunfish (Lipomis macrochirus)	Acc	1979	
Walleye (Stizostedion vitreum)	Acc	1972	Only record from Lees Ferry area, 1971-1972, may be extinct in system

KEY:

Ext – extinct – native species previously represented by viable populations, now no longer in area.

Abun – abundant – easily captured, always present in large numbers

C – common – easily captured, although not present in large numbers

FC – fairly common – occasionally captured, but not unexpected

LC – locally common – captured easily in specific areas, often present in numbers

R – rare – always unexpected; may be going extinct

Acc – accidental – one or two specimen records, isolated incidences of bait bucket releases, relatively unsuccessful transplants or individuals dispersing from Lake Mead; probably not part of breeding ichthyofauna (continued)

Arch – Archaeological remains, 4000 B.P.

FOOTNOTES:

1. Dates represent first and last published collection records. The 1979 collection records represent species taken in this study.
2. Listed in the Federal Register as an endangered species.

147

its former range; three are extinct, and one is almost assuredly doomed to become extinct in Grand Canyon [23, 35].

Ten of the 19 exotic species thus far recorded within the study area have been collected or observed so infrequently that they should be considered insignificant components of the Grand Canyon ichthyofauna. The one or two collection records for each species in the past decade probably represent isolated captures of non-breeding, accidental bait-bucket releases (golden shiner, green sunfish, bluegill sunfish), unusual upstream penetration of Lake Mead species (coho salmon, black bullhead, largemouth bass) or unsuccessful transplants by fishery management agencies (Virgin River spinedace, woundfin, walleye). The red shiner, collected at several canyon localities by Miller and Smith [14] and twice by Suttkus et al. [17], was not represented in the extensive sampling of Carothers and Minckley [25] and may be extinct.

The relative abundance of fishes in the Grand Canyon area is presented in Table II. From these data, it is obvious that although four native species are still present within the system and three of these are relatively abundant (speckled dace, flannelmouth and bluehead suckers), the exotic trout and carp are dominant in the system (combined frequency all trout species and carp = 55.5 percent) with the carp being the most commonly encountered fish within the entire system (41.6 percent of all captures). Carothers and Minckley [25] attribute the present species assemblage of Grand Canyon ichthyofauna to the combined effects of Hoover and Glen Canyon dams on the riverine habitat - changes that have been to the general benefit of exotic species and detriment of native species - and to concerted efforts by a variety of management concerns (including park managers in the 1930's and 40's) to establish a trout sports fishery in the Glen and Grand canyon area.

Exotic fishes were first introduced into the area in 1919, prior to the creation of Grand Canyon National Park. These introductions were made by the U.S. Forest Service, but neither the species nor the numbers planted were recorded [36]. The National Park Service began stocking Grand Canyon tributaries in 1920 by planting brook trout into Bright Angel Creek, and in later years, they also stocked Clear and Havasu creeks with brook trout. Rainbow trout were first introduced at Tapeats Creek in 1922 by the U.S. Forest Service [37]. Subsequently, this species was introduced into several tributaries and is now the most common fish stocked into the area. Stocking of brown trout began in 1924 at Bright Angel Creek and was later extended to other creeks.

Introduction of exotic species was stopped by the National Park Service in 1964 after a final stocking of rainbow trout

Table II. Relative abundance (expressed as percent) of fishes collected from the mainstream Colorado River and its tributaries, Coconino and Mohave Counties, Arizona, during 1977 – 1978. From Carothers and Minckley (1981).

Exotic and Native Fish	Exotic Species	Native Species
Carp – 41.6	Carp – 73.3	Speckled dace – 35.0
Speckled dace – 15.6	Rainbow trout – 23.0	Flannelmouth sucker – 30.9
Flannelmouth sucker – 13.8	Channel catfish – 1.4	Bluehead sucker – 20.8
Rainbow trout – 13.3	Striped bass – 0.9	Humpback chub – 13.1
Bluehead sucker – 9.3	Brook trout – 0.7	Razorback sucker – 0.03
Humpback chub – 5.8	Brown trout – 0.6	
Channel catfish – 0.7	Fathead minnow – 0.4	
Striped bass – 0.5	Largemouth bass – 0.2	
Brown trout – 0.3	Rio Grande killifish 0.2	
Brook trout – 0.3	Green sunfish – 0.06	
Fathead minnow – 0.2	Bluegill sunfish – 0.03	
Rio Grande killifish – 0.09	Golden shiner – 0.03	
Largemouth bass – 0.09		
Green sunfish – 0.03		
Bluegill sunfish – 0.01		
Golden shiner – 0.01		
Razorback sucker – 0.01		

149

into Bright Angel Creek. The U.S. Fish and Wildlife Service ended their introduction of exotic species after a final planting of rainbow trout into Diamond Creek in 1971.

The Arizona Game and Fish Department started an ongoing stocking program in 1964. From 1964 to 1979, 441,310 rainbow trout, 147,880 brook trout and 20,000 coho salmon and 50,000 cutthroat have been introduced at Lees Ferry. The success of fish introductions prior to 1964, and the construction of Glen Canyon Dam, was apparently limited. Rainbow trout have maintained populations in Tapeats, Havasu, and Bright Angel creeks via limited reproduction and stocking [25, 36, 37]. At present the rainbow trout is the most common species in the Glen Canyon area (McCall, this symposium) resulting in the most exciting trout trophy fishery in North America [31].

Aquatic Plants and Invertebrates

It has proven impossible to reconstruct the pre-dam aquatic plant and invertebrate community structure of the Colorado River in Glen and Grand canyons. Although a few (minimal) biological studies were conducted in the section of Glen Canyon Recreation Area to be inundated by Lake Powell, no baseline studies were conducted in Grand Canyon National Park. In Carothers and Minckley [25], Griffith Hardwick presents a general description of existing conditions relative to the present macrophyte communities in the main stem and tributaries within Grand Canyon National Park. Further, the periphytic communities of the area are described by Czarnecki et al. [21], Czarnecki and Blinn [38, 39], and Czarnecki [40]. Data on area invertebrates may be found in Carothers and Aitchison [10], Hofknecht [41], and Carothers and Minckley [25].

Riparian Vegetation

The terrestrial habitats immediately adjacent to the river below Glen Canyon Dam have become ecologically more diverse since the dam was built. Peak river flows are presently less than 30 percent of those of the pre-dam era, and the riparian vegetation has been adjusting to a new set of limiting factors. In the post-dam environment, with the threat of annual scouring gone, woody plants that were unable to undergo yearly inundation or to grow on an unstable substrate have colonized the previously inhospitable river bank. Typical desert vegetation has been moving downslope toward the river, while plant species known for their ability to grow rapidly on moist streambank soils, such as native willows (Salix sp.) and the introduced Saltcedar, have contributed to the formation of dense thickets [8, 30, 31]. Turner and Karpiscak [8] summarize the present condition of the vegetative community thus: "...We believe that in the short period of 13 years the zone of post-dam fluvial deposits has been transformed from a barren skirt on both sides of the river to a dynamic double strip of vegetation." Carothers and Dolan [31] add, "This dynamic strip of woody, perennial vegetation occurs from the dam to Lake Mead and varies in width from one to several meters depending upon local relief and rock type. The species composition of this riparian forest is a mixture of native and exotic species that has not yet reached an equilibrium, or climax, condition."

The new strips of post-dam vegetation are growing on sediments deposited pre-Glen Canyon Dam. The stability of these pre-dam deposits has been the subject of some debate and is reviewed in Laursen and Silverston [42] and Dolan [43].

Riparian Vertebrates

While the riparian vegetation of the dam-controlled Colorado River system in Grand Canyon has responded to the stabilized river regimen, the overall increase in vegetational biomass and apparent diversity has resulted in some interesting changes in the riparian animal communities. Carothers and Dolan [31] summarize these changes: "Rodents, reptiles, amphibians, and small birds now abound in a developing streamside forest where previously there was bare ground. This new riparian habitat is no replacement for the hundreds of thousands of acres that now lie under Lake Powell, but it is an unanticipated benefit. Bell's vireos, (<u>Vireo belli</u>), yellow-breasted chats (<u>Icteria virens</u>), Lucy's (<u>Vermivora luciae</u>) and yellow (<u>Dendroica petechia</u>) warblers, house finches (<u>Carpodacus mexicanus</u>), indigo (<u>Passerina cyanea</u>) and Lazuli (<u>P. amoena</u>) buntings, blue grosbeaks (<u>Guiraca caerulea</u>), and at least four species of native mice, have all benefited from a substantial increase in suitable breeding habitats along the banks of the Colorado River. Before the variable flows were stabilized, beaver were present in the system, but the food supply in the main river was limited and the annual floods surely posed a serious threat to their survival. Now, with food abundant and floods all but gone, the beavers are enjoying a habitat made to order. There is no population-density information on the animal before 1961, but today almost every beach has its family of bank-dwelling beavers. In addition the desert spiny lizard (<u>Sceloporus magister</u>), side-blotched lizard (<u>Uta stansburiana</u>), collared lizard (<u>Crotophytus collaris</u>), and whiptail lizard (<u>Cnemidophorus</u> spp.) have increased in number. Interestingly, the whiptail lizards, usually restricted to a desert domain, now forage at the water's edge for a new food source - amphipods stranded at low tide. The eventual fate of the developing forest and its inhabitants is unknown."

The increases in avian diversity and population densities are the best known of the faunal elements [Table III; 44, 45]. Still the baseline information for determining accurately the magnitude of these changes is unavailable. As mentioned previously, limited studies were conducted in the section of Glen Canyon to be inundated by Lake Powell [47]. However, an apparent lack of concern for and/or appreciation of future effects of Glen Canyon Dam on the Grand Canyon National Park riverine ecosystems resulted in a dearth of pre-dam information for the Colorado River in Grand Canyon National Park.

Table III. Obligate riparian breeding birds that have either recently colonized the river corridor in Grand Canyon or have experienced rapid population increases due to the availability of new riparian habitat.

RECENT COLONIZERS:

   Willow Flycatcher[1] (Empidonax traillii)

   Vermilion Flycatcher[2] (Pyrocephalus rubinus)

   Bell's Vireo[3] (Vireo bellii)

   Northern Oriole (Icterus galbula)

   Hooded Oriole (I. cucullatus)

   Summer Tanager (Piranga rubra)

   Great-tailed Grackle (Quiscalus mexicanus)

INCREASED IN NUMBERS:

   Yellow warbler (Dendroica petechia)

   Common Yellowthroat (Geothlypis trichas)

   Yellow-breasted Chat (Icteria virens)

   Blue Grosbeak (Guiraca caerulea)

1. Phillips et al. [46] indicated breeding along Havasu Creek and at Lees Ferry prior to 1963. Its status at that time along the river corridor in Grand Canyon was uncertain.
2. Present in small numbers during the breeding season, although actual nesting has not been confirmed
3. There is some question as to whether small numbers of Bell's Vireos were overlooked due to a paucity of earlier work or whether they did indeed colonize the river corridor after 1963 [see 45, 46].

Present National Park Service research prioritization has given major importance to determining the impacts of Glen Canyon Dam operational scenarios on the stability of the present environment. It has been generally assumed [43] that increased variation in minimum and maximum dam discharges (e.g., peaking power) can affect the rate of beach erosion and therefore influence the stability of the sediment associated riparian communities. The true relationship between the dam, the present vegetation/animal riparian community, and the fate of that community under the existing or alternative dam operational scenarios is, however, unknown.

RIVER RECREATION AND ASSOCIATED IMPACTS IN GLEN AND
GRAND CANYONS

Glen Canyon National Recreation Area is undoubtedly best known for its 180 mile long (290 km), 9 trillion gallon (34 trillion liters) reservoir, Lake Powell. The excellent recreational opportunities afforded by the lake and its 1800 mile (2900 km) shoreline attract hundreds of thousands of visitors annually, either in pursuit of water-based recreation or out of simple curiosity about one of the largest man-made reservoirs and dams in the world. Another significant recreational attraction and unique geographical feature of this National Park Service administrative unit is the 15 mile (24 km) segment of flowing river from the dam to Lees Ferry. This stretch of river attracts approximately 30,000 visitors annually, most of whom are trophy trout fisherpersons [48]. Recreation in the Glen Canyon area prior to the dam was minimal. It is doubtful whether more than 200 persons per year entered Glen Canyon for recreational or other purposes prior to the mid-1960's.

The river environment in Grand Canyon National Park has a similar history of human visitation and use. By the early 1950's only two hundred people had navigated the Colorado River through the Grand Canyon National Park. By 1967, river running had become a thriving business on the Colorado Plateau. Due partially to the relatively constant flows through Grand Canyon National Park, resulting from controlled flows through Glen Canyon Dam, more than 2,000 people annually were going down the Colorado in Grand Canyon National Park. By 1972, more than 15,000 people annually, a 700 percent increase in 6 years, were enjoying the whitewater recreation of the Grand Canyon National Park [49].

Numerous impacts, hypothesized or measured, have been linked to Colorado River recreation. These impacts include: 1) footstep induced mass wasting of alluvial sediments from recreational beaches [50], 2) contamination of beaches by human wastes [48, 51, 52], 3) water quality problems [53], 4)

degradation of aesthetic quality of beach campsites from campfire scars and litter [33, 48], and 5) trampling of vegetation
[32].

It is noteworthy that Carothers et al. [48] found significant differences in the aesthetic attributes of river campsites
between Glen and Grand canyons. Measurements of beach sand dis-
coloration by uncontrolled campfires, and litter and feces accumulation indicated that Glen Canyon beaches were far more
degraded (by a factor of 20) than Grand Canyon beaches. The
differences were attributed to 1) the apparent differences in
environmental conscientiousness between Grand Canyon users
(primarily river runners) and Glen Canyon users (primarily
fisherpersons), 2) the relative allocation of commercial versus
private users in the two areas (primarily commercial in Grand
Canyon and private in Glen Canyon), and 3) the environmental
use restrictions placed on users in both areas (stringent controls in Grand Canyon, virtually none in Glen Canyon).

CONCLUSION

Management of National Park Service areas in the Colorado
drainage is exceptionally difficult. Attempts to retain natural values in areas set aside for this purpose is next to impossible on National Park Service lands where upstream and/or
downstream water projects affect the aquatic and riparian ecology along rivers.

Challenges to resource managers and cooperating scientists
arise not only from different management agencies with conflicting programs and goals, but also within agencies. For example,
some activities conducted by the National Park Service in Glen
Canyon National Recreation Area conflict with practices in Grand
Canyon National Park. Ironically, the National Park Service is
in charge of both areas. Thus, exotic fishes are introduced in
Glen Canyon National Recreation Area where trophy rainbow trout
are the pride of fisheries managers while scientists and managers in adjacent Grand Canyon National Park fear further loss of
the almost extinct, native humpback chub to voracious, predatory
game fish. Scientists are concerned about the spread of the introduced Russian olive (_Elaegnus angustifolia_) throughout the
riparian zone along the river in Grand Canyon National Park
(where the exotic Saltcedar is already well established) while
Glen Canyon National Recreation Area managers plant _Elaegnus_ in
the Lees Ferry campground within hundreds of meters of the Colorado River in Grand Canyon National Park. In Lake Mead National
Recreation Area, just downstream, powerboats cruise at will
while amongst Grand Canyon river boatmen controversy continues
over the use of motors on boats traversing the Grand Canyon.
These unique resources management problems will continue to be

a source of dilemma for National Park Service staff as long as recreation areas, such as Glen Canyon National Recreation Area, with one set of management criteria are immediately adjacent to National Park Service areas, such as Grand Canyon National Park, set aside by Congress on entirely different use criteria with emphasis on managing the area for its naturalness.

In addition, the Colorado River in Grand and Glen canyons continues to be sought after for increased hydroelectric power production and increases in water storage and usage. These influences, which originate outside the adminstrative units and continue to plague the National Park Service, are often consumptive of natural resources with little potential for mitigation.

Dangers to natural processes within national parks which originate outside the parks are of sufficient magnitude to cause the National Park Service to develop a special "External Threats to Parks Program." This special program is designed to examine the sources, impacts, and magnitude of damage from these external threats, whether extirpation of native species, reduction in water quality, or production of acid rain.

The problems are not unique to the National Park Service. When numerous political entities share a common natural resource such as a river, airshed, or watershed, these types of problems arise. In the case of the Colorado River they have state, federal, and even international ramifications. This has resulted in various states accusing others of taking "their water" and has even led Mexico to complain that the United States is allowing her an insufficient quantity of Colorado River water and/or water which is too saline for practical use. It is hoped that the National Park Service's External Threats to Parks Program will serve as a model for examining cross-boundary problems and arriving at their solutions.

The future does not seem bright for the native aquatic and riparian ecosystems of the Colorado River in Grand Canyon National Park. However, during the last two decades, since the completion of Glen Canyon Dam, considerable progress has been made toward understanding the native (natural) elements of the ecosystem as well as the newly created, naturalized elements of these ecosystems. As recently as the 1950's, concern was expressed for Glen Canyon, to be mostly inundated by Lake Powell, while the vast impact to be thrust upon the Grand Canyon National Park by Glen Canyon Dam was ignored or unforeseen. Today scientists and managers from a variety of resource management agencies are attempting to develop means of multiple use management which would prevent further degradation of one of the Southwest's showcase and renowned natural resources, the Colorado River. A critical element in "the Colorado River masterplan" is the retention of that portion of the river through lower Glen Canyon and the Grand Canyon National Park as the only relatively free flowing whitewater segment of the Lower Colorado River, an unequaled natural-recreational resource. This can be accomplished only through research establishing a suitable information base to allow resource managers to accomplish this mission.

REFERENCES

1. C. B. Hunt, Physiography of the United States. W. H. Freeman and Co., San Francisco. 480 pp., 1967.

2. W. O. Spofford, A. L. Parker, and A. V. Kneese (eds.), Energy Development in the Southwest: Problems of water, fish, and wildlife in the Upper Colorado River Basin (Vol. II). Resources for the Future, Washington, D.C. 541 pp., 1980.

3. J. Boslough, Rationing a river. Science 81 4:26-37, 1981.

4. P. L. Fradkin, A river no more: The Colorado River and the West. A. Knopf, New York. 360 pp., 1981.

5. R. Roy Johnson, and Steven W. Carothers, Riparian habitats and recreation: Interrelationships and impacts in the Southwest and Rocky Mountain Region. Eisenhower Consort. Bull., in press.

6. W. O. Smith, C. P. Vetter, and G. B. Cummings, Comprehensive survey of sedimentation in Lake Mead, 1948-9. US Geol. Surv. Prof. Pap. 295. U.S. Govt. Print. Off., Washington, D.C. 254 pp., 1960.

7. D. C. Roden, Base fisheries data, Phase I - Boulder Canyon to Davis Dam. Contract No. 14-06-300-2705, USDI Bureau of Reclamation. Boulder City, Nev. 1978.

8. R. M. Turner, and M. M. Karpiscak, Recent vegetation changes along the Colorado River between Glen Canyon Dam and Lake Mead, Arizona. U.S. Geol. Surv. Prof. Pap. 1132. U.S. Govt. Print. Off., Washington, D.C. 125 pp. 1980.

9. C. O. Minckley, and S. W. Carothers, Recent collections of the Colorado squawfish and razorback sucker from the San Juan and Colorado rivers in New Mexico and Arizona. Southwest. Nat. 24:686-687, 1980.

10. S. W. Carothers, and S. W. Aitchison (eds.), An ecological survey of the riparian zone of the Colorado River between Lees Ferry and the Grand Wash Cliffs, Arizona. Colorado River Tech. Rpt. No. 10, Grand Canyon National Park, Ariz. 251 pp., 1976.

11. J. G. Blake, Birds of the Lake Mead National Recreation Area. LAME Tech. Rpt. No. 1, Coop. Natl. Park Res. Studies Unit, Univ. Nevada, Las Vegas. 218 pp., 1978.

12. B. W. Anderson, A. Higgins, and R. D. Ohmart, Avian use of Saltcedar communities in the Lower Colorado River Valley. Pages 128-136 in R. R. Johnson and D. A. Jones (tech. coords.), Importance, Preservation and Management of Riparian Habitat: A Symposium. USDA For. Serv. Gen. Tech. Rpt. RM-43. Rocky Mt. For. & Range Expt. Stn., Ft. Collins, 1977.

13. R. R. Johnson, Synthesis and management implications of the Colorado River research program. Colo. R. Res. Tech. Rpt. No. 17, USDI Natl. Park Serv., Grand Canyon National Park. 75 pp., 1977.

14. R. R. Miller, and G. R. Smith, Report on fishes of the Colorado River drainage between Lees Ferry and Pierce's Ferry. Unpubl. Ms. Nat. Hist. Mus., Univ. Michigan, Ann Arbor, 1968.

15. P. B. Holden, and C. B. Stalnaker, Systematic studies of the cyprinid genus Gila, in the Upper Colorado River Basin. Copeia 1970:409-420.

16. C. O. Minckley, and D. W. Blinn, Summer distribution and reproductive status of fish of the Colorado River in Grand Canyon National Park and vicinity during 1975-76. Colorado River Tech. Rpt. No. 14, Grand Canyon Natl. Park, Arizona. 17 pp., 1976.

17. R. D. Suttkus, G. H. Clemmer, C. Jones, and C. R. Shoop. Survey of fishes, mammals, and herpetofauna of the Colorado River and adjacent riparian areas of the Grand Canyon National Park. Colorado River Tech. Rpt. No. 5, Grand Canyon Natl. Park, Arizona. 48 pp., 1976.

18. A. R. Gaufin, G. R. Smith, and P. Datson, Aquatic survey of Green River and tributaries within the Flaming Gorge Reservoir Basin. Pages 139-175 in Ecological Studies of the Flora and Fauna of Flaming Gorge Reservoir Basin, Utah and Wyoming. Univ. Utah Anthro. Pap. No. 48, 1960.

19. G. A. Cole, and D. M. Kubly, Limnologic studies on the Colorado River from Lees Ferry to Diamond Creek. Colorado River Tech. Rpt. No. 8, Grand Canyon Natl. Park, Arizona. 88 pp., 1976.

20. C. O. Minckley, A report on aquatic investigations conducted during 1976-1977 on Bright Angel, Phantom and Pipe creeks, Grand Canyon National Park, Coconino County, Arizona. Unpubl. Rpt., Grand Canyon Natl. Park, Arizona. 112 pp., 1978.

21. D. B. Czarnecki, D. W. Blinn, and T. Tompkins, A periphytic microflora analysis of the Colorado River and major tributaries in Grand Canyon National Park and vicinity. Colorado River Tech. Rpt. No. 6, Grand Canyon Natl. Park, Arizona. 106 pp., 1976.

22. M. R. Sommerfeld, W. M. Crayton, and N. L. Crane, Survey of bacteria, phytoplankton, and trace chemistry of the Lower Colorado River and tributaries in the Grand Canyon National Park. Colorado River Tech. Rpt. No. 12, Grand Canyon Natl. Park, Arizona. 136 pp., 1976.

23. Alan Howard, and Robert Dolan, Geomorphology of the Colorado River in the Grand Canyon. J. Geology 89(3):269-298, 1981.

24. Carothers et al., see Carothers and Aitchison 1976, [10].

25. S. W. Carothers, and C. O. Minckley, A survey of the aquatic flora and fauna of the Grand Canyon. Final Rpt. on Contract No. 7-07-30-X0026, USDI Water and Power Res. Serv., Lower Colorado Region. Boulder City, Nev. 401 pp., 1981.

26. W. L. Minckley, and J. E. Deacon, Southwest fishes and the enigma of "endangered species." Science 159:1424-1432, 1968.

27. R. R. Miller, Origin and affinities of the freshwater fish fauna of western North America. Pages 187-222 in C. L. Hubbs (ed.), Zoogeography. Amer. Assoc. Adv. Sci. Publ. No. 51, 1958.

28. M. Molles, The impacts of habitat alterations and introduced species on the native fishes of the Upper Colorado River Basin. Pages 163-181 in W. O. Spofford, A. L. Parker, and A. V. Kneese (eds.), Energy Development in the Southwest. Resources for the Future, Washington, D.C., No. 2, 1980.

29. U.S. Geological Survey, Water resources data for Arizona, water year 1976. U.S. Geol. Surv. Water Data Rpt. AZ-76-1, 1976.

30. R. Dolan, A. Howard, and A. Gallenson, Man's impact on the Colorado River in the Grand Canyon. Amer. Sci. 62(4): 392-401, 1974.

31. S. W. Carothers, and R. Dolan, Dam changes on the Colorado River. Nat. Hist. 91(1):74-83, 1982.

32. R. R. Johnson, S. W. Carothers, R. Dolan, B. P. Hayden, and A. Howard, Man's impact on the Colorado River in the Grand Canyon. Natl. Parks and Conserv. Mag. 51(3):13-16, 1977.

33. R. Dolan, B. Hayden, and A. Howard, Environmental management of the Colorado River within the Grand Canyon. J. Envir. Mgmt. 1(5):391-400, 1977.

34. U.S. Fish and Wildlife Service, Endangered and threatened wildlife and plants. U.S. Code of Federal Regulations 50 (Part 17):52-135, U.S. Govt. Print. Off., Wash., D.C. 1980.

35. W. L. Minckley, Status of the razorback sucker (Xyrauchen texanus) in the Lower Colorado River Basin. Southwest. Nat., in press.

36. R. R. Williamson, and C. F. Tyler, Trout propagation in Grand Canyon National Park. Grand Canyon Nature Notes 7(2):11-16, 1932.

37. J. P. Brooks, Field observations. Grand Canyon Nature Notes 4(10):70, 1931.

38. D. B. Czarnecki, and D. W. Blinn, Diatoms of lower Lake Powell and vicinity (Diatoms of southwestern USA I). Biblio. Phycol. 28:1-119, 1977.

39. D. B. Czarnecki, and D. W. Blinn, Diatoms of the Colorado River in Grand Canyon National Park and vicinity (Diatoms of southwestern USA III). Biblio. Phycol. 38:1-181, 1978.

40. D. B. Czarnecki, Diatoms of northern Arizona - their distribution, ecology, and taxonomy, including notes on successional patterns in Montezuma Well and morphological variability in Navicula cuspidata (Kutz). Ph.D. Diss., Northern Ariz. Univ., Flagstaff. 126 pp., 1978.

41. G. W. Hofknecht, Seasonal community dynamics of aquatic invertebrates in the Colorado River and its tributaries within Grand Canyon, Arizona. M.S. Thesis, Northern Ariz. Univ., Flagstaff. 105 pp., 1981.

42. E. M. Laursen, and E. Silverston, Hydrology and sedimentology of the Colorado River in Grand Canyon. Colorado River Tech. Rpt. No. 13, Grand Canyon Natl. Park, Ariz. 27 pp., 1976.

43. R. Dolan, Analysis of erosion trends of the sedimentary deposits in the Grand Canyon. Unpubl. Rpt., Grand Canyon Natl. Park, Arizona, 1981.

44. S. W. Carothers, and R. R. Johnson, Recent observations on the status and distribution of some birds of the Grand Canyon region. Plateau 47:140-153, 1975.

45. B. T. Brown, P. S. Bennett, S. W. Carothers, L. T. Haight, R. R. Johnson, and M. M. Riffey, Birds of the Grand Canyon region: An annotated checklist. Grand Canyon, Ariz. 64 pp., 1978.

46. A. Phillips, J. Marshall, and G. Monson, The birds of
    Arizona. Univ. Ariz. Press, Tucson. 212 pp., 1964.

47. A. M. Woodbury (ed.), Ecological studies on the flora and
    fauna in Glen Canyon. Univ. Utah Anthro. Pap. No. 40
    (Glen Canyon Series No. 7), Univ. Utah Press, Salt
    Lake City. 226 pp., 1959.

48. S. W. Carothers, R. A. Johnson, B. G. Phillips, M. M. Sharp,
    and A. M. Phillips, III, Recreational impacts on river-
    ine habitats in Glen Canyon National Recreation Area,
    Arizona. Final Rpt. on Contract No. CX8210-0-022, Glen
    Canyon Natl. Rec. Area, Page, Arizona. 62 pp., 1981.

49. R. R. Johnson, and S. P. Martin, The Colorado River re-
    search project: A National Park Service multidiscipli-
    nary, interdisciplinary research project. Pages 29-43
    in Proceedings of the Third Resources Management Con-
    ference. Southwest Region, Natl. Park Serv., Santa Fe,
    New Mexico, 1976.

50. S. Valentine, and R. Dolan, Footstep-induced sediment dis-
    placement in the Grand Canyon. Envir. Mgmt. 3:531-533,
    1979.

51. A. B. Knudsen, R. Johnson, K. Johnson, and N. R. Henderson,
    A bacteriological analysis of portable toilet effluent
    at selected beaches along the Colorado River, Grand Can-
    yon National Park, Arizona. Pages 290-295 in USDA For.
    Ser., Proceedings of the River Recreation Management
    and Research Symp. USDA For. Serv. Gen. Tech. Rpt.
    NC-28, 1977.

52. R. Phillips, and C. Sartor-Lynch, Human waste disposal on
    beaches of the Colorado River in Grand Canyon. Colorado
    River Tech. Rpt. No. 11, Grand Canyon Natl. Park, Ariz.
    79 pp., 1977.

CHAPTER 11

THE NATURE AND AVAILABILITY OF PARTICULATE PHOSPHORUS TO ALGAE IN THE COLORADO RIVER, SOUTHEASTERN UTAH

Richard J. Watts
    Division of Environmental Engineering and
    Utah Water Research Laboratory, Utah State
    University, Logan, Utah

Vincent A. Lamarra
    Utah Water Research Laboratory, Utah State
    University, Logan, Utah

INTRODUCTION

Phosphorus is the nutrient most commonly attributed to limiting algal growth in aquatic ecosystems [1,2]. Phosphorus in freshwater systems may exist in both soluble and particulate forms. Included in the soluble forms of phosphorus are orthophosphate, polyphosphates, and organic phosphates [3]. Particulate phosphorus may be associated with biological material, adsorbed on clays, and complexed with metals such as iron, manganese, and calcium [4].

The availability of particulate phosphorus for algal growth has been the topic of thorough investigation, since this process is often the key to eutrophication associated with increased phosphorus loading from human activity. Helfrich and Kevern [5] demonstrated that the movement of phosphorus-32 from clay and its availability to algal cells is governed by pH and equilibrium reactions. Harter [6] found that when between 0 and 2.2 mg phosphorus were added to 0.1 g sediment samples from a eutrophic lake, adsorbed phosphorus was extracted as aluminum and iron phosphates, with no calcium phosphate present. He concluded that adsorption of phosphorus into the aluminum-based fraction is significant in terms of phosphorus exchange to the water.

Recent controversy concerns the availability of phosphorus associated with apatite and calcareous sediments. Golterman et al. [7] concluded that hydroxyapatite was

161

available for algal growth. Wentz and Lee [8] demonstrated that the available phosphorus in Lake Mendota, Wisconsin, was associated with the calcareous portion of the sediment. In contrast, Williams et al. [9] and Burns et al. [10] found phosphorus associated with apatite unavailable to algae.

The objective of this study was to determine the nature of the particulate phosphorus in the suspended material of the Colorado River above Lake Powell, Utah. In addition, this particulate phosphorus was evaluated regarding availability for the growth and metabolism of algae.

MATERIALS AND METHODS

The Colorado River flows through Southeastern Utah near the east-west edge of the Colorado Plateau below the Rocky Mountains on the east, and the Tavaputs, Wasatch, Fish Lake, and Paunsagant plateaus on the north and west. It is a region of nearly flat-lying sedimentary rocks, modified occasionally by gentle, broad folds and minor faults. The Colorado River was sampled above Lake Powell, approximately 100 km upstream from its confluence with the Green River. The river is relatively turbid in this region with suspended sediment concentrations of several thousand mg/l being common [11].

Samples were collected on the Colorado River 2 km upstream from the Moab bridge on US 163. Sample collection and bioassay experiments were performed monthly from July through October 1978.

Laboratory and in situ investigations were performed in the study. In situ investigations included diurnal sampling of nutrient parameters and estimation of primary productivity. Laboratory studies consisted of the US EPA Algal Assay Bottle Test, whole water algal bioassays, x-ray diffraction analysis of suspended sediment, and chemical fractionation of phosphorus associated with suspended sediment.

During monthly in situ investigations, samples were collected at three-hour intervals over a period of one day. Sampling times were 6 a.m., 9 a.m., 12 noon, 3 p.m., 6 p.m., 9 p.m., and 12 midnight. An integrated sample over the top meter of river was taken by attaching a weight to a Nalgene sample bottle and towing the bottle through the upper meter of the water column. Samples were collected at midstream. For each of the samples collected, nitrogen analyses included ammonia-nitrogen, nitrite-nitrogen, and nitrate-nitrogen. Phosphorus analyses included orthophosphate, total filterable phosphorus, acid hydrolyzable phosphorus, and total phosphorus. Chlorophyll a, temperature, pH, and secchi disc transparency also were measured. All analyses and measurements were in accordance with Standard Methods for the Examination of Water and Wastewater [3].

Estimation of net and gross primary productivity was performed using light and dark 300 ml Pyrex bottles [3]. These light and dark bottles were suspended at the surface, 0.3 m, 0.6 m, and 0.9 m in depth. Light and dark bottles were not placed below the 0.9 m depth level since secchi disc transparencies were never greater than 0.9 m.

At each depth, two sets of light and dark bottles were established--one set with river water only, and one additional set with macronutrients and micronutrients added in order to avoid nutrient limitation. These nutrient additions were performed by removing 5 ml of water from the selected light and dark bottles and adding 5 ml of concentrated mixture containing the nutrients listed in Table I.

The light and dark bottles were placed in the Colorado and Green Rivers at mid-day for four hours. Initial and then final dissolved oxygen were determined by Winkler titration with Azide modification [3] using a standard sodium thiosulfate titrant diluted five-fold to 0.005 N to increase sensitivity.

The US EPA Algal Assay Bottle Test was performed each month to determine the soluble nutrient that may limit primary production in the Colorado River [12]. Nutrient additions involved in the Algal Assay Bottle Test are listed in Table II.

The Algal Assay Bottle Test was performed in acid-washed 500 ml Erlenmeyer flasks using a 100 ml water sample which was filtered through a 0.45 μm filter. All treatments were performed in triplicate. Upon the addition of nutrients, water samples received 1000 cells per ml of the green alga Selenastrum capricornutum. An inverted beaker was placed over the mouth of the flask to avoid contamination. The cultures were placed under illumination of 1413 lux at 24°C $\pm$ 2°C. Algal growth was monitored by measuring optical density at 750 nm using a Bausch and Lomb Spectronic 70 from day three through a peak in optical density, generally at day 14.

The Algal Assay Bottle Test is a bioassay in which the sample is filtered prior to the experiment. Filtered bioassays have been shown to assess nutrients which are readily available to algae [12]. On the other hand, whole water algal bioassays, in which the sample water is autoclaved, provide an assay of total nutrient availability [13]. The

Table I. Nutrients added to light and dark bottles to avoid nutrient limitation in the estimation of primary productivity.

Nutrient	Final Concentration
Phosphate–phosphorus	0.05 mg/l
Nitrate–nitrogen	1.00 mg/l
Trace elements (c.f. algal bioassays below)	0.188 mg/l
Disodium EDTA (for metal detoxification)	1.00 mg/l

Table II. Experimental design and final concentrations of nutrients used to define nutrient limitation in the algal assay bottle test.

Sample
Sample + 0.05 mg P l^{-1} as K_2HPO_4
Sample + 1.00 mg N l^{-1} as $NaNO_3$
Sample + 0.05 mg P l^{-1} + 1.00 mg N l^{-1}
Sample + Trace Elements*

*Trace Elements Included:

Compound	Element	Concentration (μg l^{-1})
H_3BO_3	B	32.460
$MnCl_2 \cdot 4H_2O$	Mn	115.374
$ZnCl_2$	Zn	1.570
$CoCl_2 \cdot 6H_2O$	Co	0.354
$CuCl_2 \cdot 2H_2O$	Cu	0.004
$Na_2MoO_4 \cdot 2H_2O$	Mo	2.878
$FeCl_3 \cdot 6H_2O$	Fe	33.051

hard water present in the Colorado River cannot be autoclaved with assurance that nutrients will remain in solution, since precipitation may occur [3]. As an alternative to auto-claving, a modified whole water bioassay was developed which was not autoclaved. This bioassay was developed for the September and October sample dates after an original whole water bioassay proposal failed to provide meaningful data.

In the whole-water algal bioassay, triplicate whole-water samples were poured into acid-washed 125 ml Erlenmeyer flasks. Nutrients evaluated for limitation were identical to those listed in Table II. All samples were measured so that sample and nutrient additions totaled 100 ml. The flasks then received 1000 cells per ml of Selenastrum capricornutum. Aluminum foil was placed over the mouths of the flasks in a loose manner to avoid contamination. The flasks were placed on a shaker at 50 rpm to suspend particulate material and incubated under 1413 lux at $24°C \pm 2°C$. To compare whole water nutrient availability to that of soluble levels, a filtered algal bioassay was conducted simultaneously with the bioassay of whole water samples. Water samples were passed through a 0.45 μm filter and placed in 125 ml Erlenmeyer flasks. Nutrient additions were identical to whole water sample additions and treatments were performed in triplicate. Similarly, the flasks received 1000 cells per ml Selenastrum capricornutum and mouths of the flasks were covered loosely with aluminum foil. The samples were shaken at 50 rpm and placed under identical lighting and temperature conditions as the whole water bioassay. Algal growth was monitored by measuring optical density at 750 nm using a Bausch and Lomb Spectronic 70 from day three through a peak in optical density. Upon completion of the bioassay, marked by a peak in optical density, biomass was measured by gravimetric analysis and volatile suspended solids [3].

Samples which underwent x-ray diffraction analyses included suspended sediment taken from the Colorado and Green Rivers during the August, September, and October 1978 sampling trips. Samples collected during July 1978 contained too little suspended sediment for mineralogical analysis.

Each sample was ground by hand to pass a 115 mesh sieve. Two slides of each sample were prepared, one slide a random grain powder mount and the other an oriented grain powder mount. The oriented mount was prepared by making a slurry of the water evaporate. In the process, platy minerals such as clays and micas orient themselves with their basal cleavage parallel to the glass slide. After the oriented mount was scanned, it was placed in an aluminum desiccator with ethylene glycol. The desiccator was placed in an oven at 65°C for 1 hour. This process expands the smectite group minerals, enabling a distinction to be made between smectite and vermiculite. Samples were analyzed on a Stremens Krystalloflex IV x-ray generator equipped with a copper tube and nickel filter.

To determine the chemical nature of particulate phosphorus, the soil fractionation procedure of Chang and Jackson

[14] was used. This analysis was performed on the October 1978 sample, since the large amounts of calcite present in this sample raised the question of a possible calcium phosphate complex. This extraction procedure differentiates between aluminum phosphates, calcium phosphates, and iron phosphates. The Chang and Jackson [14] phosphorus extraction scheme is summarized in Table III. Suspended sediment from the Colorado River was allowed to settle and the supernatant decanted. The sediment was dried at 103°C.

A 1.0 gm sample of soil was placed in a 100 ml centrifuge tube and extracted with 50 ml of 1 N NH_4Cl for 30 minutes on a shaker at 50 rpm to remove water soluble and loosely bound phosphorus and the exchangeable calcium. The suspension was centrifuged and the supernatant solution discarded. Aluminum phosphates were extracted by adding 50 ml of neutral 0.5 N NH_4F to the NH_4-soil in the centrifuge tube and extracting the suspension on a shaker at 50 rpm for 1 hour. The suspension was centrifuged and the clear supernatant solution decanted for phosphorus analysis. The soil sample in the tube was saved for extraction of iron phosphate.

The soil sample saved after the extraction of aluminum phosphate was washed twice with 25-ml portions of saturated NaCl solution. It was extracted with 50 ml of 0.1 N NaOH on a shaker at 50 rpm for 17 hours. The soil suspension was centrifuged for 15 minutes at 2400 rpm to obtain a clear solution which was decanted into another centrifuge tube. The soil sample was then used for extraction of calcium phosphate.

Table III. Extractants and the forms of phosphorus released according to Chang and Jackson [14]

Phosphorus Fraction	Extractant	Forms of Phosphate Extractable
1. Al-phosphate	Neutral 0.5 N NH_4F	Al-phosphate completely Fe-phosphate slightly
2. Fe-phosphate	0.1 N NaOH	Al-phosphate Fe-phosphate Organic phosphorus
3. Ca-phosphate	0.5 N H_2SO_4	Ca-phosphate completely Al- and Fe-phosphate considerably

The soil sample was washed twice with 25-ml portions of saturated NaCl solution and extracted with 50 ml of 0.5 N H_2SO_4 for 1 hour on a shaker at 50 rpm. The suspension was centrifuged and the supernatant solution decanted for the determination of phosphorus.

Phosphorus was determined on all of the above extracts by the ascorbic acid method [3].

RESULTS AND DISCUSSION

Average monthly nutrient and water temperature values for the Colorado River are presented in Table IV. These values are the mean of seven samples collected diurnally for the monthly sampling day. Phosphate-phosphorus in the Colorado River was found to average well below 10 µg/l each month. Total phosphorus ranged between 81 µg/l and 105 µg/l. Nitrate-nitrogen levels were quite high with a range of 688 µg/l to 1394 µg/l. Ammonia-nitrogen levels were relatively consistent, averaging approximately 40 µg/l.

Estimates of primary productivity in the Colorado River are presented in Table V. Net primary production measured in the Colorado River ranged from 12.6 mg $C/m^2/hr$ to 501 mg $C/m^2/hr$. These values are greater than many other estimates of net primary production in large rivers. Hammer [15] found maximum net primary productivity of the Rio Negro at Manaus, Brazil to be 5.5 mg $C/m^2/hr$ to 10.9 mg $C/m^2/hr$. Similarly, the maximum estimate of primary production in the Volga River was 33.2 mg $C/m^2/hr$ [16]. Czeczuca et al. [17], using the ^{14}C method, found net primary production to range from 0.061 to 10.9 mg $C/m^2/hr$ in the River Claska, Poland. Nitrogen probably does not limit primary production in the Colorado River, since nitrate-nitrogen and ammonia-nitrogen concentrations are high.

However, Golterman et al. [7] stated that many species of algae are unable to utilize orthophosphate at levels below 10 µg/l. Since orthophosphate concentrations are generally below 10 µg/l in the Colorado River, it appears that other forms of phosphorus must be supplied in some manner to maintain these high rates of primary production.

Growth curves for the Algal Assay Bottle Test are presented in Figures 1 through 4. A randomized block design analysis of variance was used for statistical analysis of algal bioassays. Nutrient additions were classified as treatments and time in days as blocks. Duncan's multiple range test was used to evaluate significant differences between treatments at the 95 percent confidence level.

167

Table IV. Mean values of diurnal nutrient and physical
           parameters collected monthly for the Colorado
           River.

	7/78	8/78	9/78	10/78
PO_4-P (μg/l)	<10	<10	<10	<10
Total filtrable phosphorus (μg/l)	30	36	19	15
Acid hydrolyzable phosphorus (μg/l)	8	8	6	13
Total phosphorus (μg/l)	81	82	98	105
NO_3-N (μg/l)	688	780	1202	1394
NO_2-N (μg/l)	10	11	6	21
NH_3-N (μg/l)	44	38	34	40
Temperature (°C)	25	22	17	11
pH	8.1	8.4	8.2	8.1
Secchi disc transparency (cm)	55	85	60	44

Table V. Estimation of primary production and chlorophyll a
          concentrations in the Colorado River.

Month	Gross Primary Production g C/m^2/hr	Respiration mg C/m^2/hr	Net Primary Production g C/m^2/hr	Chlorophyll a g/m^3
July 1978	109.4	39.8	69.6	57.3
August 1978	833	332	501	29.0
September 1978	35.3	17.4	17.9	27.6
October 1978	35.6	23	12.6	5.7

     Growth curves and results of Duncan's multiple range
test, indicate that the sample with phosphorus and sample
with nitrogen and phosphorus grew significantly more than
samples spiked with other nutrients for all bioassays performed on Colorado River water samples. Generally, no
substantial growth was detected in the unamended sample,
sample with nitrogen, and sample with trace elements.

     Algal Assay Bottle Test results from the Colorado River
indicate that phosphorus is limiting for the sampling dates
in July, August, September, and October 1978. Phosphorus

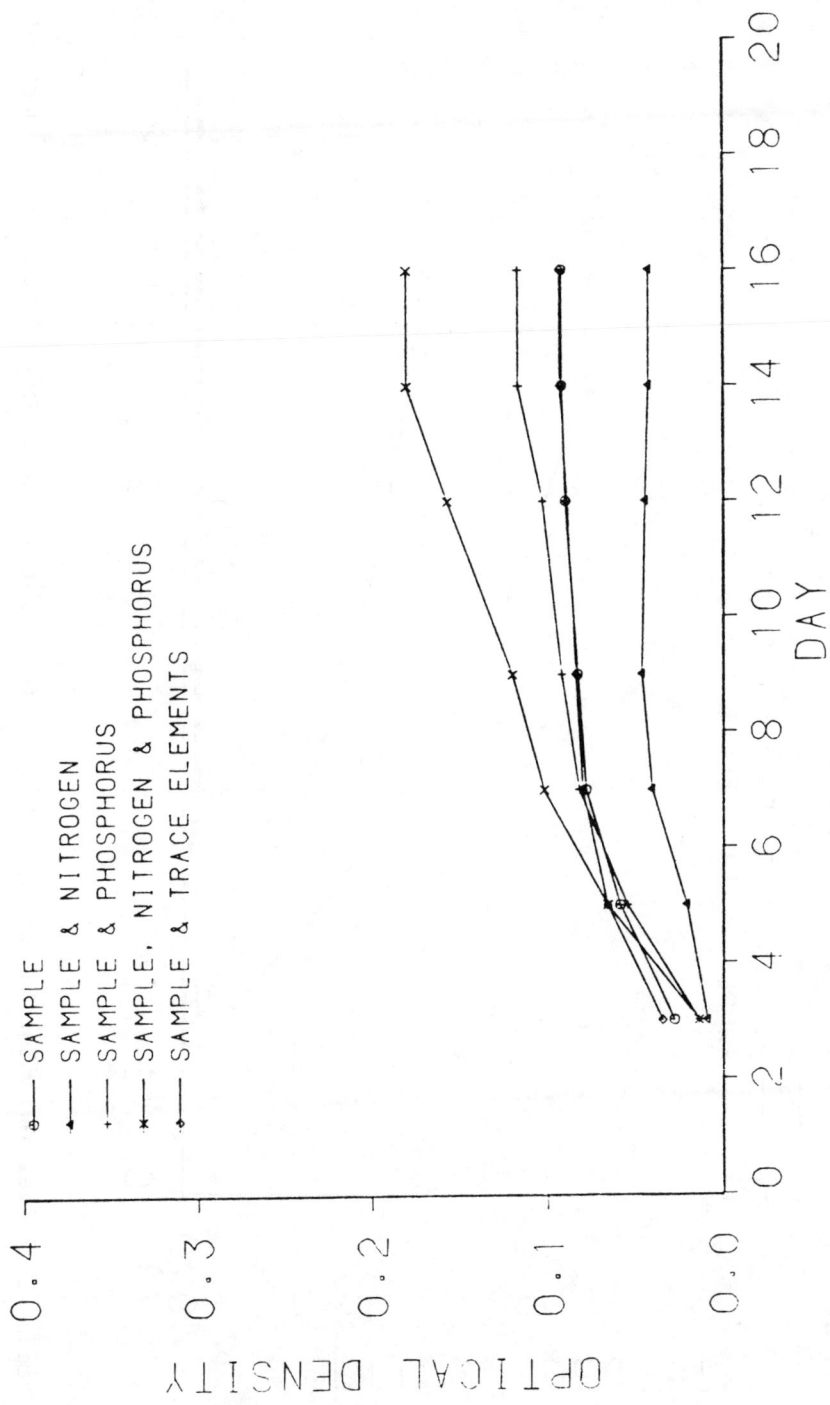

Figure 1. Algal assay bottle test growth curve for Colorado River sample collected in July 1978.

SAMPLE
SAMPLE & NITROGEN
SAMPLE & PHOSPHORUS
SAMPLE, NITROGEN & PHOSPHORUS
SAMPLE & TRACE ELEMENTS

OPTICAL DENSITY

DAY

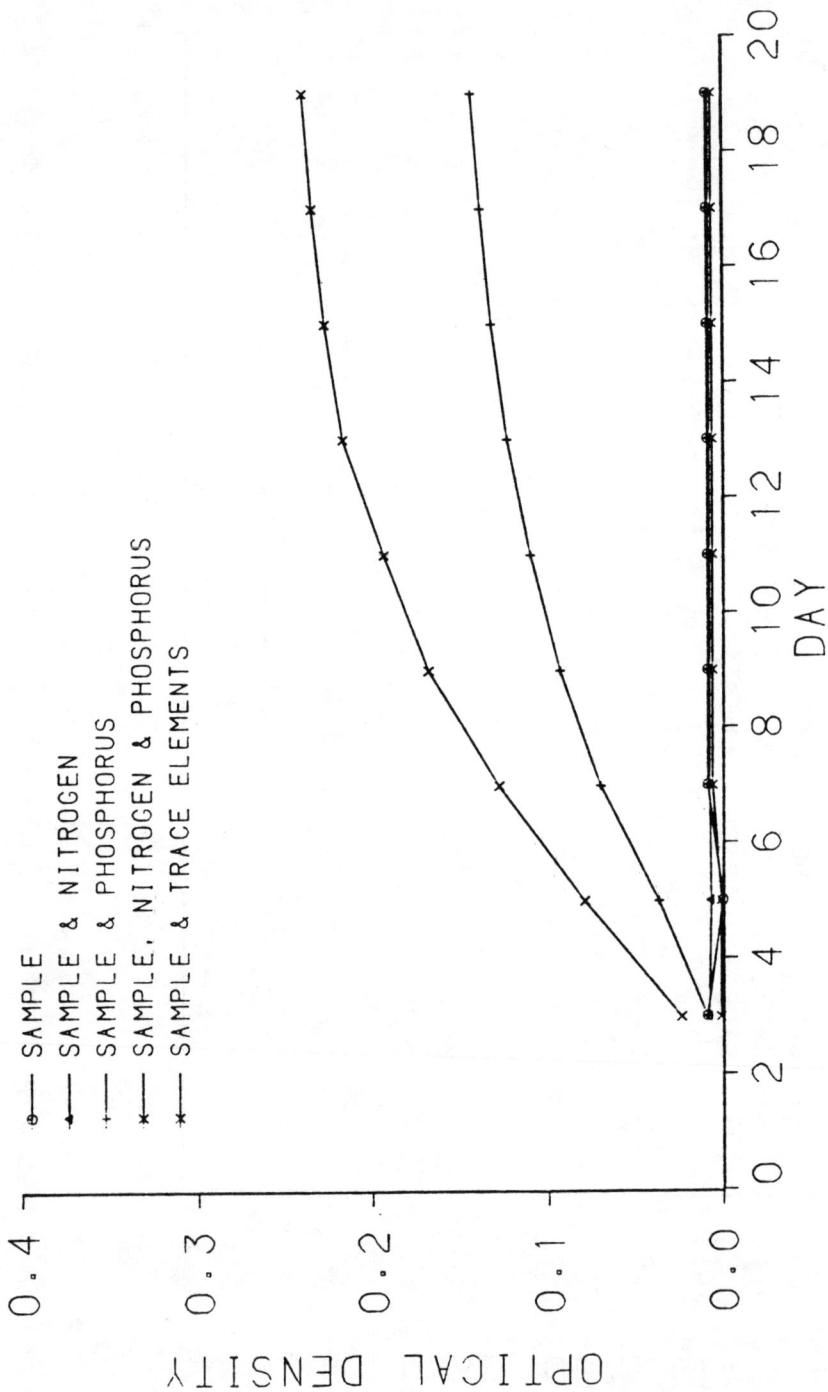

Figure 2. Algal assay bottle test growth curve for Colorado River sample collected in August 1978.

LEGEND:
- SAMPLE
- SAMPLE & NITROGEN
- SAMPLE & PHOSPHORUS
- SAMPLE, NITROGEN & PHOSPHORUS
- SAMPLE & TRACE ELEMENTS

OPTICAL DENSITY

DAY

OPTICAL DENSITY

0.4 0.3 0.2 0.1 0.0

0   2   4   6   8   10   12   14   16   18   20

DAY

SAMPLE
SAMPLE & NITROGEN
SAMPLE & PHOSPHORUS
SAMPLE, NITROGEN & PHOSPHORUS
SAMPLE & TRACE ELEMENTS

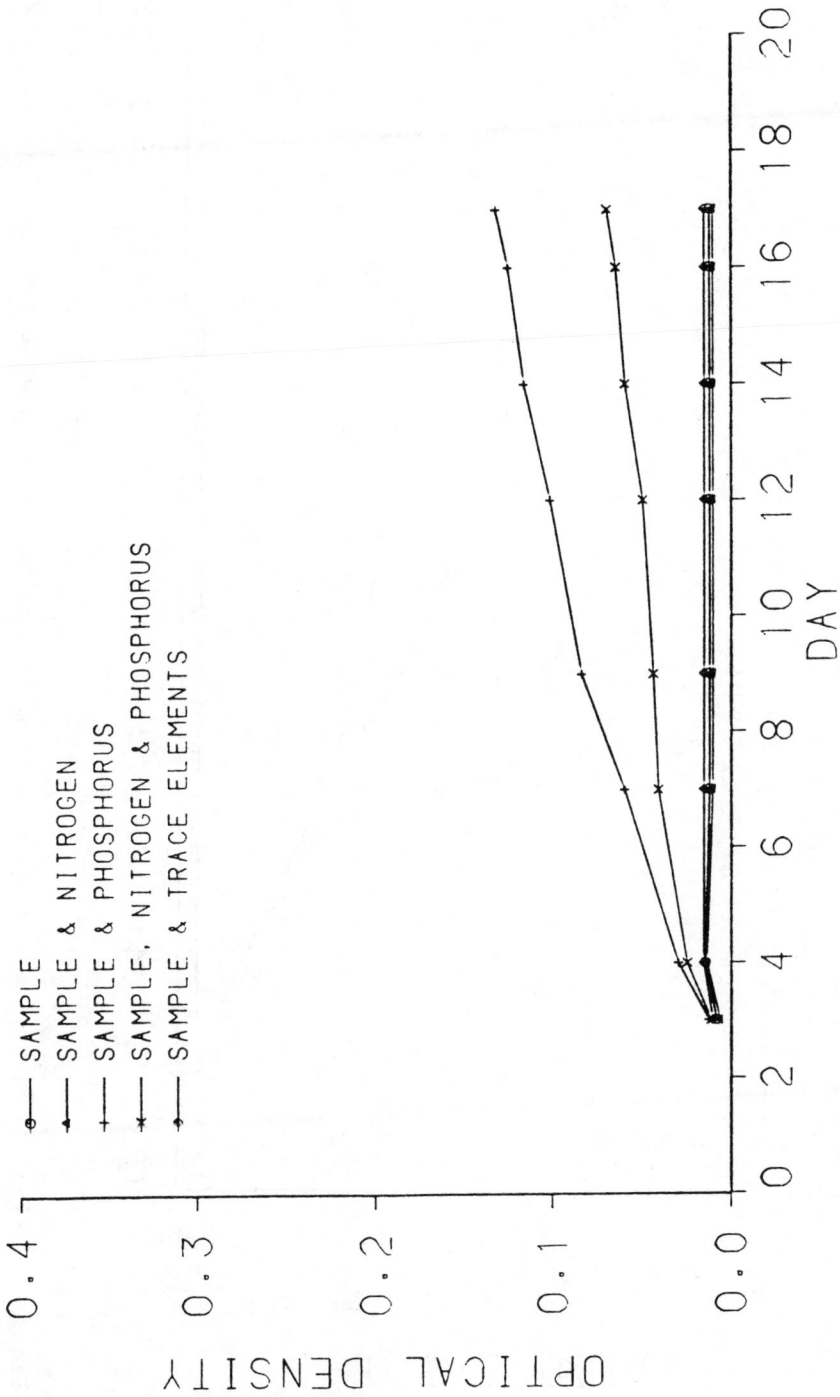

Figure 3. Algal assay bottle test growth curve for Colorado River sample collected in September 1978.

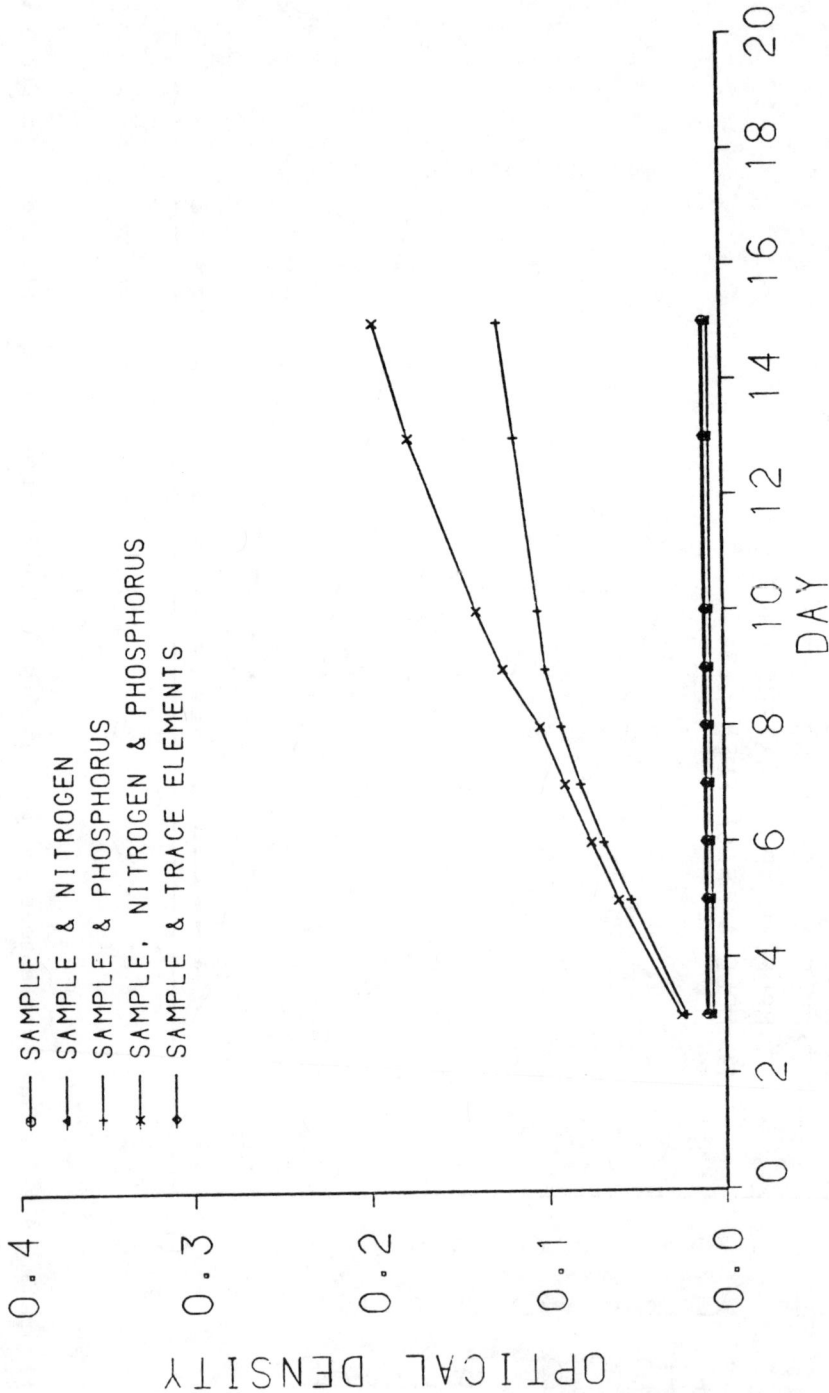

Figure 4. Algal assay bottle test growth curve for Colorado River sample collected in October 1978.

172

limitation is in agreement with the nutrient status of the Colorado River since the total soluble inorganic nitrogen to phosphate-phosphorus ratio (TSIN:PO_4-P) was on the order of 300:1. Generally, a TSIN:PO_4-P ratio of 11:1 is the border between nitrogen limitation and phosphorus limitation [12]. In all cases, the unamended sample, sample with nitrogen, and sample with trace elements were found not to be significantly different at the 95 percent confidence level by Duncan's multiple range test. None of these treatments attained significant growth, since phosphorus was limiting. However, the sample with phosphorus showed a significant increase in growth. The sample with nitrogen and phosphorus produced significantly higher growth at the 95 percent confidence level than all other nutrient additions. This addition of nitrogen and phosphorus confirms phosphorus limitation since this culture grew, with the addition of nitrogen, until it was driven to phosphorus limitation.

Monitoring of algal growth in whole water bioassays and accompanying filtered bioassays was originally conducted by the measurement of optical density at 750 nm. However, optical density data were erroneous, possibly due to the evident dumping of suspended sediment and algal cells. This dumping and precipitation was attributed to calcium carbonate precipitation associated with high pH in hard water. Due to the inability to use optical density data, the parameter used for data analysis of the modified whole water algal bioassay was maximum standing crop by gravimetric analysis and vola-tile suspended solids [3].

Maximum standing crop values and initial water chemistry data for the Colorado River modified algal bioassays are presented in Table VI. Stoichiometric interpretation of nutrient uptake may be examined using maximum standing crop (MSC) and water chemistry data. Miller et al. [12] that the average cell of <u>Selenastrum capricornutum</u> under bioassay conditions contains one part phosphorus for each 429 parts of other elements. Therefore, this alga is 1/430 phosphorus and the maximum standing crop of a phosphorus limited culture may be estimated by multiplying the phosphorus concentrations by 430. The stoichiometric calculations for particulate phos-phorus availability are outlined in Table VII. The amount of growth attributed to particulate phosphorus may be determined by subtracting the filtered sample MSC from the whole water sample MSC. Available particulate phosphorus is obtained by dividing the MSC attributed to particulate phosphorus by 430 (mg MSC)/(mg P). From this value, the percent particulate phosphorus available may be calculated by dividing by the level of particulate phosphorus in the sample times 100.

173

Table VI. Maximum standing crop and nutrient data for
modified whole water bioassays of Colorado River.

	Sept 1978	Oct 1978
Filtered Samples	Maximum Standing Crop (mg/1 Dry Weight)	
Sample	5.6	2.3
Sample + P	39.5	32.9
Sample + P + N	42.3	37.8
Whole Water Samples		
Sample	14.3	22.7
Sample + P	27.0	13.0
Sample + P + N	64.5	126.3
	Chemical Analyses (mg/1)	
Total P	0.105	0.113
Total filtrable P	0.019	0.015
Acid Hydrolyzable P	0.007	0.003
PO_4-P	0.003	0.007
NO_2-N	0.006	0.026
NO_3-N	1.265	0.394
NH_3-N	0.032	0.038
TSIN ($NO_2-N + NO_3-N + NH_3-N$)	1.303	0.459

    With particulate phosphorus available, it is likely that
phosphorus was still limiting primary production, since the
MSC of the sample with soluble phosphorus treatments was
greater than the MSC of the whole water samples. The whole
water samples for September and October grew less than the
whole water samples with phosphorus added since the samples
without nutrient addition were limited by phosphorus. Nitrogen levels were very high, negating the possibility of nitrogen limitation. Therefore, from stoichiometric analysis of
whole water and filtered bioassays, 21 percent and 49 percent
of the particulate phosphorus were available for algal
metabolism in September 1978 and October 1978, respectively.

    Suspended sediment from water samples collected in
August, September, and October, 1978, was analyzed for
mineralogical content by x-ray diffraction. These data are

Table VII. The availability of particulate phosphorus from whole water and filtered bioassays using the assumption that the alga Selenastrum capricornutum contains one part phosphorus for 429 parts of other elements.

Sample (no nutrient addition)	Maximum Standing Crop (MSC) (mg/1 Dry Weight)	Available P from Particulate Material (mg/1) $\dfrac{\text{(Whole Water MSC} - \text{Filtered MSC)(mg/1)}}{430 \frac{\text{mg MSC}}{\text{mg P}}}$	Particulate P (mg/1)	Percent Particulate P Available $\dfrac{\text{Available P}}{\text{Particulate P}} \times 100$
September – Whole Water	14.3			
September – Filtered	5.6	0.020	0.096	21 Percent
October – Whole Water	22.7			
October – Filtered	2.3	0.048	0.097	49 Percent

summarized in Table XIII. These data show that suspended sediment in the Colorado River was composed of a high percentage of quartz and calcite with small amounts of the clay minerals, smectite, illite, and kaolinite.

Phosphorus analysis was performed to determine whether phosphorus was associated with clays or calcite. Data resulting from phosphorus extractions are presented in Table IX. These data show that there was very little aluminum phosphate (i.e., phosphate adsorbed to clays) or iron phosphate. Organic phosphates are extracted with iron phosphate by 0.1 N sodium hydroxide [14]. Therefore, small amounts of organic phosphorus were present in these samples relative to other forms of phosphorus. Most extractable phosphorus was in the form of calcium phosphate. For the Colorado River, calcium phosphate represented 24 percent of the total phosphorus in the suspended sediments.

Table VIII. Results of x-ray diffraction analysis of suspended sediments in the Colorado River.

July 1978	Too low of suspended load for collection of sufficient sample.
August 1978	Approximately equal amounts of quartz and calcite, minor amounts of illite, kaolinite, dolomite gypsum, trace amounts of palygorskite.
September 1978	Dominantly quartz, lesser amount of calcite; minor amounts of illite, kaolinite, and halite.
October 1978	Approximately equal amounts of quartz and calcite; minor amounts of mortmorillonite, halite, and subequal amounts of illite and kaolinite.

Table IX. Results of phosphorus fractionation of October 1978 suspended sediment sample.

	µg P/g Dry Suspended Sediment Colorado River
Aluminum phosphate (clays)	<10
Iron/organic phosphate	<10
Calcium phosphate	350
Total phosphorus	1,460

The nature of particulate phosphorus may be discerned by viewing x-ray diffraction results in conjunction with phosphorus extraction data. Monthly x-ray diffraction data were consistent throughout the study. Particulate phosphorus may be associated with calcite, organic matter, clays, or iron [14]. It is unlikely that phosphorus is associated with quartz, since quartz is biologically and chemically inert [18]. Initially, phosphorus was thought to be associated with clays. However, x-ray diffraction data showed consistently high amounts of calcite over clay minerals. Results of phosphorus extractions indicate that extractable phosphorus was predominantly based in the calcium-phosphate fraction. Since the dominant calcium-based component of Colorado River suspended sediment was calcite, this calcium-based phosphorus is probably associated with calcite.

The precise nature of this calcite-phosphorus complex cannot be discerned from the results of this study. However, there are at least two possibilities for the nature of this calcite-phosphorus complex: Phosphorus adsorption on the surface of the calcite crystal [19], and coprecipitation of phosphorus during calcite formation [20].

Particulate phosphorus in the Colorado River has been shown to be associated with calcite, an apatite mineral. A significant portion of this particulate phosphorus was available for algal growth. These findings are in agreement with the results of many other workers (e.g. Pomeroy et al. [7], Golterman et al. [21]) in that algae can utilize some of the phosphorus present in geological substrates. However, Williams et al. [9] stated that the available phosphorus is not a constant proportion of the total particulate phosphorus but is directly related to the nonapatite inorganic phosphorus. The results of this study regarding the availability of apatite phosphorus are in agreement with recent work by Smith et al. [22] who found that ground apatite crystals supported the growth of bacteria and algae. Similar conclusions have been made by Wentz and Lee [8], Gerhold and Thompson [23], and Golterman et al. [7].

Although phosphorus associated particulate material has been shown to be available for algal metabolism in the Colorado River, turbidity may play a more important role in phytoplanktonic primary productivity. Watts [24] found that, of three chemical-physical factors studied, turbidity played a dominant role in controlling phytoplanktonic primary production and standing crop in the Colorado River. Similarly, Swale [25] found that phytoplankton of the River Lee, England, were not limited by nutrient levels, since nitrate-nitrogen ranged from 1 to 12 mg/l and phosphate-phosphorus ranged from 0.2 to 3 mg/l. Limitation of algal growth was attributed to rates of flow and high turbidity. Making similar conclusions to Swale's [1964] for the River Lee, Lund [26] stated that even if phosphorus and nitrogen were reduced to a tenth of their current concentration, a large phytoplankton community would develop if light intensity were not limiting.

The Colorado River is similar to the River Lee as described by Swale [25] and Lund [26] in that turbidity is one of the most important physico-chemical factors controlling planktonic primary production. Although orthophosphate concentrations were below 10 µg/l in the Colorado River, estimates of planktonic primary production were high if turbidity was low [24]. Therefore, the availability of phosphorus in the phosphorus-calcite complex appears to allow for high rates of primary production in the Colorado River when turbidity is not limiting light penetration. Furthermore, the availability of the phosphorus in the phosphorus-calcite complex for algal metabolism may play a critical role in the primary production levels of Lake Powell and other Colorado River system reservoirs.

REFERENCES

1. Vollenweider, R. A. 1968. Scientific Fundamentals of the Eutrophication of Lakes and Flowing Waters, with Particular Reference to Nitrogen and Phosphorus as Factors in Eutrophication. Paris, Rep. Organization for Economic Cooperation and Development, DAS/CSI/68.27. 192 pp; Annex, 21 pp; Bibliography, 61 pp.

2. Keup, L. E. 1968. Phosphorus in Flowing Waters. Water Res. 2:373-386.

3. American Public Health Association. 1975. Standard Methods for the Examination of Water and Wastewater. 14th Ed. New York, N.Y. 1193 pp.

4. Stumm, W., and J. J. Morgan. 1970. Aquatic Chemistry. John Wiley and Sons, Inc., New York. 583 pp.

5. Helfrich, L. A., and N. R. Kevern. 1973. Availability of phosphorus-32, adsorbed on clay particles, to a green alga. The Michigan Academician. 6:71-81.

6. Harter, R. D. 1968. Adsorption of Phosphorus by Lake Sediment. Soil Sci. Soc. Amer. Proc. 32:514-518.

7. Golterman, H. L., C. C. Bakels, and J. Jakobs-Mogelin. 1969. Availability of Mud Phosphates for the Growth of Algae. Verh. Internat. Verein. Limnol. 17:467-479.

8. Wentz, D. A., and G. F. Lee. 1969. Sedimentary Phosphorus in Lake Cores--Observation on Depositional Pattern in Lake Medota. Environ. Sci. Technol. 3:754-759.

9. Williams, J. D. H., H. Shear, and R. L. Thomas. 1980. Availability to Scenedesmus quadricauda of Different Forms of Phosphorus in Sedimentary Materials from the Great Lakes. Limnol. Oceanogr. 25:1-11.

10. Burns, N. M., J. D. Williams, J. M. Jacquet, A. L. Kemp, and D. C. Lam. 1976. A Phosphorus Budget for Lake Erie. J. Fish. Res. Bd. Can. 33:564-573.

11. Iorns, W. V., C. H. Hembree, and G. L. Oakland. 1965. Water Resources of the Upper Colorado River Basin--Technical Report. U.S. Geol. Surv. Prof. Pap. 441. 370 pp.

12. Miller, W. E., J. C. Greene, and T. Shiroyama. 1978. The Selenastrum capricornutum Printz Algal Assay Bottle Test. US EPA Bull. 600/9-78-018. 123 pp.

13. Weiss, C. M. 1976. Field Evaluation of the Algal Assay Procedure on Surface Waters of North Carolina. p. 29 - 76. E. J. Middlebrooks, D. H. Falkenborg, and T. E. Maloney (eds.). Biostimulation and Nutrient Asessment. Ann Arbor Science, Ann Arbor, Michigan.

14. Chang, S. C., and M. L. Jackson. 1957. Fractionation of Soil Phosphorus. Soil Science. 84:133-144.

15. Hammer, L. 1965. Photosynthese und Primarproduktion im Rio Negro. Int. Revue ges. Hydrobiol. 50:335-339.

16. Pyrina, I. L. 1959. Photosynthetic Production in the Volga and its Reservoirs. Byull. Inst. Biol. Vodakhran. 3:17-20.

17. Czeczuga, B., F. Gradzki, and E. Bobiotynski-Ksok. 1968. Primary Production in a Chosen Site of the River Ploska. Part I. Phytoplankton Production. Acta Hydrobiol. 10:85-94.

18. Brady, N. C. 1974. The Nature and Properties of Soil. Macmillan Publishing Company, Inc., New York, N.Y. 639 pp.

19. Griffin, R. A., and J. J. Jurinak. 1973. The Inter-action of Phosphate with Calcite. Soil Sci. Soc. Am. Proc. 38:75-79.

20. Otsuki, A., and R. G. Wetzel. 1972. Coprecipitation of Phosphate with Carbonates in a Marl Lake. Limnol. Oceanogr. 17:763-767.

21. Pomeroy, L. R., E. E. Smith, and C. M. Grant. 1965. The Exchange of Phosphate Between Estuarine Water and Sediments. Limnol. Oceanogr. 10:167-172.

22. Smith, E. A., C. I. Mayfield, and P. T. Wong. 1978. Naturally Occurring Apatite as a Source of Ortho-phosphate for Growth of Bacteria and Algae. Microb. Ecol. 4:105-117.

23. Gerhold, R. M., and J. E. Thompson. 1969. Calcium Hydroxy-apatite as an Algal Nutrient Source. Abs. Am. Chem. Soc. 158:Wtr. 71.

24. Watts, R. J. 1981. Some Chemical and Physical Mecha-nisms Affecting Planktonic Primary Production and Phytoplanktonic Standing Crop in the Colorado and Green Rivers, Southeastern Utah. M.S. Thesis. Utah State University. 87 pp.

25. Swale, E. M. F. 1964. A Study of the Phytoplankton of a Calcareous River. J. Ecology. 52:433-446.

26. Lund, J. W. G. 1969. Phytoplankton, pp. 306-330. In: G. A. Rohlich (Ed.). Eutrophication: Causes, conse-quences, correctives. Natl. Acad. Sci., Washington, D.C.

CHAPTER 12

CHARACTERIZATION OF YAMPA AND GREEN
RIVER ECOSYSTEMS: A SYSTEMS APPROACH
TO AQUATIC RESOURCE MANAGEMENT

Thomas C. Annear
     Wyoming Dept. of Game and
          Fish

Dr. John M. Neuhold
     Utah State University

INTRODUCTION

    Most of the significant contributions regarding the pro-
duction and utilization of energy in lotic environments have
been based either upon laboratory studies such as those by
McIntire and Phinney [1] and Whitford and Schumacher [2] or
studies conducted on low order, woodland streams of the temperate eastern United States [Fisher and Likens 3, Minshall
4, Nelson and Scott 5, and others].   Typically, these systems
are heterotrophic, relying on imported allochthonous materials
(leaf fall) for a major share of the biologically available
organic energy in the system.   More recent studies by
Minshall et al. [6] and Minshall and Wlosinski [7] on southeastern Idaho's Deep Creek showed that, where a deciduous
canopy is absent, autotrophy becomes a dominant source of
organic energy.
    Many of the ideas concerning larger streams are based on
theoretical constructs such as functional ordering by Vannote
[8] and Cummins [9].   Although increased attention has recently been focused on the characteristics of larger streams in
the western U. S., [Minshall 10, Seyfer and Wilhm 11, and
Watts 12], there remains an inadequate understanding of basic
ecosystem attributes of these systems.   This study was intended to develop an increased awareness of energy production and
utilization functions in large rivers of the semi-arid western United States and to gain insight into the potential
impacts of water development (high dams) on large river
ecosystems.

Investigations were conducted on the Yampa River in north-western Colorado and the Green River in northeastern Utah, both fifth order streams, using Strahler's [13] system of tributary classification. The rivers flow through a semi-arid region of the Colorado Plateau characterized by rolling hills with deeply incised canyons. Streamside vegetation along both streams consists largely of sagebrush (Artemisia tridentata), Utah juniper (Juniperus osteosperma), boxelder (Acer negundo), sedges, and grasses.

Four permanent sampling stations were established on the Yampa River from its confluence with the Green River eastward 90 km (Figure 1). This stream is the only major tributary in

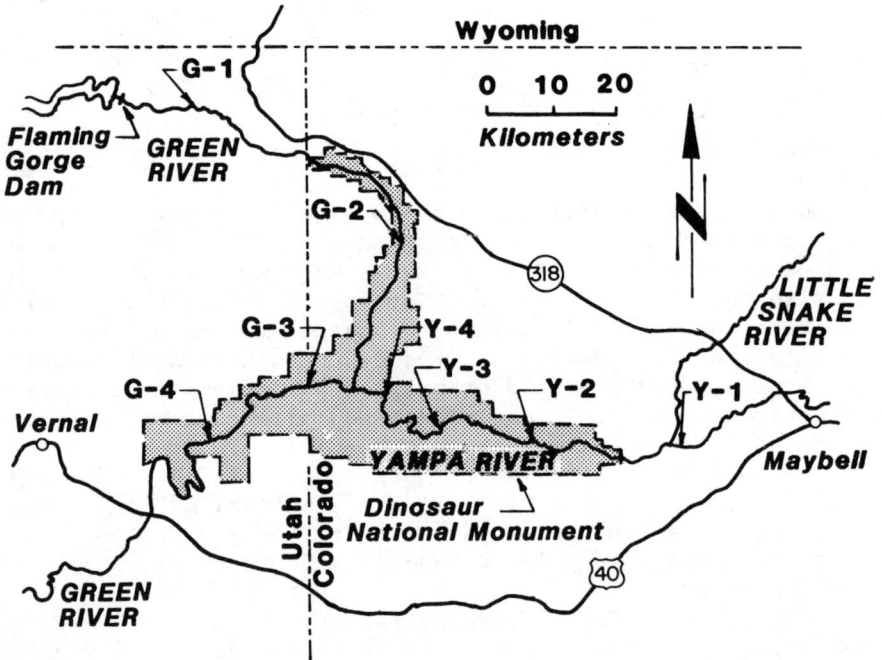

Figure 1. Location of sampling stations.

the Colorado River system which has not been developed. It exhibits high seasonal variation in discharge rates, temperature, and suspended sediment loads. According to USGS records, mean annual discharge is 58.4 m^3/sec with rates ranging from near 400 m^3/sec in late May to 5.5 m^3/sec in September. Water temperatures range from 0° C in mid winter to 21° C in late summer. Mean annual suspended sediment loads are 1.7 million tons at the river's mouth [Andrews 14].

Four permanent sampling stations were also located on the
Green River between Flaming Gorge Dam and a site near Jensen,
Utah, 135 km below the dam (KBD). Similar in size and historically exhibiting equivalent physical and biological characteristics to the Yampa River, the Green River is now strongly influenced by controlled releases from Flaming Gorge at
least to Lodore Canyon 78 KBD [Pearson 15]. The seasonal flow
regime has been stabilized (ranging from 39.9 m^3/sec in March
to 81.5 m^3/sec in June) and water temperature fluctuations
have been reduced (now varying from 4° C in mid winter to near
13° C from May to October). Sediment loads at Jensen (150
KBD) remain up to 30 percent lower than loads recorded prior
to closure of the dam in 1962.

METHODS

Each site was sampled eight times at approximate five week
intervals from August, 1978 to September, 1979. Parameters of
interest were gross production/respiration ratios (P/R) and
gross production/chlorophyll a ratios (P/C) of periphyton communities, organic energy transport rates, and spatial differences in aquatic macroinvertebrate functional groups. Substrates for P/R and P/C ratios and macroinvertebrate samples
were taken only from areas where depths and velocities were
approximately 0.4 m and 0.5 m/sec respectively.

Methods for measuring rates of periphyton production and
respiration have been established for some time in laboratory
settings [McIntire and Phinney 1], however, it is only recently that techniques have been developed for in situ investigations. In turbulent streams where periphyton is the dominant
primary producer, Bott et al. [16] found that enclosed circulating chambers monitoring changes in dissolved oxygen were
the most reliable method for measuring rates of primary production. Paired chambers similar to those used by Bott were
used to measure hourly rates of gross primary production and
respiration of natural substrates. Winkler titrations, azide
modification [APHA 17], were used to monitor changes in dissolved oxygen concentrations.

Chlorophyll a concentrations were determined by removing
all periphyton from the substrates used in the experiments and
steeping it in acetone as described by McConnell and Sigler
[18]. Optical density was converted to milligrams of chlorophyll a using Strickland and Parsons [19] trichromatic formula.

Suspended organic matter (SOM) concentrations were measured using a technique similar to that of Minshall[4]. Three
200 ml samples were filtered at each site using a millipore
syringe hand pump. Filters and filtrate were dried and then
ashed in a muffle furnace.

Aquatic macroinvertebrates were collected at all sites
only in 1979 using a modified Hess sampler and included re-

sults from concurrent studies made by Bio-West, Inc., Logan,
Utah. Four subsamples were obtained at each site and the mean
density (number/m^2) was calculated for each species. The
ecological function of organisms was determined using the
schema developed by Merritt and Cummins [20] where organisms
are classified as collectors, grazers, shredders, or predators.

Data for P/R ratios, P/C ratios, and suspended organic
matter were analyzed using a two-way factorial analysis of
variance (AOV) for each river. A randomized block design AOV
was used for analyzing differences in the mean percent abundance of macroinvertebrate functional groups. Between river
comparisons for all parameters were made using F and Student's
t-tests [Ott 21].

RESULTS/DISCUSSION

Autotrophy was an important source of organic energy for
both the Yampa and Green Rivers ($\overline{X}$=3.12, S.E.=2.49 and $\overline{X}$=3.84,
S.E.=2.39 respectively), however, significant differences
were noted both between the rivers and within discrete segments of each river.

The peak in productivity noted by other studies [Marker
22, Cushing 23, and others] to occur in mid to late spring was
not found on either river (Figure 2). The Yampa River exhibit-

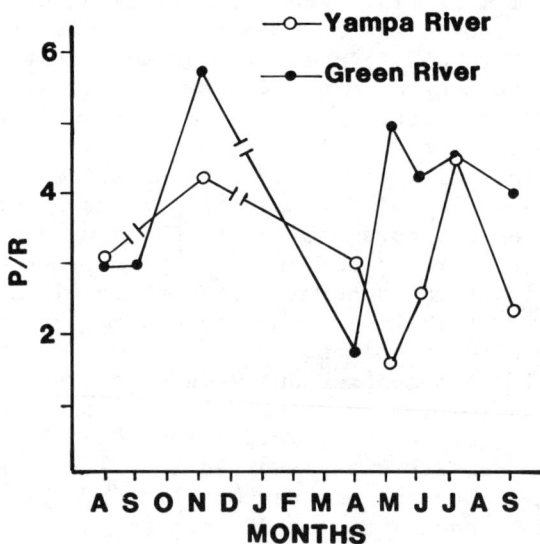

Figure 2. Mean gross production/respiration ratios for the
Yampa and Green Rivers from August, 1978 to September, 1979.

184

ed a decrease in productivity in early spring associated with cold, rising waters, high sediment loads (scouring), and turbidity. Although water discharge during May and June sampling was the same ($200m^3$/sec), productivity was 40 percent higher in June as flows receded, waters warmed, and light penetration increased. The peak noted in mid summer at Yampa River sites also contrasts sharply with other studies. In the relatively constant physical environment of the Green River below Flaming Gorge Dam, periphyton communities reached high levels of productivity in early spring which were maintained until late summer.

Where previous works have shown increased heterotrophy in late fall [Teal 24, Gumtow 25], evidence was found in this study of a similar trend in late summer. The most likely mechanism for this phenomenon in the Yampa and Green Rivers is senescence whereas the mechanism associated with this pattern in other regions is increased allochthonous inputs. In further contrast to those previously mentioned studies, both the Yampa and Green Rivers showed a second peak in P/R in late autumn. This seasonal pattern is unique to western streams where adverse environmental conditions associated with spring runoff occur later and autumnal leaf fall contributions are less than in eastern temperate zone streams.

Periphyton productivity (P/R) at Green River sites was significantly higher than the mean for all Yampa River sites ($P < .10$) (Table I). No significant difference was found be-

Table I. Mean trophic state (P/R), assimilation ratio (P/C), and suspended organic matter concentration (SOM) at Yampa and Green River sample sites.

Sample Site	River km Above Mouth	P/R	P/C	SOM (mg/1)
Y1	88.2	3.71	1.57	15.4
Y2	60.2	1.16	2.76	27.6
Y3	32.5	3.59	2.41	25.8
Y4	3.7	2.85	1.20	20.0
G1	413.3	3.32	.42	4.7
G2	391.4	3.27	.62	15.3
G3	352.0	4.96	.83	20.0
G4	329.0	3.84	.66	26.5

tween upper Green River sites (G1 and G2) and the Yampa River despite the gross physical differences in these two systems. A significant increase ($P < .05$) was noted in the Green River below the Yampa River's mouth as a likely result of increased nutrient loads and modified temperature regime caused by Yam-

pa River flows. A significant "pulse" was also noted on the
Yampa River below the mouth of the Little Snake River (Y2),
but, while the mechanism was related, the response was opposite of that noted in the Green River. According to USGS records, the Little Snake River contributes only 27 percent of
the mean annual discharge of the Yampa River below their confluence but it provides 70 percent of the suspended sediments.
This excessive scouring and turbidity likely results in the
decreased autotrophy at Y2 and is apparently attenuated at
downstream sites.

Evaluation of P/C ratio seasonal dynamics further illustrates the differences between these two systems and the
impact of spring runoff. Production efficiencies were significantly higher (P<.001, $\bar{X}=1.88$, S.E.=2.07) at Yampa River
sites than at all Green River sites ($\bar{X}=0.63$, S.E.=0.51). At
times of the year when both systems are exposed to similar environmental conditions (August through March), P/C ratios of
the two rivers are nearly the same (Figure 3). During spring

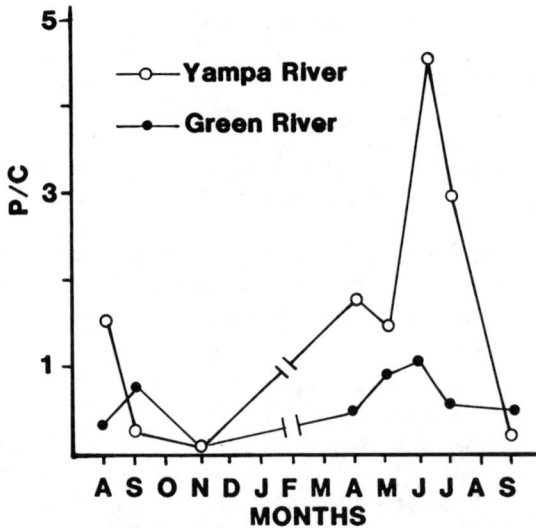

Figure 3. Mean gross production/chlorophyll a ratios for the
Yampa and Green Rivers from August, 1978 to September, 1979.

runoff, algal biomass and light penetration are lowered in the
Yampa River resulting in the dramatic increase in P/C ratios.
In the absence of perturbations associated with spring runoff,
standing crops of algae remained high and P/C ratios low in
the Green River. Although still lower than ratios in the Yampa River, energy utilization rates below the Yampa River's
confluence were greater (P<.05) than rates at upstream stations (Table I).

The late summer and autumnal peaks in SOM export rates

noted by Nelson and Scott [5] and others was not noted in either study area (Figure 4). Export rates were highest in both

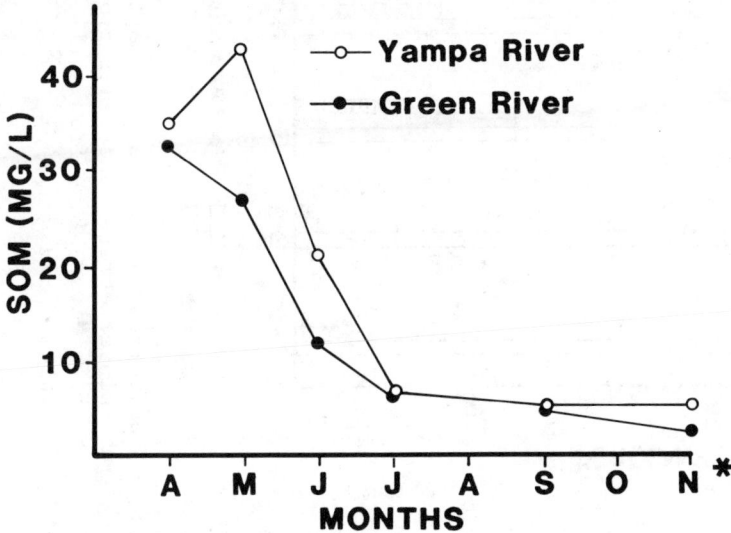

Figure 4. Mean concentration of suspended organic matter (mg/1) for the Yampa and Green Rivers from April to September, 1979.

streams during spring runoff and remained low during the rest of the year. Considering the absence of a deciduous canopy and paucity of rainfall, it is apparent that the majority of these materials are scoured periphyton and allochthonous materials are of limited importance. No significant differences were noted between Yampa River stations, but the rate of export increased significantly at downstream Green River stations (Table I).

Functional group diversity was relatively higher at Yampa River sites than at Green River sites (Figures 5 and 6) and generally resembled the community structure described by Vannote [8] for streams of this size. Most organisms in both study areas were regarded as collectors. However, high numbers of grazers were also found at Yampa River sites as might be expected considering the high turnover of periphyton. Few shredders were captured at Yampa River sites and none were captured at Green River sites supporting the hypothesis that allochthonous inputs are low and autotrophy is the primary energy source driving the system. The Green River community structure did not resemble any described by Vannote and, as that schema relates only to undeveloped systems, this lack of fit can be viewed as evidence of an unnatural perturbation in the system.

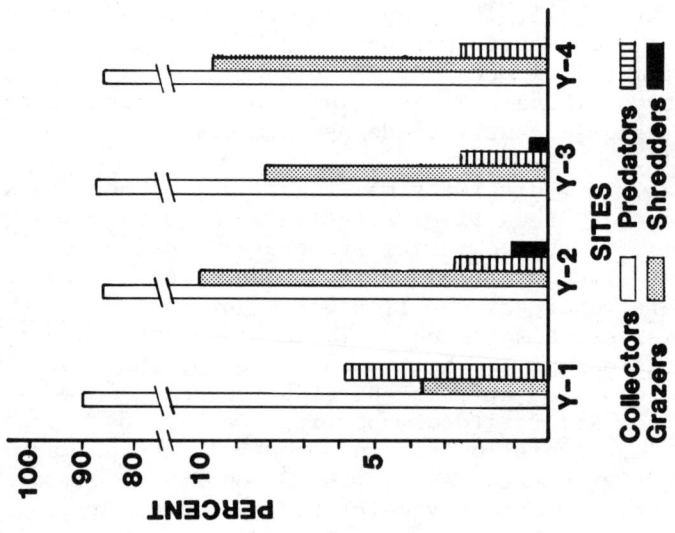

Figure 5. Mean percent abundance of aquatic macroinvertebrate functional groups at Yampa River sites from April to September, 1979.

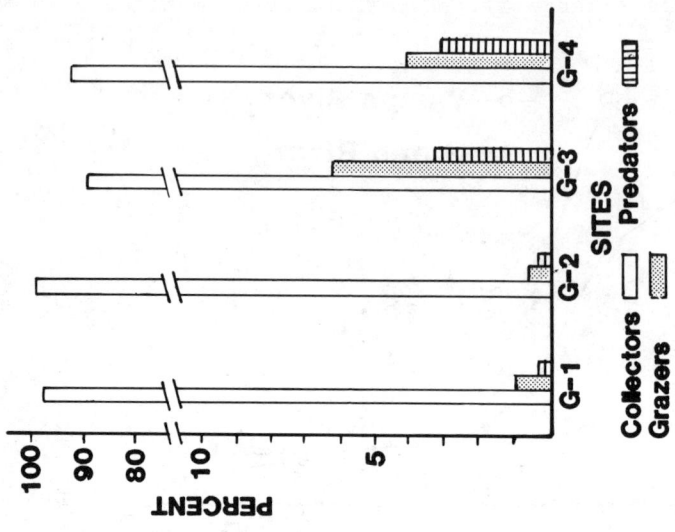

Figure 6. Mean percent abundance of aquatic macroinvertebrate functional groups at Green River sites from April to September, 1979

CONCLUSIONS

In a 1978 work Minshall [10] characterized two types of
autotrophic communities based on seasonal patterns of standing crop and energy transport: 1) low (or high) productivity
with most of the annual autochthonous contribution coming
within a short time following senescence and 2) low standing
crop with high turnover and export. The first type is composed of vascular plants and macrophytic algae and is characteristic of low gradient streams. The second is typical of
streams with little detritus accumulation, primarily periphytic production, and is associated with high gradient streams.
    The Yampa River quite clearly fits the second model, but
evidence from the Green River suggests a third type of stream
characterized by a high periphytic biomass, high production
and export, with low turnover. It should also be noted that
these two ecologically dissimilar stream segments exhibit
quite similar characteristics in terms of productivity, production efficiency, and energy transport during a large part
of the year (late summer through early spring). This phenomenon emphasizes the need to monitor systems on a regular
basis.
    The apparent similarity of upper Green River sites and
the Yampa River in terms of productivity should be reviewed
critically as these two stream segments differ greatly in
most other regards. Vannote's [8] river continuum thesis
provides at least a partial accounting as seen in a simplified representation of stream development in terms of trophic
state (Figure 7). The upper Green River and Yampa River

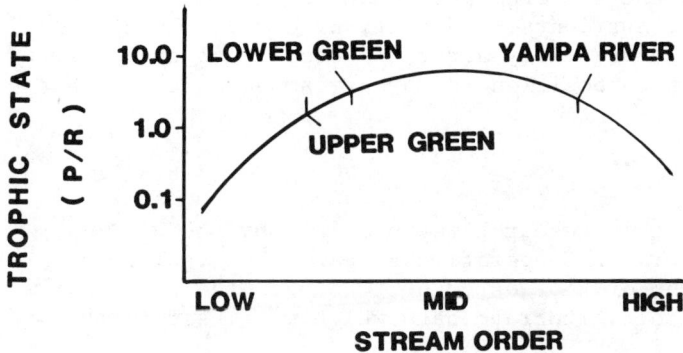

Figure 7. Conceptualization of Yampa and Green River trophic
           states.

study areas are both technically considered fifth order
streams and historically were likely characterized by similar
flora, fauna, and trophic state. Below their confluence, a

higher order stream was formed which theoretically exhibited a lower trophic state. With damming of the Green River in 1962, some ecological changes were reported [Vanicek et al. 26 and Pearson 15] and, as seen in Figure 7, it is not unreasonable for the upper Green River to have experienced a change in "stream order" without exhibiting a change in trophic state. That the mean P/R for the lower Green River is greater than the upper indicates that these river segments are not representative of high order streams and have changed. It is also evidence that the Yampa River is providing a measureable effect on Green River ecosystems below their confluence.

The parameters monitored in this study represent a meaningful method of quantifying changes in river ecosystems. From a management perspective, the approach assumes relevance in that data are more reliably collected and less subject to annual variation than fishery data (due to small sample sizes, missing year classes, variant migration or distribution patterns, etc.). Although this type of study does not provide specific information about fish communities, its synthesis with fishery studies constitutes a valuable management tool. The seasonal patterns in these parameters serve as a barometer of existing physical environmental conditions which in turn are important for "habitat" development and maintenance, food production, and/or eliciting behavioral responses (spawning).

These data provide evidence that management of the Yampa and Green River fisheries in Dinosaur National Monument should consider more than minimum flow or water temperature management. Any water development activity which significantly modifies the temporal patterning of any energy input (water quantity, nutrients, sediment, or organic load) must be expected to subsequently alter the ecological character and fish species composition in downstream segments of that river system.

ACKNOWLEDGEMENTS

Funding for this research was provided by the U. S. National Park Service Cooperative Research Unit. I also thank Paul Holden of Bio-West, Inc., Logan, Utah for providing aquatic macroinvertebrate data which were used in the analysis.

LITERATURE CITED

1. McIntire, C. and H. Phinney. 1965. Laboratory studies of periphyton production and community metabolism in lotic environments. Ecol Monogr. 35(3): 237-257.
2. Whitford, L. and R. Schumacher. 1961. Effect of current on mineral uptake and respiration of freshwater alga. Limnol Oceanogr. 6: 423-425.
3. Fisher, S. and G. Likens. 1973. Energy flow in Bear Brook, New Hampshire: an integrative approach to stream ecosystem metabolism. Ecol Monogr. 43: 421-449.
4. Minshall, G. 1967. Role of allochthonous detritus in the trophic structure of a woodland springbrook community. Ecol. 48: 139-149.
5. Nelson, D. and D. Scott. 1962. Role of detritus in the productivity of a rock-outcrop community in a piedmont stream. Linmol Oceanogr. 7: 396-413.
6. Minshall, G., J. Brock, D. McCullough, R. Dunn, M. McSorley, and R. Pace. 1975. Process studies related to the Deep Creek ecosystem. US/IBP Desert Biome Research Memorandum 75-46. Utah State Univ., Logan. 31 p.
7. _____, and J. Wlosinski. 1975. Further development of a stream ecosystem model. US/IBP Desert Biome Research Memorandum 75-48. Utah State Univ., Logan. 34 p.
8. Vannote, R. 1977. The river continuum: a theoretical construct for analysis of river ecosystems. Stroud Water Research Center, Avondale, Pa. Unpubl Typescript. 42 p.
9. Cummins, K. 1977. From headwater streams to rivers. Amer Biol Teach. 5: 305-311.
10. Minshall, G. 1978. Autotrophy in stream ecosystems. BioScience. 28(12): 767-771.
11. Seyfer, J. and J. Wilhm. 1977. Variation with stream order in species composition, diversity, biomass, and chlorophyll of periphyton in Otter Creek, Oklahoma. SW Nat. 22(4): 455-467.
12. Watts, R. 1982. Some physical and chemical mechanisms affecting phytoplankton productivity and standing crop in the Colorado and Green Rivers, Southeastern Utah, M.S. Thesis, Utah State University, Logan, Utah.
13. Strahler, A. 1957. Quantitative analysis of watershed geomorphology. Trans Amer Geophys Union. 38: 913-920.
14. Andrews, E. 1978. Present and potential sediment yields in the Yampa River Basin, Colorado and Wyoming. USGS/WRI 78-105. 38 p.
15. Pearson, W. 1967. Distribution of macroinvertebrates in the Green River below Flaming Gorge Dam, 1963-1965. M.S. Thesis. Utah State Univ., Logan. 105 p.

16. Bott, T., J. Brock, C. Cushing, S. Gregory, D. King, and R. Petersen. 1978. A comparison of methods for measuring primary productivity and community respiration in streams. Hydrobiol. 60(1): 3-12.

17. American Public Health Association. 1975. Standard methods for the examination of water and wastewater. APHA, Wash. D.C. 1193 p.

18. McConnell, W. and F. Sigler. 1959. Chlorophyll and productivity in a mountain river. Limnol Oceanogr. 4: 335-351.

19. Strickland, J. and T. Parsons. 1972. A practical handbook of sea water analysis. Fish Res Bd Can Bull. 310 p.

20. Merritt, R. and K. Cummins. 1978. An introduction to the aquatic insects of North America. Kendall/Hunt Publ Co., Dubuque, Iowa. 441 p.

21. Ott, L. 1977. An introduction to statistical methods and data analysis. Wadsworth Publ Co., Inc. Belmont, Calif. 730 p.

22. Marker, A. 1976. The benthic algae of some streams in southern England. II. The primary productivity of the epilithon in a small chalk stream. J Ecol. 64: 359-373.

23. Cushing, C. 1967. Periphyton productivity and radionuclide accumulation in the Columbia River, Washington, U.S.A. Hydrobiol. 24: 125-129.

24. Teal, J. 1957. Community metabolism in a temperate cold spring. Ecol Monogr. 27: 283-302.

25. Gumtow, R. 1955. An investigation of the periphyton in a riffle of the West Gallatin River, Montana. Trans Amer Microsc Soc. 74: 278-292.

26. Vanicek, C., R. Kramer, and D. Franklin. 1970. Distribution of Green River fishes in Utah and Colorado following closure of Flaming Gorge Dam. SW Nat. 14(3): 297-315.

PART 4

WATERSHEDS

CHAPTER 13

A HEADWATERS VIEW OF CLOUDSEEDING
IN THE COLORADO RIVER BASIN

Barbara C. Welles
     Colorado Department of Natural Resources
     Denver, Colorado

INTRODUCTION

The purpose of this paper is to explain the central
position of the State of Colorado in relation to winter
cloud seeding in the Colorado River Basin. Additionally,
the paper provides an overview of current state regulatory
procedures, commercial projects, research questions, and
public issues in Colorado.

If winter cloud seeding of orographic clouds to
increase runoff proves as successful as advocates believe
it can be, and, if pressures to augment the flow of the
Colorado River continue, the state of Colorado is
positioned for heavy impact. Colorado includes a large
percent of the seedable terrain in the upper basin as over
70 percent of the water flowing to the lower basin at Lee
Ferry comes from the State of Colorado.[1]

In 1978 the National Weather Modification Advisory
Board reported:

> In terms of scientific and technological
> underpinning, and in consideration of
> probable high economic value, it is the
> judgment of this Board, that seeding winter
> orographic storms to increase the amount of
> snow in the high mountainous watersheds of
> the West is the most advanced -- and
> closest to significant, broad scale
> operational use -- of all cloud seeding
> possibilities. A successful confirmatory
> experiment -- including, of course,
> provision for assessing the seeding results
> -- must be completed before large-scale

seeding of winter orographic storms is
considered to be an acceptable tool,
available to water resource makers.[2]

When viewed from the State of Colorado, the effects
of increased snowpack on people and lands in the Colorado
mountains must also be addressed to determine if basinwide
seeding is a viable option, both economically and
politically.

REGULATION OF WEATHER MODIFICATION IN THE STATE OF COLORADO

Weather modification is defined by Colorado law as
any attempt to change the atmosphere by artificial
means.[3]   State regulations apply to all cloud seeding
technologies, but this paper addresses only winter cloud
seeding.

The modern era of weather modification technology
began in 1946 and the first law regulating cloud seeding
activities in Colorado was passed in 1951. Two thirds of
the states in the country have passed laws controlling
cloud seeding activities within their boundaries. The
current Colorado Weather Modification Act, passed in 1972,
directs responsibility for regulation and support of
research activities to the executive director of the
Colorado Department of Natural Resources.

Although the statute recognizes the potential
economic value afforded by seeding and encourages
operations, research, and development, state funding in
most years has been minimal. Typical of a number of state
weather modification statutes, the Colorado law prescribes
a state advisory committee and requires licenses and
permits prior to conducting a specific seeding project.
Seeding without a license or permit in Colorado is a
felony.

The Colorado Weather Modification Advisory Committee

The Committee, appointed by the Governor, has ten
members. Five must have technical expertise in areas
appropriate to weather modification, and the other five
represent farming or ranching interests from the river
basin districts within the state. Members offer advice to
the Department, assist at permit application hearings, and
make recommendations on issuing of licenses and permits.

Licenses

Any individual wishing to conduct weather modification in Colorado must obtain a license. Licenses are granted for specific cloud seeding technologies and cost $100.00 each year. An applicant must meet specified education and experience requirements to qualify for a license.

Permits

In addition to a license for a particular weather modification technology, a permit is required for each specific commercial or research project. The application for a permit must indicate the objectives of the project, the dates during which the proposed operation is planned, and the specific area in which the project is to be conducted. It must also include any additional counties of the state which might reasonably be expected to be affected.

The application must list the persons or organizations for whom the project will be conducted and describe anticipated effects, if any, upon the environment. Proposed techniques to disperse seeding material into the orographic clouds must be explained. Dispersal is from manually operated ground generators which burn a solution of silver iodide and acetone in a propane flame or from seeding aircraft which dispense silver iodide flares or dry ice pellets.

Colorado permit applications also include procedures for seed-no seed decisions, and in the case of seeding from the ground, the number of generators and their proposed locations. Generators are most often placed on private property and are activated by nearby residents who are employed by the commercial firm or the research organization conducting the seeding.

As concern for public interest and comment on proposed projects is of foremost importance in Colorado regulatory procedures, a public hearing is held within each proposed project target area. Hearings provide opportunity for discussion and review of the permit application. Efforts are made to ensure that hearings receive wide publicity and attendance. Hearings are conducted by the executive director of the Department of Natural Resources or his designee, with assistance from members of the Weather Modification Advisory Committee.

In the application and testimony at the public hearing, the applicant must address the statutory requirements and the reasons for the project to be conducted. The Department has progressively required increased substantiation of anticipated economic gain expected from a project. At commercial project hearings this fall, project sponsors presented detailed analysis of economic benefit expected from the project, as well as substantiation for anticipated loss if the project is not conducted.

Individual hearing attendance in recent years has ranged from less than 20 to close to 100 people. Attendance is apt to be larger following winters with heavy snow. Attendees may speak on behalf of, or in opposition to, the proposed undertaking and comments are frequently spirited. In Colorado special effort is made to develop citizen awareness and understanding of seeding activities.

For a permit to be granted, the law dictates that the director of the Department of Natural Resources must decide, based on the application and the public hearing, that the following conditions regarding the proposed project exist:

o   The project offers promise of providing economic benefit.

o   If a commercial project, it is scientifically and technically feasible.

o   If a scientific or research project, it is designed to expand the knowledge and technology of weather modification.

o   It does not involve a high degree of risk of substantial harm to land, people, health, safety, property, or the environment.

o   It will not adversely affect another project.

o   It is designed to minimize risk and maximize scientific gain or economic benefits to residents of the area or the state.

o   It is designed to include adequate safeguards to prevent substantial damage to land, water rights, people, health, safety, or the environment.[4]

Safeguards and minimal risk of harm, therefore, are essential to any weather modification permit granted in Colorado. The basic source of data regarding environmental impacts of winter cloud seeding is the often quoted report entitled "Ecological Impacts of Snowpack Augmentation in the San Juan Mountains, Colorado". The report presents the findings of the San Juan Ecology Project conducted by natural scientists from Colorado State University, the University of Colorado and Ft. Lewis College in conjunction with the Bureau of Reclamation's five year Colorado River Basin Pilot Project.[5]

Overall findings from the San Juan studies are that winter cloud seeding, properly conducted, in the high mountain ranges of the West, will have minimal effect on mountain ecosystems. State and privately funded projects are not subject to National Environmental Protection Act environmental impact statement requirements. But results from federally funded environmental studies are almost always cited to substantiate the minimal environmental impacts anticipated. Although studies to date indicate nomimal affects, considerable concern and support exist for the need to monitor environmental implications of seeding over a longer time period than has been done to date.

Through experience, permits in Colorado now include safeguard conditions and requirements pertinent to each project. These usually include:

o   Specific meteorological and cloud physics parameters under which the project may operate (e.g., cloud temperatures, wind speeds, wind directions, and cloud moisture content).

o   A requirement that the operator communicate with a state government representative in the Department of Natural Resources before any seeding is conducted.

o   Suspension criteria to reduce the likelihood of spring flooding from the melting of additional snow developed by cloud seeding. Suspension is based on a percent of the average snowfall each month as recorded by the United States Department of Agriculture, Soil Conservation Service, Snow Survey Unit. Reports are distributed from January to June. Readings are now also readily available from many automated SNOTEL snow measuring sites.

199

o   Suspension is also required when the United
        States Forest Service, Rocky Mountain Avalanche
        Forecast Center, issues an "extreme caution for
        all travelers" warning within the project area,
        when the State Highway Department experiences
        difficulty with road maintenance, and when the
        State Division of Wildlife has concerns
        regarding deer and elk herd migrations and
        availability of browse.

A project local review committee is appointed by the
Department of Natural Resources for each project in the
state. Membership includes representatives of predominant
interest groups within the project area which are likely
to be affected or who believe that they may be affected.
Local project committees are kept well-informed of project
status and members are asked to report citizen concern in
their area to the Department when need occurs.

PROJECT ACTIVITY

The State statutory provisions and regulatory
procedures just described govern all cloud seeding
projects conducted in Colorado. Winter seeding has been
financed by the state, by federal agencies, and by
numerous public and private entities. Requirements differ
for commercial and research projects. Numbers of licenses
and permits granted recently in Colorado are identified in
Table I.

Table I. Numbers of licenses and permits granted in each
          of the last five years in Colorado

	Licenses	Commercial Permits	Research Permits
1981	6	3	1
1980	5	3	0
1979	6	3	1
1978	8	4	1
1977	6	3	0

Federally Funded Research Projects

Major winter seeding research projects have been
conducted in Colorado since 1960. These include Climax I
and Climax II, the Colorado River Basin Pilot Project, the

San Juan Ecology Project and studies conducted by Colorado
State University to determine cloud processes and
dispersal of seeding materials. Major funding for these
studies has come from the Bureau of Reclamation, the
National Science Foundation, and the National Oceanic
Atmospheric Administration. Information resulting from
these research projects is the basis for seeding
techniques and anticipated results for current commercial
and research projects in Colorado.

State Funded Projects

Concerned by very dry conditions in the winter of
1976-77, the Colorado legislature appropriated funds for
seeding in late winter of 1977 and again for the winter of
1977-78. Funding for the first of the two drought relief
projects conducted in late winter of 1977 was $251,200.
The second project, authorized in the spring of 1977 and
conducted from November, 1977 to May, 1978 received state
appropriations of $300,000 for seeding and $50,000 for
evaluation. State funds were augmented by $600,000
drought relief funds from the Bureau of Reclamation and
supplemented by local funds of $16,500 for a total program
approaching $1,000,000.

During the winter of 1977-78 seeding was conducted in
most mountain areas of the state above 2,743 m. (9,000
ft.). Public hearings for the project were held in four
mountain communities. Due to the prevalent dry conditions
at that time considerable support existed for the
statewide project. Opposition was expressed in the San
Luis Valley and in the San Juan Mountains where seeding
had been conducted recently for the Colorado River Basin
Pilot Project.

Commercial Projects

Commercial projects sponsored by public and private
entities usually include a variety of sponsors who
contribute differing levels of funding to the programs.
Increased confidence in the potential of the technology
has resulted from the research mentioned above. In
addition, perceived economic gain from either snowpack or
snowmelt run-off is so great in comparison to cost that
cloud seeding is considered by some as low cost
insurance.

Commercial project sponsors willing to fund winter seeding include a variety of entities such as:

o Ski areas
o Summer recreation groups, e.g., river runners
o Small ditch companies
o Large municipal water providers
o Conservancy districts
o Towns and counties benefitting from tourism
o Resorts
o Stores and commercial operations

Where Seeding Occurs

Winter cloud seeding in Colorado is generally conducted to target elevations at above 2,896 m. (9,500 ft). These regions have been identified as the major areas for cloud seeding because snowfall accumulates in winter above that elevation and produces ten inches or more annual runoff per square mile. [6]

Commercial programs are being conducted in the winter of 1981-82 in the vicinity of major ski areas and in water district collection areas. The projects are for the mountains where snow is produced and for the drainage areas where snow melts in the springtime and runoff is collected and diverted to the eastern slope of the continental divide.

CONTROVERSY

Public reaction to cloud seeding is almost always controversial. Those believing in the potential of the technology, who stand to receive economic benefit, are the ones who fund commercial projects and were listed above as commercial project sponsors.

Seeding Proponents

Water users are strong proponents of winter seeding in order to maintain adequate soil moisture and storage reservoir water levels. The water users, however, are not concerned with early program starting dates. As long as snow arrives in time to melt in the spring, when it falls during a season is insignificant.

On the other hand, ski areas sponsoring cloud seeding are most interested in early winter seeding. They request projects beginning in November to provide favorable skiing conditions early in the season. Many ski areas have invested heavily in snowmaking equipment particularly to

augment snowpack on lower slopes, but still maintain
strong interest in early season seeding to avoid economic
losses from delayed or poor conditions at opening.

Colorado Ski Country, the ski area operators'
promotional organization, and the organization of county
commissioners, Colorado Counties, Inc., have many members
who support winter seeding projects. However, due to the
nature and variety of membership in these organizations,
they do not necessarily develop unified positions
regarding seeding. In contrast, the Colorado Water
Congress, which represents most of the major water
interests in the state, served recently as a catalyst to
generate support for a state funded weather modification
research project.

Opponents and Skeptics

Not everyone approves of winter weather
modification. Leading the opposition in Colorado have
been residents of high mountain communities who normally
experience considerable annual snowfall, residents of
communities where ranching is a major economic factor, and
those who believe man should not "fool around with nature".

County commissioners from some high mountain
communities, while supporting a need for additional
research to determine seeding effects, have claimed that
their counties incur additional heavy snow removal costs
as a result of cloud seeding programs. They seek
financial assistance in meeting those additional costs.
At some future date sufficient data may exist on which to
negotiate the question of economic loss from snow
resulting from cloud seeding, but so far the problem has
not been adequately addressed to produce a basis for
calculations.

At a recent permit application public hearing a
number of members of the American Association of Retired
People from one of Colorado's higher mountain communities
were present to protest increased snowfall. They argue
that mountain winters are difficult enough without
increased snowfall. The state granted the permit but
excluded seeding activity which might have been likely to
affect the concerned community.

Those engaged in ranching frequently oppose seeding
on several grounds. Many believe snow will be increased
on their pasture land, rather than in the higher areas
targeted for snow increase. They fear programs allowed to

203

begin in early November will make pasture land less accessable for their cattle and will create need for earlier-than-otherwise supplemental hay feeding. In the spring ranchers do not wish heavy snowfall during calving time, which can result in numerous calf losses. To meet ranching concerns, state permit dates for which seeding is allowed are frequently delayed until December in ranching areas where concern exists. Those programs are also terminated well ahead of normal calving season.

The final groups of opponents are those skeptics who do not believe that seeding works and those who feel strongly that man should not "fool around with nature". Both groups are vocal and the state endeavors to consider their claims with the same care received by other opponents.

Fluctuations in Cloud Seeding Interest

While controversy is rarely totally absent in cloud seeding discussions, major interest in the subject frequently rises and falls with the weather. Calls for seeding wax in times of drought and wane in times of ample moisture. In times of drought, however, fewer seedable clouds appear overhead as targets for seeding than during more normal winters. Some people advocate limited seeding on a regular basis in all but very wet years to maintain higher average water supply.

As expected, legislators reflect the concerns of their constituents. When their constituencies include both proponents and opponents of cloud seeding, legislators tend to avoid taking sides. In periods of extended drought, however, more legislative support for seeding can be expected despite limited seedable opportunities.

EMERGING ISSUES

Significant political documents, coupled with sizable population increases in the West, have caused increasing demands for water from the Colorado River. The significant documents include the Colorado River Compact, signed in 1922, which apportioned the River waters between the upper and lower basin states; the Boulder Canyon Project Act of 1928, which divided the water among the lower basin states; the Upper Colorado River Basin Compact of 1948, which divided the waters among the upper basin states; the Mexican Treaty of 1944, which guaranteed

Mexico 1.5 million acre feet of water annually; and the Colorado River Basin Project Act of 1968, which authorized the Central Arizona Project.

Most significant for the application of winter cloud seeding in Colorado, the Colorado River Basin Project Act, Public Law 90-537, directed the Department of Interior to determine the best means of augmenting the flow of the River. In testimony before the House Subcommittee on Water and Power Resources on March 25, 1981, Bureau of Reclamation Commissioner Robert Broadbent testified:

> One of the most promising areas we have been investigating for augmenting water supplies in the Colorado River Basin is weather modification. The technology of cloud seeding for significantly enhancing snowpack is scientifically possible, appears to be the most cost-effective means for increasing water supplies, and is within sight. Several authoritative studies indicate that the flow of the Colorado River could be increased at least 10 percent by snowpack augmentation in the mountainous areas of the Basin States. Presently, Water and Power is considering a proposal for an accelerated Colorado River Augmentation Demonstration Project which could result in Basinwide operational cloud seeding by 1991. Expectations are that the best means will prove to be winter cloud seeding.[7]

Assuming that winter seeding claims of a 10 to 20 percent increase over a winter season prove to be valid, the costs of producing additional runoff by this method in the Colorado River Basin are relatively low. As previously mentioned, the state of Colorado contains over 70 percent of the seedable terrain of the upper basin. Thus, as pressures continue to build to augment the flow of the river, increased interest in seeding winter clouds over Colorado can be expected. Such development will bring to the fore a number of key issues.

Proof of Augmentation Effects

First there is need to develop proof-positive of augmentation effects. This points to the importance of developing continuing support for the Bureau of Reclamation Project Skywater confirming project currently proposed. Careful assessment of presently operating

205

commercial projects must also be encouraged. Commercial
project assessment may be accomplished with Federal-State
cooperation or with private funding. The specific
locations of seeding results and the elevations where they
will occur must be more accurately documented.

Advantages and Disadvantages

A second issue is improved understanding of
advantages and disadvantages. While some people gain from
cloud seeding, others incur additional costs. Studies are
needed to quantify the economics of weather modification.
In Colorado winter cloud seeding permit applications and
hearings, proponents are currently providing increasingly
sophisticated analyses of economic benefits from augmented
snowfall. These include anticipated gains received by ski
areas enabled by seeding to open at the beginning of the
season with sufficient snow, gains anticipated by farmers
who quantify increased crop production due to augmented
water supplies, and increases to utilities from generation
of electricity. These claimed benefits are of
considerable economic magnitude.

More difficult to finance is the collection of
information on costs that may be incurred as a result of
increased snowfall. These can include removal of
increased snowfall from streets, highways, and railroad
tracks, compensation to ranchers for earlier-than-usual
supplemental feedings if weather modification snow is the
cause, and increased payment to farmers or ranchers for
damages caused by wildlife forced to seek feed. Need must
also be addressed for plans to develop long-term
monitoring of potential ecological impacts.

Perhaps if seeding activity is carefully regulated
and monitored, snowfall can be maintained within the
bounds of natural variability and consequently create no
ecological impacts. This possibility has not yet been
demonstrated.

Public Information

A third issue is citizen understanding. Grass roots
education is essential if wide-spread cloud seeding is to
be politically acceptable. In Colorado, effective
vehicles for informing the public have included public
hearings on permit applications, state and local project
advisory committees, news media releases, local decisions
on project start-up dates dependent upon conditions in a
specific area, meetings to discuss cloud seeding in

geographical areas of particular concern, and a public
official open door policy.

It is Colorado policy to keep the public well
informed of all proposed cloud seeding or cloud seeding
related activities and to encourage public participation
and interest. If questions and concerns raised by the
area citizens are met in a straight forward and timely
fashion, chances will be improved for seeding activity to
continue and to be studied. Inadvertent lapses in
communications about seeding plans and activities can turn
off public acceptance and create a negative political
attitude toward seeding, requiring years to regain trust
and acceptance.

Many legal questions still surround "developed" water
and are raised as additional key issues for study. Water
developed for seeding for a long time to come likely will
be subject to the laws of the river as opposed to being
claimed, stored, and put to use by the individual or
entity who pays for the seeding.

A look at cloud seeding from the headwaters of the
Colorado River Basin reveals a promising subject of lively
interest for Colorado and other Western States.

REFERENCES CITED IN TEXT

[1]   Water and Related Land Resources: Colorado River
      Basin, Colorado Water Conservation Board and U.S.
      Department of Agriculture, October 1963.

[2]   Weather Modification Advisory Board, The Management
      of Weather Resources, Volume I: Proposals for a
      National Policy and Program, Report to the Secretary
      of Commerce (Washington: Government Printing Office,
      1978) 229 pages, p. 184.

[3]   The Weather Modification Act of 1972, Article 20,
      Colorado Revised Statutes, 1973.

[4]   Ibid.

[5]   Steinhoff, Harold W., and Jack D. Ives, Ecological
      Impacts of Snowpack Augmentation in the San Juan
      Mountains, Colorado, Department of Watershed Science,
      Colorado State University, prepared for Office of
      Atmospheric Water Resources, U.S. Bureau of
      Reclamation, Department of Interior, under contract
      No. 14-06-D-6292, August, 1970.

[6] "The Colorado Pilot Project: Design
     Hydrometeorology", Patrick A. Hurley, Journal of the
     Hydrolics Division Proceedings, Proceedings of the
     American Society of Civil Engineers, May 19, 1972.

[7] U.S. Congress, House, Subcommittee on Water and Power
     Resources, 97th Congress, First Session, Hearing on
     the Augmentation of Colorado River Water Supplies,
     Statement of Robert Broadbent, Commissioner, Water
     and Power Resources Services, Department of Interior,
     March 25, 1981, 6 pp., p. 5.

                        OTHER REFERENCES

Colorado Department of Natural Resources, Draft,
Cooperative Agreement Report, Prepared for: Department of
Atmospheric Resources Research, U.S. Bureau of
Reclamation, Department of Interior, Denver, Colorado
under Contract and Master Document No. 9-07-85-V0027,
October, 1981.

Colorado River Basin Project Act, Public Law 90-537.

Fischer, Ward H., Appendix-"Legal Considerations",
Generalized Criteria for Verification of Water Developed
through Weather Modification, M.W. Bittinger and
Associates, Inc., Ft. Collins, Colorado, August, 1975.
Final Report to Office of Water Research and Technology,
U.S. Department of Interior Contract 14-31-0001-9020.

Hemel, Eric I. and Holderness, Clifford G., An
Environmentalist's Primer on Weather Modification,
Stanford Law Society, September, 1977, 106 pp.

Upper Colorado Region Appendix VI, Land Resources and Use,
Land Use and Management Workgroup, Upper Colorado Region
State-Federal Inter-Agency Group, Pacific Southwest
Inter-Agency Committee, Water Resources Council, June,
1971, 142 pages.

Ibid, Appendix I, History of Study, and Appendix II, The
Region, 80 pp.

U.S. Congress, Senate, Committee on Commerce, Science, and
Transportation, Weather Modification Programs, Problems,
Policy, and Potential printed for the use of the Committee
on Commerce, Science, and Transportation, 95th Congress,
Second Session,(Washington: U.S. Government Office, 1978),
746 pages.

                             208

U.S. Department of the Interior, Bureau of Reclamation, Atmospheric Water Resources Management, Project "Skywater", Denver, Colorado, Precipitation Management and the Environment: An Overview of the Skywater IX Conference, September 1977, 223 pp.

U.S. Department of Interior, Bureau of Reclamation, and Department of Natural Resources, State of Colorado, Water for Tomorrow, Colorado State Water Plan, Phase I-Appraisal Report, February, 1974.

U.S. Department of Interior, Bureau of Reclamation, Final Environmental Statement, Project Skywater, A Program of Research in Precipitation Management, 1977, Three volumes.

CHAPTER 14

POTENTIAL ECOLOGICAL IMPACTS
OF SNOWPACK AUGMENTATION IN
THE UINTA MOUNTAINS, UTAH

Kimball T. Harper
    Department of Botany and
      Range Science
    Brigham Young University
    Provo, Utah

The Utah Division of Water Resources in conjunction
with the Bureau of Reclamation began in 1976 to gather
information for an environmental impact statement for winter
cloud seeding to augment snowpack in the upper Colorado
River Basin. Hypotheses related to the potential impact of
increased snowpack on regional vegetation were to be
formulated and tested. The required field studies were
conducted by Brigham Young University.

Earlier research programs funded by the Bureau of
Reclamation concentrated on the potential impact of winter
snow augmentation in the Medicine Bow Mountains of
southeastern Wyoming and the San Juan Mountains of
southwestern Colorado. The results of those studies have
been reported by Knight [1] and Steinhoff and Ives [2]. The
Utah study was to focus on the Uinta Mountains, northeastern
Utah. The relationship of late lying snow to the plant
cover of four major ecosystem was to be studied during the
snow melt and growing season for four years. Also, the
relationship of snowpack to runoff in nine major watersheds
was to be analyzed.

The study was not designed to evaluate all possible
impacts of cloud seeding aimed at augmentation of winter
snowpacks in mountains. Prior studies had demonstrated that
several potential impacts were not likely to be deleterious
or could be circumvented by judicious management of seeding
operations. It was decided that it was unnecessary in this
study to further investigate potential impacts that had been
considered in sufficient detail by others. Such potential
impacts are discussed in the paragraphs that follow.

PRIOR STUDIES

Effectiveness of Cloud Seeding

The first aspect of winter orographic cloud seeding
that demanded attention was whether or not the technology
was capable of producing additional precipitation. Several
independent evaluations have been made of winter orographic
weather modification programs in western United States.
Foehner [3] summarized the results of snow augmentation
programs in the Sierra Nevada Mountains of California:
workers there have consistently concluded that precipitation
was enhanced by cloud seeding. Grant [4] concluded that "a
practical technology now exists for augmenting precipitation
from some wintertime orographic clouds. Increases in
overall precipitation on the order of 5-20%, depending
primarily upon the location, can be expected with some
confidence." The Utah Division of Water Resources [5] has
recently completed an evaluation of the central and southern
Utah weather modification program for the period 1974-78
(five years of treatment). The Utah report showed an
overall increase in January-March precipitation of about 13%
in the treated area for the five year period: the
probability that the effect was real was computed to be
0.862. The observed increase was larger (15%) above 2438 m
elevation than below (8.8%).

Snowpack and Streamflow

Research in Utah by Jeppson et al. [6] and Hill et al.
[7] demonstrated that larger snowpacks on watersheds
consistently result in greater runoff. Leaf's [8] work in
Colorado also found a very strong correlation between
heavier snowpacks and greater runoff. Our studies in the
Uinta Mountains show that increased water content in the
April 1st snowpack is directly correlated with increased
streamflow [9]. Our results predict a 13% increase in
annual streamflow if the water content of the April 1
snowpack is increased 10% over the long-term average: a 20%
increase in snowpack should generate a 26.6% increase in
runoff.

Big Game Winter Range

Ward et al. [10], working in The Medicine Bow Mountains
of Wyoming, have concluded that "increased snow
accumulations on elk range below 2743 m (9000 ft) would have
some detrimental effects, particularly if they came early in
the season ... or during the calving season." Sweeney and
Steinhoff [11] concluded that winter range useable by elk

would decrease about 5% given an increase of 15% in the average snowpack of the San Juan Mountains of Colorado. Wyckoff [12] observed that snow cover on big game winter ranges is positively correlated with larger snowpacks on higher elevation snowcourses in central Utah. Strickland and Diem [13] evaluated the impact of snow on mule deer in the Medicine Bow Mountains of Wyoming. They considered that increased snowpack below 2743 m reduced available winter range.

Since it seems likely that seeding of winter clouds will produce additional precipitation below 2743 m (9000 ft) on treated mountains, careful monitoring of snowpacks on big game winter ranges must be coordinated with cloud seeding operations. Even though expected increases in precipitation due to seeding can be expected to be less below 2743 m than above [5], there is still a need to minimize the effects of additional snow on winter ranges. Suitable start-stop criteria for seeding are commonly used to avoid undesirable effects of seeding. In the case of winter ranges, criteria could be chosen that would prohibit seeding before natural storms had pushed big game from higher elevation ranges and onto the winter range. Once on the winter range, other criteria could be used to terminate seeding should snowpack on the area exceed seasonal norms by a given amount. Carefully chosen criteria would nearly eliminate the possibility that cloud seeding would adversely affect wildlife.

Silver as a Potential Toxicant

Although some environmentalists have considered silver iodide (the preferred ice nucleating agent in most cloud seeding operations) to be potentially dangerous to soil and microorganisms and bacterial symbionts of ruminant digestive systems, research has not supported that fear. As early as 1970, Cooper and Jolly [14] reviewed the literature on silver iodide and concluded that silver was not likely to concentrate to harmful levels through either terrestrial or aquatic food chains. More recently, Klein [15] summarized the results of numerous experiments designed to evaluate the impacts of silver iodide on soil microbiological processes, higher plants, aquatic invertebrates, fish, sewage treatment processes, rumen and caecum microorganisms in vertebrate animals, and human physiology. His conclusion was that silver iodide (and all other ice nucleating agents commonly considered for routine use) represented negligible environmental hazards. Klein [15] considered that use of ice nucleating agents would not involve unacceptable risks.

213

Effects of Additional Snow on Nutrient Leaching

Knight and Kyte [16] felt that an increased loss of essential elements for life could be expected from watersheds where snowpacks were larger. In contrast, Lewis and Grant [17] reported that biologically active elements such as nitrogen, phosphorus and potassium were lost significantly faster from a high elevation (2900 m) watershed in the Colorado Rockies in a year of light snowpack than in a year of heavy snowpack. They considered that soil frost was greater when snow cover was incomplete and frost stimulated the loss of nutrients by interfering with biological processes that would otherwise have resulted in the uptake of greater amounts of those elements and reduced losses from the terrestrial environment. More research is needed in this area, but such work was beyond the budgetary capabilities of this study.

THE UINTA MOUNTAINS

The Uinta Mountains lie in the northeastern corner of Utah (Figure 1). Unlike other major North American mountain ranges, the long axis of the Uintas is oriented east and west rather than north and south. The range is large in both areal extent and elevation. Over 13,620 km^2 (5.26 x 10^3 mi^2) of land lie above the 2135 m (7000 ft) contour in the range. Twenty-six peaks rise above 3965 m (13,000 ft) in the Uintas. Kings Peak in the center of the range rises to an elevation of 4117 m (13,498 ft) and is the highest point in Utah.

The Uintas give rise to four of Utah's major rivers: the Bear, Duchesne, Provo and the Weber. About 1.8 km^3 (1.6 x 10^6 acre ft) of surface runoff water are produced annually by the Uintas [6, 18]. About 68 percent of that water feeds into the Green River, a major tributary of the Colorado [6]. The Uinta Mountains are thus a major watershed of the Upper Colorado River Drainage Basin.

Climate

Maximum precipitation in the Uintas is about 102 cm (40.0 in) per year. Along the foothills of the range, annual precipitation averages about 54 cm (about 21 in). Precipitation increases approximately 7.5 cm (3.0 in) per 305 m (1000 ft) rise in elevation. Approximately, three-quarters of the total annual precipitation falls between October 1 and July 1 [9]. The northwestern quarter of the range generally receives more winter precipitation at a given elevation than other parts of the range. In summer, the east end of the range receives more rainfall than the west end [9].

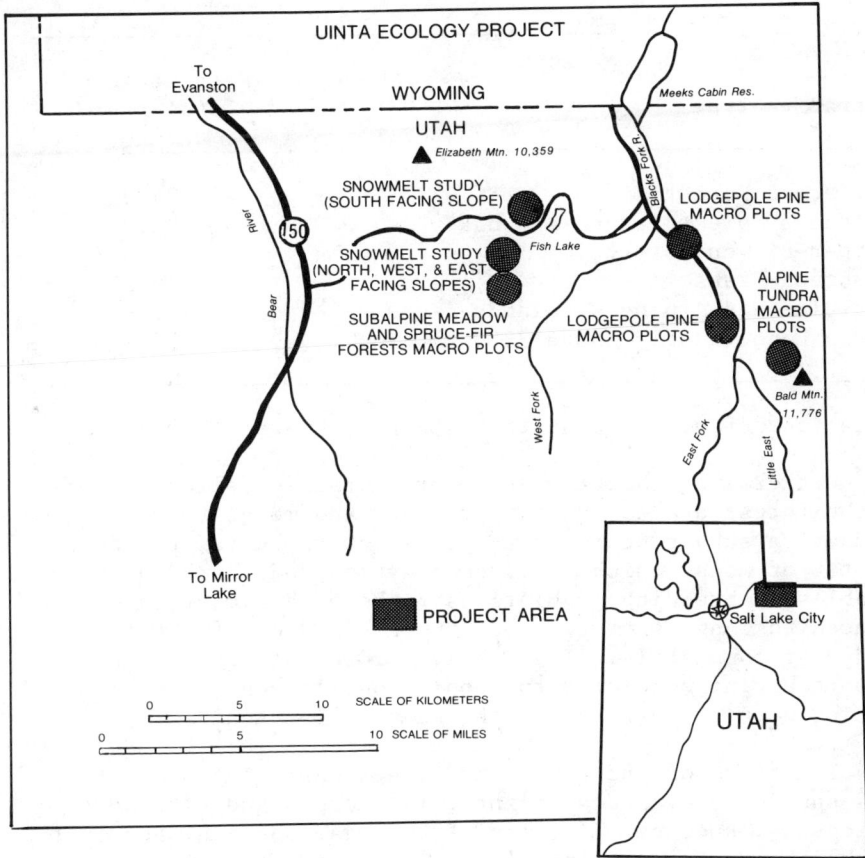

Figure 1. Location of the major study sites in the Uinta Mountains.

Vegetation

The vegetation of the Uintas above 2135 m is forested (Table I) with coniferous forest prevailing above 2740 m. Roughly 60% of the coniferous forest is dominated by lodgepole pine, the most widespread vegetation type in the Uintas. Large areas of Engelmann spruce and subalpine fir also occur in the range. Ponderosa pine is less common in the area and is largely confined to the east and south slopes of the range. Douglas fir forms stands locally throughout the range, usually on limey outcrops.

215

Table I. Vegetation of the Uinta Mountains. Only the area
          above 2135 m is considered.*

Vegetative Type	Percentage of Total Area
Coniferous Forest	34.9
Deciduous Forest (aspen and oak)	31.5
Juniper-Pinyon Forest	2.4
Sagebrush-Grass	16.3
Subalpine and Alpine Herblands	9.8
Barren Land (mostly above 3350 m)	5.1

*Data taken from a map based on Landsat imagery [19].

    Wet drainage bottoms and poorly drained valley bottoms
in the forested areas support verdant and moderately
productive subalpine meadows and willow thickets, the chosen
habitat of moose which have recently invaded the area and
established breeding populations. Above 3350 m, forests are
reduced to tangles of twisted, shrub-sized conifer trees or
disappear completely. The gentler slopes at high elevations
support alpine herblands that only locally consist of a
species complex that merits the name alpine tundra.

    The flora of the Uintas is rich, consisting of over
1050 species of vascular plants, 140 mosses and liverworts,
at least 40 species of lichen, 235 fungal species, and over
800 known species of algae [20]. No endangered plants are
known in the Uinta Mountains, but six plants that have been
legally designated as threatened [Federal Register 40 (125),
part V, pages 27880-27883, July 1, 1975] do occur in the
area. Species designated as threatened are:

> Cryptantha stricta (Osterh.) Payson (Erect Cryptantha)
> Mertensia viridis var. cana (Rydb.) L. O. Williams
>   (Green Bluebell)
> M. viridis var. dilata (A. Nels.) L. O. Williams (Green
>   Bluebell)
> Parrya rydbergii Botsch. (Rydberg's Parrya)
> Penstemon acaulis L. O. Williams (Stemless Penstemon)
> P. uintahensis Pennell (Uinta Penstemon)

Welsh [21] continues to recognize these taxa as threatened
even though he has recommended removing 36 entities from the
endangered and threatened lists published for Utah in 1975

by the Fish and Wildlife Service. Since these taxa are found at low elevations or on windswept ridges, snow augmentation is not likely to harm them.

Animal Resources

The invertebrate animals of the Uintas are not well known. The mountain pine bark beetle (<u>Dendroctonus monticola</u> - family Scolytidae) may exert a greater economic impact on the Uinta Mountains than any other wild animal occurring there. Vertebrate animals of the area are well known and include 22 species of fish, 8 amphibians, 15 reptiles, 82 mammal species and 186 species of birds [20]. Domestic animals that regularly forage on the range include cattle, sheep and horses.

Big game hunting is a major activity on the Uintas. The range supports the largest of Utah's three moose herds [22, 23] and large elk and deer herds. In 1976, 45 bull moose were harvested on the Uintas [24]. The range also provided a harvest of 791 elk in 1976, roughly one-third of the elk harvested in Utah that year. During the period 1970-76, an average of 1500 mule deer per year were harvested on the Uintas. The deer harvest was made by an average annual force of 4300 hunters [24].

Sage grouse, ruffed grouse, and the blue grouse have sizeable populations in the Uinta Mountains. In 1976, in an attempt to enrich the upland game bird resource of Utah, the Division of Wildlife Resources reintroduced the ptarmigan into the alpine zone of Painter Basin on the Uinta North Slope. The population is apparently reproducing in its new home [25].

Timber Resources

About 44% of the saw timber now harvested in Utah comes from the Uintas [26]. The total volume of timber harvested on the range amounts to over 1.41×10^5 m^3 (5×10^6 ft^3) per year. Lodgepole pine is the primary contributor to the harvest at 58.6% of the total; Engelmann spruce contributes almost 31% of the total [20].

Grazing Resources

The Uintas provide summer forage for domestic grazers for an equivalent of 119,709 animal unit months. The grazing is allotted about equally to cattle (52.1%) and sheep (45.7%). Horses account for 2.2% of the permitted grazing pressure [20].

Recreation

The Uintas are a major recreation ground for the
population centers along the Wasatch Front in Utah. If one
includes visitors to the Flaming Gorge Recreation Area and
Strawberry Reservoir, both on the edges of the study area,
total recreation use on the Uinta Mountains is currently
close to 3×10^6 visitor days per year. The use
distribution among the National Forests that manage the bulk
of the Uinta Mountain area is roughly as follows: 50% on
the Ashley, 42% on the Wasatch, and 8% on the Uinta (the
Uinta National Forest manages only a small portion of the
Uinta Mountains). The three major contributors to the
visitor day total appear to be camping (33%), fishing (20%)
and hiking (13%). Hunting probably contributes about half
as many visitor days as fishing [20].

Nonbiological Resources

Mineralization is generally thought to be minimal or
absent in the Uintas [18]. The deep gravel deposits around
the fringes of the range have been little used because of
their remoteness from markets. There is a coal withdrawal
on Currant Creek near the Strawberry Reservoir on the south
slope. The cuestas that form the shoulders of the entire
perimeter of the Uintas have proven to be productive of oil
and gas. Currently, producing wells occur all around the
range and active exploration continues [18, 20].

THE STUDY AREAS

Detailed evaluations were made on the impacts of late
lying snow on the plant components of four major terrestrial
ecosystems of the Uinta Mountains. Studies were made in
lodgepole pine forests, spruce-fir forests, subalpine
meadows, and alpine herblands. The geographic locations,
elevations, average precipitation, and number of macroplots
(each with an area of 0.02 ha or 0.05 acres) sampled in each
ecosystem are given in Figure 1 and Table II. As the data
show, the study is based on an array of sites that vary
widely in respect to climate and elevation.

Tables II. Location and general characteristics of sites
            considered in this study.

Ecosystem	No. Macroplots Sampled	Legal Description of the Area
Lodgepole Pine	16	T.2N.,R.12E.,SE¼, Sect 3
	9	T.2N.,R.12E.,NW¼, Sect 25
Spruce-Fir	36	T.2N.,R.11E.,SW¼, Sect 15
Subalpine Meadow	29	T.2N.,R.11E.,SW¼, Sect 15
Alpine Herbland	21	T.2N.,R.13E.,S ½, Sect 30

	Ave. Elevation (m)	Ave. Annual Precipitation (cm)
Lodgepole Pine	2760	59.4
	2846	63.3
Spruce-Fir	3203	73.0
Subalpine Meadow	3203	73.0
Alpine Herbland	3373	82.5

METHODS

In each ecosystem, sites were sought out where a
gradient in snow depth developed each winter across an area
of uniform topography, soils and plant lifeform. At all
sites, snow drifting patterns were responsible for the
observed gradient in snow depth. At three of the sites
(lodgepole pine, spruce-fir and subalpine meadow), an abrupt
forest edge adjacent to a meadow and at right angles to the
prevailing winds caused drifts to form parallel to the
forest edge. At Elizabeth Ridge, the site of both the
spruce-fir and subalpine meadow studies, the prevailing
winds began to drop their burden of snow more than 100 m in
front of the forest edge. Beyond the edge, the drifts also
extended several meters into the forest. At the point of
maximum depth (usually just inside the forest), the drift
was often over twice as deep as along its leading edge. At
the alpine herbland site, drifts owe their existence to a

slight change of topography. There on Bald Mountain, the prevailing winds strike the gently convex ridgetop at right angles and scour snow from the windward side and deposit it in drifts beyond the crest and in the lee of the ridge. Thus within a kilometer, snowpacks vary significantly in depth without a major change in topography or plant lifeform.

At each study site, circular macroplots (0.02 ha in area) were established in a stratified block design across the gradient in snow depth but within a common vegetation type. In each macroplot, 25 subsamples (each 0.25 m^2 in area) were uniformly spaced. Plant cover and above ground production were taken annually at each subsample during the last week in July or the first week in August. In forested plots, 100% surveys were made of trees to establish density, canopy cover, and growth rates. Tree seedlings and shrubs were inventoried in 4.0 m^2 quadrats. See Harper [27] for sampling details.

RESULTS

Water content of the April 1 snowpack varies strongly among the four ecosystems considered in this report. Using the early snow-free date for comparison, the lightest snowpacks observed (average 7.1 cm water content) occurred on the alpine herbland site, while the heaviest packs (58.2 cm water) were found in the spruce-fir forest (Table III). In both cases, the values reflect the action of wind. On the alpine herbland site, wind sweeps away all snow not sheltered by rocks or clumps of vegetation, while the heavy snowpacks in spruce-fir forest are partially built from snow blown off the adjacent meadow.

The data demonstrated that the natural plant communities of the Uintas are remarkably resistant to change in the face of progressively heavier snowpacks on April 1. None of the population parameters for trees differed significantly between snow-free date zones even though April 1 water content of snowpacks increased 39% between the early and late snow-free date zones in lodgepole pine forests and 148% in the spruce-fir forests. Forest understory was more affected by that increase in snowpack, but even there, the changes were desireable in the lodgepole pine forests: the data show more species/0.25 m^2, more cover, and more above-ground production on the sites with heavier snowpacks [28]. In spruce-fir, both species diversity and understory cover declined significantly along the gradient of progressively heavier snowpacks; above ground production in the understory also declined, but within zone variance was

Table III. Response of the plant cover to progressively later snow-free dates in four Uinta Mountain plant communities. Tree data are for lodgepole pine in the zone named for that species and for Engelmann spruce in the spruce-fir forest. Trends for less common tree species were similar to those for lodgepole and spruce. Values followed by the same letter are not significantly different between snow-free date categories (5% level of probability).

Lodgepole Pine

Characteristic	Snow-Free Date		
	Early	Moderate	Late
No. study plots	8	13	4
Ave. amount water stored in Snowpack on April 1 (cm)	14.8a	15.8a	20.6b
Ave. snow-free date (May)	23.1a	25.2b	25.5b
Ave. Tree Canopy Cover (%)	45.8a	41.8a	37.6a
Ave. tree age (years)	130.0a	145.7a	119.3a
Ave. tree height (m)	15.9a	16.3a	14.4a
Ave. tree diameter (cm)	10.6a	13.0a	8.8a
Ave. No. tree reproductions/0.1 ha	113.7a	250.0a	295.5a
Ave. No. species/0.25m^2	4.8a	5.8a	8.5a
Ave. understory cover (%)	27.2a	32.4a	51.9b
Ave. understory production (g/m^2)	19.7a	17.5a	31.9a

Spruce-Fir

Characteristic	Snow-Free Date		
	Early	Moderate	Late
No. study plots	14	16	6
Ave. amount water stored in snowpack on April 1 (cm)	58.2a	69.9a	144.3b
Ave. snow-free date (June)	17.9a	25.6b	July 14.9c
Ave. tree canopy cover (%)	15.4a	19.5a	24.0a
Ave. tree age (years)	118.8a	133.1a	185.1a
Ave. tree height (m)	12.8a	12.0a	13.7a
Ave. tree diameter (cm)	9.7a	10.7a	9.9a
Ave. no. tree reproductions/0.1 ha	238.7a	361.2a	318.2a

Table III. Continued.

Ave. no. species/0.25m^2	9.0a	7.4a	5.9a
Ave. understory cover(%)	56.3a	52.9a	35.8a
Ave. understory production (g/m^2)	35.8a	33.1a	29.2a

Subalpine Meadow

	Snow-Free Date		
Characteristic	Early	Moderate	Late
No. study plots	11	10	8
Ave. amount of water stored in snowpack on April 1 (cm)	42.7a	56.9b	80.1c
Ave. snow-free date (June)	12.6a	16.6b	24.8c
Ave. no species/0.25m^2	12.0a	11.7a	11.8a
Ave. herbaceous cover (%)	73.2a	79.4b	67.2c
Ave. herbaceous production (g/m^2)	96.5b	107.3a	91.7b

Alpine Herbland

	Snow-Free Date		
Characteristic	Early	Moderate	Late
No. study plots	4	8	9
Ave. amount of water stored in snowpack on April 1 (cm)	7.1a	9.1a	21.1b
Ave. snow-free date (June) May	28.0a	0.5a	8.4b
Ave. no. species/0.25m^2	8.5a	10.8ab	12.5b
Ave. herbaceous cover(%)	53.5a	63.5b	67.3b
Ave. herbaceous production (g/m^2)	70.4a	97.6a	80.8a

sufficiently high that the zone means could not be shown to differ significantly ($p > 0.05$) by analysis of variance [27].

   In the subalpine meadow and alpine herbland studies, the plant communities again proved to be resistant to adverse effects from late lying snow (Table III). Although April 1 water content averaged about 88% greater in snowpacks in the late versus the early snow-free date zone of the subalpine meadow, we could demonstrate no statistically significant differences among zones in respect to species diversity (number of species/0.25 m^2). Plant cover and plant production, however were significantly depressed in the late snow melt zone [29]. In the alpine herblands, late lying snow was actually associated with greater species diversity and plant cover than on sites kept relatively free of snow by wind. Plant production did not differ significantly across the snow-free date gradient in the alpine herblands (Table III).

Since weather modification planners often state the anticipated increases in precipitation as a percentage of some base value, the expected change in two critical variables (changes in snow-free date and forage production) for three levels of snowpack augmentation were computed (Table IV). The results show that the direction and magnitude of expected changes are heavily influenced by the initial amounts of water stored in pretreatment snowpacks. Where initial snowpacks are light (as in the lodgepole pine and alpine herbland ecosystems), augmentation of the average pretreatment snowpack by as much as 25% can be expected to have positive (desireable) impacts on plant production and species diversity (Tables III and IV). The data show that the effects are strongest at the lowest elevations. That result is reasonable, because temperatures are higher, plant transpiration stresses should be greater, and growing seasons are longer at lower elevations. Consequently, in habitats where meltwater from the unmodified snowpack is not likely to completely saturate the soil within the plant root zone, any additional moisture should prolong the period of vigorous plant growth and enhance production. Both lodgepole pine study areas have deep, moderate textured soils that would not be fully recharged by the average April 1 snowpack observed in the early snow-free date zone (water content averaged 14.8 cm) [28].

At the higher elevations of the alpine herblands considered, soils are shallow, moderate textured and highly skeletal [30]. In addition, temperatures are cooler, growing seasons are shorter, and the likelihood of growing season precipitation is greater [9]. As a result, augmentation of the winter snowpack is unlikely to greatly alter the incidence of late growing season water stress on plants. Perhaps the desireable effects of additional snow in the alpine herblands is as much related to protection of plants from periodic exposure to dessicating winds and alternate freezing and thawing in winter as to greater availability of soil moisture in late summer.

Table IV. The effects of additional snow on snow-free date and understory (all annual growth below a height of 1.5 m in forests) or herb layer production in four plant communities of the Uinta Mountains. Increases in snow are stated in terms of the average water content of the April 1 snowpack in the early snow-free zone of each community. See Table III for actual values considered in each community.

Percentage Increase in April 1 Snowpack	Delay in Snow-Free Date (Days)	Change in Plant Production (%)
Lodgepole Pine		
5	0.3	+ 10.0
15	0.8	+ 30.1
25	1.3	+ 50.1
Spruce-Fir		
5	0.9	- 0.6
15	2.6	- 1.7
25	4.3	- 2.8
Subalpine Meadow		
5	0.9	- 2.3
15	2.6	- 4.4
25	4.3	- 10.8
Alpine Herbland		
5	0.4	+ 0.2
15	1.4	+ 0.7
25	2.3	+ 1.1

On Elizabeth Ridge, location of the subalpine meadow and spruce-fir plots, heavy snowpacks occur in even the earliest snow release zone. Such snowpacks are fully capable of completely saturating the shallow, stoney soils at the site. Thus supplemental snow in the pack can do nothing to relieve late season plant water stress, but does have the undesireable effect of retarding snow-free date and thus shortening the already scant growing season characteristic of those elevations [27, 29]. Accordingly, increases in the size of the April 1 snowpack produced negative impacts on plant production in both the subalpine meadow and spruce-fir forest (Table IV). Nevertheless, the magnitude of the adverse effects is small (e.g., 4.4% decline in plant production in the subalpine meadow and 1.7% decline in spruce-fir understory production given a 15% increase in snowpack).

DISCUSSION

Assuming that the average water content in the early snow release date zone (Table III) is a fair estimate of the

224

water content of the April 1 snowpack across the entirety of
each of the four ecosystems studied, the weighted average
water content of the April 1 snowpack can be computed across
the all four ecosystems. Such an estimate requires
knowledge of the area covered by each ecosystem.
Fortunately, those areas are known (Table V). Now, by
computing the proportion of the entire study area
contributed by each ecosystem, multiplying the average water
content of the April 1 snowpack in that ecosystem by that
proportional value, and summing those products across the
four ecosystems, a weighted average is obtained for water
content of the snowpack across all ecosystems. The
resultant weighted average water content of the snowpack on
April 1 is 23.5 cm. Given a 10% increase, the average
snowpack across the four ecosystems would contain 25.9 cm or
an extra 2.4 cm of water (0.9 in).

Earlier work in the Uintas suggested that a 10%
increase in regional snowpacks on the west end of the Uintas
resulted in a 13% increase in streamflow [9]. If 85% of the
additional water content of the snowpack becomes runoff
water (an assumption justified by data presented by Harper
et al. [9]), surface flow would be increased by about 2.54
cm (1.0 in) per unit area. Given a value of $20 per acre
foot for runoff water (a value used by the Utah Division of
Water Resources [5]), the additional runoff (2.54 cm) would
have a value of $4.12/ha ($1.67 per acre).

Table V. Areal extent, average above-ground production in
the herb layer, and predicted change in production
given a 10% increase in average snowpack in each
community. Basic data are from Harper [27].

Community	Area (ha)	Average Production kg/ha	Predicted Change (%)	Absolute Change (kg/ha)
Lodgepole Pine	294,000	205	+ 20.1	+ 41.2
Spruce-Fir	122,940	335	− 1.2	− 4.0
Subalpine Meadow	16,832	989	− 3.4	− 33.6
Alpine Herblands	124,241	852	+ 0.5	+ 4.3
	558,113			

Since the influence of additional snow had an unequal
impact in the four ecosystems studied, it would be helpful
to know what the integrated effect of heavier snowpacks
would be across all ecosystems. The data compiled in Table
V can be manipulated to show that weighted average above-
ground production in the herb and understory layers of the
four ecosystems of concern is 401 kg/ha. In combination,

the four ecosystems dominate about 558,113 ha in the Uintas.
Given a 10% increase in average snowpack, weighted above-
ground, herbaceous production is predicted to be 422 kg/ha
across the area considered above. Thus a 10% increase in
snowpack would increase herb production about 5.2% across
the entire area even though reduced production is expected
locally where normal snowpacks are already sufficient to
saturate the root zone on melting.

Assuming that all 21 kg/ha of the additional herb
production resulting from a 10% increase in snowpack is
usable as forage, the extra forage is sufficient to feed one
animal unit (one cow or five sheep) for 2.6% of a month
(considering an animal unit to require 720 lbs/acre or 807
kg/ha per month). In 1981, the federal government charged
$2.31/Animal Unit Month (AUM): thus 21 kg/ha of forage is
worth $0.06.

Using data from Table IV and weighting procedures as
outlined above, one can also predict the average delay in
snow-free date across the entire study area due to a 10%
increase in snowpack. That procedure yields an estimate of
0.9 day delay in snow-free date.

It is possible to manage timber harvesting and grazing
so as to harvest all resources that might be declared
saleable by managers despite a delay of 0.9 day in the
snow-free date. Nonwinter sports and recreation activities
may, however, be handicapped by a delay in snow melt.
Winter sports, on the other hand, would be benefitted by
late lying snow, but there is currently little winter sports
activity in the Uintas. If winter sports are neglected and
the nonwinter recreation use is assumed to average three
million man days per year across the entire Uinta Mountain
region (area of 13,620 km^2), one can again compute a
weighted impact factor for recreation. For the purpose of
this study, a conservative assumption was made that summer
recreation use is spread uniformly across the Uintas (it is
actually concentrated on the margins of the range) and
throughout the snow-free season (it is actually concentrated
in July and August). The weighted average snow-free date
across the study area (558,113 ha or 41% of the entire Uinta
mountain area above 2135 m) was estimated to be May 30. The
nonwinter sports season was assumed to terminate on October
26 (the end of the hunting season). Thus recreational
season is considered to include 150 days. The economic
value of a recreation day is assumed to be $3.00, a value
currently used by the U.S. Forest Service.

Assuming 1,230,000 visitor days are spent in the four ecosystems of concern, there are 2.2 visitor days per hectare during a 150 day period. The economic value to society of that recreation use would be $6.60/ha. If 0.9 day of the 150 day season were lost without compensation, the economic loss would be 0.6% of $6.60/ha or $0.04/ha.

Assuming that supplemental runoff water can be generated at a cost of $1.00/acre foot (about $0.82/ha dm), a cost shown to be realistic by the Utah Division of Water Resources [5], one can use that value and the foregoing costs and benefits to calculate a first approximation benefit/cost ratio for cloud seeding in the Uinta Mountains. In doing so, timber production is neglected in the analysis, but that omission is justified by the failure of tree parameters (Table III) to show any consistent trends due to snow-pack variations. The results show a benefit/cost ratio of 16.7 (Table VI).

Table VI. Summary of estimated economic benefits and costs associated with a 10% increase in April 1 snowpack in four higher elevation ecosystems in the Uinta Mountains. See preceding paragraphs for assumptions on which the values in this table are based.

Benefits		Costs	
2.54 cm runoff water =	$4.12/ha	Generation of 2.54 cm	
21.0 Kg forage/ha =	0.06/ha	water = $0.21/ha	
Total	$4.18/ha	Loss 0.9 days recreation	
		=	$0.04/ha
		Total	$0.25/ha

SUMMARY

Results of studies on the impacts of late-lying snow on four ecosystems (lodgepole pine and spruce-fir forests, subalpine meadow and alpine herbland) are reported. An increase of 10% in the average snowpack is estimated to retard snow-free date 0.6-1.8 days in the four ecosystems considered here and to increase runoff from the combined study areas by 13%. Ten percent more snow could not be shown to alter tree growth or reproduction in the forests studied. A 10% increase in snowpack tended to increase above-ground herb growth in ecosystems that normally have light snowpacks (lodgepole forest and alpine herbland) and to decrease herb production in zones of heavy snowpacks (spruce-fir forest and subalpine meadow). All changes in herb layer production were small. A 10% increase in

227

snowpack is estimated to produce an overall increase of 5.5%
in herb production since communities with smaller snowpacks
are more widespread than those with heavy snowpack in the
study area. Impacts on recreation should be slight.

LITERATURE CITED

1. Knight, D. K. (ed.). 1975. The Medicine Bow Ecology
    Project. Final Report to U.S. Bureau of Reclamation,
    Division of Atmospheric Water Resources Management,
    Denver, Colorado. 397 pp.

2. Steinhoff, H. W. and J. D. Ives. 1976. Ecological
    impacts of snowpack augmentation in the San Juan
    Mountains, Colorado. Final report of the San Juan
    Ecology Project to the U.S. Bureau of Reclamation,
    Division of Atmospheric Water Resources Management,
    Denver, Colorado. 489 pp.

3. Foehner, O. H. 1977. Weather modification: a major
    resource tool. Proceedings of the 45th Annual Western
    Snow Conference. 13 pp.

4. Grant, . D. 1977. Scientific and other uncertainties
    of weather modification, pp. 7-20. In W. A. Thomas
    (ed.), Legal and scientific uncertainties of weather
    modification. Duke University Press, Durham, N.C. 155
    pp.

5. Utah Division of Water Resources. 1981. Evaluation of
    five years of operation of a central and southern Utah
    weather modification program, 1974-1979. Utah Division
    of Water Resources, Salt Lake City. 66 pp.

6. Jeppson, R. W., G. L. Ashcroft, A. L. Huber, G. V.
    Skogerboe, and J. M. Bagley. 1968. Hydrologic atlas
    of Utah. Publication No. PRWG35-1 from the Utah Water
    Research Laboratory, Utah State University, Logan. 306
    pp.

7. Hill, G. E., N. E. Stauffer, Jr. and H. K. Woodward.
    1975. Assessment of cloud seeding programs and
    evaluation techniques in the State of Utah. Utah
    Division of Water Resources Publication, Salt Lake
    City. 61 pp.

8. Leaf, C. F. 1975. Watershed management in the central
    and southern Rocky Mountains: a summary of the status
    of our knowledge by vegetation types. USDA Forest
    Service Research Paper RM-142. 28 pp.

9. Harper, K. T., R. A. Woodward and K. B. McKnight. 1980. Interrelationships among precipitation, vegetation, and streamflow in the Uinta Mountains, Utah. Encyclia 57:58-86.

10. Ward, A. L., K. Diem, and R. Weeks. 1975. The impact of snow on elk, pp. 105-133. In D. H. Knight (ed.), The Medicine Bow Ecology Project. Final report to the Bureau of Reclamation, Division of Atmospheric Water Resources Management, Denver, Colorado. 397 pp.

11. Sweeney, J. M. and H. W. Steinhoff. 1976. Elk movements and calving as related to snow cover, pp. 415-436. In H. W. Steinhoff and J. D. Ives (eds.), Ecological impacts of snowpack augmentation in the San Juan Mountains, Colorado. Final Report, San Juan Ecology Project.

12. Wyckoff, J. W. 1980. Evaluating mule deer winter range relationships through Landsat satellite imagery. Ph.D. Dissertation, University of Utah, Salt Lake City. 106 pp.

13. Strickland, M. D. and K. Diem. 1975. The impact of snow on mule deer, pp. 137-174. In D. H. Knight (ed.), The Medicine Bow Ecology Project. Final Report to the Bureau of Reclamation, Division of Atmospheric Water Resources Management, Denver, Colorado. 397 pp.

14. Cooper, C. F. and W. C. Jolley. 1970. Ecological effects of silver iodide and other weather modification agents: a review. Water Resources Research 6:88-98.

15. Klein, D. A. (ed.). 1978. Environmental impacts of artifical ice nucleating agents. Dowden, Hutchinson, and Ross, Inc. Stroudsburg, PA. 256 pp.

16. Knight, D. H. and C. R. Kyte. 1975. The effect of snow accumulation on litter decomposition and nutrient leaching, pp. 215-223. In D. H. Knight (ed.), The Medicine Bow Ecology Project. Final Report to the Bureau of Reclamation, Division of Atmospheric Water Resources Management, Denver, Colorado. 397 pp.

17. Lewis, W. M., Jr., and M. C. Grant. 1980. Relationship between snow cover and winter loss of dissolved substances from a mountain watershed. Arctic and Alpine Research 12:11-17.

18. Wasatch National Forest. 1976. Environmental statement, North Slope land use plan. USDA Forest Service, Intermountain Region, Ogden, Utah. 161 pp.

19. Ridd, K. 1978. Vegetation of the Uinta Mountains above 7,000 feet (a map). Department of Geography, University of Utah, Salt Lake City.

20. Harper, K. T., W. K. Ostler, and D. C. Anderson. 1982. History, environments and resources of the Uinta Mountains, Utah. Great Basin Naturalist 42: (in press).

21. Welsh, S. L. 1978. Endangered and threatened plants: reevaluation. Great Basin Naturalist 38:1-18.

22. Wilson, D. W. 1971. Carrying capacity of the key browse species for moose on the north slope of the Uinta Mountains, Utah. Masters thesis, Utah State University, Logan. 57 pp.

23. Babcock, W. H. 1977. Continuing investigations of the Uinta North Slope moose herd. Utah Division of Wildlife Resources Publication No. 77-19. 76 pp.

24. John, R. T. and J. S. Fair. 1977. Big game harvest report - 1976. Utah Division of Wildlife Resources Publication 77-4. 96 pp.

25. Hall, D. E. (ed.) 1978. Information needed on ptarmigan. The Wildlife Report, June 12, 1978, p. 2. Published by the Utah Division of Wildlife Resources, Salt Lake City.

26. Setzer, T. S. and T. S. Thorssell. 1977. Utah timber production and mill residues, 1974. USDA Forest Service, Research Note INT-234. 5 pp.

27. Harper, K. T. (ed.) 1981. Potential ecological impacts of snowpack augmentation in the Uinta Mountains, Utah. Final report submitted to the Bureau of Reclamation, Office of Atmospheric Resource Management, Engineering and Research Center, Denver, Colorado 80225 and Utah Division of Water Resources, 231 E. 400 S., Salt Lake City. 291 pp.

28. Harper, K. T., C. W. Morden, D. L. Hunter and K. B. McKnight. 1982. Effects of heavier snowpacks on plant parameters of the lodgepole pine-bluegrass habitat type in the Uinta Mountains, Utah. (in review).

29. Ostler, W. K., K. T. Harper, K. B. McKnight and D. C.
    Anderson. 1982. The effects of increasing snowpack on
    a subalpine meadow in the Uinta Mountains, Utah. J.
    Arctic and Alpine Research 14: (in press).

30. Harper, K. T., W. K. Ostler, K. B. McKnight and D. L.
    Hunter. 1982. Effects of late-lying snow on an alpine
    herbland in the Uinta Mountains, Utah. Encylcia 59:
    (in press).

CHAPTER 15

SIMULATED IMPACT
OF WEATHER MODIFICATION
ON THE COLORADO RIVER

John C. Lease
Division of Atmospheric
Resources Research
U.S. Bureau of Reclamation
Denver, Colorado

BACKGROUND

The natural water supply of the Colorado River is rapidly becoming inadequate to meet the needs of the Colorado
River Basin states. This demand is directly related to the
historically arid conditions of the Basin and the lack of
adequate streamflow to meet existing conditions. Projections
have been made which indicate that with future development
and growth, water shortages will begin to occur as early as
1990 in the Lower Basin and will become increasingly more
severe thereafter. Although large amounts of water presently
in reservoir storage will temporarily meet Basin demands, and
excess flows may occur with normal precipitation over the
next 10 to 20 years, projected requirements would exceed
natural flows by several billion cubic meters annually after
that time. Increased water supplies will be needed due to:
rapidly expanding population and industry in the region;
salinity control and other water quality problems; irrigation; reservoir and streamflow maintenance for environmental,
wildlife, and recreational benefits; and hydroelectric power
generation. Another significant factor affecting water and
energy requirements in the region will be the rate at which
oil shale, coal, and oil reserves in the Basin are developed.
Additionally, over half of the population of 11 western
states depend to a large extent on the Colorado River as a
source of water, with a greater percentage of water exported
from the Colorado than from any other major river system in
the United States.

The Bureau of Reclamation's DARR (Division of Atmospheric
Resources Research) is currently in the planning phase for a
program designed to help alleviate this impending water
shortage. This program, the Colorado River Enhanced Snowpack

Test, will use weather modification as a method of increasing precipitation within the Basin. Weather modification, in the form of seeding wintertime orographic cloud systems to augment mountain snowpack, has been developing over the last 30 years. Experimental results suggest that seeding under favorable conditions should increase seasonal precipitation by 10 to 15 percent. To estimate the potential impact and benefit of weather modification on the Colorado River Basin, the DARR conducted a series of simulations using the CRSS (Colorado River Simulation System) model. Such mathematical modeling was conducted by using both modified and unmodified historical streamflows, and the resulting differences were examined in terms of water shortages, salinity reductions, international agreements with Mexico, reservoir response, hydroelectric power production, and economic benefit.

AUGMENTATION POTENTIAL

Most of the flow of the Colorado River originates from seasonal snowpack in alpine and subalpine watersheds above the 2,750-m (9,000-ft) level, where winter precipitation amounts are high and evapotranspiration losses low. Figure 1 shows five major runoff-producing areas within the Upper Basin, totaling about 34,000 km^2 (13,000 mi^2), which produce 75 percent of the Basin runoff while accounting for only 12 percent of the Basin area.

To investigate the potential of weather modification in the Upper Colorado and other river basins, the Bureau of Reclamation sponsored a comprehensive study of 20 years of precipitation, runoff, and storm data to provide an estimate of potential annual increases in runoff. This study (1), conducted by North American Weather Consultants, used a combination of meteorological and hydrological analyses to estimate incremental runoff resulting from cloud seeding. The meteorological analysis consisted of entering detailed data on atmospheric soundings, terrain features, seeding sources, and cloud tops into a numerical model which determined the potential for precipitation increases by calculating the difference between modified and unmodified precipitation rates. Meteorological information was obtained from those stations within or closest to the subbasin of interest. Information on terrain features was obtained by examining the mountain massif of each subbasin and determining a representative topographic profile and crest orientation. The results of the model were adjusted to fit historical data witin the subbasin, and final precipitation values were calculated as a function of distance from the massif.

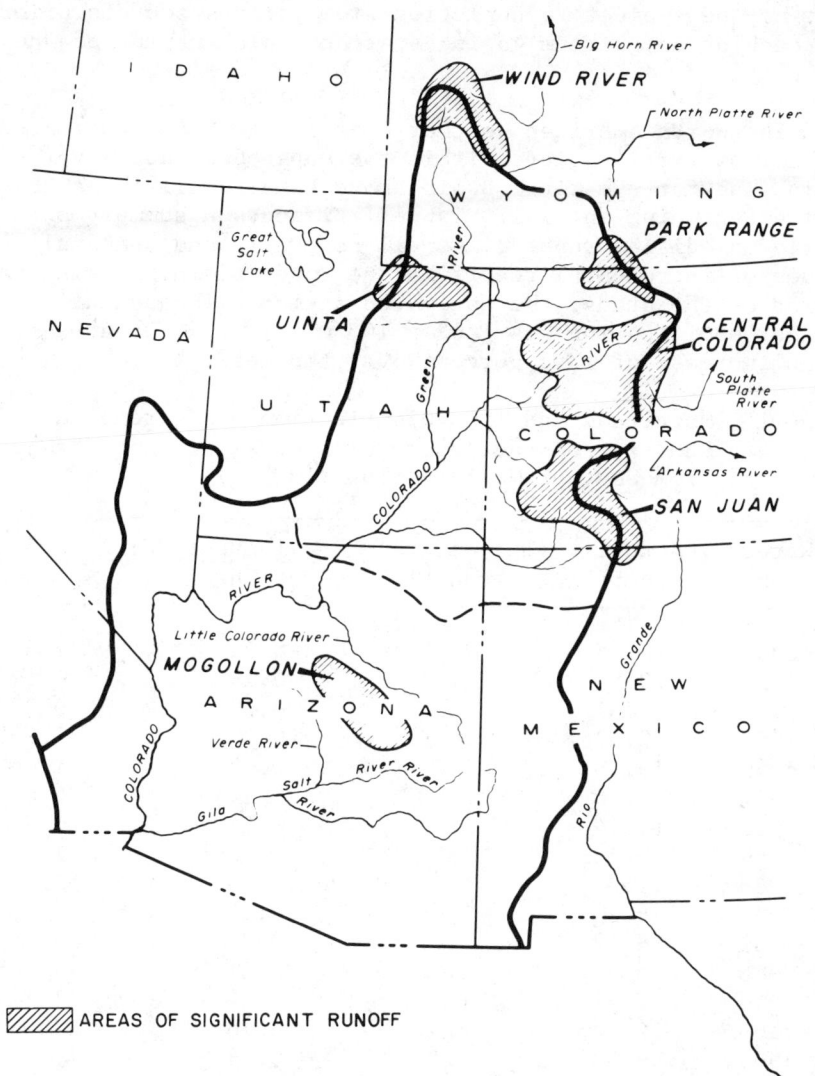

Figure 1. Major runoff-producing areas in the Colorado River
Basin (Mogollon Area not included in study),

To convert the incremental precipitation quantities
derived in the meteorological analysis into runoff, sophisticated precipitation-runoff relationships were developed by
massif and applied seasonally by elevation zones to the previously derived precipitation values. Runoff amounts were
accumulated to represent the total additional water available
for each subbasin. The results represent additional water

that may be subject to depletion from various sources prior to reaching downstream gaging stations which represent the subbasin.

The North American Weather Consultants study estimated streamflow for 11 points within the Upper Colorado River Basin, and for the total Basin above Lake Powell for a 20-yr period beginning in 1951. Table I presents a summary of yearly unmodified runoff, as well as yearly and seasonal values of incremental runoff in the Upper Basin. During the period of the study, the estimated augmentation potential from weather modification was 1.62×10^9 m^3 (1.32×10^6 acre-ft), or an increase of 13.2 percent over the mean flow.

Table I. Magnitude and Time Distribution of Incremental Runoff from the Upper Colorado River Basin Resulting from Weather Modification (10^9 m^3)

Water year	Actual runoff	Incremental runoff	Incremental runoff Oct–Mar	Incremental runoff Apr–Sep
1952	22.15	2.53	0.21	2.32
1953	10.84	1.31	0.19	1.12
1954	7.53	1.16	0.19	0.97
1955	8.99	1.29	0.20	1.09
1956	10.78	1.66	0.20	1.46
1957	21.36	2.32	0.24	2.08
1958	17.54	2.01	0.27	1.74
1959	8.32	1.26	0.18	1.08
1960	11.33	1.53	0.23	1.30
1961	8.20	1.34	0.19	1.15
1962	18.22	2.02	0.26	1.76
1963	6.21	1.21	0.21	1.00
1964	7.51	1.26	0.14	1.12
1965	16.12	2.09	0.18	1.91
1966	9.63	1.17	0.24	0.93
1967	9.54	1.45	0.19	1.26
1968	11.70	1.64	0.18	1.46
1969	13.59	1.64	0.19	1.45
1970	13.57	1.84	0.25	1.59
1971	12.54	1.71	0.26	1.45
Average	12.28	1.62	0.21	1.41

SIMULATION PROCEDURES

To simulate the impact of additional runoff produced by weather modification, the results of the North American

Weather Consultants study for the Upper Colorado River Basin
were used as input to the CRSS model. The CRSS model (2, 3,
4, and 5), developed by the Bureau of Reclamation, is a
research model of the River which reflects: water availability; salinity; demands on water by municipal, energy, and
agricultural users; and other water demands.

The CRSS model is structured so that the Basin is divided
into a series of subbasins called nodes. These nodes, which
represent specific geographic areas, are further subdivided
into individual points at which inflow of water, demands for
water, and reservoirs can be located. Based on projected time
and water demand relationships made for each demand point, the
impact on river water supply and quality can be analyzed. In
the Upper Basin, each demand point generally represents these
uses: current agricultural, municipal, industrial, and
export. In the Lower Basin, each demand point generally represents a specific user, such as the MWD (Metropolitan Water
District) of California or the CAP (Central Arizona Project).

As input information describing the demand, the CRSS
model requires reservoir operation data and hydrologic data
(see Fig. 2). A set of historic hydrologic flows (1933-1974)
were subjectively selected and used as the unmodified flow
for the period 1979 through 2020. The data from the North
American Weather Consultants study were used to synthetically
generate flows which would represent those due to weather
modification.

In order to allow for yearly flow variation and suspension due to excess snowpack, avalanche control, and big game
hunting, two corrections were necessary to duplicate actual
weather modification activity. Adjustments for yearly variations in snowfall and subsequent runoff were made by employing
linear regression techniques to establish a relationship
between unmodified runoff and incremental increases in flow
for the period 1952 to 1971. A similar technique was employed
for computing the salt loading adjustment.

Operational suspension periods were established from data
obtained during the San Juan Pilot Project (6). Values were
computed from the actual suspension periods observed between
1971 and 1975. It was determined that on 18.5 percent of the
operational days, activities were suspended due to excess snow
or avalanche danger during 2 of the 5 years of the program.
In the remaining 3 years, suspensions were made for avalanche
control on 2.3 percent of the days. Suspensions were made at
the beginning of each seeding season to avoid big game hunting. This resulted in an additional reduction in possible
seeding days of 10.3 percent. The adjustments listed above
were applied for excess snowfall and avalanche control during

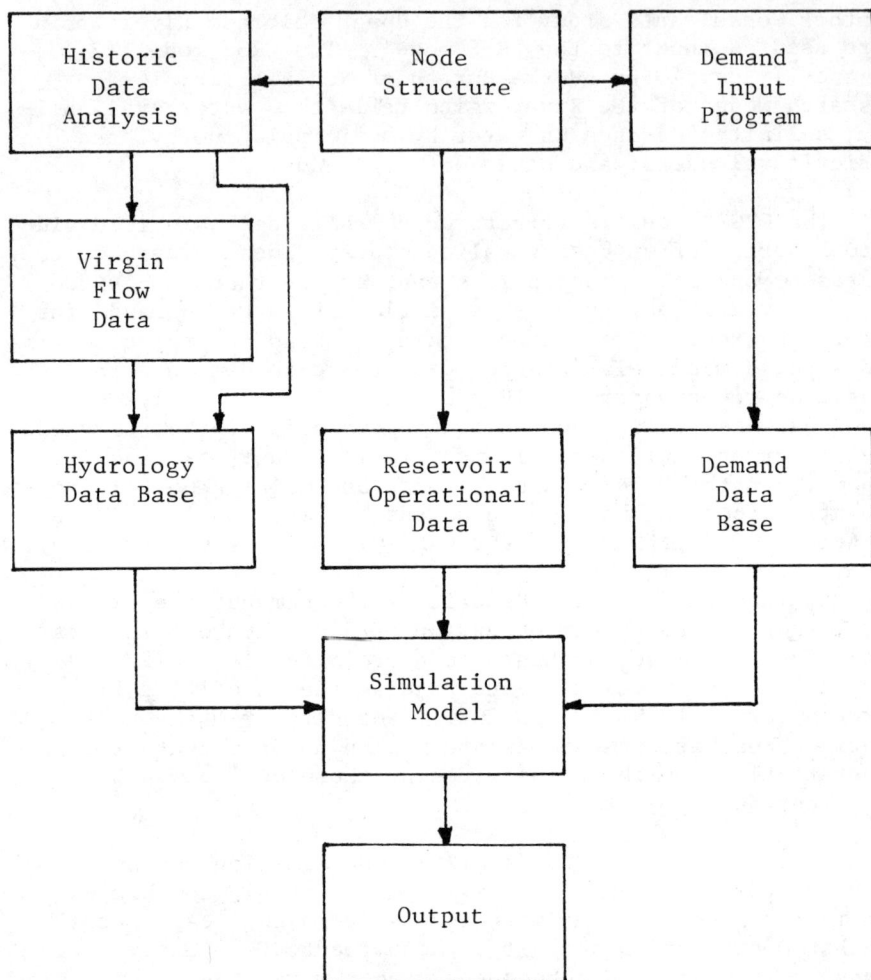

Figure 2.   Colorado River Simulation System model block
            diagram.

the 8 highest flow years, and for avalanche control only in
the remaining 12 years. Uniform adjustments were made
throughout the entire 20-year period to account for suspension due to hunting. Again, linear regression techniques
were used to develop flow and salt loading relationships.

    Final derived relationships vary from subbasin to sub-
basin, but the general tendency for both flow and salinity is
for incremental increases to be lower in years with above-
average flows and greater in below-average flow years. In
terms of flow into Lake Powell, values ranged from a 19 percent increase with 6 x 10^9 m^3/yr to 11 percent with 21 x
10^9 m^3/yr.

RESULTS

Historic augmented and unaugmented flows were individually run through the CRSS model and the results compared (7). Specific factors evaluated were reduction of shortages, changes in salinity, flows to Mexico, flow response, reservoir response, and hydroelectric power production. This evaluation was conducted by assuming a Basin-wide operational weather modification program from 1991 through 2020.

Water Shortages

Shortage messages for the Upper Basin were generated by the CRSS model to provide information on specific demands. Consumptive needs, water availability, and volume-short water values were listed by month for demands not satisfied. Shortages not satisfied with NWM (no weather modification) totaled 7.2×10^9 m^3 (5.9×10^6 acre-ft). These shortages were reduced to 6×10^9 m^3 (4.9×10^6 acre-ft) with WM (weather modification). Shortages of consumptive use water were reduced over 15 percent with weather modification for an average yearly shortage reduction of 41×10^6 m^3 (33×10^3 acre-ft). Releases below Hoover Dam provided a minimum flow at Imperial Dam of 6.7×10^9 m^3 (5.4×10^6 acre-ft) with NWM and 6.8×10^9 m^3 (5.5×10^6 acre-ft) with weather modification after 1991.

Salinity Reduction

A reduction of the concentration of salts in terms of total dissolved solids occurred with WM, as illustrated in Table II. However, at Lee's Ferry, the increased flow due to WM caused an increase in total tonnage being carried by the River, although the concentration of salts decreased. Yearly average total dissolved solids concentration with WM varied at Lee's Ferry from 623 mg/L to 816 mg/L. Below Hoover Dam, the concentration of salts with WM ranged from 798 mg/L to 954 mg/L.

Mexican International Agreements

The CRSS model is programmed to deliver required flows to Mexico, as stipulated in United States-Mexico agreements. Therefore, 1.85×10^9 m^3 (1.5×10^6 acre-ft) are always delivered to Mexico, but water delivery intended for the CAP and the MWD of California can be reduced. With NWM, the minimum yearly delivery of water to MWD after 1991 was 0.68×10^9 m^3 (0.55×10^6 acre-ft), while during the same period, the minimum delivered to CAP was 0.49×10^9 m^3 (0.4×10^6 acre-ft), with WM, the minimum delivery to MWD remained the same, while

239

Table II. Changes in Salinity and Flow from 1991-2020

Source	Lee's Ferry			Hoover Dam			Imperial Dam		
	Flow 10^9 m^3	TDS* mg/L	10^6 tons/yr	Flow 10^9 m^3	TDS mg/L	10^6 tons/yr	Flow 10^9 m^3	TDS mg/L	10^6 tons/yr
NWM	10.60	764	8.94	10.72	934	11.04	6.83	1169	8.82
WM	11.94	700	9.22	11.49	867	10.99	6.98	1078	8.30
Difference	1.34	-64	0.28	0.77	-67	-0.05	0.15	-91	-0.52

* Total Dissolved Solids

deliveries to CAP more than doubled to over 1.0×10^9 m^3 (0.85×10^6 acre-ft). During the 30-year period studied, more than 3.7×10^9 m^3 (3.0×10^6 acre-ft) of additional water were delivered to MWD as a result of WM, while an additional 14.8×10^9 m^3 (12.0×10^6 acre-ft) were delivered to CAP.

Flow Response

Based on the 1952 to 1971 virgin flows into Lake Powell, the average yearly incremental increase from WM was 13.2 percent. The average yearly increase from WM calculated from CRSS results was 14.1 percent. This increase is greater, even with suspension periods because, as shown in Table III, the average flow for the period was almost 1.2×10^9 m^3 (1.0×10^6 acre-ft) less than that during the historic source period used to establish flow coefficients. This reflects the trend described earlier that below-average flows tended to produce higher than average yearly incremental runoff. The actual magnitude of the yearly increase from historic flows was 1.62×10^9 m^3 (1.32×10^6 acre-ft), as compared to average yearly incremental increases of 1.57×10^9 m^3 (1.27×10^6 acre-ft) generated by the model.

Reservoir Response

Reservoir response was examined for both Lake Powell and Lake Mead by investigating how water was transferred with respect to the dam and the storage and water balance over the simulation period. In the CRSS model, water can be transferred from above the dam to below by three different methods: spilled over the top of the dam, passed through diversion works but not through the powerplant, and through the powerplant. Results of the model show that there were no spills at Hoover Dam during the entire period with either WM or NWM. Two spills totaling 1.5×10^9 m^3 (1.2×10^6 acre-ft) did occur with WM at Glen Canyon Dam. Flows which passed through diversion works as a result of WM totaled 6.8×10^9 m^3 (5.5×10^6 acre-ft) higher than flows generated with NWM. Diversion through Hoover Dam did not occur. All remaining flow, including that to Mexico, passed through the powerplants and generated electricity.

The volume of water stored at Lake Powell and Lake Mead was reduced during the 1991 to 2020 period with NWM. The volume in Lake Powell at the beginning of the period was 26.8×10^9 m^3 (21.7×10^6 acre-ft). This volume fluctuated, as shown in figure 3, until it reached a final volume of 24.2×10^9 m^3 (19.6×10^6 acre-ft). Lake Mead, on the other hand, experienced a more drastic reduction, as shown in figure 4, going from 27.1×10^9 m^3 (21.9×10^6 acre-ft) in 1991

Table III. Flow into Lake Powell (10^9 m^3)

Year	No Weather Modification	Weather Modification	Potential Increase
1991	11.95	13.36	1.41
1992	9.13	10.39	1.26
1993	12.85	14.80	1.95
1994	13.56	15.11	1.55
1995	13.41	15.74	2.33
1996	11.25	12.74	1.49
1997	10.30	11.63	1.33
1998	18.39	10.49	2.10
1999	8.94	10.14	1.20
2000	6.20	7.14	0.94
2001	7.38	7.88	0.50
2002	9.66	9.30	−0.36
2003	16.77	19.91	3.14
2004	13.26	16.30	3.04
2005	6.53	8.05	1.52
2006	8.74	10.13	1.39
2007	7.74	7.33	−0.41
2008	14.08	16.29	2.21
2009	7.83	8.54	0.71
2010	7.89	8.27	0.38
2011	15.19	17.85	2.66
2012	9.42	11.68	2.26
2013	8.99	9.27	0.28
2014	10.41	11.98	1.57
2015	11.80	13.85	2.05
2016	12.58	14.66	2.08
2017	11.82	14.61	2.79
2018	10.12	11.48	1.36
2019	16.20	18.34	2.14
2020	11.02	13.16	2.14
Average	11.11	12.68	1.57

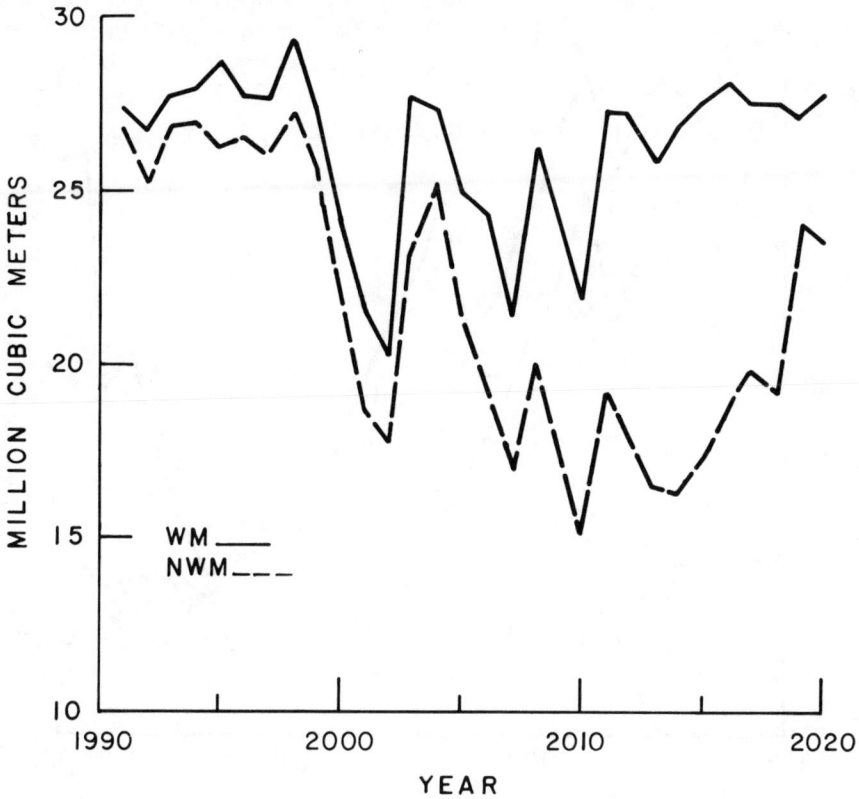

Figure 3. Lake Powell storage.

to 14.1 x 10^9 m^3 (11.4 x 10^6 acre-ft) in 2020. With the use
of weather modification, the reservoir situation changes significantly. Lake Powell, instead of decreasing by 2.6 x
10^9 m^3 (2.1 x 10^6 acre-ft), now increased by 0.3 x 10^9 m^3
(0.2 x 10^6 acre-ft). Lake Mead, instead of decreasing by 13.0
x 10^9 m^3 (10.5 x 10^6 acre-ft), decreased by only 1.2 x
10^9 m^3 (1.0 x 10^6 acre-ft). This resulted in a net increase
in storage at the end of the 30-yr period of 14.7 x 10^9 m^3
(11.9 x 10^6 acre-ft), and was directly attributable to weather
modification. In addition, minimum yearly flows at Lee's
Ferry were exceeded by releases of water at Glen Canyon Dam
during 19 years of WM, while with NMW, they were exceeded only
in 6 years.

Power Production

    The yearly hydroelectric power production at both Glen
Canyon Dam and Hoover Dam decreased significantly during the
period of the study with NWM; however, with increased flows

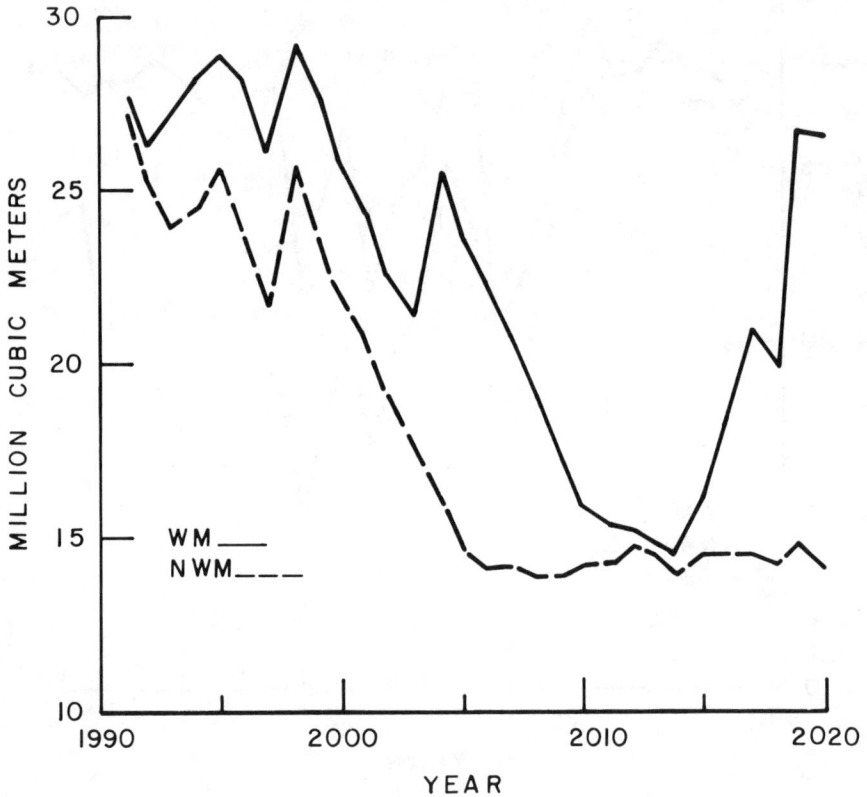

Figure 4. Lake Mead storage.

due to WM, production increased. The greater generating
capacity at Hoover Dam, which precluded the need for any non-
power releases, and the increased hydraulic head resulting
from WM, produced increases in power productions that were
greater than those at Glen Canyon Dam. The range of total
power generated (Table IV) with WM varied from a low of 9.9 x
10^9 kWh to 13.6 x 10^9 kWh with a yearly average of 11.8 x
10^9 kWh, or a mean yearly increase of 1.45 x 10^9 kWh.

ECONOMIC IMPACT

A dollar value for the increased streamflow resulting
from weather modification is difficult to determine. Water
in the Colorado River Basin is used and reused many times
before it flows into Mexico. The Colorado supplies water for
cities, industry, agriculture, energy development, and the
natural environment, each of them placing a different value
on water. Hydroelectric and recreational facilities are

Table IV. Average Annual Power Production (10^6 кWh)

Source	Hoover Dam	Glen Canyon Dam	Total Basin
NWM	3,709	3,837	10,366
WM	4,233	4,492	11,811
Difference	524	655	1,445

useful by-products of reservoirs along the River. Finally, the esthetic value of water cannot be measured in monetary terms.

Recognizing these difficulties, no attempt was made to assess all possible benefits and disadvantages of weather modification. Rather, dollar values were assigned to those areas which seemed to have the greatest impact. Specific areas addressed were: power production; salinity reduction; and additional value of consumptively used water, flow-through water, and stored water.

Revenues produced by increased power production were determined for Hoover Dam, Glen Canyon Dam, and the remainder of the Basin. Values were computed on coal replacement costs in thermal generating powerplants which were set at 20 mills per кWh. With an average annual increase in power production of 1.45×10^9 кWh, additional revenue of $29 million could be realized.

Salinity reductions in the Lower Basin appeared to pro-duce the highest economic return of any potential benefit. Computations were based on the dollar value of salinity reduc-tion (8) to agricultural, municipal, and industrial interests. This value was based on an average of $450,000 per mg/L at Imperial Dam; using the 91-mg/L average reduction in salinity resulted in a potential benefit of $40,950,000 annually.

The value of additional water depends upon its use. Consumptively used water in the Upper Basin was assigned a value of $12/$10^3$ m^3 ($15/acre-ft). A 41×10^6 m^3 (33 x 10^3 acre-ft) average annual reduction of consumptively used water would result in a value of $495,000. Additional water flowing through the system is used by municipalities, agri-culture, and industry. The average yearly flow below Hoover Dam, attributable to weather modification, was 0.78×10^9 m^3 (0.63×10^6 acre-ft). The majority of this water, an average of 0.5×10^9 m^3 (0.41×10^6 acre-ft), went to the CAP, while the MWD received an average of 0.13×10^9 m^3

$(0.11 \times 10^6$ acre-ft). At a value of $\$6.50/10^3$ m^3 ($\$8$/acre-ft) the additional water below Hoover Dam was worth $\$5,040,000$. Finally, water created by weather modification and stored in the reservoir system became available for use, and therefore had a value. Although increased storage takes place in upstream reservoirs, only Lake Mead storage was analyzed. During the 30-yr period, additional water stored averaged 0.49×10^9 m^3 (0.4×10^6 acre-ft) annually. This water would be continually available for municipal drought relief, irrigation, or substitution for other sources. A value of $\$9.75/10^3$ m^3 ($\$12$/acre-ft) was assigned to this water, resulting in a benefit of $\$4,764,000$.

CONCLUSIONS

Weather modification is a viable method for augmenting the flow of the Colorado River. Simulation, using historic meteorological and hydrological data as input to the CRSS model, indicates that streamflow can be increased by an average of over 14 percent, or 1.6×10^9 m^3 (1.3×10^6 acre-ft) over a 30-yr period. Furthermore, the value of this additional water totals approximately $\$80$ million annually for the following sources: $\$29$ million from increased hydroelectric power production, $\$40$ million from salinity reduction, $\$5.5$ million from water supplies to reduce deficits in Arizona and California, and $\$5$ million from water available for new uses.

REFERENCES

1. Elliott, R. D., J. F. Hannaford, and R. W. Shaffer, May 1973, Twelve Basin Investigation, Volumes I and II, North American Weather Consultants.

2. Elliott, R. D., R. W. Shaffer, A. Court, and J. F. Hannaford, October 1976, Colorado River Basin Pilot Project, Comprehensive Evaluation Report, Aerometric Research, Inc.

3. Glenwright, E. T., June 1976, Colorado River Simulation System Documentation, Part C, Bureau of Reclamation.

4. Kleinman, A. P., and F. B. Brown, December 1980, Colorado River Salinity, Bureau of Reclamation.

5. Lane, W. L., June 1973, Colorado River Simulation System Documentation, Part E, Bureau of Reclamation.

6. Lane, W. L, June 1976, Colorado River Simulation System Documentation, Part B, Bureau of Reclamation.

7. Main, R. B., August 1976, Colorado River Simulation
   System Documentation, Part A, Bureau of Reclamation.

8. Novak, E. M., July 1980, Application of the Colorado
   River Simulation System in Evaluation of Weather
   Modification Activity, Bureau of Reclamation.

CHAPTER 16

ENVIRONMENTAL AND SOCIAL
IMPLICATIONS OF CLOUD SEEDING
IN THE COLORADO RIVER BASIN

Edward R. Harris
        Division of Atmospheric
        Resources Research
    Bureau of Reclamation
    Denver, Colorado

Three years ago, the national Weather Modification
Advisory Board completed its congressionally-mandated study
of the state of scientific knowledge concerning weather
modification. The Board's report to the Secretary of
Commerce stated: "a usable technology for systematically
and extensively enhancing rain and snow by seeding certain
types of clouds is scientifically possible and within sight."
(1) However, this positive evaluation of technological
prospects was qualified by the Board's admonition that public
acceptance of precipitation augmentation will be tied directly
to the level of public confidence that environmental and
social results will be acceptable.

The Bureau of Reclamation has been assigned an instrumental role in the national research program to develop a
reliable precipitation augmentation technology. Reclamation
scientists agree that present-day understanding opens the
door to a reliable cloud seeding technology. Reclamation
also believes that research programs must seek the information that is needed to resolve environmental and social
issues. Careful study of the remaining technological,
environmental, and social questions is required to analyze
the risks and define the techniques by which benefits can be
maximized and risks minimized.

PERSPECTIVE

The requirement of the Colorado River Basin Project Act
of 1968 that the Secretary of the Interior find a means of
providing additional and adequate water supplies in the
Upper and Lower Colorado River Basin as well as satisfy the

requirements of the Mexican Water Treaty has resulted in study of cloud seeding as an augmentation tool. The Colorado River Basin Pilot Project investigated the seeding potential of winter orographic clouds in Colorado's Rocky Mountains from 1970 to 1975. Even though the project scientists were hampered by a lack of sophisticated equipment that would permit precise, daily forecasting of cloud types and conditions, the project did indicate that candidate storms were available that could be seeded for precipitation increases. Increased snowpack in the high-country "reservoir" watersheds of the Upper Colorado River Basin is extremely valuable as a greater percentage of water is exported from this Basin than from any other major river system in the United States. A great section of the western and south-western United States runs on Colorado River water.

Commissioner of Reclamation, Robert Broadbent discussed the augmentation potential of cloud seeding in recent congressional hearings. (2) He noted that use of one option will not foreclose other options for augmentation, and that cloud seeding appears to be one of the most promising of the new technologies.

We are approaching a decision point on the future of the Federal Government's precipitation augmentation activites. What future goals will be set? Can we expect a transition toward a greater reliance upon cloud seeding to augment Basin water supplies?

THE REGION

The arid-to-semi-arid climate of the region features notable variations in the precipitation caused primarily by atmospheric moisture supply and topography. Most of the precipitation is provided by Pacific air masses that move inland from the west. Canadian arctic air can occupy the northern portion of the region during winter months. Since the region is distant from major sources of moisture and the air masses cross numerous mountain ranges en route to the area, precipitation is sparse except in high mountain areas. Average annual precipitation varies from less than 152.4 mm (6 in.) in the lowest valleys to more than 1270 mm (50 in.) in the higher mountains. Average precipitation in the valleys and agricultural areas is from 254 to 508 mm (10 to 20 in.) per year. Snow accumulations can exceed 254 cm (100 in.) at the higher elevations and do not completely melt until late summer.

Approximately 60 percent of the Upper Basin land areas is under Federal administration, 15 percent is Indian Trust land and 25 percent of the area is State, corporate and private land. Grazing and crop production are two of the most important land uses. Forests cover about one-third of the area. Extensive outdoor recreation use is made of both public and private lands.

Historically, the Basin has been sparsely populated. However, the population began to increase significantly about 1965. Since the 1980 census, the increases have been attributed to the so-called Sunbelt phenomena. However, the most dramatic local population growth patterns are related to expanding energy development industries. Service industries benefitting the energy and recreation communities also account for a significant portion of the population increase.

COOPERATIVE TECHNOLOGY DEVELOPMENT

Since completion of the Colorado River Basin Pilot Project analysis, Congress has given the Bureau of Reclamation funds to continue planning and data collection in the Basin. Much of this money has been spent in cooperative undertakings with the Basin states. State governments and universities have invested cooperative funds in investigations of regional storm types, equipment development, and environmental and social studies. Cooperative agreements now exist between the Bureau and the states of Colorado and Utah to continue to examine the usefulness of cloud seeding as a water management tool. Reclamation also supports the study of winter storms in Colorado's Park Range by Colorado State University. Cooperative agreements with the United States Forest Service call for collection of meteorological data in high country forest areas and for the study and evaluation of avalanches. General meetings have been held at least once each year with representatives of state agencies, water management and user organizations, federal agencies, and environmental groups to review the status of cloud seeding activities and the prospects for continued development of the technology in the Colorado River Basin.

A majority of the western states have enacted legislation regulating weather modification activity. The Division of Atmospheric Resources Research has assisted state legislative committees that drafted new state laws. Although state requirements vary widely in complexity and degree of regulation, many states require weather modification companies to show scientific and engineering competence to obtain an operator's license. Generally, they also may require a

251

permit for the conduct of each field project as well as public hearings or publication prior to cloud seeding activity.

The Bureau of Reclamation has supported the idea that the state level is the appropriate regulatory level for non-Federal cloud seeding decisions. In a cooperative spirit, the Bureau's policy has been to insure that its research project and project contractors have followed the spirit of state law requirements.

THE STATE OF ENVIRONMENTAL AND SOCIAL ANALYSIS

A significant amount of knowledge about the environ-mental effects and social, or community, responses to increased precipitation, (or the idea of increasing precipi-tation) has accrued since the 1960s. One result of this ongoing evaluation has been a moderation of expectations about the effect of cloud seeding. The "Noah's Ark" syndrome has faded as the public more clearly understands the likely cloud seeding scenarios as well as the suspension criteria that are incorporated into operational plans because it is good science, and good sense, and because of state regulations.

Over the course of time, the public is beginning to understand that large storms and those that are naturally efficient will not be seeded; that, in fact, it would be counter-productive to seed them. Also, cloud seeding can be stopped at any time a desired amount of precipitation is reached during any season. The ability to select which storm will be seeded and which one will be permitted to pass by unseeded is one of the plusses of cloud seeding.

The Chief of the Division of Atmospheric Resources Research put the question of public understanding of weather modification into focus recently: "Weather modification suffers from a major dilemma which tends to hinder progress; claims for beneficial increases are treated with skepticism until "nine yards" of proof is produced, while the mere suggestion or suspicion that seeding is associated with dis-beneficial weather events is readily accepted without any call for proof. Weather modification scientists should be and are looking for the truth, both the positive and negative impacts of cloud seeding. The public should maintain a balanced and fair perspective by making the criteria for acceptance of both the positive and negative effects equal." (3)

It should be pointed out that the major portion of

environmental and social analysis has applied to the precipitation augmentation research mode. However, all environmental and social evaluations of the effects of the technology have emphasized the distinction between short-term research effects and the potential for larger-scale and unknown effects resulting from long-term precipitation management. There is a consensus that operational cloud seeding analysis must look at offsite downstream impacts. The eventual use of the water is the key factor in an array of questions which include: the effect on urban growth and its attendant problems, potential for energy production, effects on farming and irrigation practices, effect on water quality in the Basin, and similar effects in the adjoining carryover Missouri, Arkansas and Rio Grande Basins.

The key scientific questions to be resolved in order to prepare an environmental evaluation of an operational program are: how much additional snowpack and streamflow will be produced, in which watersheds, and in what time frame?

The key policy questions to be resolved for the most efficient preparation of an environmental evaluation of an operational program are: where will the water be used, and how?

The Weather Modification Advisory Board stated: "Experience to date indicates a clear need for information and education programs if weather modification is to be accepted and used effectively. These, again, are legitimate costs to be weighed against benefits of the particular weather-modification technology concerned." (1, p. 131)

The public may view the entry of the Federal government into an operational precipitation management program as an action or major and significant environmental and social impact. The program could be controversial in some areas. The extent and degree of the program's controversy will relate directly to, and diminish in proportion to, the effectiveness of the project's public involvement and education efforts.

Experience in the Bureau's atmospheric research programs demonstrates that public access to and participation in project environmental and societal assessments as well as access to the decisionmaking process is important in shaping public confidence in cloud seeding as a water resource management option.

LEGAL ISSUES

The closer we come to a point of transition from
research to operational mode precipitation augmentation by
the Federal government, the greater the public interest in
resolution of intertwined legal and policy issues.

These questions divide into four categories: water
rights, general liability concerns, social costs, and
regulation. Although it is not possible to forecast how
quickly and in which order these issues will mature, we
believe the social costs issue will have to be addressed in
the immediate to mid-term in order to secure and maintain
public acceptance of operational technology in the project
watersheds. Resolution of the other issues, while no less
important, appears to be of less immediate concern.

A reliable evaluation of how much additional snow and
water can be produced by cloud seeding, its timing, and
locations are the critical factors in the maturation of these
issues.

Social Costs

Under a Federal operational precipitation management
program, a definite public demand will develop for a com-
pensation/reimbursement system for people and communities
in the project areas to cover actual damages from increased
precipitation. People in the affected areas and environmental oversight groups will perceive a Federal operational
program as a major and significant action impacting the
environment and their lives. A widespread perception is
that the people in the project areas will incur the incon-
veniences and costs of increased snowpack, while the downstream areas will enjoy the "real" benefits.

Compensation claims can be expected in areas such as:
increased costs of snow removal, structural damage due to
increased snow loading, livestock losses, loss of the use of
grazing areas for indefinite periods in the late autumn and
early spring, "haystack losses" from changes in big game
feeding patterns, mitigation needs to protect wildlife from
loss of winter range or more severe winters, and delayed
access to mining sites. The development of a compensation
system is not only a legal-policy question; it will be a
primary factor in community acceptance of the technology.
State and local governments are sensitive to complaints
about disadvantages that may be caused by increased snowpack.

254

Water Rights

Under the research mode, the Bureau of Reclamation and the Department of the Interior position has been that any additional water produced by cloud seeding is "incidental" to the learning process, and that this incremental amount of water simply enters the system for free distribution among users according to existing water rights.

Many Western states have enacted weather modification laws which include a statutory claim of rights to all water produced in a state through weather modification technology. If the Federal government undertakes an operational precipitation management program to produce water to meet needs involving the Federal government, the question of a potential Federal water right claim to the new water will concern the states and other water rights owners. An eventual Federal policy decision regarding a legal claim by the Federal government will be desired by state and local interests as they develop their own policy and attitudes about a Federal program.

A great deal of discussion and study has been, and will continue to be, devoted to the question of who can, or should, capture the economic value inherent in the "new" or "incremental" water. It appears the larger the mutuality of interest that can be identified, the easier the road to agreement will be. It appears that the largest community of interest that could be realized through cloud seeding would be accomplished by declaring cloud seeding increases to be an augmentation to the river, and therefore, subject to existing water doctrines of the western states.

General Liability Concerns

The general rules of liability for harm resulting from a person's or entity's actions or negligence have always applied to weather modification activity. To date, no legal action concerning weather modification activities has resulted in a favorable decision for the plaintiff due to the difficulty in establishing the necessary cause and effect relationship. As operational activity increases and scientific certainty of effect improves, this situation will change. The Weather Modification Advisory Board Report, in discussing liability under expanded operational scenarios, examined the theory of absolute liability as well as liability for fault. The report tended to favor the absolute or "no-fault" liability theory. The interest in more definitive approaches to liability questions at the Federal level

will increase with the advent of a Federal operational
program.

Regulation

As stated previously, there are no Federal regulations
governing operational cloud seeding activities except for
the requirement of Public Law 92-205 that weather modification activities in the United States and its territories
shall be reported to the Secretary of Commerce. Many state
statutes govern precipitation management activities and, even
though these do not apply to Federal activity, Reclamation
research programs have sought to comply with the spirit of
state laws as a matter of intergovernmental cooperation and
good will. The weight and urgency of public demand for a
Federal regulatory system might increase as we move toward
a Federal operational program.

PUBLIC ACCEPTANCE

Case studies of public controversies relating to weather
modification reveal several commonalities associated with the
emergence of opposition. A technology study of hail suppression (4) noted the following factors: local heterogeneity
of weather needs, occurrence of drought periods during cloud
seeding efforts, lack of scientific consensus about the
readiness of the technology for operational application,and
general lack of public participation in decisions to adopt
hail suppression. The same issues are found in relation to
cloud seeding to enhance precipitation.

Over the longer term, social acceptance of the precipitation augmentation component of weather modification will be
tied directly to the scientific certainty about the results
of seeding the storms of a particular region. The greater
the uncertainty about target area and downwind effects due
to seeding, the greater the perception of risk by the
community. In line with its perception that some degree of
risk is connected with cloud seeding, the public has taken
an active interest in the decisionmaking process. Public
hearings held as part of the state regulatory process have
resulted in altered project designs by changing project
starting dates (delays to avoid possible conflict with big-
game hunting seasons), changing the project area (locating
project generators to avoid seeding storms crossing certain
mountain ranges), and changing project ending dates (sus-
pending seeding activity when the snowpack reaches a certain
percentage above the "average" or when unexpected wildlife
feeding problems arise).

Expanded dialog between atmospheric scientists and citizens of project areas is necessary. Whatever degree of risk is found in cloud seeding ultimately, it is understandable that the public should wish to take its own risks rather than to have such decisions made for them. Consequently, scientists and project managers at the Bureau of Reclamation believe the degree of public understanding of the technology and the degree of public participation in the decisionmaking surrounding the project will influence its acceptability in the community.

The optional aspect of cloud seeding as a water resource management tool should be emphasized. The decision to seed is reviewable each year, depending upon moisture and water supply conditions. The decision can change if water management objectives change. In addition, the treatment strategy may be changed to accommodate each year's needs. The possible mix of seed and no-seed options available over an extended period of time underscores the necessity for confidence in the reliability of the technology. Certainly, the matter of predicting or discerning impacts upon the environment or society that may result from cloud seeding is complicated by the flexibility of its potential use. An additional complicating factor is the difficulty in separating responses due to varying amounts of precipitation from other human or natural environmental influences. Other management practices, cultural preferences, and inadvertent weather modification may intensify or, perhaps, counteract the effect of cloud seeding.

As Cooper and Jolly stated, it cannot be expected to be easy to ascertain ecological change due to deliberate weather modification in the short run. Ecological studies show that plant and animal communities respond in the long run to mean climatic conditions. The length of time required for such adjustments may depend upon the variability of the premodification precipitation regime.

"Since weather modification will be a pervasive, not a localized phenomenon, it will be linked in its effects to a variety of other environmental processes. For this reason, it is almost impossible to separate research needed to predict ecological effects of weather modification from other aspects of environmental research. This creates a problem for mission-oriented federal agencies accustomed to sponsoring investigations aimed at providing answers to specific problems in the shortest feasible time." (5)

If a new Federal cloud seeding effort is authorized in the Colorado River Basin, we intend to begin work on its

257

environmental and societal aspects with this caveat in mind. A careful and long-term analysis and monitoring effort will be needed to provide the answers.

REFERENCES

1. Weather Modification Advisory Board, 1978, "The Management of Weather Resources, Volume I, Proposals for a National Policy and Program," Report to the Secretary of Commerce, Department of Commerce, Washington, D.C., p. 30.

2. Broadbent, Robert N., March 25, 1981, Statement by the Commissioner of Reclamation before the House Sub-committee on Water and Power Resources Hearings on the Augmentation of Colorado River Water Supplies.

3. Silverman, Bernard A., October 9, 1981, Personal Communication with Dr. John Koch, Miles City Community College, Miles City, Montana.

4. Farhar, Barbara C. et al., 1977, "Hail Suppression and Society," Summary of Technology Assessment of Hail Suppression, Illinois State Water Survey, Urbana, Illinois.

5. Cooper, Charles F. and William C. Jolly, 1969, "Ecological EFfects of Weather Modification: A Problem Analysis," The University of Michigan School of Natural Resources, Department of Resource Planning and Conservation, Ann Arbor, p. 125.

PART 5

OIL SHALE DEVELOPMENT

CHAPTER 17

AN ECOSYSTEM APPROACH TO ENVIRONMENTAL MANAGEMENT

John Carter and Vincent Lamarra
    Co-Directors
    Ecosystems Research Institute
    Logan, Utah 84321

ABSTRACT

The central problem in the preparation of the Detailed Development Plan for the White River Shale Project, was the preparation of a long- term environmental management/monitoring program. The program had to satisfy regulations of the various governing agencies, and be flexible and economical.

The major difficulty encountered was analyzing and interpreting six years of environmental data in a manner which would allow the management of the White River Shale Oil Corporation to make logical and correct decisions about program needs and which would reflect environmental processes occurring in the region of development. This difficulty was overcome by use of a conceptual ecosystem model as the organizing principle for the analysis. Use of the model has allowed the establishment of statistically valid quantitative relationships and pointed out areas of insufficient data or lack of coordination. The 1981-82 program was designed to complete the important ecosystem relationships which will be used to guide the long-term monitoring program and aid in evaluation of project impacts and mitigation.

INTRODUCTION

The federal Oil Shale Leasing Program was created in response to President Nixon's 1971 request to the Department of the Interior for a plan to develop the extensive oil shale resources in the United States. Six tracts of land were selected for lease in the prototype program: two each in Colorado, Utah, and Wyoming. The tracts are in the Green River Formation, where oil yield

may average 30 gallons per ton of shale (Figure 1).
Competitive bidding on the tracts took place from January
to June of 1974. Bids were received for the Colorado and
Utah tracts only. Both of the Utah leases became
effective on 1 June 1974.

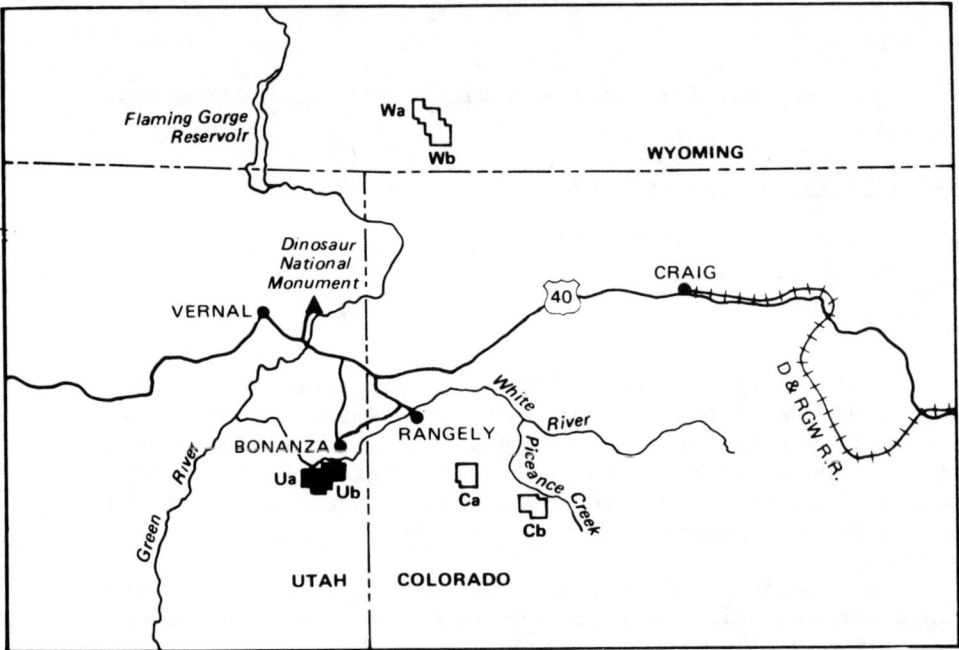

Figure 1. Location of Federal Prototype oil shale tracts.

During the baseline data collection period,
1974-1976, an extensive data base for air, climate, soils,
biology, and water resources was developed. Effort was
reduced substantially after this period while the many
legal problems surrounding the project were addressed.
The interim program, 1977-1980, concentrated on
terrestrial biology (plant and animal data) with less
effort in the other areas. Because oil generated from oil
shale in now projected to be economically feasible,
efforts are again underway to begin development with shale
oil production scheduled for 1988. The environmental
program is now designed to fill gaps in the data base
while concentrating on ecosystem relationships which can
be used to evaluate impacts of development and aid in
determining necessary mitigation measures.

The central problem for the White River Shale Oil
Corporation environmental managers is to ensure the

environmental integrity of the developed areas and to develop technology for environmental protection. Given the complexity and variability of the environment, how could managers make logical decisions from such a highly variable data base? We felt that organizing the data base, its analysis, and interpretation using an ecosystem framework would provide tools which were usable and understandable.

The following environmental management program is being developed for use on the federal oil shale tracts Ua and Ub. These tracts are leased by the White River Shale Oil Corporation. The reclamation goal is to restore the tracts to the pre-mining condition. The management/monitoring program is designed to identify ecosystem components most affected by oil shale mining and to restore and guide reclamation by tracking ecosystem changes afterward. We discuss here the application of this program to management of the aquatic resources in the White River basin. The same approach is being integrated into the entire environmental program for the Ua and Ub tracts.

FACTORS IMPORTANT TO THE DESIGN OF THE ENVIRONMENTAL MANAGEMENT SYSTEM

In addition to the need for a system that leads to logical, correct, data based decisions, several other factors were important in the design of the environmental management program for White River Shale Oil Corporation. Among these were regulatory requirements for monitoring and expected impacts due to the industry.

The Federal Oil Shale Supervisor's Office (OSO) indicated that a primary goal of the Federal Prototype Oil Shale Program is, To insure the environmental integrity of the affected areas and at the same time develop a full range of environmental safeguards and restoration techniques that will be incorporated into the planning of a mature oil shale industry should one develop. The Oil Shale Lease Environmental Stipulations require the lessees to conduct a monitoring program before, during, and after development operations. To achieve these goals, OSO published a set of monitoring guidelines to be used in developing monitoring programs (OSO, 1979). Included in these guidelines are the following:

1. Describe existing environmental conditions.

2. Identify candidate potential parameters to monitor during the initial development phase.

3. Select potential parameters to monitor based on likelihood of impact, degree of impact, importance, legal requirements, measurability, interpretability, and cost effectiveness.

4. Design statistical procedures for detecting and evaluating degree of impact.

5. Develop a quality assurance program.

6. Build into a computer program appropriate threshold values for specific parameters.

7. Design a contingency plan.

Oil shale development may have many direct and indirect effects upon the organisms and physical environment of tracts Ua and Ub. It is necessary to consider these potential effects of development so that monitoring parameters can be selected based upon detection and evaluation of these impacts. A brief summary of development related activities which can result in environmental change follows:

1. Surface disturbance by site preparation, road and corridor construction

2. Noise and high levels of activity

3. Air emissions and dust

4. Leaching of compounds into surface and subsurface water

5. Accidental spills

6. Habitat enhancement/reclamation

Other factors important in the design of the program are timeliness of information and cost. For these reasons the determination of ecosystem parameters that have the greatest importance and utility for long-term monitoring and assessment of effects is given emphasis. In order to accomplish the required ends, data collection must incorporate flexibility. It is important that data can be used to evaluate more than one impact and will be useful in the design, implementation, and evaluation of mitigation and reclamation measures.

With these goals in mind, it is necessary to use a conceptual view of the ecosystem which can identify the dominant static and dynamic variables (structure, function, and rate process) which can depict ecosystem "health" and identify departures from normality. These departures can then be used in contingency planning as the trigger for management decision making guided by a logic flow model.

CONCEPTUAL VIEW OF THE AQUATIC ECOSYSTEM

The ecosystem view recognizes that the biological components of the system exist within, interact with and are constrained by the physical environment (Figure 2).

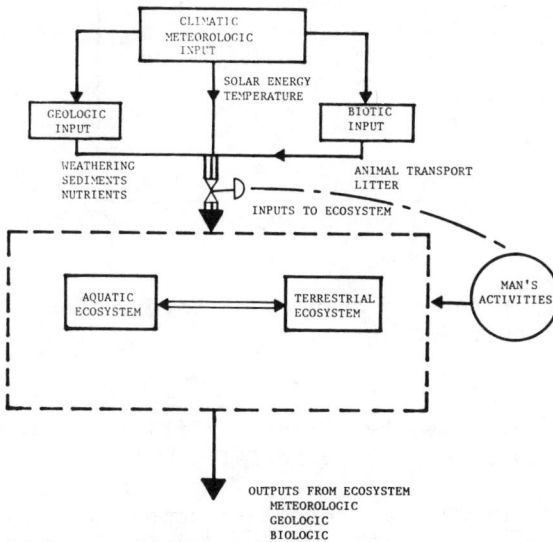

Figure 2. Conceptual view of ecosystems relative to external influences.

In addition, this view recognizes that the activities of man may have a large influence on these ecosystem components changing inputs and outputs across system boundaries and the structure and functioning of the biological components within individual ecosystems (Figure 3).

The primary producers (plants and algae) use solar energy and input materials to generate carbohydrates which are used for growth and reproduction (increase in biomass, or primary production). Production then has two basic

265

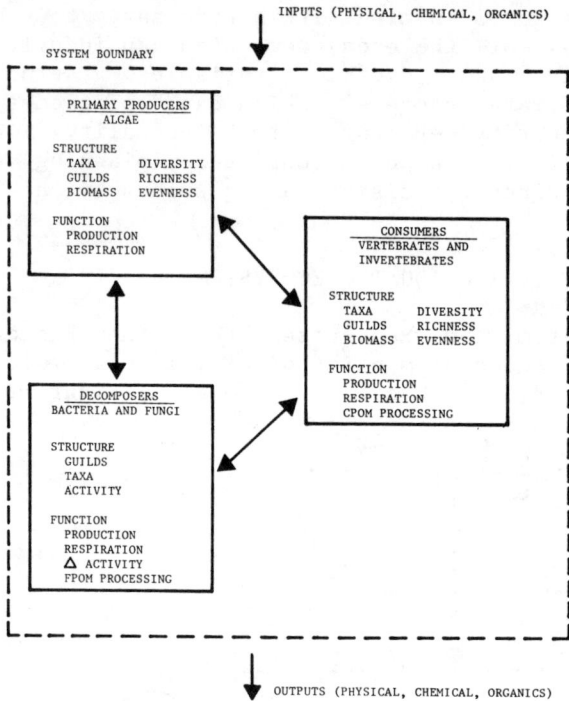

INPUTS (PHYSICAL, CHEMICAL, ORGANICS)

SYSTEM BOUNDARY

PRIMARY PRODUCERS
ALGAE

STRUCTURE
TAXA   DIVERSITY
GUILDS   RICHNESS
BIOMASS   EVENNESS

FUNCTION
PRODUCTION
RESPIRATION

CONSUMERS
VERTEBRATES AND
INVERTEBRATES

STRUCTURE
TAXA   DIVERSITY
GUILDS   RICHNESS
BIOMASS   EVENNESS

FUNCTION
PRODUCTION
RESPIRATION
CPOM PROCESSING

DECOMPOSERS
BACTERIA AND FUNGI

STRUCTURE
GUILDS
TAXA
ACTIVITY

FUNCTION
PRODUCTION
RESPIRATION
Δ ACTIVITY
FPOM PROCESSING

OUTPUTS (PHYSICAL, CHEMICAL, ORGANICS)

Figure 3. Major components and interactions
of the aquatic ecosystem.

fates: (1) it is consumed by animals or (2) it enters the
decomposer cycle as dead plant material.

Within the animal or consumer component, plant
material is used for growth and reproduction. Growth is
termed secondary or tertiary production depending upon
whether the animal is an herbivore (secondary producer) or
a carnivore (tertiary producer). The animals ultimately
die and enter the decomposer cycle.

Within the decomposer component all dead plants and
animals are ultimately processed by invertebrates, fungi
and bacteria. These organisms through mechanical or
metabolic (chemical) activities break the large organic
molecules into smaller inorganic and organic molecules
which can be used by the primary producers.

Within each component there are structural and
functional parameters which affect the way energy flows
through the ecosystem. These parameters are defined at a
point in time and therefore are instantaneous values, that
is, they are mere snapshots of a continually changing

266

milieu. Examples of these parameters are species or guild abundance, richness and diversity.

The functional, or dynamic, aspects of the system can be described by such parameters as production, respiration, and nutrient cycling (decomposition pathway). Interactions between system components can best be described in terms of processes or rates which reflect energy flow and the interaction within the biotic components as well as with the driving environmental variables. Thus, processes as well as structural factors must be considered as quantitative descriptors of the system.

Physical or environmental variables are seldom constant, but may oscillate over time (Figure 4). Biological components of the system exist within the constraints of physical factors and respond with oscillations in populations, production, and other processes (Figure 5). Irregardless of these system behavioral changes, the trajectory of this behavior can be predicted, and one can identify a stability zone within which the variance in system behavior is expected to lie. This behavior can be used as a quantitative descriptor of the ecosystem.

EXTERNAL PHYSICAL FACTORS

There are four major sources or external inputs into the White River in the study area: (a) direct atmospheric contribution, (b) surface flow from the adjacent watershed, (c) upstream contribution, and (d) accrual (springs). The impact of these external sources of water, materials, and energy is both direct and indirect. Various substances enter the White River system through the air-water interface (Table 1).

The introduction of these products from the atmosphere as well as from the watershed can provide the driving forces upon which this river system operates. For example, during a series of storms in March and April, 1979, 1.02 inches of precipitation fell. As a result, flows in the White River increased from 1,000 cfs to 1,800 cfs and total suspended solids increased from 200 mg/l to 4,500 mg/l, thus decreasing light penetration and decreasing net photosynthesis by the benthic algae. In general, one of the dominant overriding, external factors regulating the river is its arid climate and seasonal

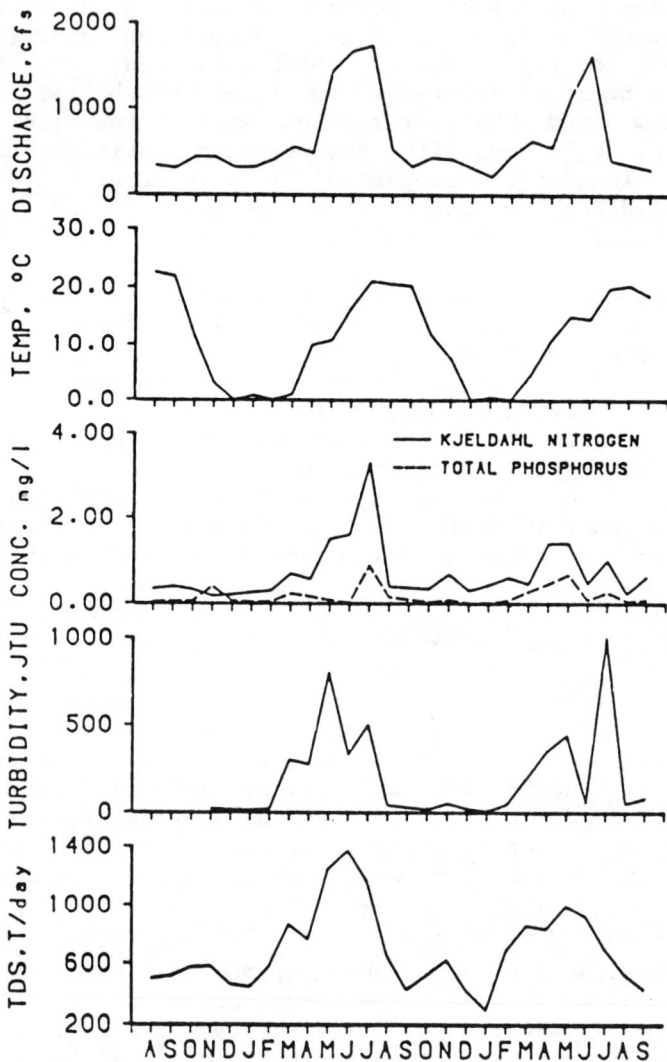

Figure 4. Variation in physical and chemical conditions in the White River during a two-year baseline period. Final Environmental Baseline Report, White River Shale Project, 1976.

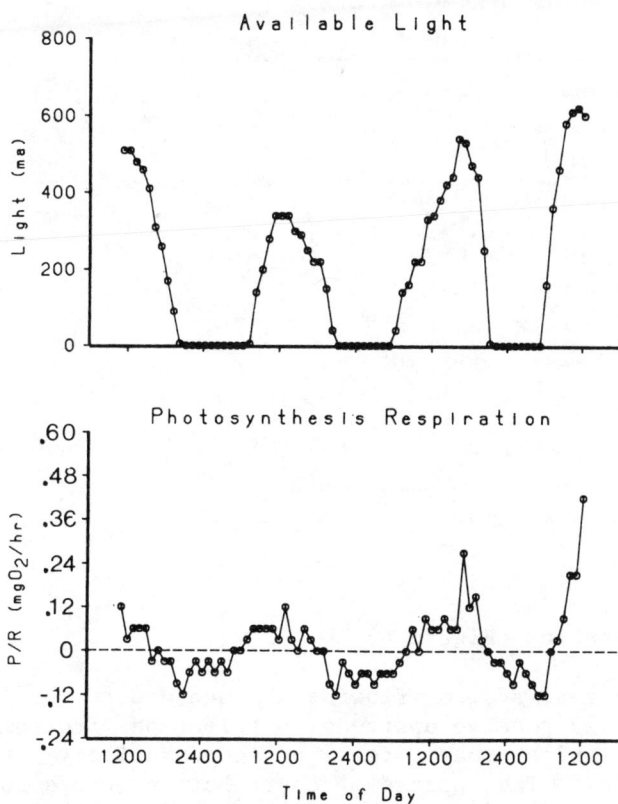

Figure 5. Light and production-respiration (P/R) data for a three-day period June 29 – July 2, 1981 in the White River, Utah. Data are continuously recorded by remote sensing from sealed, continuously recirculating P/R chambers.

Table 1. The major external physical factors affecting the White River System.

_____

A. Direct atmospheric contribution

    (1) Precipitation (dissolved, suspended substances).
    (2) Dry fall (nitrogen, phosphorus, metals, etc.).
    (3) Light (solar radiation, photoperiod, intensity).
    (4) Heat (temperature).

B. Surface flow from adjacent watersheds

    (1) Water.
    (2) Dissolved constituents (nutrients, metals, etc.).
    (3) Suspended constituents.

        (a) Inorganic (sediment, silt, nitrogen, etc.).
        (b) Organic (detritus, insects, plants, etc.).

C. Upstream contribution

    (1) Water.
    (2) Dissolved constituents (nutrients, metals, etc.).
    (3) Suspended constituents.

        (a) Inorganic (sediment, silt, nitrogen, etc.).
        (b) Organic (detritus, insects, plants, etc.).

    (4) Heat (temperature).

D. Evaporation - crystallization process

_____

discharge patterns (Figure 4).

Major stream systems flowing through desert or arid regions usually receive upstream contributions from higher elevations which may receive comparatively high precipitation. The upstream contribution occurs during three periods. The first is spring melting of lowland snow packs (lower basin runoff). Lower basin runoff represents only a small portion of the spring runoff, but the nutrient content of this source water is high. This factor, in combination with the low quantity of water, results in increasing nitrogen and phosphorous levels in the White River at a time when temperatures are increasing (Figures 4 and 6). The second major water input into the White River is the snow melt from high elevations (upper basin runoff). The third is baseflow.

The importance of discharge (seasonal or storm events) on the biological components are numerous and important because of the potential for large biological changes brought about by rapid changes in the physical and chemical environment (habitat).

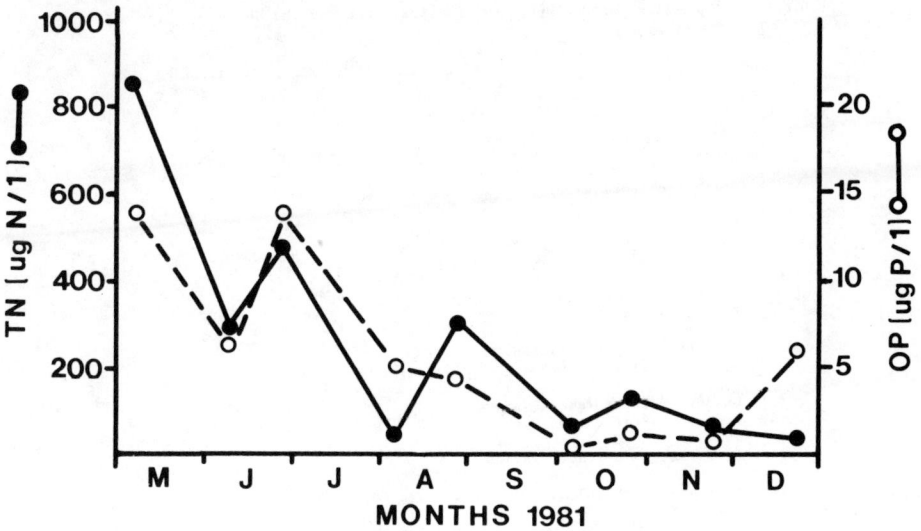

Figure 6. The concentration of orthophosphate and total inorganic nitrogen in the White River during 1981. The data represents 5-day averages (monthly) at Southam Canyon.

INTERNAL PHYSICAL FACTORS

For over 500 years, earth scientists have viewed rivers as a link in the hydrologic cycle. These systems were the logical center of erosion, transportation, and deposition of dissolved, suspended, and tactively carried geologic materials. The concept of rivers as dynamic systems combined with the basic principles of fluid dynamics have formed the nucleus of modern fluvial geomorphology. The description of the White River in terms of internal physical factors must include consideration of channel geomorphology and fluvial (water) characteristics (Table II).

Channel Geomorphology

Leopold and Maddock (1953) showed empirically that water depth, width, and velocity were functions of the load transported by a river and that one could thus predict the effects of changes in load supplied by side streams upon the entire geometry of the main channel. Structural changes caused by external factors are important considerations. For example, Wolman (1964) noted that a local increase in sediment load imposed by a tributary stream upon a river immediately began to alter the width to depth ratio of the river channel and siltation occurred on the river bed. In both cases, the

271

Table 11.   The major internal physical factors affecting the White
             River System.

A.   Channel geomorphology

     (1) Elevational change (slope).
     (2) Cross sectional depth distributions.
     (3) Cross sectional water velocity distributions.

          (a) Current.
          (b) Turbulence.

     (4) Substrate.

          (a) Size.
          (b) Texture.
          (c) Chemical composition.
          (d) Stability.

     (5) Channel meander rate.
     (6) Pool to riffle ratio.

B.   Fluvitile (water) characteristics

     (1) Heat (temperature)
     (2) Chemistry.

          (a) Dissolved gases.
          (b) Dissolved substances.
          (c) pH.

     (3) Turbidity and silt load.

---

ambient physical system was predictably modified, with concurrent effects on the biological community.

Changes in current and turbulence can shift and restabilize the bed in the White River. This dynamic process can produce molar action that may physically damage animals and modify habitat types (pools, runs, riffles). Periods of high discharge or storm events will produce greater habitat modifications than normal, baseline water level currents. It should be noted that in sections of the White River where shifting stream bottoms are encountered, the configuration of canyon walls or large boulders may be important in creating permanency in deflecting currents in a predictable manner.

Substrate stability is dependent on current velocity (Bull, 1981). Biologically, substrate stability relates to the organisms which use the benthic region for attachment (periphyton, macroinvertebrates). The current-determined substrate size also determines the size of organic particles (food for higher organisms) caught within pore spaces (Farnworth, 1979). The substrate type (size, texture, and chemistry) may at times influence the distribution of benthic algae and invertebrates in streams (Whitton, 1975). Large, stable, rocky substrates have

272

thus far been observed to have a higher standing crop of periphyton and macroinvertebrates than small substrates, such as gravel and sand (Figure 7).

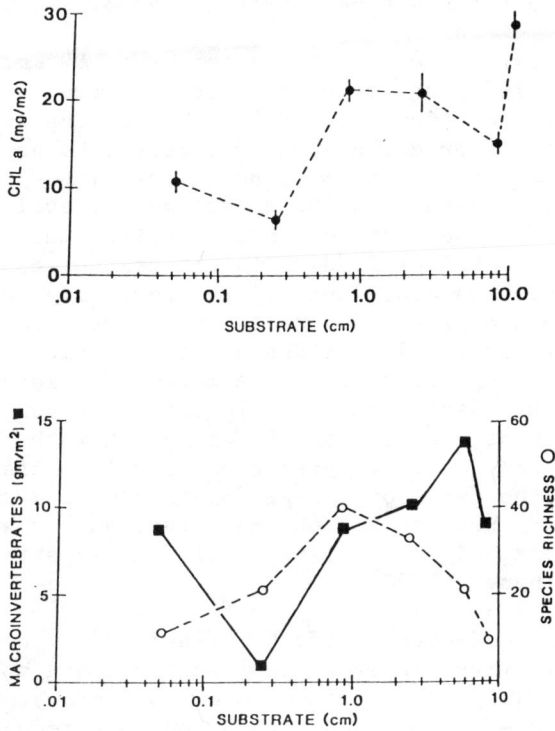

Figure 7. The distribution of biological parameters by substrate sizes in the White River, Utah. Bars denote 95% confidence intervals.

Even among the rocky substrates, large differences have been observed in algal standing crop between rock types, probably due to textural and mineral differences. Thus, the distribution of habitat types (pools and riffles) which result from the dissipation of energy within the study sections should be mapped and the size, texture and chemistry of substrate types determined.

The response of primary producers is often stochastic, depending upon random changes in flow and, hence, sediment movement. Overall, increased levels of suspended sediment reduce light intensity and limit photosynthesis; however, the presence of associated nutrients may ameliorate this effect to some extent. Species composition of primary producers also is influenced by suspended and deposited sediment. Changes in species and biomass are related to changes in suspended solids concentrations, rate of deposition, bed-load movement, and substrate type (Whitton, 1975).

Sediment influences macroinvertebrates by modifying habitats and inducing movement out of areas with high rates of deposition (Cummins, 1975). These high rates of deposition often reduce the number of benthic organisms by burial of sessile organisms and alteration of the substrate. Increased suspended sediment levels often result in an increased rate of insect drift out of the affected area. Filter-feeding invertebrates, especially certain molluscs and crustaceans, are adversely affected by high levels of suspended solids over extended periods of time. There is little evidence for sediment-induced mortality for groups other than the molluscs. Recovery by mobile invertebrates is rapid, following the resumption of normal conditions. Recovery depends upon how the habitat was modified, the substrate preference and life history of the organism, the types of refugia available, and the mode of migration. Recolonization may involve downstream drift, upstream migration from within the substrate, and reproduction (Hynes, 1970).

Suspended sediments affect fish by increasing mortality, altering rates of reproduction, and modifying growth rates. Suspended sediment does not appear to be lethal for juvenile and adult fish, but may reduce their resistance to disease and also damage gill tissues. Fish may avoid localized areas of increased sediment concentrations. The deposition of clay to sand-size particles adversely affects reproduction in some groups, particularly the salmonids. This results from reducing the flow of water and, hence, the renewal of oxygen to deposited eggs. Habitat modification, particularly the loss of cover for juvenile fish, and removal of suitable spawning sites may reduce certain fish populations. The endemic fishes found in the White River have special, unique adaptations to overcome the natural turbidity of this ecosystem.

The amount of solar energy entering the White River ecosystem is important when attempting to predict the effect of changes in other physical characteristics. The light climate of each major habitat type may explain the benthic community responses to changes in flow, silt load, temperature, and chemical alterations. It is evident that net production and respiration closely follow available

274

light at the substrate (Figure 5). Such results are easily quantified.

Because most aquatic organisms are poikilothermic, temperature is an essential parameter to be measured. Wlosinski (1979) indicated in the validation of his Desert Stream Ecosystem model (DSEM) that temperature was the most important physical parameter affecting the community metabolism. Experiments conducted in the White River indicate that temperature is highly significant in regulating decomposition (Figure 8).

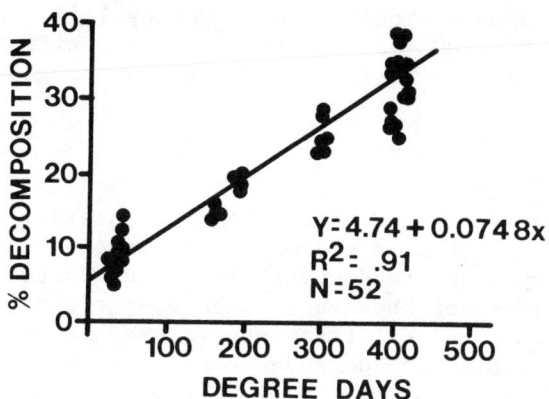

Figure 8. Leaf decomposition rates as percent weight loss. Studies conducted in the White River (October, 1981–April, 1982).

Fluvial Characteristics

The second major set of characteristics needed to define the abiotic structure of the White River is the fluvial characteristics (Table II). The chemical load of the river is derived from atmospheric precipitation, from mineral matter dissolved by the surface and ground waters, and from atmospheric gases (previously described in Table I, as external inputs).

Gibbs (1970) has distinguished three basic origins for dissolved substances in surface waters. These are atmospheric precipitation, rock weathering, and evaporation-crystallization processes. The processes affect the relative concentration of the major macro-elements found in freshwater ($Na+$, $K+$, $Ca+^2$, $Mg+^2$,

275

SO_4^{-2}, Cl^-, HCO_3^-, CO_3^{-2}). Because of the sources and mechanisms of weathering, the air-water-terrestrial interface is critical in the movement and translocation of chemical constituents into the White River. For example, a decrease in the pH of rainfall to 5 or 6 from a normal 7 to 8 can result in soil nutrient leaching. At lower pH values rainfall can leach 10-15 thousand years of accumulated exchangeable soil nutrients in periods as short as 30 to 100 years. These soil nutrients, dominantly monovalent and divalent cations, could move into the river system, increasing the concentration of dissolved solids and the potentially growth limiting nutrients, nitrogen and phosphorus. During 1981, diurnal variations of 200-300% in orthophosphate and nitrate were documented for the White River during a single storm event.

BIOTIC FACTORS

The purpose of the following discussion is to describe dominant biotic processes of stream communities which may be important to document for the White River. The present state of knowledge concerning the biological structure and function of stream ecosystems is based on four factors: (1) the dependence of consumer organisms for a large portion of their energy supply on allochthonous organic material from the terrestrial system watershed; (2) the utilization of organic input during the fall-winter period of lowest temperatures (prior to spring runoff); (3) the dominance of some stream segments by primary producers and (4) the relationship between the allochthonous and autochthonous organic inputs to the system.

One technique for characterizing the White River ecosystem is to determine the input, processing, and export rates of particulate detritus. A brief summary of the processing that occurs for "particulate organic material" (POM) is as follows. After a piece of coarse particulate organic material (CPOM) enters the stream (twigs, branches, leaves, etc.), two processes occur rapidly (Cummins, 1973). First, the soluble fractions are leached from the coarse material. This leaching occurs within the first 24 hours and can account for a 5-30% weight loss. The second process is the colonization of the coarse particles by microorganisms. The major groups are: cells and spores of bacteria, aquatic hyphomycete fungi, and protozoans. The large colonized particles are reduced to smaller sizes (FPOM - fine particulate organic matter) by abrasion (current and turbulence), animal feeding, and microbial assimilation. The

interrelationships between the organisms which process the
coarse particles (called shredders) and the microbial
flora which colonize the particles have been demonstrated
(Cummins, 1973). This processing rate (by structural
components) is a quantifiable behavioral characteristic of
any river system which is dependent upon temperature,
turbulence, pH, water hardness and ionic constituents
(Lush and Hynes, 1973).

The organisms which feed on the FPOM produced are
called collectors. They aggregate the small particles by
their feeding activities. A preliminary analysis of the
macroinvertebrate data collected on the White River
indicates that the dominant types of invertebrates are
collectors dependent upon the heterotrophic food chain
(Figure 9).

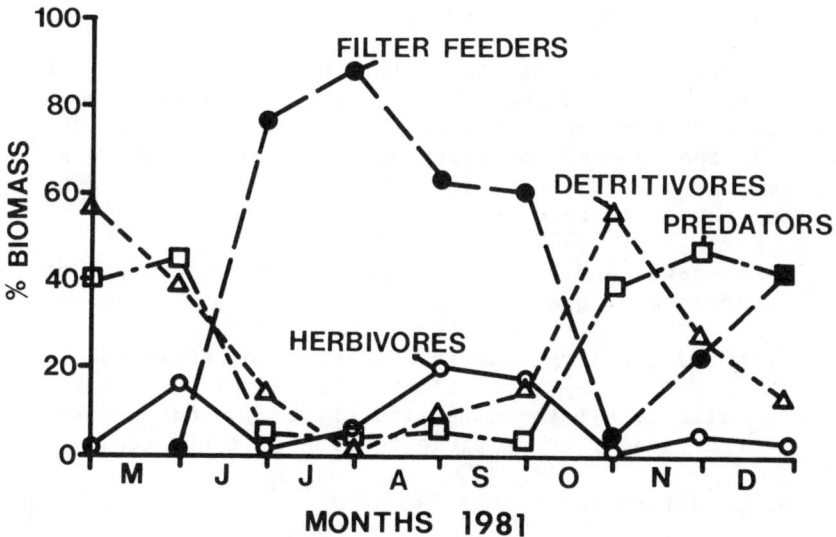

Figure 9. The distribution (% composition) of the major
functional groups of macroinvertebrates in a riffle habitat
at Southam Canyon in the White River during 1981.

Respiration rates and absence of invertebrates from
decomposing leafpaks indicate microorganisms are very
important in the regeneration of this allochthonous
detritus for nutrients (Figure 8).

The impact of physical-chemical factors on the
distribution of macroinvertebrates is usually indirect,
with the major controlling factor being the quantity or
quality of food. In general, invertebrates are adapted to

the particular environmental conditions present (such as stream temperature, velocity, oxygen concentration, light intensity, water hardness or alkalinity). It should also be noted that within a specific stream reach, a habitat type (pool or riffle) will also affect the distribution of macroinvertebrates. Hynes (1970) has extensively documented the truly unique or specialized macroinvertebrate adaptations found in riffle or pool systems. Distributional patterns, within stream systems of the same size, in terms of occurrence and density, are influenced by water movement and the nature of the surface available for colonization (substrate particle size). It has been previously discussed how velocity and turbulence limit the ranges of particle size, and thus macroinvertebrate distribution. Data as to the role of the above factors on macroinvertebrate distributions and abundances in the White River are being collected in the ongoing 1981 environmental program.

Undoubtly, food is the ultimate determinant of macroinvertebrate distribution and abundance in non-perturbed running waters. Cummins (1973) has described the type of community based on food habits. An analysis of the White River macroinvertebrate data indicates the presence of the following groups:

1. grazers and scrapers-herbivores feeding on attached algae.

2. shredders-large particle feeding detritivores.

3. collectors-both suspension (filter) and deposit (surface) fine particle feeding detritivores.

4. predators-carnivores

The role that the primary producers play in the stream system is related to the previous discussion on CPOM and FPOM. If the major primary producer is a micro-algae, its biomass may be reduced by direct predation (i.e., by macroinvertebrates, functionally called scrapers) or be sluffed off into the FPOM pool. However, if it is an aquatic hydrophyte, or a moss, it may enter the food chain, almost exclusively during die off, therefore following the same general CPOM pathway as terrestrial inputs.

There are several important factors that must be studied in the White River if natural fluctuations in the primary producers are to be fully understood. Depth and

278

velocity are important factors. It has been noted that the faster the current, the more the loosely attached species will be washed downstream. Velocities greater than about 5 m/sec erode all but the most tenacious periphyton. Depth is important as it interacts with turbidity and available light to limit photosynthetically active light at the stream bottom (Whitton, 1975). Some benthic algae are very tolerant of shade. Whitton (1972) found that a quarter of the species taken from the River Wear, England, and incubated in the dark for two months, were able to grow when given favorable conditions. On the other hand, Whitton (1973) has shown some species to be tolerant only of high light conditions. The shallow, lower velocity, high light areas in the White River are always higher in chlorophyll a than the deeper, faster areas (Figure 10).

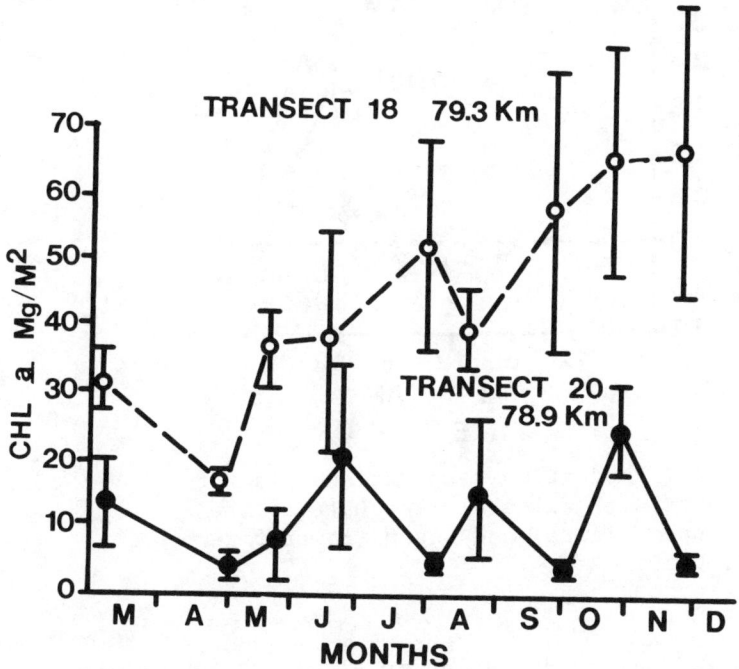

Figure 10. The temporal change in periphyton biomass at a pool and a riffle transect in the White River at Southam Canyon.

In preliminary experiments, oil shale leachate has been experimentally introduced into P/R chambers in situ. Low-level leachate additions either depressed or stimulated the community responses depending upon the stream studied (Figure 11).

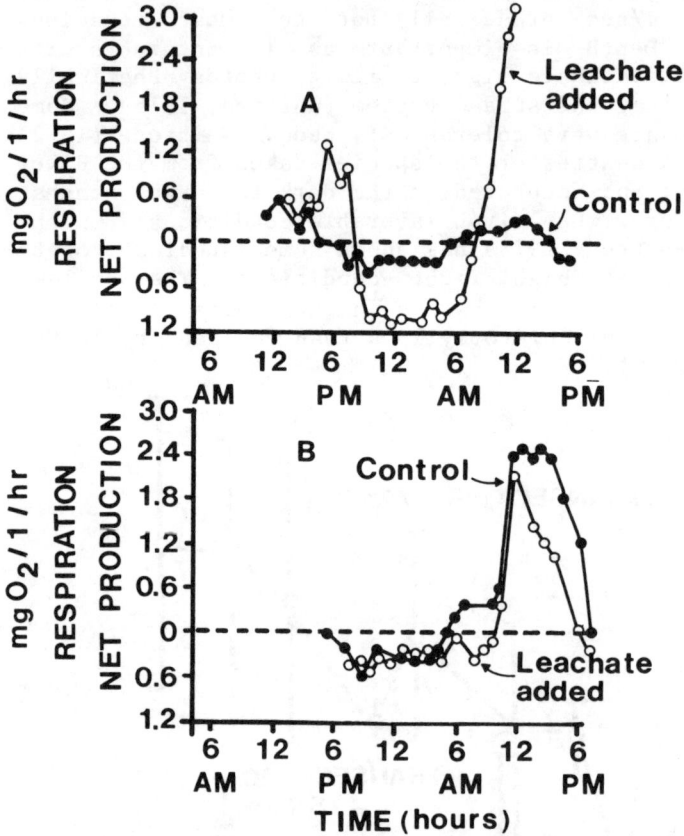

Figure 11. Production-respiration data for experiments with oil shale leachate in the White (upper figure) and Logan (lower figure) Rivers, Utah.

These results indicated that an in situ technique using a community-level parameter (P/R) was more sensitive than laboratory bioassays using single species algal cultures. These results quantify the fact that different stream ecosystems respond differently to perturbations and that laboratory bioassays did not predict this result.

When determining rates of production, respiration, and processing of organic material in the White River, the source of the organic material (allochthonous vs. authochthonous) is critical. A stream heavily dominated by terrestrial input will have high respiration values (heterotrophy) and low photosynthesis (autotrophy). However, if autotrophy (high primary production) is greater than heterotrophy, then the ratio will be greater than 1. Thus, continuous monitoring of production and respiration rates through the seasons will allow a direct determination of the balance between primary production and consumption of organic material. It is this balance which will be a measure of the relative health of the system from which departures may be identified and related to the system structural components. This mechanism of classifying the White River system is of great importance in determinimg potential structural or functional changes in the biotic component in a timely manner.

RATIONALE OF THE ECOSYSTEM APPROACH

Ecosystem resolution can be carried from the most gross features (macrostructure/function) to the most miniscule (microstructure/function). What must be kept in mind with the management/monitoring program, therefore is a reasonable perspective based on logic, economics, regulatory requirements, ecosystem knowledge and the expected pathway or mechanism of effects.

Based on the previous view of the aquatic ecosystem, we can begin, for example, at the macro-level by establishing relationships between primary production and environmental physical factors such as light, temperature, water chemistry and so forth. For the invertebrate or consumer populations we can establish relationships between environmental factors, terrestrial organic input, primary production, and production of the consumers. At the decomposer level we can measure indices of microbial community dynamics such as enzymatic activities coupled to community metabolism and relate this to stream physical environment, organic matter and water chemistry. Overlaying all these subsystems we can determine production/respiration (P/R) and decomposition rates as related to environmental factors through the year to arrive at a temporal, quantitative picture of gross community behavior.

By considering an ecosystem in terms of energy or carbon flow, measuring production and consumption between each subsystem and for the overall system, and calculating mass balances, we will arrive at a quantitative description of the ecosystem. This quantitative description then becomes a statistically based valid reference with which to compare future system behavior for determination of departures from normality. If the relationships under consideration are valid, departures

281

from this state by any component should be reflected by an associated change in one or more related components or sub-components.

This method has distinct advantages. If valid statistical relationships between biological/physical, biological/biological and physical/physical parameters are established for the oil shale tracts, the long-term monitoring effort will be reduced in scope and cost by not having to monitor all possible components through time. In addition, basing the monitoring program on these ecosystem relationships is a diagnostic procedure leading the environmental manager closer to the probable cause of departure from normality by statistical analysis of observed versus expected relationships. Once probable cause is determined, intensive effort may be expended in a smaller area to elucidate cause and determine corrective action.

LOGIC OF THE MONITORING PROGRAM

Using our conceptual model and the rationale described above, the management program is able to use logic which leads to the appropriate contingency plan, mitigation or change in reclamation in the event of an adverse effect, or no further action in the event the state is normal. An integral component of the following procedure, the computer program "MONITOR" was developed to streamline the analytical process.

Prior to operation of the development-level monitoring program, baseline data collection and pre-development monitoring are used to measure environmental (biological/physical) parameters and construct the ecosystem model (Figures 2 and 3). This model, which is a statistical and mathematical description of the ecosystem, describes inter- and intra- component structure, functional relationships, and interactions with driving variables. These relationships then become the definition of ecosystem "health," i.e., state variables against which future data are compared. This comparison takes place as shown in Figure 12, a flow model of contingency planning using MONITOR. The analysis employed should follow logically through the various levels of resolution described below.

Level I. The on-going monitoring effort will measure those ecosystem parameters and relationships which were determined significant during baseline and predevelopment data collection. These might include parameters such as chlorophyll a, invertebrate functional group biomass, and

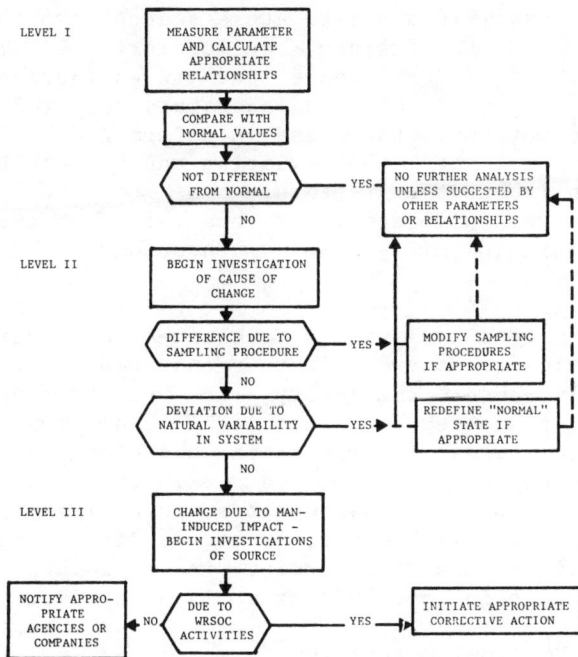

Figure 12. Flow model of environmental analysis and contingency planning.

light vs. primary production among others. These data are used in MONITOR. Data which fall outside the allowable range (acceptable confidence limits) or cause a significant change in the particular ecosystem relationship of interest are flagged. For those parameters satisfying the model requirements of normality, no further action would be taken. If a parameter did not satisfy the criteria for normality, the next step in contingency planning should occur.

Level II. For relationships deviating from the allowable range as defined by the model, the next level of analysis begins. In this case, subsets of the structural and functional characteristics of ecosystem (biological/physical) parameters are considered. These are compared with the baseline data base and the ecological literature. If the deviation can be explained by a change in a driving variable, a co-relationship with other ecosystem parameters, or from the literature, the model can be refined to include this relationship. If the deviation cannot be explained in this manner, the last step of analysis occurs.

Level III: Having failed to explain the deviation by

detailed data analysis and literature search, hypotheses must be generated to explain the deviation. These hypotheses will form the basis for investigations to determine the problem. Field and laboratory research will then be carried out until the cause is determined. If, by gathering additional data, the deviation can be attributed to natural environmental factors, the model will be refined and analysis will stop. If, however, the cause is industry related, mitigation methodology can be implemented.

Application of this logic to reclaimed or modified areas is identical. However, in this instance, the model is used to measure similarity to pre-development conditions or projected goals and legal requirements for the area. Effectiveness of mitigation, reclamation, or habitat enhancement will require evaluation of each important relationship by knowledgeable investigators as well as consultation with the governing regulatory agency at appropriate intervals to determine if reclamation is complete and mitigation or enhancement is effective.

EXAMPLE OF THE MONITORING APPROACH

To illustrate the application of this logic, consider the following example. Assume the following represent a set of primary relationships describing a subset of the ecosystem and that primary production (mg $O_2/m^2/hr$) is the community-level parameter being monitored.

Net Primary Production = f(light, chlorophyll, community respiration)

Light = f(season, time of day, turbidity)

Chlorophyll = f(season, storm events, nutrients)

Community Respiration = f(standing crop of invertebrates, microbes, organics)

If at Level I we enter the most recent data into MONITOR and the relationships still hold, that is, the confidence limits of the relationship curves are not violated, the system is normal and we need proceed no further. If, however, we observe net primary production to be lower than expected by our model and that correspondingly, community respiration is higher than expected, while light and chlorophyll are normal, MONITOR outputs a red flag for net primary production (NPP) and community respiration (CR).

NPP ↓   CR ↑

   We then proceed to Level II to determine if CR
changes can be explained. We explore the relationships
determing CR, i.e.:

   CR = f(biomass of invetebrates, microbes,
         organics)

If we determine the standing crops of invertebrates and
microbes to be normal, the relationship has not
explained the change. Therefore, other factors which
could be causing the change must be considered. We
enter Level III where other hypotheses are formulated to
explain the change.

   We may hypothesize that invertebrate or microbe
respiration has increased, thus, respiration must be
directly measured by utilizing respirometers developed
for the project. If microbial respiration is normal but
invertebrate respiration is above normal, the deviation
in invertebrate respiration must be determined. Perhaps
leached metals or organics caused sublethal stress. We
then analyze properly preserved samples of stream
invertebrates for these compounds. Concurrently,
process waste stream records are inspected for abnormal
levels of the suspected agent. If levels are above the
baseline, then leachate is implicated and the pathway
and appropriate mitigation must be determined.

   In summary, community level parameters and
ecosystem-based relationships with sublevels can be used
to narrow the path of investigations of an abnormal data
point. This can then lead to a logical management
decision and action. It should be noted that many of
the activities presented as level I, II, and III could
be occurring concurrently to reduce reaction time.

FUTURE DEVELOPMENT OF THE MANAGEMENT/MONITORING PROGRAM

   Data gathered in the baseline and monitoring
programs for Tracts Ua and Ub will be incorporated into
an ecosystem model which will be continually refined as
more data are accumulated. Emphasis is being placed on
the preproduction period to complete the majority of
ecosystem relationships for the model and refine
techniques to be used in the long-term program.

As quantitative relationships are developed and proven valid, the model can be refined. Accumulated information may then be used in the contingency plans to aid in solving problems as they arise. Those parameters which have limited relevance to expected effects can be dropped from continuous monitoring. With this approach, as the model is refined, the number of parameters measured and the frequency of measurement can be reduced or changed, maximizing the cost effectiveness of the monitoring and reclamation programs.

This management program has been developed and is being phased into operation, integrating the efforts of a team of environmental managers and consultants covering all environmental disciplines. It is expected to save time and money for the White River Oil Shale Corporation, allowing monitoring of the White River ecosystem without measuring all possible factors, and lead to mitigation when necessary. It is logical, intuitively appealing, simple, and manageable. We believe it is a beginning to logical monitoring and correction of industrial impacts.

Acknowledgements

The authors express appreciation to James W. Godlove, Environmental Affairs Coordinator and Rees C. Madsen, Vice President of Administration for White River Shale Oil Corporation. Their insight into the environmental problems associated with oil shale development has allowed this experiment in environmental management based upon ecosystem principles to begin.

LITERATURE CITED

1. Bull, William B. 1981. Soils, geology and hydrology of deserts. In: Daniel D. Evans and John L. Thames [eds.]. Water in Desert Ecosystems. Vol. II. US/IBP. Synthesis Series. pp. 42-58.

2. Cummins, K.W. 1973. Trophic relations of aquatic insects. Ann. Rev. Ent. 18:183-206.

3. Cummins, K.W. 1975. Macroinvertebrates. In: River Ecology. B.A. Whitton, [ed.]. Universtiy of California Press. pp. 170-198.

4. Farnworth, E.G., M.C. Nichols, C.N. Vann, L.G. Wolfson, R.W. Bosserman, P.R. Hendrix, F.B. Golley and J.L. Cooley. 1979. Impacts of sediment and nutrients on biota in surface waters of the United States. U.S. Department of Commerce, National Technical Information Service PB80-1299588. 314 pp.

5. Gibbs, R. 1970. Mechanisms controlling world water chemistry. Science 170:1088-1090.

6. Hynes, H.B.N. 1970. The ecology of running waters. University of Toronto Press. 555 pp.

7. Leopold, L.B. and T. Maddock, Jr. 1953. The hydraulic geometry of stream channels and some physiographic implications. U.S.G.S. Prof. Paper 252. 27 pp.

8. Lush, D.L. and H.B.N. Hynes. 1973. The formation of particles in freshwater leachates of dead leaves. Limnol. Oceanogr. 18:968-977.

9. OSO. 1979. Guidelines for monitoring the environmental effects of development in the federal prototype lease tracts. Oil Shale Supervisor, Grand Junction, Colorado. 26 pp.

10. Whitton, B.A. 1972. Environmental limits of plants growing in flowing waters. In: Conservation and productivity of Nat. Waters Symp. R.W. Edwards and Garrod [ed.]. Zool. Soc. London 29:3-19.

11. Whitton, B.A. 1975. Algae. In: River ecology. B.A. Whitton [ed.]. University of California Press. pp. 81-105.

12. Whitton, B.A. 1973. Freshwater plankton. In: The biology of blue green algae. N.G. Carr and B.A. Whitton [eds.]. Blackwell-Oxford. 676 pp.

13. Wlosinski, Joseph H. 1979. Predictability of stream ecosystem models of various levels of resolution. PhD Dissertation, Utah State University, 138 pp.

14. Wolman, M.G. 1964. Problems posed by sediment derived from construction activities in Maryland. Report to the Maryland Water Pollution Control Commission, Annapolis, Maryland. 125 pp.

CHAPTER 18

ENERGY DEVELOPMENT AND
RIPARIAN MANAGEMENT IN A COLD
DESERT ECOSYSTEM

C. Val Grant
    Bio-Resources, Inc.
    Logan, Utah

The Upper Colorado River Basin is under a siege from energy, agricultural, and urban development and what we are seeing now is only the tip of the iceberg. Those charged with monitoring and reclaiming disturbed ecosystems should be concerned at the very least. The situation is aggravated by the lack of adequate baseline data on plant and animal communities. The flora and fauna of this arid region which includes eastern Utah, western Colorado, and southwestern Wyoming, are most similar to those of the Great Basin Desert but this area was not considered part of the North American deserts by MacMahon [1]. Given the current state of knowledge of this ecosystem this exclusion is not surprising. Ecosystem management based on inadequate data can only be managed inadequately.

This paper reports recent information collected on vertebrate communities in four terrestrial habitats adjacent to the White River, Utah. We hope to illustrate how this information will help meet needs of monitoring and reclaiming disturbed ecosystems in the future. The first step in meeting these needs is to identify critical ecosystems, i.e., ecosystems of both high wildlife and economic value that are most likely to be affected adversely during development. In the arid west, these are usually riparian habitats. Second, these critical habitats must be monitored to evaluate the state of biotic resources. This includes an assessment of the physico-chemical environment -- the context in which organisms (including man) must exist -- which typically includes studies of soils, geology, hydrology and weather, and an analysis of how the biota interrelates with the physical and chemical environment.

Historically, wildlife specialists have emphasized game
birds and mammals and the vegetation with which these species
are associated. More recently raptors, threatened and
endangered species, and other species of special interest
have been considered. Mitigation, reclamation, and management
strategies are then planned around these animals. Non-game
birds, mammals, and reptiles have been largely ignored by
managers, apparently on the assumptions that non-game animals
have little to offer a wildlife monitoring scheme and that
game animals are an accurate measure of environmental health.
Both of these assumptions are incorrect. While game animals
have their place in a comprehensive managing program, hunting
pressure and, for some species, great mobility make them less
than perfect indicator species. Raptors and endangered
species occur in such low densities that effects of development are frequently difficult to measure. On the other hand,
low mobility, high site tenacity, and lack of hunting pressure
make most non-game animals attractive indicators of environmental health.

As an illustration of the application of monitoring non-
game animals we present six years of data from the Utah Oil
Shale Tracts, Ua and Ub, in the Uintah Basin of northeastern
Utah. The work was funded by the White River Shale Project,
a joint venture of Phillips Petroleum, Sohio, and Sun Oil Co.

The Uintah Basin is a broad, asymmetric basin near the
northern boundary of the Upper Colorado River Basin (Figure
1). The Uintah Basin overlies alluvial sediments from a pre-
historic depositional lake basin. The major land features
bordering the basin are the Uintah Mountains on the north,
the Wasatch Range on the west, the Roan and Book Cliffs on
the south, and the highlands associated with Douglas Creek
subsurface arch along the eastern edge at the Utah-Colorado
state line.

The Green River, flowing generally southwestward, is the
major drainage within the basin. The majority of the water
in the Green River comes from the westward flowing Yampa and
White Rivers (headwaters in the Rocky Mountains of Colorado)
and the eastward flowing Duchesne River (headwaters in the
Uintah Mountains). Oil shale, tar sands, and other oil and
gas development are scheduled for this area.

The region near the tracts, including the area extending
southward to the Roan Cliffs and westward to the Green River,
is a gently north sloping highly disected plateau. The
plateau is characterized by steep wall canyons with ephemeral
dry streams or washes. Within the tracts, the landscape is
composed of north-south trending valleys separated by narrowly

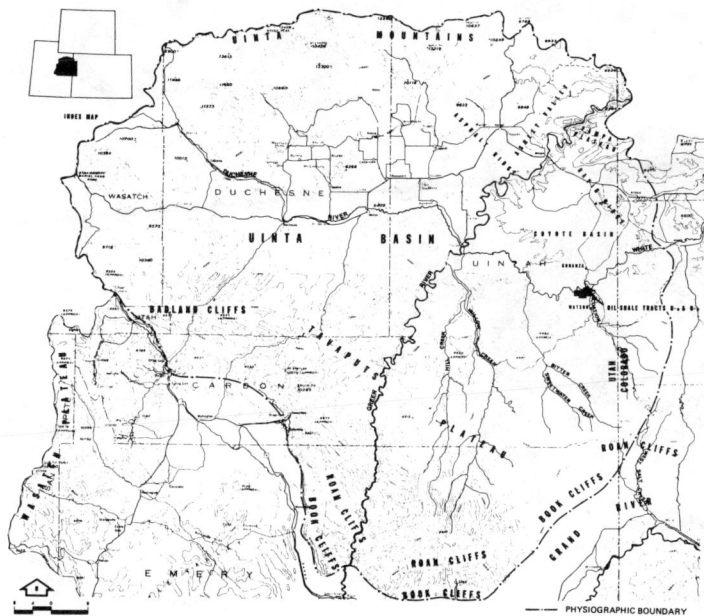

Figure 1.   Location of the Oil Shale Tracts in the Uinta
Basin of Northeastern Utah.

elongated mesas.   Elevation on the tracts ranges from 1500 to
1900 meters.   Tracts Ua and Ub are bounded by the White River
to the north, Hell's Hole Canyon to the east, upland areas at
slightly less than 1900 meters to the south, and Asphalt Wash
to the west.   Most of the upland soils or lack of them are
from sandstone of the Uinta Formation.

The White River is the only perennial water on the
tracts; however, Evacuation Creek contains a low flow for
several months of the year which is mainly brackish due to
groundwater discharge from the Bird's Nest Aquifer.   Perennial
water also exists in small ponds in Asphalt Wash, and tributaries of Evacuation Creek and White River are all ephemeral.

From 1975-1980 wildlife in Greasewood, Shadscale,
Juniper and Riparian habitats on the White River and Evacuation Creek (Figure 2) were sampled at two sampling sites
per habitat.   Amphibians, reptiles, birds and mammals were
sampled using line transects one kilometer long.   Transects
were walked in February, April, June, August, and October
each year.   Bats were sampled with mist nets at a pond in
Asphalt Wash during June and August for four years.   Rodent
populations were sampled in the above months with transect
trapping and during August rodents were sampled on large
trapping grids (144 traps).

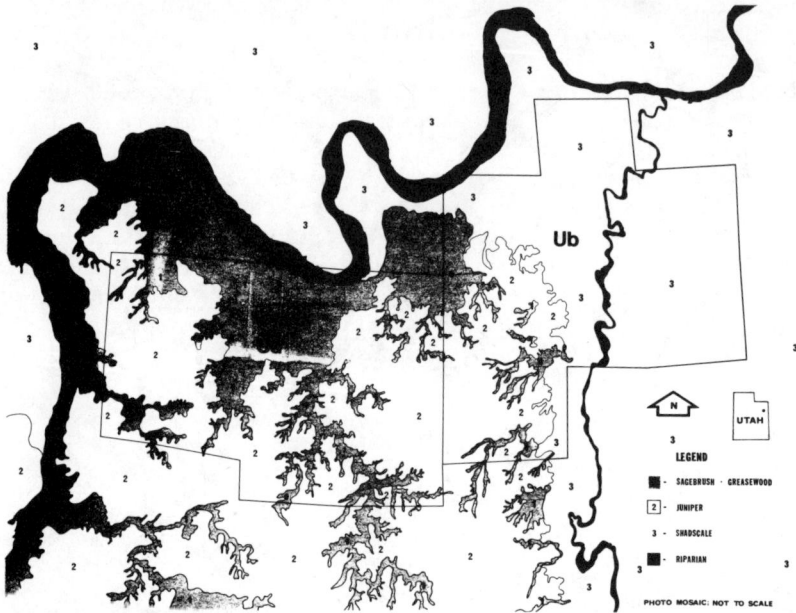

Figure 2. Vegetation map of Oil Shale Tracts, Ua and Ub.

In 1975, a very wet winter and spring resulted in a desert bloom of annual plants (Figure 3). In 1977, the area experienced a region-wide drought. Since 1977 there has been

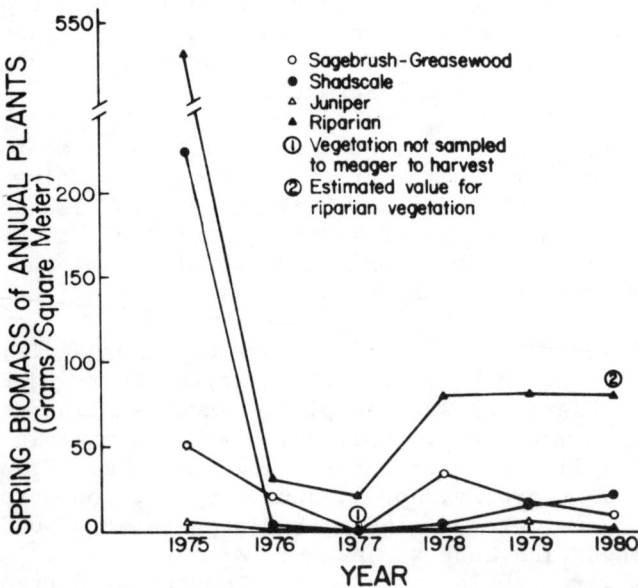

Figure 3. Annual plant biomass from 1975-1980 on Tracts Ua and Ub.

292

a gradual increase in annual plant biomass and in precipitation (Figure 4).

In order to manage Riparian systems we first need to know what constitutes the Riparian fauna and compare it to the fauna in adjacent habitats. Among the amphibians there are four Riparian species found: two toads and two frogs. Only the toads occur in the upland habitats at scattered stock ponds and ponds formed by deep well discharge. Eleven species of reptiles occur in the area. Nine species occur in Riparian and upland habitats with a difference in composition among four species. Racers (Coluber constrictor) and western terrestrial garter snakes (Thamnophis elegans) are restricted to Riparian whereas short-horned lizard (Phrynosoma douglassi) and western rattlesnakes (Crotalus viridis) are found in upland habitats. Although western rattlesnakes are found along Evacuation Creek, a Greasewood-Tamarisk habitat, none have been encountered along the White River.

Amphibians are low value in terms of management due not only to low species richness but also their abundance is too low to be quantified. Reptiles are useful in Riparian settings because of the consistency in species composition; however, their distribution and abundance present some problems. The four species of snakes in Riparian occur at low abundance and are restricted to dense vegetation of cottonwoods and grasses.

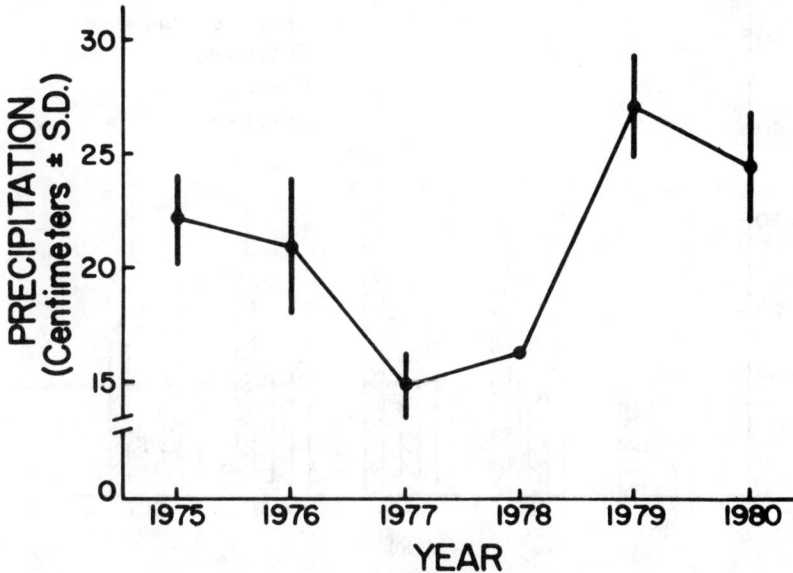

Figure 4. Annual precipitation from 1975-1980 on Tracts Ua and Ub.

Of the five lizard species, three are restricted to areas of
sparse vegetation of shrubs with little to no understory.
The other two species, the tree lizard (Urosaurus ornatus)
and eastern fence lizard (Sceloporus undulatus), present
difficulties in measuring their abundance due to their vertical distribution in the cottonwood trees.

The value of amphibians and reptiles in Riparian management appears to be limited to presence/absence relationships.
There are, however, some parameters among the lizards which
aid in management decisions. First, species diversity [2]
among lizards is significantly higher (p<0.95) in structurally diverse wooded habitats (Riparian and Juniper) than in
the less complex shrub habitats (Greasewood and Shadscale).
Second, lizard abundance is significantly higher (p<0.95) in
habitats with the least vegetative ground cover (Juniper and
Shadscale). The percentage of bare ground correlates positively with lizard abundance (r=0.88, p<0.99). A decrease
in lizard diversity and/or an increase in lizard abundance
may well be the best indicator(s) of trampling effects of
humans and livestock in desert Riparian systems.

Riparian is by far the most important habitat for one
class of wildlife, the birds. Species richness is highest
in Riparian compared to upland habitats both seasonally
(p<0.99, Figure 5) as well as annually (p<0.99). Riparian

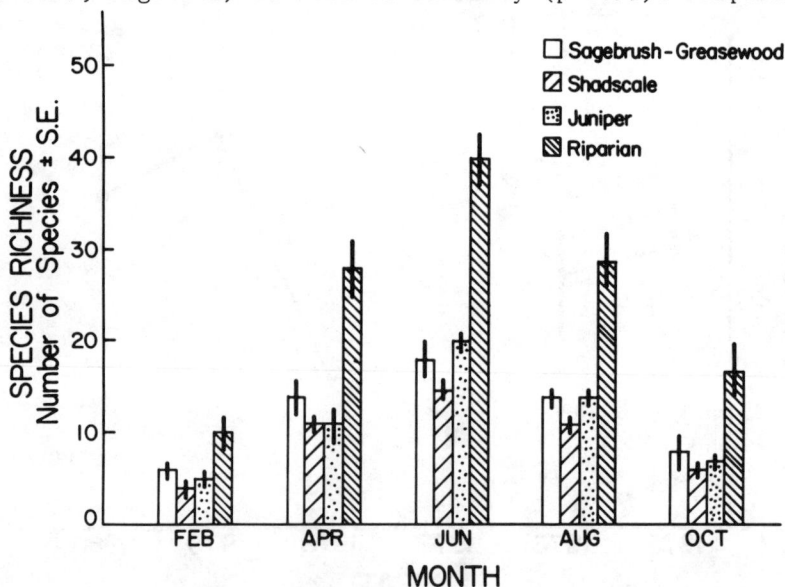

Figure 5.   Seasonal avian species richness in four habitat
            types on Tracts Ua and Ub.

also supports the highest annual abundance of birds (p<0.99, Figure 6). Loss or change in Riparian habitat will influence the avian community more than other wildlife groups.

The problem with dealing with such a wide range of birds, hummingbirds to geese, is selecting important species which will indicate change. Based on a recent analysis of this avian community (Vander Wall, Steele and Grant, in prep.), use of birds in management is best served by identifying feeding guilds and tracking annual and seasonal changes in the guilds rather than using selected species. Three major guilds represent most of the avian community - the granivores, omnivores and insectivores. Changes in precipitation and primary productivity appear to influence Riparian granivores and omnivores but not the insectivores. Dividing the insec-tivores into sub-guilds, i.e., ground gleaners, foliage gleaners, air cruisers, air hawkers and bark gleaners and probers, show that the air hawkers (Tyrannidae) and bark gleaners (Troglodytidae, Sittidae and Certhidae) and probers (Picidae) are the least important sub-guilds in terms of abundance. Air cruiser abundance (primarily Hirundinidae) appears to be positively correlated with spring flow in the White River, suggesting that high flows which flood the cottonwood stands result in more breeding areas for insects, hence, an increase in the air cruiser's food resource.

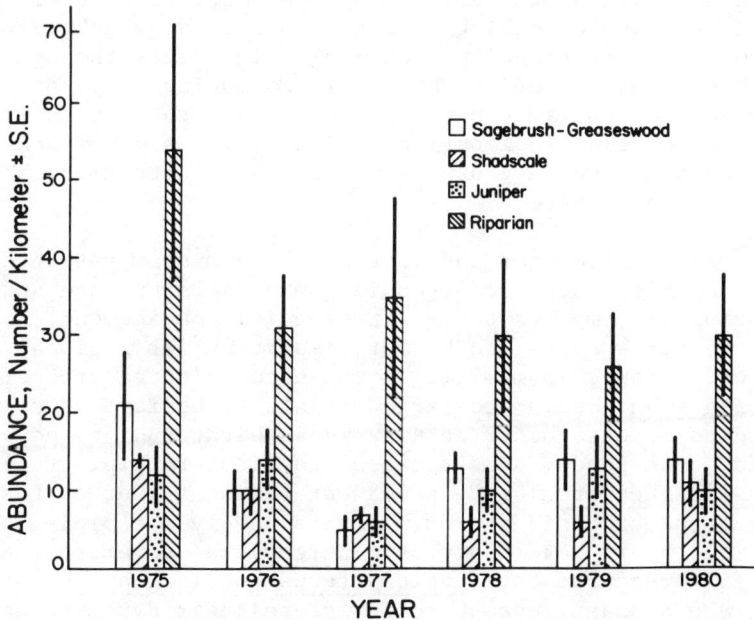

Figure 6. Annual avian abundance in four habitat types on Tracts Ua and Ub.

295

Annual changes in Riparian ground and foliage gleaners increased with declines in precipitation and increases in temperature. Both these avian sub-guilds plus the insectivorous
lizards reached peak abundance during the 1977 drought, a
period when the other avian guilds and groups of wildlife
decreased significantly.

Waterfowl and raptors are important in any riparian management scheme; however,decisions based on changes in either
group present some problems, especially among the raptors.
Both groups occur at low abundance; hence , it is difficult
to measure change. Both are dependent on biotic and abiotic
factors which can be precisely measured, e.g., prey base and
water quality and quantity. Establishing the needed relationships between these large birds and their environment and
using the small birds as the key to avian management provides
a quantitative base for decision making which more closely
tracks changes in Riparian habitat. In an extensive wetlands
habitat, the ease of measuring waterfowl populations would
preclude the need to measure non-game birds. Since wetlands
constitute a small percentage of desert rivers, avian management is best achieved through the non-game species.

The major drawback with using non-game birds is that
most are part-time residents; hence, changes in their composition and abundance are influenced by conditions on their
wintering grounds and during migration. This drawback can
be overcome by selecting Riparian monitoring sites throughout
a river basin and comparing changes in community structure
annually and seasonally. Until we understand the variability
of avian communities in Riparian habitats, decisions which
affect the most diverse group of wildlife in North America
are more art than science.

The mammals are, for the most part, permanent residents
with low mobility ; hence changes in mammalian abundance and
distribution are important for site-specific management,
whereas the birds appear to be more important for regional
management. Change in mammals is measured in three groups:
bats, rodents and medium-to-large mammals (rabbits to deer).
Nine species of bats occur in Riparian habitat and at the
scattered stock ponds. One species, the western pipistrelle
(Pipistrellus hesperus) is a permanent resident,whereas the
others are summer residents or migrants. Only the pipistrelle, the hoary bat (Lasiurus cinereus), a summer resident,
and the silver-haired bat (Lasionycterus noctivagans), a
migrant, are abundant enough to integrate their dynamics into
a management program. All are insectivores and like the
avian insectivores,their annual changes are puzzling. Pipistrelle abundance appears to be dependent on the severity of

296

winter, increasing after mild winters and declining after
harsh winters. Since these small bats (3-6 g) dwell in caves
and crevices in the sandstone, cold winter temperatures would
definitely affect their survival. Hoary bats, a tree-dwelling
species, besides being the largest bat in Utah (23-39 g), are
also the most abundant. Changes in their seasonal and annual
abundance as well as sex ratio suggest that the deserts in the
Intermountain West are the prime sites for hoary bats' par-
turition and breeding, contrary to previously suggested distribution [3]. Where the pipistrelles and hoary bats change
seasonally and annually, silver-haired bats occur only in
June at an almost constant abundance through four years.
Annual changes in bat abundance followed much the same pattern
as abundance of one sub-guild of insectivorous birds, the air
cruisers. Like the birds, knowledge of bat community on a
regional scale can integrate another part of our wildlife into
management decisions.

The thirteen species of rodents found in Riparian and
adjacent upland habitats are vivid illustrations of the
effects of soil types on rodent distribution. Riparian
supports the lowest diversity (p<0.95), lowest richness
(p<0.95), and lowest biomass (p<0.95) of rodents compared to
Greasewood, Shadscale and Juniper. Riparian and the other
wooded habitat, Juniper, support the lowest rodent densities
(p<0.95). Concerning the soils where they are deep and well-
drained (Greasewood and Shadscale), rodent densities are
highest through six years. Where soils are well-drained
(Greasewood, Shadscale, Juniper) rodent biomass is high.
Where soils are shallow, well-drained and vegetation structure complex (Juniper), rodent richness and diversity are
highest. The silty, wet soils in Riparian, which freeze
every winter, support one species at moderate densities, the
deer mouse (Peromyscus maniculatus), with low densities of
bushy-tailed woodrats (Neotoma cinerea) who reside above
ground in abandoned woodpecker holes, and western harvest
mice (Reithrodontomys megolotis) who reside in sandy deposits
near the river's banks.

The rodents also illustrated the ameliorating effects of
Riparian during a severe drought. The effects of the 1977
drought resulted in a significant decline in rodent densities
in all habitats (Figure 7). In the dry upland habitats, deer
mice became locally extinct in Shadscale and Juniper, white-
tailed antelope squirrels (Ammospermophilus leucurus) became
extinct in Shadscale through 1980, as did desert woodrats
(Neotoma lepida) in Shadscale and Greasewood. Deer mice remained at low densities in the upland habitats through 1980;
however, their densities in Riparian began increasing in 1978
and remained at median density through 1980.

297

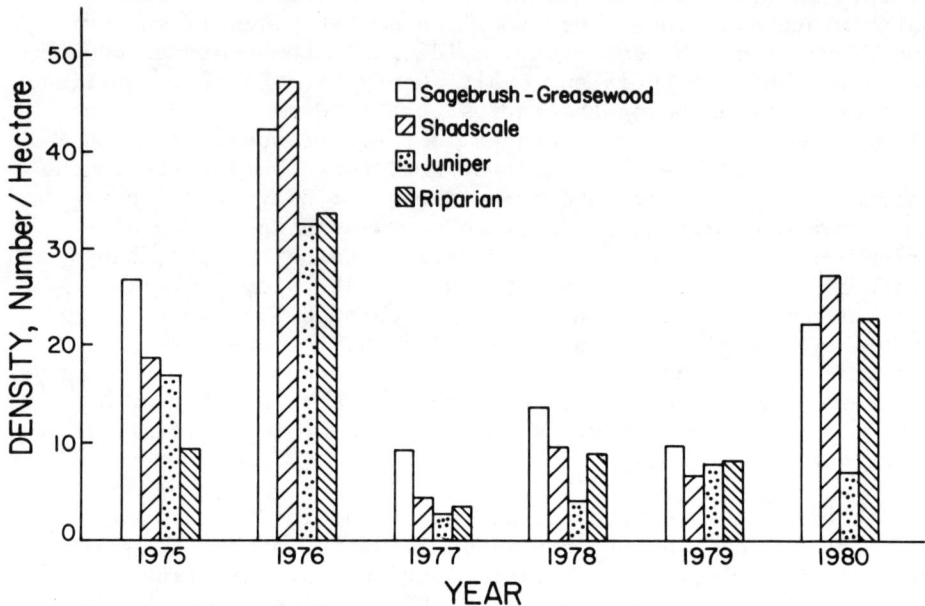

Figure 7. Annual rodent density in four habitat types on Tracts Ua and Ub.

The drought had a lasting influence on the herbivorous and omnivorous rodent guilds (Cricitidae and Sciuridae) in all habitats but Riparian. The one rodent guild which was not affected by either the bloom or the drought was the granivores (Heteromyidae). The two species in this guild remained at a stable density through 1979. In 1980 Ord's kangaroo rat (Dipodomys ordii) densities increased significantly (p<0.95) in Greasewood and Shadscale and they invaded Juniper and Riparian. Apache pocket mice (Perognathus apache), the other granivore, increased in density in Shadscale and remained near their median density in Greasewood in 1980.

The rodents provide the manager with site-specific input on the most numerous and diverse permanent residents among all wildlife species. Besides providing an indicator to soil conditions and distribution, rodent density and distribution reflect the influence and magnitude of weather changes and changes in primary productivity. Regressing annual plant biomass and an indicator to perennial shrub growth on Cricetidae densities from the following year, i.e., a one year lag time, resulted in a significant correlation (r=0.96, p<0.99) suggesting primary productivity is an adequate predicator of the herbivorous rodent guild.

298

The rodents also demonstrate as do the birds and reptiles that management based on one year of information is essentially useless. Some species do become extinct, others invade different habitats, and very few maintain an unchanging abundance. Without knowledge of these changes, environmental management to determine site-specific or regional impacts or to measure the effectiveness of reclamation must relegate wildlife to a status of little or no importance in the decision-making process.

The one group of wildlife which receives the most attention plus economic support are the medium to large mammals. These include the game mammals, the furbearers and the carnivores. Current wildlife management practices judge the value of land according to economic rather than biological input. Although the desert Riparian habitat along the White River is the mainstay for mule deer (Odocoileus hemionus), beaver (Castor canadensis) and coyote (Canis latrans), the number residing along a 20 mile reach of the river are low and difficult to assess. Using these mammals as a basis for monitoring impacts or measuring reclamation success necessitates subjective rather than objective input for the decision-making process.

The only medium-sized mammal, which is a game species, that provided adequate input for management is the desert cottontail (Sylvilagus audubonii). Cottontails are most abundant in Riparian and Greasewood (p<0.95) and their annual abundance changed initially in the same manner as the rodents, i.e., at median abundance during the floral bloom of 1975 and increasing by 500% in some habitats in 1976 (Figure 8). The 1977 drought did not radically reduce their abundance as it did the rodents. The reason it did not is attributed to hunting pressure in 1976 and 1977. By 1978, cottontails maintained high abundance in Greasewood only. In 1979 and 1980 the only habitats which supported measurable numbers of cottontails were Juniper and Riparian.

Desert Riparian habitats overall support the highest diversity and biomass of wildlife and should represent the monitoring key to management and mitigation in the Upper Colorado River Basin. If Riparian loss is to be mitigated, a need to replace apples with apples exists. To assume that one mitigates losses of Riparian habitat with increased waterfowl production on an impoundment makes some sense economically, but not biologically. Our current practice emphasizes replacing a diverse system with a monoculture. Although monocultures are definitely easy to manage, they are also easy to lose. A vital need exists to maintain the present level of diversity until we have the needed infor-

299

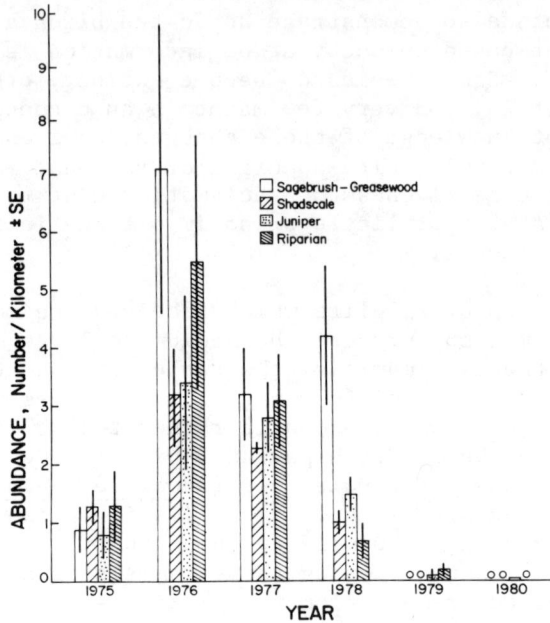

-Figure 8. Annual cottontail abundance in four habitat types
            on Tracts Ua and Ub.

mation to assess change on a site-specific as well as regional basis.

In conclusion, four points require recognition and
clarification for a management scheme:

1. Physical, chemical, and biological components must
   be integrated when assessing impacts on an ecosystem. The physics, chemistry and biology of any
   system are not mutually exclusive.

2. Since Riparian habitat is of prime importance in a
   desert from the viewpoint of terrestrial vertebrates, efforts should be concentrated on monitoring
   and managing the rivers in the Upper Colorado River
   Basin.

3. Since the response time by small animals is faster
   than that of large animals, efforts should concentrate on the small animals in the system. Currently
   our work is being expanded to include invertebrates
   to make sure the most reliable indicators of
   environmental change are being monitored.

4. Intensive monitoring of any area for its biological

300

components should be a continual process. If industry or government regulatory agencies relied on data collected for a single year, the results would be nonsensical. This is sufficiently illustrated by the extreme differences in wildlife population dynamics from a six year data base.

LITERATURE CITED

1. MacMahon, J.A. 1979. North American Deserts: their floral and faunal components. In Arid-Land Eco-Systems: Structure, Functioning and Management. Cambridge Univ. Press, Great Britain. Vol. 1:21-82.

2. Shannon, C.E. 1948. A mathematical theory of communication. Bell Syst. Tech. J. 27:379-432.

3. Findley, J.S. and C. Jones. 1964. Seasonal distribution of the hoary bat. J. Mam. 45:461-470.

CHAPTER 19

OIL SHALE DEVELOPMENT: POTENTIAL
EFFECTS ON THE BIRDS NEST AQUIFER
IN THE VICINITY OF FEDERAL LEASE
TRACTS Ua-Ub, UINTA BASIN, UTAH

Michael P. Donovan
Mark W. Bulot
    VTN Consolidated, Inc.
    Irvine, California

ABSTRACT

The economic development of oil shale from eastern Utah
will be largely dependent upon the availability of suitable
fresh water in the vicinity of the development. The Birds
Nest Aquifer could represent a supplementary source of water
for oil shale development where surface water resources may
have limited availability. However, specific activities
associated with oil shale mining could detrimentally affect
the aquifer and accordingly decrease its usefulness as a
water supply source to augment present surface water supplies.

The proposed oil shale activities that could potentially affect the aquifer include underground mining, surface disposal of retorted shale, surface water impoundment
and ground water withdrawal.

The potential effects from underground mining, surface
disposal of retorted shale and water impoundments are considered inconsequential compared to the effects caused by
water withdrawal for supplementing surface water supplies.
Such effects could include significant dewatering of the
aquifer (change in storage) which in turn could affect
annual discharges and recharge rates of adjacent surface
streams. The magnitude and long-term effects from ground
water withdrawal will be dependent on the management of
water resources in this region.

INTRODUCTION

The economic development of oil shale resources in the
Uinta Basin of northeastern Utah will require suitable quantities of fresh water for oil shale processing and associated needs. The purpose of this report is to present basic
existing information on the area and the potential effects
of proposed mining activities on the Birds Nest Aquifer,
which may become a significant source of water for future
oil shale development. Mining and associated development of
ground water supplies could have a significant impact on the
existing ground water systems. The Birds Nest Aquifer, a
major ground water system in proximity to the proposed mining activities, could be detrimentally affected by underground mining activities, ground water withdrawal, and surface disposal of retorted oil shale.

Presently, little information is available to characterize the potential availability of usable quantities of
water from the Birds Nest Aquifer to determine the initial
and long-range effects of oil shale development on the
aquifer. Most studies in the area have been conducted near
Federal Lease Tracts Ua and Ub[1] with subsequent studies
conducted by the United States Geological Survey[2].

Oil Shale Development

The various present-day technologies for oil shale
development include true in-situ, modified in-situ, and surface retorting techniques. Oil shale mining technologies
include open-pit mining, underground mining, and underground
retort development. Generally, underground room-and-pillar
mining and surface retorting are the proposed methods, as in
the Detailed Development Plan (DDP) for Federal Lease Tracts
Ua-Ub[3].

Independent of the technology, a prime concern for
development of oil shale in the region will be the quality
and availability of water. The U.S. Water Resources Council[4] has suggested that synthetic fuels development in the
upper Colorado region will consume an average of 2.4 barrels
of water per barrel of oil (based on a 50,000 barrel per day
plant). Accordingly, the development of a synthetic fuels
industry in northeastern Utah, producing 1 million barrels
per day, will require almost 312 acre-feet per day.

Oil shale development in the White River region could
be limited by the development of water from major surface
reservoirs or ground water aquifers in the region. The
present surface water supplies are subject to seasonal low

304

flows, droughts, and the acquisition of surface water rights
by oil shale developers. The development and maintenance of
ground water resources in the region may play an important
role for some oil shale projects. The Birds Nest Aquifer
may be able to accommodate some of the high consumption, low
quality water requirements for these oil shale developments.

The following oil shale development activities represent possible sources of effects to existing ground water
systems in the Uinta Basin:

1. underground mining;
2. surface disposal of retorted shale and water
    impoundments;
3. ground water withdrawals

The proximity of mining activities to the Birds Nest
Aquifer could affect both the aquifer and surface drainages
with which there is a system interaction.

PHYSICAL ENVIRONMENT

Physiography

The area described in this report is in the southeastern Uinta Basin, located in northeastern Utah and
northwestern Colorado (see Figure 1). The area encompasses
approximately 7,770 square kilometers (3,000 square miles)
and is part of a broad asymmetric basin located on the
northeastern edge of the Colorado Plateau. The land's
surface elevation varies from 1,707 meters (5,600 feet) at
its lowest point on the White River to 2,895 meters (9,500
feet) in the Roan Cliffs at the southern edge of the basin.

The most prominent regional surface feature is the
northwesterly slope of the portion of the basin which is
traversed by the White River. The area immediately adjacent
to the river consists of moderate to extremely steep canyons
(slopes from 20% to greater than 100%). The rock outcrops
vary in their resistance to erosion and exhibit bench and
slope features on the canyon walls, with the resistant
layers forming the benches. The topography of the upland
portions of this area (above the river and ephemeral stream
channels) is gently rolling, following the regional slope.
The climate of the basin is semi-arid with occasional
localized thunderstorms occurring during the generally hot,
dry summers. Winters are cold, with snow accumulation
usually confined to higher altitudes.

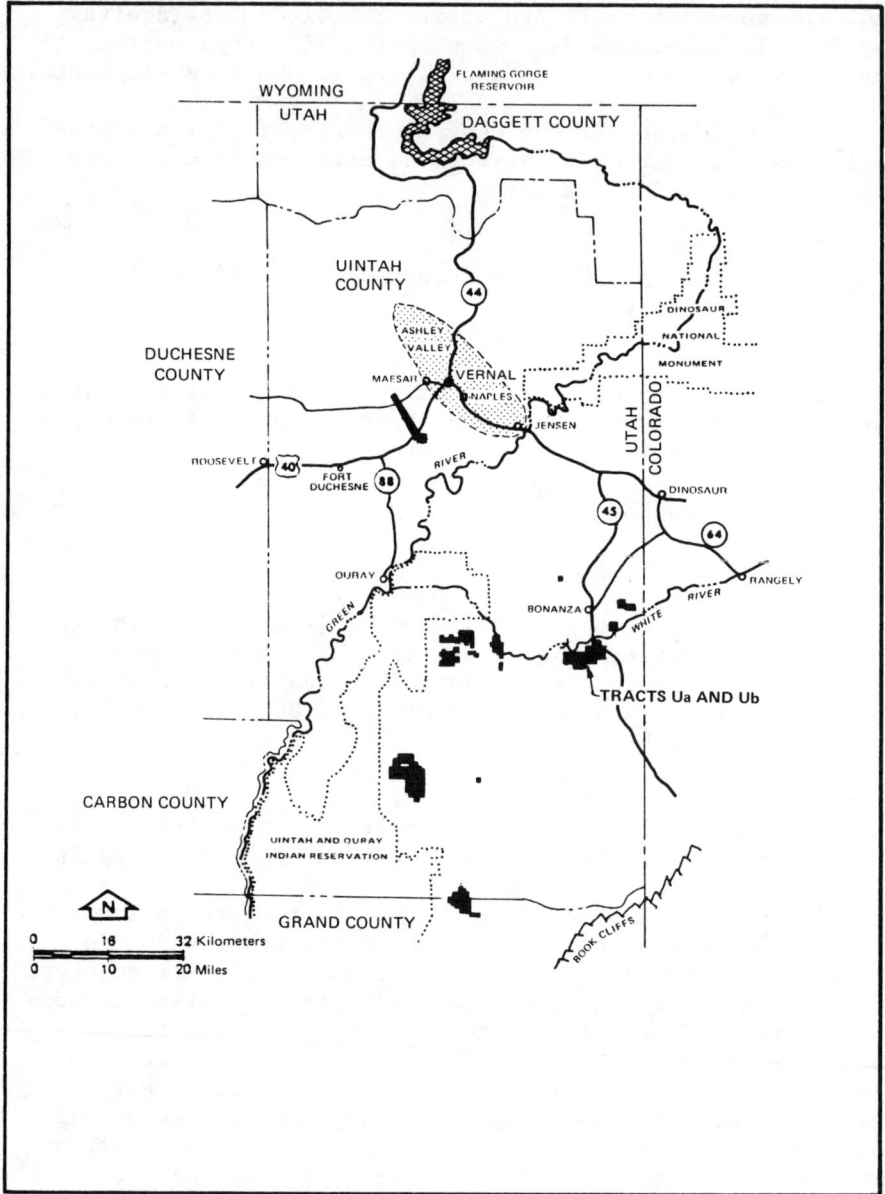

Figure 1. Location of Oil Shale Tracts Ua/Ub, Uinta Basin, UT

306

Geologic Setting

The Uinta Basin is part of the remnant Lake Uinta which covered the entire Uinta and Piceance Creek Basins during the Eocene Epoch. During that Epoch, a thick sequence of deep sediments derived from the surrounding highlands was deposited, forming the present-day sedimentary rocks. The extensive Tertiary sedimentary rocks located in the basin have been described by Cashion[5] where a detailed discussion of the geology of the area was provided. The two geological formations of current interest are the Green River and Uinta Formations.

Structural geology

The beds of the Green River and Uinta Formations exhibit a dip of less than 5` to the north and northwest. This is typical of the structure in this portion of the basin. Minor folding of the strata is extremely small in scale and differs little from the regional dip. No faults have been mapped in the vicinity of Tracts Ua-Ub although a large number of near-vertical joints occur throughout the area. The major joint set is oriented in a northwest direction.

Lithology

Green River Formation

The Green River Formation is composed primarily of marlstone and oil shale, with minor amounts of interbedded sandstone, siltstone, and thin volcanic tuff beds. It is comprised of three members: Douglas Creek, Garden Gulch, and Parachute Creek (Figure 2). The total thickness of the Formation is approximately 488 meters (1,600 feet) in the vicinity of Tracts Ua-Ub[1].

The Douglas Creek Member is primarily composed of interbedded shale, sandstone, and siltstone. The rocks are light-brown to gray in color and contain occasional beds of limestone. The member varies from approximately 107 to 244 meters (350 to 800 feet) in thickness in the vicinity of the study area.

The Garden Gulch Member consists primarily of gray and brown marlstone strata with minor amounts of siltstone, sandstone, and thin beds of oil shale. The Garden Gulch Member is of limited areal extent compared to other rock units of the basin. It is the stratigraphic equivalent of upper Douglas Creek Member beds found at the southern end of the basin. It is more than 31 meters (100 feet) thick in

307

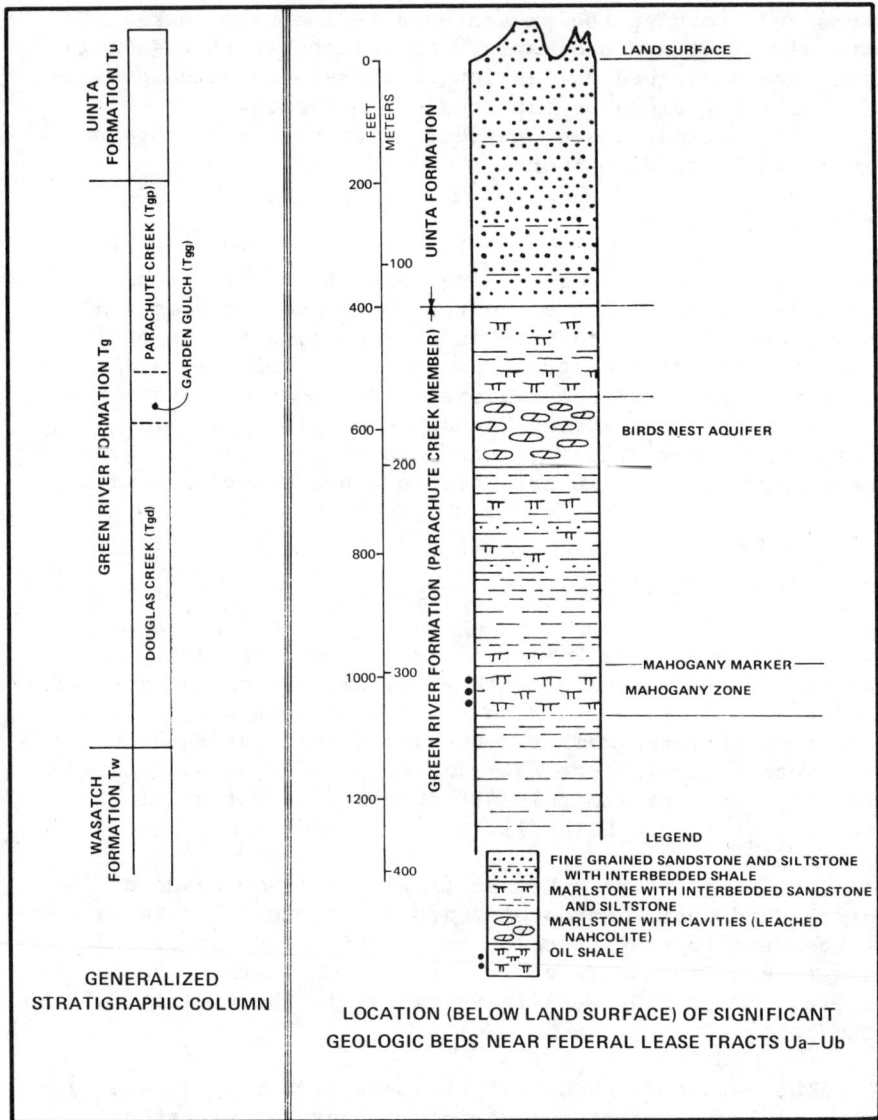

Figure 2. Stratigraphic Cross Section of Geologic Beds
Near Federal Lease Tracts Ua-Ub

the study area and thickens to the northeast, to almost 91 meters (300 feet) along the White River in western Colorado.

The Parachute Creek Member is located in the upper portion of the Green River Formation and is predominantly marlstone and dolomite with minor beds of siltstone, sandstone, and altered volcanic tuff. The member contains the most economically important sequences of oil shale deposits within the Green River Formation.

The Birds Nest Zone is found near the top of the Parachute Creek Member. It is a zone of strata approximately 35 meters (115 feet) thick[1] situated approximately 122 meters (400 feet) above (stratigraphically) the richest oil shale sequence in the area (the Mahogany Zone). The Birds Nest Zone is composed primarily of marlstone and shale but has significant water storage and transmitting capabilities as a result of secondary permeability and porosity accomplished through the leaching out of nodules and stringers of the soluble mineral nahcolite (sodium bicarbonate) in localized areas. The leached-out cavities resemble birds' nests in outcrops of the zone, explaining the origin of its name[5].

Uinta Formation

The Uinta Formation, which overlies the Green River Formation, is a fluvial fine-grained sandstone and siltstone interbedded with minor quantites of shale and congolomerate. Like the Green River Formation, it is of Eocene Age and generally found exposed throughout the basin. Maximum thickness of the Uinta Formation in the area has been estimated at 305 meters (1,000 feet)[5]. In the vicinity of Tracts Ua-Ub, the contact between the Uinta and Green River Formations averages about 122 meters (400 feet) beneath the land surface.

Hydrologic Setting

Detailed information on the Birds Nest Zone in the area has been acquired by extensive hydrogeologic work on Tracts Ua-Ub[1,2] and southwest of Ua-Ub by the U.S. Geological Survey. The saturated portion of the Birds Nest Zone is described in the literature as the Birds Nest Aquifer[6]. These previous studies have indicated that the quality of the water contained in the Birds Nest Aquifer is poor, and that the quantity available may be limited to regional recharge areas. Evidence of recharge to the aquifer from the White River in this area is found in similarities in water temperature between ground water and river water near the northern boundary of Tract Ua[7]. Other than monitoring wells on Tracts Ua-Ub, no wells producing water from this zone in the southwestern Uinta Basin are documented in available literature.

Transmissivity values for the aquifer vary widely, as the hydraulic properties of the zone depend upon the proximity to areas of recharge, as well as the numbers and sizes of cavities and fractures intersected by the test well. Transmissivity values measured from pump tests varied from less than 0.1863 meters per day (15 gallons per day per foot) to more than 93.15 meters per day (7,500 gallons per day per foot)[6].

Flow rates during well drilling on Tracts Ua-Ub ranged from zero to 2,650 liters per minute (700 gallons per minute)[6]. Aquifer saturation ranged from zero near the southeast boundary of Tract Ub to full saturation under approximately 73 meters (250 feet) of artesian head near the northern boundary of Tract Ua. Variability in well yield was a result of the degree of local aquifer saturation and potential head.

A detailed discussion of ground water quality in the Uinta Basin is provided by Austin and Skogerboe[8]. Localized ground water quality for Tracts Ua-Ub is provided by VTN Colorado[1]. A table summarizing the general characteristics is provided in Table I. The aquifer yields sodium-bicarbonate-sulfate type water and is generally considered poor quality for most industrial and domestic uses.

There is substantial potential for development of the Birds Nest Aquifer as a water supply resource for oil shale development. A number of test wells drilled into the aquifer have produced significant quantities of water. However, variability in production from the zone could put limitations on specific well sites. Proper well development techniques may result in ground water wells which produce suitable quantities of water for supplementing oil shale development requirements.

POTENTIAL EFFECTS ON THE BIRDS NEST AQUIFER

Potential effects on the Birds Nest Aquifer will be determined by the scale of oil shale development in the region and the technologies proposed for such development. Slawson[7] has suggested various activities by which potential pollutants could reach ground water systems at Tracts Ua-Ub. Activities that are common to most oil shale developments in the region include underground mining, surface

310

disposal of retorted shale and surface impoundments, and
ground water withdrawal (Figure 3). These activities and
their potential effects on the Birds Nest Aquifer are
discussed below.

Table I. Partial Summary of Ground Water Quality
for Birds Nest Aquifer Near Ua-Ub

	No. of Samples	Maximum	Mean	Minimum
Total Dissolved				
Solids (mg/l)	41	6000	4030	1430
pH, field	41	11.4	7.4	7.3
Alkalinity as				
CaCO$_3$ (mg/l)	47	1970	616	216
Calcium (mg/l)	47	210	130	5.7
Magnesium (mg/l)	44	260	170	67
Potassium (mg/l)	43	13	7.2	2.3
Sodium (mg/l)	42	1800	890	310
Chloride (mg/l)	47	280	75	36
Sulfate (mg/l)	46	3300	2150	180
Sulfide (mg/l)	43	150	19	0
Fluoride (mg/l)	43	11	1.7	0.2
Boron (mg/l)	43	78	5.5	0.03
Iron (mg/l)	43	4.6	0.68	0.02
TOC (mg/l)	31	66	20	4.6
Strontium (mg/l)	40	19	7.1	2.2

Source: VTN Colorado[1].

Underground Mining

    Generally, underground mining activities in the region
will include shaft-sinking (production and surface ventila-
tion shafts) and oil shale extraction from the Mahogany
Zone. Dewatering of the Birds Nest Aquifer adjacent to
the shafts will probably be required during construction.
Information obtained from exploration holes on Tracts Ua-Ub
indicates that substantial downward leakage from the Birds
Nest Aquifer into the mine zone is unlikely[3]. However,
if subsidence were to occur from mine failure or collapse,
appreciable dewatering of the aquifer into the mine zone
would probably occur. The overall result would be a decrease in water storage in the aquifer near the area of
subsidence.

311

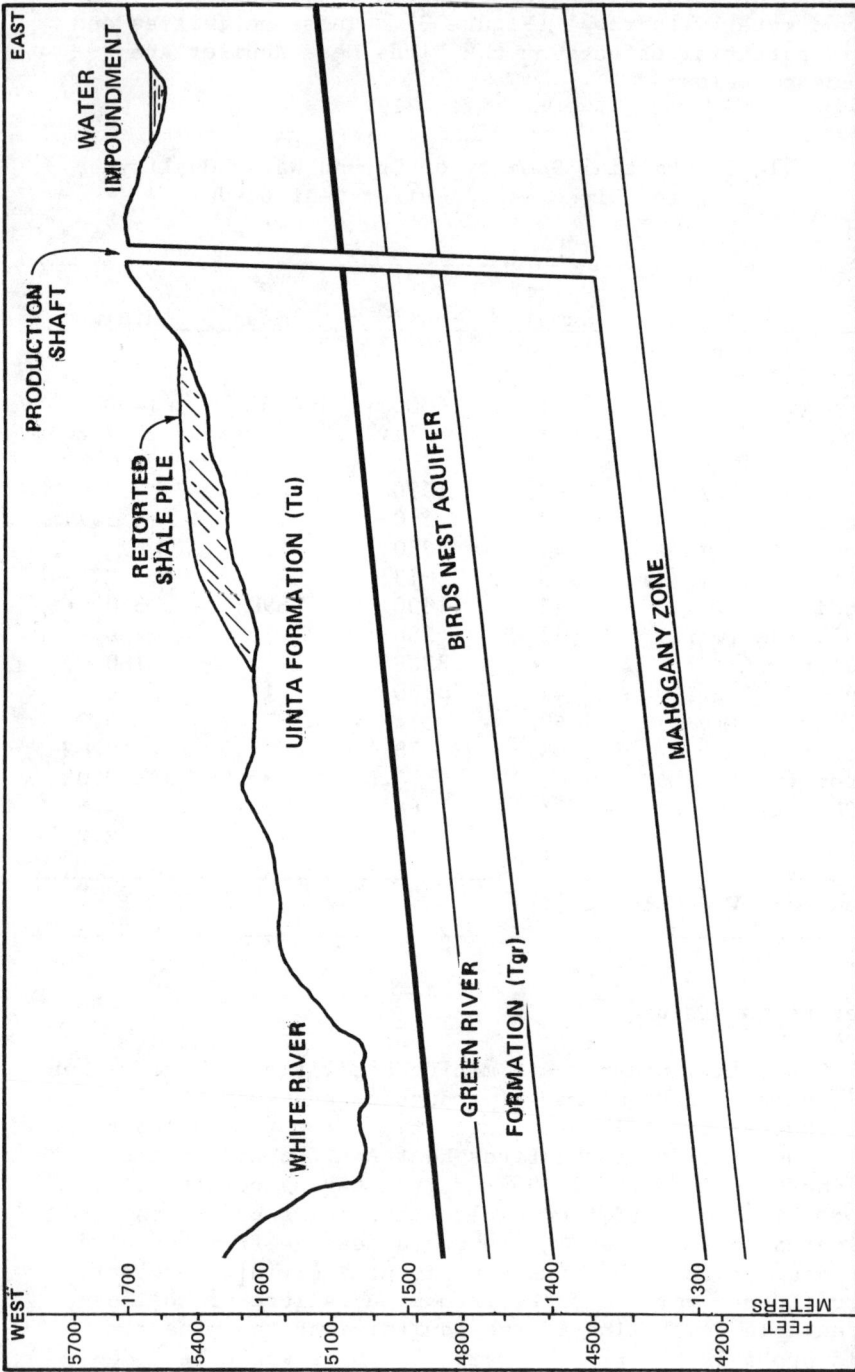

Figure 3. Generalized Cross-Section of an Oil Shale Project in the Vicinity of the White River, Utah

Surface Disposal of Retorted Shale and Surface Water
Impoundments

A possible long-term effect on the Birds Nest Aquifer
is likely to be associated with the surface disposal of
retorted shale, as leachate derived from water percolating
through the pile could eventually reach the aquifer. However, Slawson[7] has suggested that, under saturated conditions, a moderately compacted effort on the spent shale
pile (575,000 joules per cubic meter [12,000 ft-lbs per
cubic foot], 152 meters [500 feet] thick) would require 36
years in order to transmit leachate through the entire
vertical column of the pile. It would take a correspondingly greater amount of time during unsaturated flow conditions. In either case, 122 to 152 meters (400 to 500
feet) of Uinta sandstones and Green River marlstones would
then have to be penetrated to reach the Birds Nest Aquifer
on Tracts Ua-Ub.

It is unlikely that substantial amounts of highly
saline leachate would be able to reach the Birds Nest
Aquifer without some type of conduit (i.e. abandonded
wells, mine workings, fractures). However, if the aquifer
received leachates in an area adjacent to a stream and then
subsequently discharged that water, localized degradation
of a surface stream could eventually result. Regional
studies have indicated that the Birds Nest Aquifer contributes only minor amounts of water to existing surface water
supplies[7].

Surface wastewater impoundments could constitute another
source of contaminants due to percolation. However, the contaminants would have to breach any impermeable liners and
then migrate through 122 to 183 meters (400 to 600 feet) of
sedimentary strata in order to reach the aquifer.

Water Withdrawal

Presently, there has been no proposal to develop the
Birds Nest Aquifer as a potential water supply source in
association with an oil shale project. As previously
stated, uncertain availability and poor water quality are
significant barriers. However, in instances where water
quality does not play a significant role for a specific
purpose, such as for retorted shale cooling and dust suppression, substitutions for high quality water could be
made. If ground water withdrawal were employed to complete
the total water demand for an oil shale project, the Birds
Nest Aquifer could experience substantial dewatering. Such
dewatering could reduce ground water discharge to surface

streams, enhance ground water recharge from surface streams, or completely dewater the aquifer in the vicinity of withdrawal.

VTN Colorado[6] indicated that the baseflow from Evacuation Creek on Tract Ub is partially sustained by discharge from the Birds Nest Aquifer. Ground water contours substantiate that the incision of the upper Green River Formation by both the White River and the perennial and ephemeral stream channels near Tracts Ua-Ub, has allowed the Birds Nest Zone to drain adjacent to its outcrop in the canyon walls. Additional dewatering could reduce the water availability to these surface tributaries.

Recharge to the Birds Nest Aquifer from the White River, where a hydraulic connection exists, could be increased through dewatering of the aquifer down-gradient from the recharge point.

Ground water development distant from recharge sources could lead to complete dewatering of portions of the aquifer, because of the overall low transmissivity.

SUMMARY

It is believed that the Birds Nest Aquifer has significant potential for augmentation of surface water supplies in oil shale development. The most significant effects on the aquifer would most likely be due to its development as a water resource, which may result in widespread decrease in storage within the aquifer. Potential effects related to mine dewatering during shaft construction and oil shale extraction are expected to be minor and localized in comparison. Potential changes in ground water quality of the aquifer from surface disposal of retorted shale and surface water impoundments is considererd unlikely.

REFERENCES

1. VTN Colorado, Final Environmental Baseline Report, White River Shale Project, Denver, Colorado, 1977.

2. W.F. Holmes, Results of Test Drilling for Ground Water in the Southeastern Uinta Basin, Utah and Colorado, U.S. Department of the Interior, Geological Survey, Water Resources Investigations 80-951. Salt Lake City, Utah, 1980.

3. White River Shale Project, Detailed Development Plan for Federal Oil Shale Leases Ua-Ub, 1976.

4. U.S. Water Resources Council, Synthetic Fuels Development in the Upper Colorado Region, Technical Report, Washington, D.C., 1981.

5. W.B. Cashion, Geology and Fuel Resources of the Green River Formation, Southern Uinta Basin, Utah and Colorado, U.S. Department of the Interior, Geological Survey Professional Paper 548, Washington, D.C., 1967.

6. VTN Colorado, White River Shale Project, First Year Environmental Baseline Report, Volumes 1 and 2, Denver, Colorado, 1976.

7. G.C. Slawson, Jr., Groundwater Quality Monitoring of Western Oil Shale Development: Identification and Priority Ranking of Potential Sources, EPA-600/7-79-023, U.S. Environmental Protection Agency, Environmental Monitoring and Support Lab, Las Vegas, Nevada, 1979.

8. L.H. Austin and G.V. Skogerboe, Hydrologic Inventory of the Uintah Study Unit, Utah State Division of Water Resources, PRWG 40-5, Utah Water Research Laboratory, Salt Lake City, Utah, 1970.

CHAPTER 20

FEDERAL PROTOTYPE OIL SHALE PROGRAM:
AN ENVIRONMENTAL SAMPLING APPROACH

Donald R. Dietz
U. S. Fish & Wildlife Service
Grand Junction, Colorado

Roger A. Tucker
U. S. Geological Survey
Grand Junction, Colorado

The U. S. Department of the Interior, U. S. Geological Survey's Oil Shale Office (Grand Junction, Colorado)* has the responsibility to determine the environmental impacts associated with development of four 5,000-acre lease tracts --two each in Colorado and Utah. This paper discusses the present environmental monitoring program for the Federal Prototype Oil Shale Lease Tracts and suggests some apparent problems and possible solutions.

The U. S. Department of the Interior established the Prototype Oil Shale Leasing Program and leased four 5,000-acre tracts for commercial oil shale development in 1973. This program was designed to determine the feasibility of producing shale oil commercially and the environmental impacts and costs of commercial operation. Two years of baseline data and 5 additional years of precommercial phase development monitoring have been completed. The lessees have submitted Detailed Development Plans (DDPs) and Monitoring Programs for several mining technologies. One DDP is currently in a review status and the others have been approved subject to specific conditions including stipulations that the monitoring and mitigation programs be both flexible and dynamic.

* Now (1/82) part of newly established Minerals Management Service, U. S. Department of the Interior.

The Colorado lease tracts are in the Piceance Creek Basin of northwest Colorado and the Utah tracts are in the adjoining Uinta Basin (Figure 1). The two basins are separated by the Douglas Creek Arch along the Colorado-Utah border. Physiographically the basins are arid to semiarid plateaus that have been greatly eroded and are intricately dissected by intermittent streams. Broad, rounded divides characterize the region and drop off over float-covered slopes to narrow alluvial valleys. Alluvial fans spread from the confluences of many tributaries to the major stream channels.

The principal stratigraphic unit in both basins is the Green River Formation of Eocene age which contains more than 1.8 trillion barrels of shale oil in the form of kerogen-rich marlstone [1].

The oil shale region is characterized by warm summers, cold winters, abundant sunshine, low relative humidity, light precipitation, and large diurnal temperature variation. The growing season averages about 3 months in the higher mountains nearby and about 5 months at lower elevations. Temperatures average about 7°C and range from 37°C to -46°C. Annual precipitation varies from about 20 cm to 80 cm according to elevation. Both air quality and visibility are generally good.

A large mule deer herd inhabits the Piceance Creek Basin. This herd is world famous with sportsmen. Mule deer feed on mountain brush species such as bitterbrush, mountain-mahogany, and serviceberry. The higher basin rims are forested with aspen, pine, and fir; however, basin-wide the predominant vegetation is pinyon-juniper and sagebrush. In the more xeric parts of the Piceance Creek Basin, and much of the Uinta Basin, desert and salt-desert shrubs predominate along with several species of juniper. Grease-wood grows along drainage channels in both basins.

The basin ranges provide grazing for cattle and sheep as well as for feral horses. Many species of birds and mammals frequent the region as residents or as migrants. Fish and other aquatic resources are not abundant, but several threatened and endangered fish species inhabit portions of the three major rivers, the Colorado, Green, and White [2].

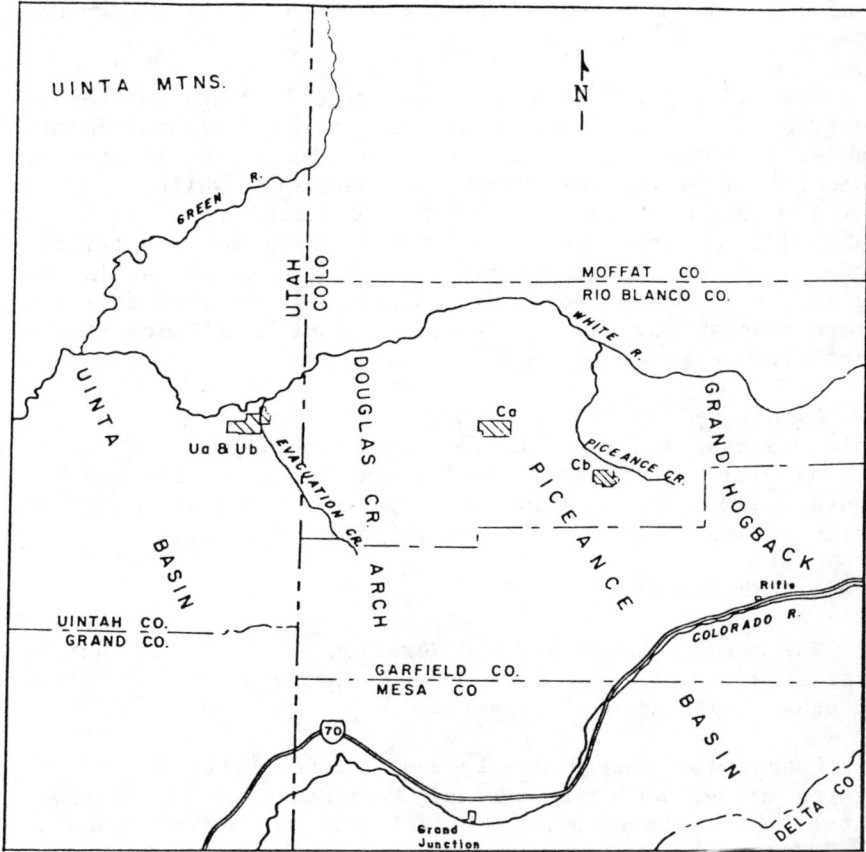

Figure 1.   Location of Prototype Oil Shale Tracts C-a, C-b,
            U-A, and U-b [2].

319

DEVELOPMENT

The success of monitoring depends upon the degree of knowledge of the mining development processes. The two Federal tracts in Colorado are being developed by both above-ground and modified in-situ (MIS) retorting processes, with raw shale feed derived from open pit and room-and-pillar mining. The two Federal tracts in Utah will be developed by room-and-pillar mining with above-ground retorting.

The MIS process involves underground mining to remove about 20 to 40% of the oil shale to provide a natural retort chamber. Remaining shale is then retorted in place, producing a desirable refinery feedstock while leaving behind a confined rubble pile of processed shale. Mined shale will be processed above ground in a surface retort. Surface retorting requires massive handling of raw shale and the resultant processed shale. Shale oil produced from any retort process may be upgraded in a special refinery at the plant site or at some other location.

Associated with the mining, retorting, and processed shale disposal are several factors which have a potential for affecting both the on-site and off-site environment. Figure 2 shows development actions associated with an MIS operation and their potential environmental effects.

IMPACT ASSESSMENT

The nature, location, and duration of an impact must be forecast before planning monitoring programs for wildlife and other environmental factors.

Leases for energy development on federally owned land require either an Environmental Assessment or an Environmental Impact Statement. The environmental stipulations of the Federal oil shale leases state: "The lessee shall conduct the monitoring program to provide a record of changes from conditions existing prior to development operations, as established by the collection of baseline data." Conditions for approval of the lease-required Detailed Development Plan (DDP) also state: "The environmental monitoring plan shall be revised as needed, based on the analysis of the final baseline report--submitted for review and approval by the (Federal) Mining Supervisor prior to commencement of commercial development" [4].

320

DEVELOPMENT-PERTURBATION MATRIX

DEVELOPMENT ACTIONS	Creates Surface Disturbance	Creates Noise	Creates Odor	Forms Barrier	Causes Collisions	Creates Dust/Particulates	Emits Gases	Alters Runoff	Creates Erosion	Alters Surface Water	Alters Groundwater	Alters Water Quality	Creates Chemical Leaching	Creates Water Effluents	Creates Seismicity	Creates Landslides/Slumps	Creates Subsidence	Alters Human Activity
MINING																		
Shaft Sinking	●	●				●				●			●					
Drift Development		●				●	●			●			●					
Retorts		●				●	●			●			●		●			
Post Retort						●				●	●	●	●	●		●		
PROCESSING																		
Retorting		●	●			●	●			●			●		●			
Gas Treatment	●	●	Q				●				●	●	●					
Steam Generation	●	●	●			●	●				●		●					
Water Treatment	●	●	●				●			●	●		●					
Oil Treatment	●	●	●				●				●		●					
Cooling Towers	●	●				●	●				●	●	●					
DAMS & IMPOUNDMENT	●			●		●		●	●	●	●	●	●					
OVERBURDEN & SHALE																		
HANDLING & DISPOSAL																		
Conveyor Belts	●	●		●		●												
Trucks		●	●		●	●	●											
Overburden & Topsoil	●					●		●	●	●	●	●	●				●	
Raw Shale Pile	●					●	Q	●	●	●	●	●	●			●		
Processed Shale Pile	●			●		●	Q	●	●	●	●	●	●		●	●		
TANK STORAGE	●			●			●				●							
WASTE MANAGEMENT																		
Raw Sewage	●		●			●		●	●	●	●	●						
Trash	●		●			●	●		●	●	●	●						
By-product Waste	●		●			●	●			●	●	●						
POWER GENERATION	●	●				●	●					Q	Q	Q				
UTILITY CORRIDORS																		
Pipe Lines	●																	
Power Lines	●			●	●													
TRANSPORTATION																		
Roads	●			●	●	●		●	●	●	●							●
Vehicles		●			●	●	●											●
MINE WATER																		
Dewatering	●	●	●				●		●	●	●	●	●	●	●		●	
Reinjection	●	●								●	●							
Mine Waste Water											●		●					
Surface Discharge	●	Q	●	●			●	●	●	●	●	●	●	●				
Evaporation		●	●			●			●		●							
SURFACE FACILITIES																		
Buildings	●					●	●											
Work Areas	●	●	●	●	●													●

Legend: ● = Potential Effect   Q = Questionable

Figure 2. The development-perturbation matrix shows potential perturbations to environment from development for modified in-situ retorting [5].

321

The Federal oil shale lease required each lessee to conduct a 2-year environmental baseline program before submitting a DDP and beginning any construction. The 2-year baseline program was intended to establish baseline conditions in the natural environment from which significant development-related perturbations could be measured. It soon became evident that treatment and control designs were necessary to separate development effects from random natural changes. Careful design is especially necessary for variables like faunal populations [2].

The prototype program is unique in that data collected on about 20,000 acres may be used to predict the potential environmental impacts on several million acres of potentially leasable oil shale lands in Colorado and Utah. Accordingly, these monitoring programs must address much more than specific site impacts and mitigation. The wide scope of the program allows the Department of the Interior to project regional impact for a mature oil shale industry.

MONITORING APPROACH

The environmental monitoring conducted by the lessees has been described in previous papers [2, 6]. Basically each program involved the following steps:

1. Describe baseline conditions.

2. Determine the sequence of major development actions in time and space.

3. Estimate important potential perturbations to environmental factors caused by development actions.

4. Select biotic and abiotic response factors for monitoring.

5. Design experimental programs to identify and test statistical significance of development-caused effects.

6. Develop an interrelationship matrix to consider interactions, related variables, cause-and-effect, etc.

7. Incorporate concepts of treatment and control, biological threshold values and mitigation based on degree, direction, and duration of environmental effect.

8. Develop systems models that permit forecasting impacts based on simulation analysis and designing integrated mitigation programs that incorporate results obtained from optimization models.

The present monitoring program approved for the Federal Prototype Oil Shale Lease Tracts is based on (1) compliance monitoring for State and Federal standards and (2) statistical null hypothesis testing for significant changes from baseline data and data from control sites. Currently, impacts affecting interrelationships are monitored more qualitatively than quantitatively.

The U. S. Geological Survey Oil Shale Office (OSO) has prepared a monitoring guideline manual to assist in the review and management of the lessees' monitoring program [7]. The manual presents the rationale for OSO's management program for environmental monitoring. These guidelines reflect numerous meetings to reach a consensus among the multidisciplinary groups of specialists in:

1. Describing and delimiting the most probable development actions, i.e., breaking down engineering actions to their component parts over time and space.

2. Forecasting the specific environmental components that would probably be perturbed (affected enough to cause a change).

3. Selecting the most important and/or relevant biotic and abiotic parameters to detect a perturbation and/or combination of perturbations.

Three matrices were prepared to facilitate the use of the guidelines. The first matrix (Figure 2) compared development actions with environmental perturbations.

The second matrix compares potential environmental perturbations with selected biotic and abiotic response parameters. While the entire matrix is too large to present here, the monitored parameters include the following:

Fauna
- deer distribution and density
- feral horse distribution and density
- small mammal composition and density
- avifauna distribution and density
- threatened and endangered species observations
- general wildlife observations

Flora
- plant condition and stress
- plant cover and density
- plant production and utilization
- plant chemical composition
- threatened and endangered species occurrence
- vegetation types and state of succession

Aquatic
- benthos biomass and diversity
- periphyton biomass and diversity
- fish (contingency)

Soil-Site
- elevation, aspect, percent slope
- soil series
- soil depth - A and C horizons
- soil temperature
- soil moisture
- soil pH and conductivity
- soil elements, major, minor, and trace

Hydrology
- stream and spring/seep flow and quality
- alluvial water levels and quality
- upper and lower aquifer levels and quality
- sediment characterization

Geology
- subsidence

Meteorology/Climatology
- air temperature and relative humidity
- precipitation and evaporation
- barometric pressure
- wind speed and direction
- inversion
- noise

Air Quality
- sulfur dioxide, nitrogen dioxide, carbon monoxide
- nonmethane hydrocarbons, ozone
- total suspended particulates
- size distribution
- trace elements
- gross radioactivity
- carcinogens
- visibility
- plume transport and diffusion

The third matrix compares the aforelisted parameters with each other in mirror image format. This facilitates a rapid scan of the many intrarelationships and interrelationships that could be affected by development.

The matrices are the basis for environmental monitoring and modeling efforts. Cause-and-effect relationships can often be predicted by the use of the matrices coupled with the lessee's Detailed Development Plans. This screening identifies those key factors that must be incorporated into experimental designs for statistical evaluation.

The matrices indicate the environmental effects of one or more development actions or, conversely, identify which actions could affect an individual parameter.

For example, construction of roads (a development action) could cause the following perturbations:

1. Create a surface disturbance,
2. Form barriers to animals,
3. Cause collisions between animals and vehicles,
4. Create dust and particulate matter,
5. Alter water runoff,
6. Create erosion,
7. Alter surface water location and movement,
8. Alter ground water, and
9. Alter human activity.

Each of the above can be monitored by studying the changes in various response parameters.

One might pose the question - How might roads and vehicular traffic on the lease tract affect trout in Piceance Creek (a stream north of Tract C-b)? The development action/perturbation matrix (Figure 2) shows nine possible perturbations from road construction and several from

subsequent vehicular use.   The environmental perturbation/
response parameter matrix (not shown but available from OSO)
shows that the following perturbations are likely to affect
fish on and/or below the lease tracts:

1.   Water effluents,
2.   Chemical leaching,
3.   Alterations in surface water, and
4.   Water quality.

While initial construction of the road system may cause
all four perturbations, presence and use of the road would
mainly alter surface water flows by affecting natural drainage patterns.   Improper road maintenance and vehicle misuse
could create erosion and thus contribute sediment and
perhaps toxic chemicals to the runoff water.

Fish populations are closely linked to their
environment.   The mirror image matrix discloses that fish
populations are related to the following monitored
parameters:

1.   Benthos,
2.   Periphyton,
3.   Stream flow,
4.   Stream quality,
5.   Spring and seep flow and quality, and
6.   Sediment characteristics.

Other habitat characteristics that may be closely
related to trout populations will be monitored in the future
on the lease tracts if deemed important.

Statistical evaluation tests for detecting significant
impacts must address several important factors:

1.   Adequacy of baseline data,
2.   Suitability of control sites,
3.   Biological threshold levels of organisms
     monitored,
4.   Accuracy and precision of the sampling technique,
5.   Ability to interpret the test results, and
6.   Translation of analytical results into effective
     control and/or mitigative actions.

A typical experimental design and data analytic method for detecting impacts upon aquatic biota was presented in "C-b Shale Oil Project Development Monitoring Program" [8]. The objective was to test the following null hypothesis:

H_O: No significant change exists in periphyton biomass over time and

H_O: No significant difference exists in periphyton communities at control stations versus development stations from baseline data, recognizing the differences during baseline.

Samples are collected in triplicate on a monthly basis and hypothesis testing is done at an alpha level of 0.10. Analysis of variance will test significance of difference.

If an impact is indicated by hypothesis testing on periphyton, then, as part of a contingency monitoring program, fish shocking will be carried out to test if fish populations are also impacted.

The relationship of periphyton population to water flow and quality will be studied using correlation analysis.

Aquatic models using data collected from both aquatic biology and hydrology programs are being investigated as a means of ecosystem analysis and management prediction.

Interrelationship monitoring is a prerequisite to development of quantitative prediction and other systems models.

If we seek the most likely impacts on the most important biota, then our candidate for intensive monitoring is the mule deer and its habitat as affected by oil shale development. Mule deer is the species that draws the most public attention. The Piceance Creek Basin has historically attracted sportsmen from around the world to hunt its famous mule deer herds.

The largest site-specific impact on mule deer will be the destruction of its habitat. This is not a difficult sampling problem since X acres of surface destruction = X acres of habitat lost, hence X numbers of deer days use are lost. The effect on mule deer of the plant operation is more a "one shot" research problem than a routine monitoring problem because one mainly needs to determine the

327

distance that a deer will keep from disturbance areas and the major characteristics of the avoidance area.

The impact most difficult to monitor will be associated with airborne pollutants that will likely have an insidious and long-term effect on habitat and which will also have both direct and indirect effects on faunal and floral species. The effect will probably be chronic in most cases but could be acute in some instances.

Designing a flora and fauna monitoring sampling network to detect effects of airborne pollutants is complicated by the lack of unaffected control areas. The Piceance Creek and Uinta Basins have complex surface wind fields dominated by upslope flow during the day (anabatic) and downslope flow at night (katabatic). The prevailing wind direction varies from WSW in Colorado to SSE over the Utah tracts. Where the synoptic pressure gradients are weak, the area is dominated by locally generated winds near the surface. These winds vary widely depending on the local terrain features (Table 1).

The wide variability in wind speed and direction renders suspect any data extrapolation from air-quality-monitoring trailers to surrounding terrain, and hence habitat. Possibly combining plume studies utilizing SF_6 (sulfur hexafluoride) tracer with air quality and meteorology data from the air quality-meteorological trailers will provide an adequate basis for placement of floral and faunal sampling stations.

Statistically it may be possible to employ only rather gross correlation studies to this problem. Without the use of on-site analysis and sampling apparatus at the biota sampling sites, "cause-and-effect" studies relating specific chemicals to biota via methods such as multiple regression and principal component analysis may not be possible.

Determining the levels of chemicals and chemical combinations that affect threshold tolerance of biota in the laboratory and determining the morphological and/or physiological indicators manifested in the plants and animals can possibly provide the basis for modifying the field sampling program for biota.

Table 1. Prevailing Wind Direction/Speed (in m/s)
on the Utah Tracts during 1980.
(White River Shale Oil Corp., 1981. [9]
Site locations shown on Figure 3)

1980	Site			
Month	4	6	11	13
Jan	W/2.5	W/2.2	NW/2.2	WNW/1.8
Feb	W/2.2	SE/2.0	NW/2.0	WNW/1.3
Mar	W/3.1	SSE/3.1	SE/3.1	SE/2.4
Apr	SE/3.4	SE/3.6	SE/3.4	WNW/2.7
May	S/3.8	SSE/3.6	SSE/3.1	SE/2.7
Jun	S/4.5	SSE/4.5	S/4.0	S/3.4
Jul	WNW/3.1	SSE/3.6	SE/3.4	SE/2.7
Aug	WNW/3.6	SSE/3.8	SE/3.6	SE/2.9
Sep	WNW/2.9	SSE/3.1	SE/3.1	SE/2.2
Oct	W/2.4	SE/2.7	SE/2.7	SE/1.6
Nov	W/2.4	SE/2.4	SE/2.2	SSE/1.3
Dec	W/2.0	SE/2.2	SE/2.2	SSE/1.3
Annual	W/3.0	SSE/3.1	SE/2.9	SE/2.2

IMPACT—MONITORING RATIONALE AND RECOMMENDATIONS

The Federal Prototype Oil Shale Leasing Program, though
still young (about 8 years), has undergone considerable
growing pains. In particular, the environmental monitoring
has changed greatly since its inception and is still
changing in response to new techniques and to newly
identified needs.

The program began as an outgrowth of the Environmental
Impact Statement for the Prototype Leasing Program [10]
required by the National Environmental Policy Act (NEPA).
The early data were mostly descriptive and of the laundry
list variety collected for a wide array of tract components.
Such data, though serving to describe baseline conditions,
were more qualitative than quantitative.

The baseline data, even though satisfying the environmental stipulations of the oil shale lease, were not
necessarily suitable for impact monitoring. Some of the
problems were:

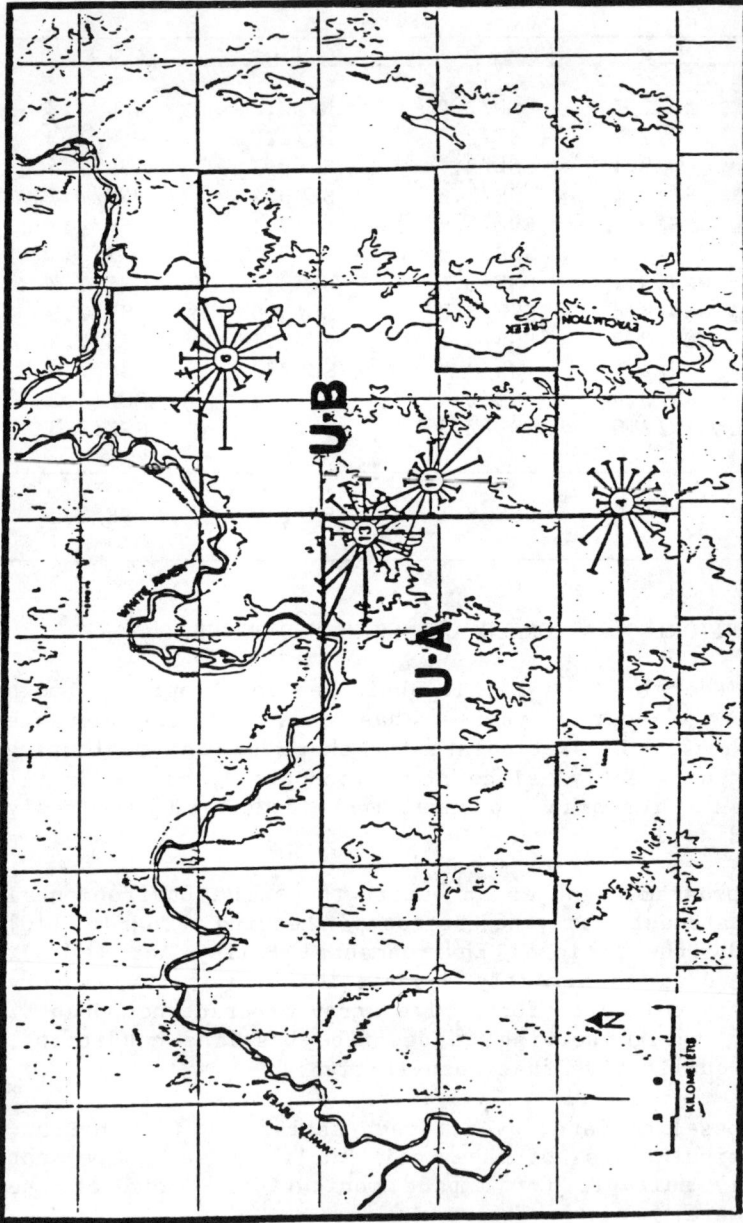

Figure 3. Directional wind roses for January, 1980, at four monitoring stations on Utah tracts U-a and U-b. The length of each bar represents the frequency of winds from the direction towards which the bar points [9].

1. When baseline programs were started, the complete
    engineering design was in flux,

2. Rigorous statistical designs utilizing the
    treatment and control concept were not addressed,
    and

3. Neither statistical hypothesis testing nor eco-
    system analysis was a major consideration.

Hence, many analytical requirements were not met such as
random selection of sites, adequate replication of treatment
and control stations, and rigid quality control specifying
allowable sampling variance and error. Fortunately, since
the baseline data collection was completed, many of these
problems have been corrected. Because full-scale commercialization is still some time in the future, current
monitoring is mostly in the baseline mode. Although
considerable land disturbance has taken place for mine
development, water and air quality have scarcely been
affected, and so habitats and their inhabitants have not
been significantly disturbed from pollutants. The interim
period before full-scale development will permit a continual
evaluation and refinement of the in-place environmental
monitoring network and procedures on the Federal Prototype
Lease Tracts.

As previously stated, the OSO staff will use the
present precommercial development period to evaluate
effectiveness of the existing environmental monitoring
program approved for the lessees.

Before the current monitoring program began, the
lessees and the OSO mutually decided on statistical
hypothesis testing as the basis of the monitoring program.
Yet this approach has not been without criticism. Hakonson
and White [11], among others, have recommended that the
monitoring program be based more on ecosystems. Several
statisticians have suggested that classical experimental
designs may not be valid on the lease tracts, because
several basic statistical rules are violated [2]. The
problem is well illustrated by Eberhardt [12], who stated:

The main pitfall, however, is that the essential
inferences in present day experimental design rest
firmly on the foundation of random assignment of
experimental "treatments" and "controls" to a

331

considerable number of test plots, animals, batches of product, or whatever. In the circumstances considered here, there is just <u>one</u> treated area and there may or may not be one or more "controls" while randomization applies only to subsampling.

Thus, there is an obvious problem in applying classical inferential methods to preoperational and postoperational data on <u>one</u> mine site. Relying solely on ecosystem models probably holds even greater problems. Common criticisms of relying on ecosystem models to detect environmental effects are:

1. They lack the sensitivity for early detection of site-specific impacts,

2. Models tend to proliferate experimental error,

3. Most ecological models have no statistical basis, and

4. Many quantitative models are never validated (or invalidated) nor can they be.

Eberhardt [12], in discussing the use of models in impact studies, stated:

My chief concerns with present day modeling of ecological processes have to do with our present day lack of knowledge of structural details in ecological systems, the almost nonexistent attention to sampling and estimation problems in most modeling efforts, and the general tendency to suppose that our models are presently useful as predictive devices.

The problem facing OSO and all groups attempting to determine the environmental effects of a development action is how to marry the two disciplines in such a way to minimize the problem areas. There is some truth in the story about the specialist knowing everything about almost nothing and the generalist knowing almost nothing about everything. We desire to know enough about the really important aspects to make the proper decisions concerning control and mitigation of detrimental impacts to the environment and natural processes.

Several recent reference books discuss the determination of impacts to the environment [13, 14, 15]. From
our own experiences with the Prototype Leasing Program and
the recommendation of others [12, 13, 14, 15, 16, 17], we
propose the following guidelines in designing environmental
impact studies:

1. Field Survey--Do a field reconnaissance of the area to
    be affected. In general terms describe major abiotic
    and biotic components. Prepare field maps depicting
    major landforms, drainages, cultural features, soil
    type, surface geology, vegetation types, and other
    obvious features. If possible, describe general
    ecosystem types and apparent interactions [15, 16].

2. Development Plans--Prepare detailed maps depicting the
    location of all facilities, roads, utility corridors,
    and point sources for potential pollutants. Delineate
    all major development actions sequentially in time and
    space [5, 7, 18, 19].

3. Literature Search--Do a search on major facets of the
    program including resource base, historic and present
    land-use practices, climatic records, ecology, poten-
    tial environmental impacts of the technology planned,
    biological threshold tolerance constraints, control
    mechanisms for preventing and/or alleviating environ-
    mental hazards, possible mitigation for impacts, rele-
    vant environmental monitoring techniques, and
    ecosystem models prepared in similar habitats.

4. Problem Analysis--Prepare an in-depth analysis con-
    sisting of:

    a. Breakdown of development actions into separate,
    sequential operations quantified as to volumes, surface
    areas, distances, etc.

    b. Preparation of a matrix relating development
    action to environmental perturbation.

    c. Listing of abiotic and biotic parameters that are
    candidates for monitoring and a mirror image matrix to
    help focus on related biotic and abiotic parameters.

    d. An interdisciplinary team workshop to formulate a
    crude systems model. This conceptual word model will

333

provide internal communications among the program
managers, mining and processing engineers, environmental specialists (air, water, biology, and
reclamation), and data management and analysis group.
The word model should show what and where vital
information is still needed, what each specialist needs
from the others, and what the specialist can supply.
The quantity and quality of existing data and other
information should be evaluated.

5. Monitoring Program--Design an operational and environ-
    mental monitoring program to meet all regulatory
    requirements as well as address critical operational
    and ecological problems. Some essential objectives of
    this program are:

    a. The baseline data collection program should
    effectively describe the environment to meet NEPA and
    lease requirements and should provide statistically
    valid data on control and on predevelopment (treatment)
    sites.

    b. This pretreatment phase should be designed to
    collect data on those key or critical parameters and
    systems identified in the conceptual model. This will
    require integrated and coordinated sampling sites and
    time periods among disciplines. Monitoring effort
    should concentrate on the most critical functions that
    permit the tract ecosystem to maintain stability.

    c. A data base management system and quality assur-
    ance program should be designed and started to handle
    and validate the collected data. It should be designed
    to standardize as much of the data as possible from the
    various lease tracts and to facilitate rapid analysis
    of integrated data. This will require the acquisition
    of necessary hardware, software, and operator/analysts.

    d. The precommercial development period where either
    pilot-size demonstration plants or small commercial-
    sized works are built is the opportune time to test and
    adjust the monitoring program. The mitigation program
    must also be coupled to the monitoring for evaluation
    purposes [6].

    The monitoring program will be centered on
    statistical analyses such as hypothesis testing or

334

quality control but must also provide the means for detecting unforeseen problems or opportunities.

e.   The early commercial period of development should also be used for model development. As data on impacts are accumulated, the opportunity exists for making long-range predictions concerning both operational results and environmental consequences. Simulation models and analysis can be powerful tools for the manager whose decisions affect both future operations and long-range impacts. Optimization models can also provide a means for both minimizing adverse impacts and maximizing development.

The monitoring concept presented here, we believe, offers a step forward in determining, controlling, and mitigating adverse environmental impacts from development programs irrespective of the type of development. It attempts to make baseline programs an integral part of development monitoring. Marrying systems analysis to inferential statistical design should let both disciplines do what they do best, i.e., assure that the big picture is looked at with special focus on critical functions and that site-specific/ action-specific impacts are properly tested statistically.

The skillful use of these powerful tools will give the manager the best available methods for rapid and effective decision-making.

LITERATURE CITED

1. Weeks, J. B., et al. 1974. Simulated effects of oil-shale development on the hydrology of Piceance basin, Colorado. U. S. Geological Survey Professional Paper 908.

2. Dietz, D. R., et al. 1978. Assessment of oil-shale development--A problem in statistical design. In Gardner, D. A., and Tykey Truette, compilers-editors, Proceedings of 1977 Department of Energy Statistical Symposium (Oct. 26-28, 1977, Richland, Wash.). Oak Ridge National Laboratory, CONF-771042, p. 87-109. Oak Ridge, Tennessee.

3. U. S. Bureau of Land Management. 1975. Oil Shale Lease, Tract C-a, Serial No. C20046.

4. Peter A. Rutledge, Area Oil Shale Supervisor, written communication, August and September 1977.

5. National Research Council. 1979. Mining and processing of oil shale and tar sands. In Appendix B of Surface Mining of Non-coal Minerals--A Study of mineral mining from the perspective of the Surface Mining Control and Reclamation Act of 1977. National Academy of Sciences. Washington, D.C.

6. Dietz, D. R. 1979. Interdisciplinary developmental and ecological assessment: A prerequisite to effective design of environmental mitigation. In The Mitigation Symposium--A national workshop on mitigating losses of fish and wildlife habitats, Colorado State University, Fort Collins, July 16-20. U. S. Department of Agriculture, Forest Service, Technical Report RM-65.

7. U. S. Geological Survey. 1979. Guidelines for monitoring the environmental effects of development on the Federal Prototype Oil Shale Lease Tracts. Oil Shale Office. Grand Junction, Colorado.

8. Occidental Oil Shale, Inc. 1979. Development monitoring program for Oil Shale Tract C-b, C-b Shale Oil Project (Feb. 23, Grand Junction, Colo.).

9. White River Shale Oil Corporation. 1981. Progress report--Environmental programs, White River Shale Project, Federal Prototype Oil Shale Leases U-a and U-b (Aug. 1981, Grand Junction, Colo.).

10. U. S. Department of the Interior. 1973. Final environmental statement for the Prototype Oil Shale Leasing Program. Washington, D.C., U. S. Government Printing Office, v. 1.

11. Hakonson, T. E., and G. C. White. 1981. Ecological effects of oil shale development--Problems, perspectives and approaches. In Oil Shale: The Environmental Challenges, Vail, August 11-14, Proceedings. Colorado School of Mines Press, p. 105-123. Golden, Colorado.

12. Eberhardt, L. L. 1976. Quantitative ecology and impact assessment. Journal of Environmental Management, v. 4, p. 27-70.

13. Green, R. H. 1979. Sampling design and statistical methods for environmental biologists. John Wiley & Sons. New York.

14. Holling, C. S., editor. 1978. Adaptive environmental assessment and management. John Wiley & Sons. New York.

15. Ward, D. V. 1978. Biological environmental impact studies: Theory and methods. Academic Press. New York.

16. States, J. J., et al. 1978. A systems approach to ecological baseline studies. Colorado, U. S. Fish and Wildlife Service, Biological Services Program. Fort Collins, Colorado.

17. Thomas, J. M. 1977. Factors to consider in monitoring programs suggested by statistical analysis of available data. In W. Van Winkle, editor, Proceedings of the Conference for Assessing Effects of Power-Plant-Induced Mortality on Fish Populations (May 3-6, Gatlinburg, Tennessee). Pergamon Press, p. 243-255.

18. Leopold, L. B., et al. 1971. A procedure for evaluating environmental impact. U. S. Geological Survey Circular 645.

19. Moore, Russell, and Thomas Mills. 1977. An environmental guide to western surface mining--Part 2: Impacts, mitigation, and monitoring. U. S. Fish and Wildlife Service, Biological Services Program. Fort Collins, Colorado.

PART 6

ENERGY IMPACTS

CHAPTER 21

POTENTIAL EFFECTS OF PROCESSED OIL SHALE
LEACHATE IN THE AQUATIC ENVIRONMENT

Judith G. Dickson
    Utah Water Research Laboratory, Utah State University,
    Logan, Utah

Mary L. Cleave
    Houston Space Center, NASA, Houston, Texas

David L. Maase
    Biospherics Inc., Rockville, Maryland

V. Dean Adams
    Utah Water Research Laboratory, Utah State University,
    Logan, Utah

Donald B. Porcella
    Tetra Tech, Lafayette, California

INTRODUCTION

The water pollution potential of spent oil shale residues is a major concern regarding the environmental impact of oil shale development [1,2,3,4]. Salts, trace elements, and organics, including carcinogenic polycyclic aromatic hydrocarbons (PAH), that may leach out of spent shale as a result of weathering, and into surface and groundwater, pose a serious problem to public health and natural resources.

At the Utah Water Research Laboratory, several projects have been directed toward characterizing the spent shale leachates, in terms of their potential pollutants, and the effect of these materials on aquatic organisms. The creative use of bioassays and chemical analysis has allowed conclusions to be made regarding the impact of spent shale disposal and, also, recommendations to be used to direct future research in this area. Three of these studies will be presented. For more details concerning these projects, refer to the original work [5,6,7,8].

EFFECTS OF SALINITY AND OIL SHALE
LEACHATE ON PHYTOPLANKTON PRODUCTIVITY

Introduction

There are several sources of poor quality, high salinity water associated with oil shale processing and disposal. It is estimated that a 50,000 barrel/day operation in Utah would produce a low quality waste stream to the process shale disposal site of between 0.4-0.7 ft^3/s [9]. An additional source of poor quality water results from leaching spent shale with good quality water in order to establish vegetation [10]. Studies show that the process will leach salinity ions into the water on a continuing basis [11,12]. Increasing the salinity in the Colorado River system due to diversion of water for shale processing is also of concern. There will be a steady increase in salinity concentration at Imperial Dam (up to 14 mg/ℓ in 1990) as a result of this practice.

There is very little published literature describing the effects of increasing salt concentration on the growth of freshwater phytoplankton. The work that is available indicates that at low concentrations of total dissolved solids (TDS), the effect is stimulatory [14]. This positive effect on productivity is mediated through an increase in the availability of nutrients. In addition, many of the major cations are required for algal growth, and therefore, may increase productivity as TDS increases. The stimulatory effect of TDS on production increases with increasing salt concentration to a certain point (1400 mg/ℓ for Alberta Lakes) [14], after which a further increase in TDS in these waters tends to inhibit production. The cause of growth suppression is unknown. It is speculated that particular elements may reach toxic levels, or that increased osmotic pressure of the solution may alter normal cell functioning [15].

The present study [5] was undertaken to determine the effect of relatively high levels of salinity on freshwater phytoplankton productivity. The ultimate purpose was to evaluate the potential impact of oil shale development given that: 1) spent shale leachate, a highly saline waste stream, will enter the Colorado River system, and/or 2) the diversion of water for shale processing will significantly increase salinity downstream.

Methods

Both standard and modified algal assay procedures were used. Standard algal assay procedures were conducted to provide data which would be comparable to other investigations [16,17]. The modifications to the standard algal assay procedure were made using the general guidelines presented in Standard Methods [18]. The standard algal assay organisms utilized, Selenastrum capricornutum, Printz, and Anabaena flosaquae (Lyngb) De Brebisson, were secured from the National Eutrophication Research Program. These algae were maintained in AAM, a synthetic algal nutrient medium [16]. The indigenous algae utilized, Synedra delicatissima var. angustissima and Scenedesmus bijuga, were isolated from samples collected at the Wahweap station of Lake Powell [19]. These algae were isolated and maintained in Lake Powell Synthetic Medium (TDS = 780 mg/ℓ) which is AAM modified by the addition of the major cations and anions measured in Lake Powell [5,20]. Unialgal cultures of all four of the test algae (hereafter referred to as Selenastrum, Anabaena, Synedra, and Scenedesmus), were maintained according to standard algal assay procedures, except for the media modification already described.

The initial bioassays (bioassays #1-8, Table I), were conducted to determine the effects on algal growth of several selected salts (eight or ten) assayed alone and in various combinations, at several concentrations (0.004-0.3 N as NaCl). Algal growth parameters used in comparisons were growth rate and maximum standing crop. The standard test alga, Selenastrum, was used in the first three of these bioassays. In bioassay #4, the test organism was Selenastrum that had been cultured for six months in Lake Powell Synthetic medium (LPS). This experiment was designed to test whether an acclimated culture could tolerate higher concentrations of salt in its environment. A similar set of bioassays was conducted using Synedra, an alga which is indigenous to Lake Powell as the test organism (bioassays #5-8, Table I). Bioassay #11 was conducted using equal numbers on three different algal phyla, represented by Anabaena, Synedra, and Scenedesmus, as the test organisms. This bioassay attempted to identify if the compounds under study could potentially alter the algal community composition. After the salt effects on algal growth of this species had been measured, the oil shale extracts were tested (bioassays #10, and 11). In one of these (bioassay #11), shale leachate was assayed along with its matching salt solution. This experiment was designed to determine those effects on algal growth which are due to the non-salt constituents of processed shale leachate.

Table I. Summary of algal bioassays conducted [5].

Bioassay Number	Test Organism(s)	Medium Perturbations[a]
1	*Selenastrum*	Single salt additions to AAM of 8 salts, equivalent to NaCl normalities of 0.05-0.13 N.
2	*Selenastrum*	Single salt additions to AAM of 8 salts, equivalent to NaCl normalities of 0.004-0.05 N.
3	*Selenastrum*	Two salt additions to AAM, all possible combinations of 10 salts, equivalent to NaCl normality of 0.03 N.
4	*Selenastrum* acclimated 6 mos. to high salinity	Single salt additions to LPS of 10 salts, equivalent to NaCl normalities of 0.05, 0.10, and 0.30 N.
5	*Synedra*	Single salt additions to LPS of 10 salts, equivalent to NaCl normalities of 0.05-0.30 N.
6	*Synedra*	Two salt additions to LPS of all possible combinations of 10 salts, equivalent to NaCl normality of 0.05 N.
7	*Synedra*	Three and four salt additions to LPS of all possible combinations of 10 salts equivalent to NaCl normality of 0.05 N.
8	*Synedra*	Multiple salt additions (5 or more) to LPS of all possible combinations of 10 salts equivalent to NaCl normality of 0.05 N.
9	*Synedra, Scenedesmus, Anabaena*	Single salt additions to LPS of 10 salts, equivalent to NaCl normality of 0.05 N.
10	*Selenastrum* acclimated 6 mos. to high salinity	Shaker extracted elutriates of shales CR, CP, DR, and DP (R=Raw, P=Processed) tested at four concentrations of additions to AAM.
11	*Synedra*	Shaker extracted elutriates of shales AR, AP, BR, and BP (R=Raw, P=Processed) and leachate of shale AP and matching salt solutions in LPS.

[a]AAM = Algal Assay Media [16].
LPS = Lake Powell Synthetic Media
Salt additions: NaCl, KCl, $MgCl_2$, $CaCl_2$, Na_2SO_4, K_2SO_4, $MgSO_4$, $CaSO_4$ (Bioassays #1 and 2), with $NaHCO_3$ and $KHCO_3$ for the remaining bioassays.

Both raw and processed shales were extracted with water (10 percent by volume). After a period of constant shaking, the elutriate was collected and sterilized by filtration. The composition of elutriate collected from two shales before and after processing is shown in Table II. Note the tremendous increase in TDS following processing. The oil shale are identified by an alphabetic code because these are unhistoried samples from prototype processes and there-fore may not be representative of a full-scale operation.

Results

Salinity Additions

Using the standard test alga, Selenastrum, it was found that growth depression occurred at the lowest salt concen-tration assayed (0.004 N) and this magnitude of the depres-sion increased with increasing salt concentration. There were significant differences in the inhibitory effect of

Table II. Summary table for the characteristics of oil shale elutriates [5].

| | | ELUTRIATE | | |
| | | TYPE "A" | | TYPE "B" | |
		RAW	PROCESSED	RAW	PROCESSED
Cations, meq/ℓ					
Na		0.592	35.74	0.117	4.545
K		0.284	1.396	0.041	0.190
Mg		0.199	20.215	0.369	4.782
Ca		0.606	32.63	0.808	12.121
	Total	1.681	89.981	1.335	21.638
Anions, meq/ℓ					
Cl		0.085	0.851	0.040	0.200
HCO_3		1.011	2.953	0.416	2.814
SO_4		0.583	89.55	0.773	18.275
CO_3		0.238	0.515	0.040	---
	Total	1.917	93.873	1.269	21.289
TOC, mg/ℓ		7.38	1.985	0.153	11.3
pH		9.04	8.84	8.85	8.33
TDS, mg/ℓ		121	7056	101	1518

cations. The divalent cations depressed the growth rate more than did the monovalent cations.

The acclimated Selenastrum tolerated higher concentrations of salts than did the standard alga. The effect of the salt addition was to decrease the standing crop; however, the growth rate for most additions was not significantly different than that in the control flasks.

The indigenous alga, Synedra, was also negatively affected by salt additions, but was not inhibited as much as either the acclimated or unacclimated Selenastrum. Growth depression occurred at 0.05 N and increased as the concentration increased. There were also differences in the ionic toxicity response of Selenastrum and Synedra to single salt additions. The monovalent cations (Na, K) were more effective in reducing the growth of Synedra than were the divalent cations (Ca, Mg). The opposite occurred with Selenastrum. When Synedra was grown in multiple salt solutions, growth was depressed, but there was no relationship for monovalent to divalent toxicity and no way to predict the interactions of these various salts on algal response.

When the three algal species (Synedra, Scenedesmus, and Anabaena) were grown in mixed culture, in the Lake Powell synthetic mean there was a reduction in biomass observed as salt concentration increased. It was noted, however, that these species were more tolerant to salinity than any other test algae. One response observed under increased salinity was a decrease in the number of nitrogen-fixing heterocysts in the blue-green alga, Anabaena.

Oil Shale Elutriates and Leachates

All six of the elutriates (raw and spent) had an equal effect in reducing the growth rate of acclimated Selenastrum, and all but one elutriate from raw shale significantly reduced the final biomass. On the contrary, the addition of many of the spent shale elutriates and leachates stimulated the growth of the indigenous green alga, Scenedesmus. The concentration effects of these additions did not provide consistent conclusions, except that the degree of stimulation appeared to be dependent upon the process applied to the shale. In general, the extracts from the spent shales stimulated growth more than did the extracts from the raw shales. This stimulation was not caused by the addition of any of the salt compounds because the matching salt controls did not cause a stimulation of growth. The processing of the shale may make growth stimulating compounds more available to Scenedesmus. The compounds may be similar to the low molecular weight aromatic hydro-carbons in other petroleum products that were found to stimulate algal growth [21].

Conclusions

In general, the sensitivity to salt addition occur-red in the following order: Selenastrum > acclimated Selenastrum > Synedra > the three combined species. It is for this reason that the authors [5] support the use of an indigenous algal or an acclimated test algal culture in this type of toxicity bioassay.

The results of this study indicate that an increase in the salinity of Lake Powell may inhibit the growth of the naturally occurring species, Synedra. However, the level of salinity required to suppress the growth of Synedra (0.05 N NaCl or 1150 mg/ℓ TDS as NaCl) would be a large increase in the salt content of the receiving water. The increased cost to the agricultural users of this water would probably prevent the salinity from attaining such a level.

Leachates from oil shale sites may increase the produc-tivity of Scenedesmus more than leachates from the raw shale. However, runoff containing material leached from raw shale could also stimulate the growth of Scenedesmus.

AN EVALUATION OF THE CARCINOGENIC POTENTIAL
OF PROCESSED OIL SHALE LEACHATES

Introduction

Processed oil shales generally contain between 0.02 and 0.2 percent benzene leachable material [22]. A portion of this material consists of polycyclic aromatic hydrocarbons, PAH (Table III), some of which are known to be carcinogenic to animals, including humans. Despite the relatively low concentration of the benzene-soluble organics in spent shale, the large quantities of spent shale that will be produced during a day of full production (estimated at one facility, the White River Shale Project, to be 86,000 metric tons [4]), will contain 17-170 metric tons of benzene-soluble materials per day. Schmidt-Collerus [22] found that 20-40 percent of the total benzene-soluble organic matter in shale can be leached by water and that most of the carcinogens in the shale were dissolved and concentrated in the salt residue. The PAH content of this percolate could be as much as 3 to 4 orders of magnitude higher than that of the pristine ground or surface water in the area. The

Table III. Polycyclic Aromatic Hydrocarbons Identified in Spent Oil Shale Extracts

Compound Name	Symbol	Investigator DRI[a]	Investigator DRI[b]	Investigator Colony[c]	Carcinogenic Potential[d]
Acridine		X			-
Anthanthrene (dibenz(c,d,j,k)pyrene)		X	X	X	-
Benz(a)anthracene	BA	X	X	X	+
Benzo(ghi)perylene	BP		X		+
Benzo(a)pyrene	BaP	X	X	X	+++
Carbazole		X			-
Dibenz(a,j)acridine		X			+
Dibenz(a,h)anthracene	DBA		X		++
7,12 Dimethylbenz(a)anthracene	DMBA	X	X	X	++
Fluoranthene	F	X	X	X	-
3-Methylcholanthrene			X	X	++
Perylene	P	X	X	X	-
Phenanthrene		X	X	X	-
Phenanthridine		X	X	X	?
Pyrene		X	X	X	-

[a]Denver Research Institute [22].

[b]Denver Research Institute [3].

[c][23].

[d]Relative carcinogenic activity on mouse skin, +++ = high, ++ = moderate, + = weak, - = inactive, ? = unknown [24].

probability that compounds having blastomogenic or mutagenic properties will enter into the environment from this source warrants an assessment of the carcinogenic potential of spent oil shale leachate.

There were two techniques used in the present study to attempt to accomplish this objective. These were chemical analysis and bioassay. Each approach has its advantages and limitations. It was therefore assumed that by applying both techniques simultaneously, more information could be obtained.

The recent advent of sophisticated chemical techniques and instrumentation (GC/MS, HPLC) has allowed researchers to identify the individual components of a sample and estimate the hazard of the material based on the number and concentration of known and suspected carcinogens. However, there are very few compounds for which the carcinogenic activity is known. A reliable bioassay which is sensitive to carcinogenic compounds can provide the needed information in this instance. Several microbial mutagenicity bioassays have recently been found that detect carcinogens as mutagens with good accuracy [25]. This has led to their use in the preliminary screening step prior to animal bioassays to confirm the activity of suspected chemical carcinogens.

Methods

Extraction of Organics

There were two approaches taken to obtain extracts of spent oil shale [6]. The first method employed soxhlet extraction using various organic solvents and separation techniques for easier identification and bioassay. Based on initial investigations with soxhlet operation conditions, the standard procedure was to extract spent shale samples for three days in benzene, followed by three days in methanol. Four hundred gram samples and 1.2 liters of leaching solvents were used for each sample. The concentrated methanol extracts were used directly in the Ames test bioassay. An attempt was made to redissolve the benzene extracts in dimethylsulfoxide, because benzene is toxic to the bioassay organism. This procedure was unsuccessful; therefore, a limited number of the extracts could be assayed.

The second approach utilized several methods to determine the potential for water to leach mutagens from shale. Extended contact of water with shale was provided in both an upflow column, and in a large teflon™*-lined drum equipped with a motor-driven mixer. The extracted organic compounds were either concentrated on an exchange resin and eluted with organic solvents, or extracted directly with an organic solvent (liquid-liquid extraction). Both the water samples and the organic solvent extracts were screened in the mutagenicity assay.

Chemical Analysis of Extracts

Organics in the concentrated extraction samples were identified by computerized gas chromatography/mass spectrometry (GC/MS). A Hewlett-Packard Model 5985A GC/MS system was used for GC separation (10 m glass capillary column coated with SP2100) and mass spectrograph identification and quantification of peaks. PAH standards were used to develop GC retention times and mass spectra for comparison with samples.

Mutagenicity Testing of Extracts

The Ames/Salmonella mutagenicity assay [26] was used to screen proposed oil shale extract for potentially carcinogenic chemicals. This assay has previously been shown to correctly identify more than 90 percent of the carcinogens tested as mutagens with a low level of false negatives (less than 20 percent) [27]. In addition, several researchers have used the bioassay to detect carcinogens/mutagens in energy-related mixtures, including petroleum products, by-products, and effluents [28,29,30,31,32,33,34].

The samples resulting from spent shale extraction were assayed using the standard plate incorporation assay over a wide concentration range (representing from 0.1 to 10 g of spent shale per plate), with and without the addition of the microsomal enzyme fraction, S-9 [7]. For comparison, the dose-response of several known PAH standards, some of which had previously been identified in spent shale extracts, were determined.

Since the spent shale extracts that were assayed were known to contain a complex mixture of chemicals, an investigation was made to determine whether the magnitude of the response in the Ames test is indicative of the total complement of mutagens and non-mutagens. Several assays were

---

*Registered trademark of E.I. duPont de Nemours and Company, Inc., Wilmington, Delaware.

conducted in which two chemicals of known mutagenic activity were combined. These solutions were either 1:9 or 1:1 mixtures with a combined total of 10, 20, and 50 μg/plate. The response was compared to that of the single compound at the same concentration to which an appropriate amount of pure solvent blank was added.

Results

PAH Composition

More than 100 organic compounds from processed oil shales were identified by GC/MS, including several carcinogens (Table IV). Four and 5 ring PAH were found to be benzene extractable from processed oil shales in concentrations ranging from less than one to greater than 50 ppb (mass PAH:mass shale). These PAH were available in water extracts below their respective solubilities.

Mutagenic Activity

The results of Ames test assays indicate that mutagens are present in each of the four types of spent shale and that mutagens can be obtained by a variety of different extraction methods (Table V). In general, these mutagens require metabolic activation (addition of enzyme-induced rat liver homogenate) for their detection as mutagens. The repeated response of TA strains 1537, 1538, and 98, indicate that many of these mutagens cause frame-shift mutations in bacterial DNA. There were fewer base-pair substitution mutagens detected in these samples (TA strains 1535 and 100). Aqueous extracts of spent shale exhibited marginal or no mutagenic response, presumably because the techniques for extraction and concentration of trace quantities of these nearly water-insoluble compounds are inadequate.

The results of assays on one-to-one mixtures of two mutagens which exhibited different dose response curves when assayed separately indicated the response to the mixture was non-additive. Furthermore, in the majority of cases, the response to the mixture was determined to be statistically indistinguishable (chi-square analysis) from the dose response curve of one of the mutagens. This masking effect was found to persist for one strong mutagen (benz(a)pyrene) even when it composed only 10 percent of the mixture.

Conclusions

In conclusion, we can say that PAH, many of which are certainly carcinogens, are present or at least can be formed

349

Table IV. Summary of Quantification of Organic and Water Developed Shale Samples; ppb Except as Indicated [6]a

Standard Species	Organic Developed[b]			Water Developed[c]			Solubility
	Max.	Mean	Min.	Max.	Mean	Aqueous Concentration (μg/1)	(μg/1)
Carbazole	69	32	0.16	NS	NS	--	--
4-Azafluorene	43	15	0.25	NS	NS	--	--
Acridine	50	18	0.22	NS	NS	--	--
2-Aminofluorene	T	NS	NS	NS	NS	--	--
Dibenzothiophene	134	50	NS	20	9.3	~10	--
Anthracene	62	20	NS	NS	NS	--	50
Phenanthrene	483	165	9.9	61	50	~10	1250
Fluoranthene	85	22	1.1	10	3.5	~1.0	250
Pyrene	97	34	1.6	10	3.4	~1.0	150
Thianthrene	T	NS	NS	T	--	--	--
Aminopyrene	T	NS	NS	T	--	--	--
Triphenylene	} 54	10	0.69	1.0	0.5	(0.01 - 1.0)	10
Benz(a)anthracene							
Chrysene							5
Benzo(e)pyrene	} 56	1.0	0.13	0.17	0.1	(0.01 - 0.1)	--
Perylene							
Benzo(a)pyrene							--
13-H-dibenzo(a,i)carbazole	T	T	NS	NS		--	--
7,12 dimethylbenzo(a)anthracene	T	T	NS	NS		--	--
Benzo(ghi)perylene	T	T	NS	NS		--	--
Dibenz(a,h)anthracene	T	T	NS	NS		--	--

aUnion, Tosco and Paraho shales included; ppb, parts species indicated per billion parts dry shale sample.

bSix samples.

cMinimum for all species sought was below detection limit (NS not shown).

Note: NS is no show (i.e. below detection limits); T is trace.

Table V. Results of Ames Mutagenicity Testing of Spent Shale Extracts [7][a]

Spent Shale Type and Sample No.	TA Strain 1535	1537	1538	98	100	Extraction Procedure[b]
Type A						
#1	(+)[c]	+	(+)	(+)	–	Sox. Ext. one day with φH followed by one day with MeOH, KD concentration, in MeOH
#2	–	–	–	Δ[d]	(+)	Sox. Ext. four days with φH followed by four days with MeOH, RE concentration, in MeOH
#3	++	–	–	–	(+)	Sox. Ext. four days with φH followed by four days with MeOH, KD concentration, in MeOH
#4	N.T.[e]	+	+	–	–	Samples #2 and #3 combined
#5	N.T.	Δ	Δ	N.T.	Δ	Sox. Ext. four days with φH–MeOH solution 1:5 by volume, KD concentration, in φH–MeOH
Type B						
#1	+	+++	+	+	–	Sox. Ext. three days with MeOH, KD concentration, in MeOH
#2	N.T.	++	+	N.T.	(+)	Same as #1
#3	–	(+)	–	–	–	TLC on silica gel of sample #1, eluted with MeOH:p-Dioxane solution 1:1 by vol.
#4	+	++	–	–	–	Same as #3 except eluted with MeOH:p-Dioxane solution 1:4 by vol.
#5 w/S-9	N.T.	++[Δ]	+[Δ]	+[Δ]	Δ	Sox. Ext. three days with φH:MeOH solution 1:5 by vol., KD concentration in φH:MeOH
w/oS-9[f]	N.T.	++[Δ]	++[Δ]	Δ	Δ	
#6	N.T.	Δ	Δ	Δ	Δ	Sox. Ext. one day with φH followed by one day with MeOH, RE concentration, in MeOH
#7	N.T.	(+)	(+)	–	–	Sox. Ext. one day with φH followed by two days with MeOH, in MeOH
#8[g]	N.T.	(+)	–	–	–	Sox. Ext. one day with Pentane, KD concentration, TLC on silica gel eluted 5 fractions w/MeOH
Type C						
#1	N.T.	+++	+++	++	–	Sox. Ext. three days with φH followed by five days with MeOH, RE concentration, in MeOH
#2	N.T.	+++	+++	++	(+)	Same as #1 except KD concentration
#3	–	–	–	–	–	TLC on silica gel of sample #2, eluted fractions with MeOH
#4	N.T.	Δ	Δ	Δ	Δ	Separation of sample #2 on Al2O3, eluted with φH
Type D						
#1	N.T.	++	+++	+	(+)	Sox. Ext. one day with MeOH, KD concentration, in MeOH
#2	N.T.	–	–	–	–	TLC on silica gel of sample #1 eluted fractions with MeOH
#3	N.T.	(+)	–	Δ	–	AQ. Leach. filtered and passed through Sephadex gel, eluted with MeOH
#4	N.T.	–	–	–	–	AQ. Leach. used in agar media preparation
#5	N.T.	(+)	(+)	–	–	AQ. Leach. filtered and passed through XAD-2 resin eluted with MeOH

[a] Results of plate incorporation assays with rat liver homogenate (S-9) except where indicated, from at least two replicates.

[b] Sox. Ext. = Soxhlet Extraction   KD = Kuderna Danish
φH = Benzene   RE = Roto-evaporation
MeOH = Methanol   TLC = Thin layer chromatography
Vol. = Volume   AQ. Leach. = aqueous leachate

[c] Symbols are used to indicate relative mutagenic strength.

Symbol	Interpretation
–	No mutagenic response
(+)	Questionable mutagenic response
+	Weak mutagenic response
++	Moderate mutagenic response
+++	Strong mutagenic response

[d] Δ indicates toxicity at high concentrations of sample.

[e] N.T. = not tested.

[f] w/o S-9 = assay performed without rat liver homogenate.

[g] Of 5 fractions eluted from silica gel, only one R_f = 0.18 showed any mutagenicity.

from precursors in processed shale and that they can be extracted in organic solvents through soxhlet extraction. In addition, many of these compounds are also extractable by water, indicating the potential of spent oil shale to be a source of carcinogens to the environment by leaching.

The results also point out the limitation of the Ames test when applied to mixed environmental samples. In the majority of experiments, the mutagenic response of the Ames test was found to be non-additive. Therefore, the magnitude of the response may underestimate the total composition of mutagens/carcinogens in the sample.

EVALUATION OF MICROCOSMS FOR DETERMINING
THE FATE AND EFFECT OF BENZ(A)ANTHRACENE
IN AQUATIC SYSTEMS

Introduction

In an earlier section of this paper, the potential for spent oil shale to be a source of carcinogens was presented. Polycyclic aromatic hydrocarbons (PAH) were suspected to be one of the groups causing the mutagenic/carcinogenic response. Identification of carcinogenic PAH in these extracts of spent oil shale was confirmed by GC/MS [6].

The mobility of these pollutants through spent shale disposal piles and into the aquatic environment is a topic for further study. However, the results of shale leaching experiments and mass transfer calculations performed by Amy and colleagues [35], supports the speculation that spent shale could be a long term source of organic pollutants to groundwater.

Several factors contribute to the worldwide presence of PAH in the aquatic environment. These include: overall magnitude of production, efficient dispersal, and ineffective treatment methods for removal from waste streams [36]. Due to their hydrophobic nature, they partition to particulate matter and are deposited to the sediments, where they tend to accumulate because they are quite resistent to degradation [37,38]. Contaminated sediments provide a continual source of dissolved PAH to the overlying water. The effect of this low-dose, long-term exposure to the aquatic organisms and the potential for bioaccumulation to occur is unknown.

Three phase microcosms were used to study this problem as it pertains to one PAH, benz(a)anthracene (BA). This compound was chosen as a model for carcinogenic PAH because it is a typical constituent of petrochemical effluents and it is one of the higher molecular weight PAH (which most of the carcinogenic PAH are). In addition, BA is potentially an easier species to study because it is reasonably soluble in water and is classified as a mild carcinogen.

At the concentrations tested (1 μg/ℓ), BA has not been found to be toxic, nor has it been shown to inhibit photosynthesis [36]. Its effect was expected to be subtle, so these experiments were designed to observe the overall system response, in terms of community structure and function. Thus, the study had the following objectives: 1) to validate the predictions made for the fate of BA in a microcosm simulating a freshwater reservoir; 2) to characterize the effect of a long-term (60 day) exposure of this carcinogen on the development of the reservoir biotic community; 3) to assess the potential for bioaccumulation; and 4) to evaluate the particular microcosm apparatus for its application to environmental fate and effect studies.

Methods

Microcosms

The microcosm used in this experiment (Figure 1) was a closed system and consisted of reservoir sediment (2.5 ℓ), synthetic media (11.5 ℓ), and gaseous (0.9 ℓ) phases. The experimental apparatus was constructed with two ports for sampling the liquid phase and for supplying fresh media. Provision for capturing and measuring gas production and consumption was made by the addition of a small volume, low displacement acid-trap and manometer. The sampling of headspace gas composition was accomplished through a teflon stopcock attached to the gas trap. Microcosms were initially filled with synthetic nutrient media [40] which was replaced on a semi-continuous basis at a rate of approximately 1 ℓ per day. A mixture of reservoir plankton was used to innoculate the microcosms.

Two microcosms (one treatment and one control) were placed in complete darkness in order to simulate hypolimnetic conditions. The other six systems were subjected to a summer-like diurnal cycle of 16 hrs lighting and 8 hrs of darkness. Lighting was provided by fluorescent bulbs which emitted 1.2 $W \cdot m^2$ of irradiation in the visible spectrum at a position midway down the length of the upright microcosm, and less than 35 $mW \cdot m^2$ in the near ultraviolet.

The temperature was maintained at 17° and 19° $\pm$ 2°C, for dark and diel light microcosms, respectively.

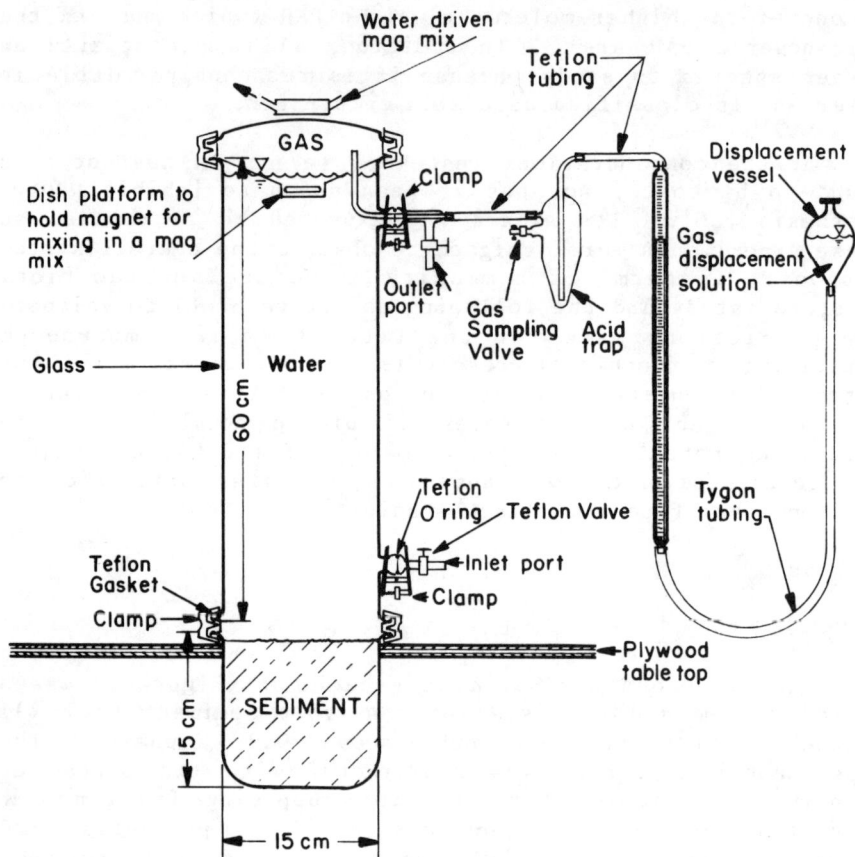

Figure 1. Schematic design of microcosm.

Analytical Procedures

Metabolic activity was determined by mass balance calculations [41] based on daily measurement of gas volume (converted to volume at STP), and weekly analyses of headspace gas composition, and corrected for inputs and outputs of dissolved gases.

Water quality analyses of microcosm media were conducted weekly. The analyses performed were: pH, alkalinity, dissolved oxygen, ammonia, nitrite, nitrate, orthophosphorus, and total phosphorus. Biomass was measured by weekly samples analyzed for total organic carbon. Plankton species composition was determined by microscopic examination of samples.

A total of 2 mg of BA was introduced into each of the treatment microcosms by contaminating a thin layer of surface sediment prior to closing the system. This procedure was found to effectively provide a small quantity of BA to the aqueous phase and was felt to reasonably simulate sorption, settling, and accumulation of BA in reservoir sediments.

The fate of BA in these microcosms was determined by weekly extraction of microcosm media (liquid-liquid extraction with hexane), and final analysis of sediment and biota (soxhlet extraction with methanol and benzene). Computerized gas chromatography/mass spectrometry (GC/MS) was used to identify and quantify BA and degradation products (Hewlett-Packard Model 5985A GC/MS System).

Results

The results of these microcosm experiments indicated that there were no significant differences in community response between the microcosms that had received BA treatment and the control systems. Even at the relatively high levels of BA concentration (3 μg/ℓ), there was no evidence of any effect, either stimulative or detrimental, on oxygen production, net photosynthetic activity, nutrient utilization, species composition (algal or invertebrates), nor on biomass accumulation. (For more information refer to the original work [8]).

The mass balance analysis of BA indicated that under conditions which exclude photolysis, the majority of the compound remained in the sediment, with some migration to deeper layers (Figure 2). A small concentration of the parent compound was maintained in the dissolved state through desorption from these sediments. The biota were also found to accumulate BA. Algae contained a small portion (partition coefficient, K_p = 1.1) which indicated adsorption was passive. Flies were found to contain a larger quantity of this PAH; however, the concentration factor was still small (K_p = 10).

Conclusions

The removal processes that were observed for BA in these microcosm experiments are consistent with those presented in the literature [37,38] with one exception. In the environment where the radiation level is higher than that presented in this experiment, greater rates of photooxidation of dissolved BA would occur and the compound would be somewhat less persistent. Radiation capable of causing photolysis of BA was essentially eliminated in these experiments in order

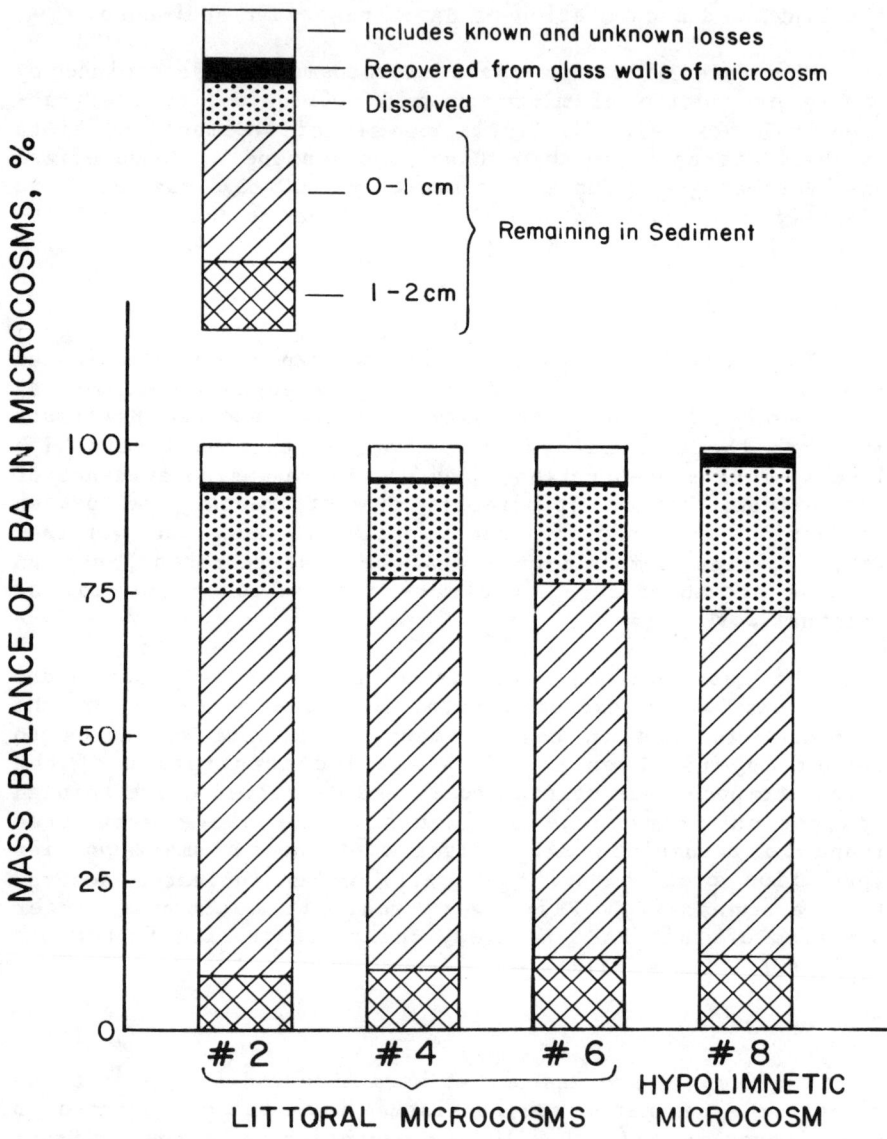

Figure 2. Mass balance of benz(a)anthracene in littoral (#2, 4, and 6) and hypolimnetic (#8) treatment microcosms.

to test the "worst case" response of the biological organisms which, as discussed previously, was neither detrimental nor stimulative.

The results of the present microcosm experiment support the speculation [36] that BA (and probably other higher molecular weight PAH) are relatively unavailable for bio-magnification in aquatic food chains. The potential pathways to humans still include the consumption of shellfish recently exposed to PAH-containing effluents, and drinking of water that has originated from the hypolimnion of a reservoir containing PAH-contaminated sediments.

While the microcosms used in this experiment may not be particularly suited to allow the detection of sublethal responses of aquatic organisms, it does appear to be appropriate for the study of the fate of relatively insoluble organic contaminants. In the recent symposium on the use of microcosms in ecological research [42] the types of micro-cosms, their appropriate use, and experimental limitations are discussed. In general, microcosms having structurally simple ecosystems are used to best advantage in the determination of chemical fate (mechanisms, rates) or chemical toxicity. Complex ecosystems are more suitable to gain information about the effect on the biological community of the contaminant (i.e., structure and function changes) and about the interactions that organisms have which affect compound fate (i.e., bioaccumulation).

The results of the present study are in line with the above generalization. The system had two trophic levels (primary producer and herbivore). The compound was not toxic and no significant differences in the biological response could be detected. The best data were obtained regarding the fate of the compound given the particular contamination procedures, physical environment, and compound recovery schemes.

CONCLUSIONS

The results of batch bioassays indicate that algal growth is inhibited by salt additions; however, the amount of salinity increase required to affect the algal population in Lake Powell is quite high. This level of salinity will probably never be reached because of the downstream agricultural interest. This study also found that releases of leachates from the shale disposal sites may be biostimulatory to the algae of Lake Powell. The agent(s) responsible for this effect is (are) probably not salts.

The procedures used to characterize the organic residue of spent shale were successful in identifying PAH, some of which are certainly carcinogenic. The combined use of the chemical and biological analysis techniques in this characterization study could have been more complementary. It is recommended that the sample be chemically fractionated prior to mutagenicity testing. This serves to reduce the chance that chemical interactions occur which may mask the detection of mutagenic activity. In addition, chemical identification and quantification of mutagenic fractions may make better use of GC/MS time and may lead to a reliable procedure for comparing the carcinogenicity of various environmental samples.

The results of the microcosm study indicated that higher molecular weight PAH that enter the aquatic environment will persist there sorbed to sediments and biota. This behavior has been fairly well documented in other microcosm studies. The presence of dissolved PAH does not appear to alter normal ecosystem structure nor function, at least during a 60 day exposure to relatively high concentrations (up to 3 $\mu g/\ell$). The experimental results also supported the speculation that bioaccumulation of higher molecular weight PAH in food chains is probably not significant.

REFERENCES

1. Ward, J. C. 1971. Water pollution potential of spent oil shale residues. Prepared for Environmental Protection Agency, Grant No. 14030EDB.

2. Dassler, G. L. 1976. Assessment of possible carcinogenic hazards created in surrounding ecosystems by oil shale development. MS Thesis, Utah State University, Logan. 99 p.

3. Schmidt-Collerus, J. J., F. Bonomo, K. Gala, and L. Leffler. 1976. Polycondensed aromatic compounds and carcinogens in the shale ash of carbonaceous spent shale from retorting of oil shale, p. 115-116. In T. F. Yen (Ed.). Science and technology of oil shale. Ann Arbor Science Publishers, Inc., Ann Arbor, Michigan. 226 p.

4. Slawson, G. C., Jr. (Ed.). 1979. Groundwater quality monitoring of western oil shale development: Identification and priority ranking of potential pollution sources. EPA-600/7-79-023. January.

5. Cleave, M. L., V. D. Adams, and D. B. Porcella. 1979. Effects of oil shale leachate on phytoplankton productivity. Utah Water Research Laboratory, Utah State University, Logan, Utah. UWRL/Q-79/05.

6. Maase, D. L. 1980. An evaluation of polycyclic aromatic hydrocarbons from processed oil shales. PhD Dissertation, Utah State University, Logan, Utah. 202 p.

7. Dickson, J. G., and V. D. Adams. 1980. Evaluation of mutagenicity testing of extracts from processed oil shale. Utah Water Research Laboratory, Utah State University, Logan, Utah. UWRL/Q-80/01.

8. Dickson, J. G. 1981. Predictive testing of environmental carcinogens. PhD Dissertation, Utah State University, Logan, Utah. 241 p.

9. Conkle, N., V. Elizey, and K. Murthy. 1974. Environmental considerations for oil shale development. EPA, NTIS PB241942, Washington, D.C. 114 p.

10. Bloch, M. B., and P. D. Kilburn (Eds.). 1973. Processed shale revegetation studies, 1965-1973. Colony Development Operation. Atlantic Richfield Company, Denver, Colorado. 208 p.

11. Colorado State University. 1971. Water pollution potential of spent oil shale residues. EPA, U.S. Government Printing Office, Washington, D.C. 116 p.

12. Ward, J. C., and S. E. Reinecke. 1972. Water pollution potential of snowfall on spent oil shale residues. Bureau of Mines Open File Report 20-72. Colorado State University, Fort Collins, Colorado. 51 p.

13. Siggia, S., and P. C. Uden. 1974. Report of the conference-workshop entitled analytical chemistry pertaining to oil shale and shale oil. National Science Foundation, Washington, D.C. 194 p.

14. Kerekes, J., and J. R. Nursall. 1966. Eutrophication and scenescence in a group of Prairie-Parkland Lakes in Alberta, Canada. Proceedings International Association of Theoretical and Applied Limnology 16(1):65-73.

15. Stewart, W. D. P. (ed.). 1974. Algal physiology and biochemistry. Botanical monographs, Vol. 10. Blackwell Scientific Publications. Oxford, England.

16. United States Environmental Protection Agency. 1971. Algal assay procedure: Bottle test. National Eutrophication Research Program. Pacific Northwest Environmental Research Laboratory, Corvallis, Oregon. 81 p.

17. Miller, W. E., J. C. Green, and T. Shiroyama. 1978. The Selenastrum capricornutum Printz algal assay bottle test. EPA-600/9-78-018. Corvallis Environmental Research Laboratory, Corvallis, Oregon.

18. American Public Health Association, American Water Works Association, Water Pollution Control Federation. 1975. Standard methods for the examination of water and wastewater. 14th ed. American Public Health Association, Washington. 1193 p.

19. Stewart, A. J., and D. W. Blinn. 1976. Studies on Lake Powell, USA: Environmental factors influencing phytoplankton success in a high desert warm monomictic lake. Arch. Hydrobiol. 78(2):139-164.

20. Medine, A., and D. B. Porcella. 1980. Heavy metal effects on photosynthesis/respiration of microecosystems simulating Lake Powell, Utah/Arizona. In R. A. Baker (Ed.) Contaminants and sediments. Ann Arbor Science, Ann Arbor, MI

21. Dunston, W. M., L. P. Atkinson, and J. Natoli. 1975. Stimulation and inhibition of phytoplankton growth by low molecular weight hydrocarbons. Mar. Biol. 31:305-310.

22. Schmidt-Collerus, J. J. 1974. The disposal and environmental effects of carbonaceous solid wastes from commercial oil shale operations. First Annual Report to NSF, NSFGI 34282 x 1, Denver Research Institute. 169 p.

23. Atwood, M. T., and R. M. Coomes. 1974. The question of carcinogenicity in intermediates and products in oil shale operations. Unpublished Colony Development Operation Paper. The Oil Shale Corporation, Rocky Flats Research Center.

24. Hoffman, D., and E. L. Wynder. 1976. Environmental respiratory carcinogenesis. pp. 324-365. In C. E. Searle (Ed.). Chemical carcinogens, ACS Monograph 173. American Chemical Society, Washington, D.C. p. 788.

360

25. Purchase, I. F. H., E. Longstaff, J. Ashby, J. A. Styles, D. Anderson, P. A. Lefevre, and F. R. West-wood. 1976. Evaluation of six short term tests for detecting chemical carcinogens and recommendations for their use. Nature 264:624-627.

26. Ames, B. N., W. E. Durston, E. Yamasaki, and F. D. Lee. 1973. Carcinogens are mutagens: A simple test system combining liver homogenates for activation and bacteria for detection. Proc. Natl. Acad. Sci. USA 70(8):2281-2285.

27. McCann, J., N. B. Spingarn, J. Kobori, and B. N. Ames. 1975. Detection of carcinogens as mutagens: Bacterial tester strains with R factor plasmids. Proc. Natl. Acad. Sci., U.S.A. 72(3):979-983.

28. Rubin, I. B., M. R. Guerin, A. A. Hardigree, and J. L. Epler. 1976. Fractionation of systematic crude oils from coal for biological testing. Environ. Res. 12:356-365.

29. Voll, M. J., J. D. Isbistee, L. I. Isaki, and M. D. McCommas. 1977. Mutagenic potential of petroleum byproducts in Chesapeake Bay waters. Water Res. Research Center, Univ. of Maryland, College Park, Md. Completion Report A-034-Md 14-34-0001-6021. Tech. Report #39.

30. Rao, T. K., et al. 1977. Correlation of mutagenic activity of energy related effluents with organic constituents. 8th Annual Meeting, Environ. Mutagen. Soc., Colorado Springs, Colo. p. 47-48.

31. Chrisp, C. E., G. L. Fisher, and J. E. Lammert. 1978. Mutagenicity of filtrates from respirable coal fly ash. Science 199(4324):73-75.

32. Epler, J. L., F. W. Larimer, T. K. Rao, C. E. Nix, and T. Ho. 1978. Energy-related pollutants in the environ-ment: Use of short term tests for mutagenicity in the isolation and identification of biohazards. Environ. Health Perspect. 27:11-20.

33. Epler, J. L., J. A. Young, A. A. Hardigree, T. K. Rao, M. R. Guerin, I. B. Rubin, C.-H. Ho, and B. R. Clark. 1978. Analytical and biological analyses of test materials from the synthetic fuel technologies. I. Mutagenicity of crude oils determined by the Salmonella typhimurium/microsomal activation system. Mutat. Res. 57(3):265-276.

34. Epler, J. L. 1979. Evaluation of mutagenicity testing of shale oil products and effluents. Environ. Health Perspect. 30:179-184.

35. Amy, G. L., A. L. Hines, J. F. Thomas, and R.E. Selleck. 1980. Groundwater leaching of organic pollutants from in situ retorted oil shale. A mass transfer analysis. Environ. Sci. Technol. 14(7):831-835.

36. Neff, J. M. 1979. Polycyclic aromatic hydrocarbons in the aquatic environment. Appl. Sci. Publ. Ltd., London. 262 p.

37. Smith, J. H., W. R. Mabey, N. Bohonos, B. R. Holt, S. S. Lee, T.-W. Chou, D. C. Bomberger, and T. Mill. 1977. Environmental pathways of selected chemicals in fresh-water systems. Part I: Background and experimental procedures. EPA-600/7-77-113.

38. Smith, J. H., W. R. Mabey, N. Bohonos, B. R. Holt, S. S. Lee, T.-W. Chou, D. C. Bomberger, and T. Mill. 1978. Environmental pathways of selected chemicals in fresh-water systems. Part II: Laboratory studies. EPA-600/7-78-074.

39. National Academy of Sciences. 1972. Particulate polycyclic organic matter. NAS, Washington, D.C. 361 p.

40. Medine, A. J. 1979. The use of microcosms to study aquatic ecosystem dynamics--methods and case studies. PhD Dissertation, Utah State University, Logan, Utah. 369 p.

41. Porcella, D. B., V. D. Adams, P. A. Cowan, S. Austrheim-Smith, W. F. Holmes, J. Hill IV, W. J. Grenney, and E. J. Middlebrooks. 1975. Nutrient dynamics and gas production in aquatic ecosystems: The effects and utilization of mercury and nitrogen in sediment-water microcosms. PRWG121-1, Utah Water Research Laboratory, Utah State University, Logan, Utah.

42. Giesy, J. P., Jr. (Ed.). 1980. Microcosms in ecological research. DOE Symposium Series 52 CONF-781101.

CHAPTER 22

PROGNOSIS FOR WATER CONSERVATION AND THE
DEVELOPMENT OF ENERGY RESOURCES AT THE
SALTON SEA - DESTRUCTION OR
PRESERVATION OF THIS UNIQUE ECOSYSTEM?

Glenn F. Black
    California Fish and Game
    Chino Fisheries Base
    Chino, California

INTRODUCTION

This paper presents information on the fish, wildlife,
and recreational values associated with the Salton Sea and
discusses the possible impacts of proposed energy development and water conservation measures on these values. Furthermore, possible methods for the perpetuation of the fish,
wildlife, and recreational resources of the Sea are examined;
strangely enough, they are closely linked to some of the
very proposals that could drastically alter this unique
ecosystem.

Setting

The Salton Sea is located in southeastern California
approximately 216 km (135 mi) southeast of Los Angeles, 56
km (35 mi) north of Baja California, and 80 km (50 mi) east
of the Colorado River (Figure 1). It is California's
largest inland water, 58 km (36 mi) long and 14 km (9 mi) to
22 km (14 mi) wide, covering 932 km^2 (360 mi^2) of surface
area having 176 km (110 mi) of shoreline. The Sea lies in a
sink more than 85 m (278 ft) below sea level. It is a
veritable "desert oasis" in an environment that averages
only 4.9 cm (1.9 in) of rainfall a year [1].

The Sea, as we know it today, was formed in 1904-07 as
a result of the Colorado River overflowing its banks and
flowing unchecked into the Salton Sink. The river filled
the sink to a depth of more than 24 m (80 ft) and inundated
more than 1295 km^2 (500 mi^2) of desert land [2]. By 1920

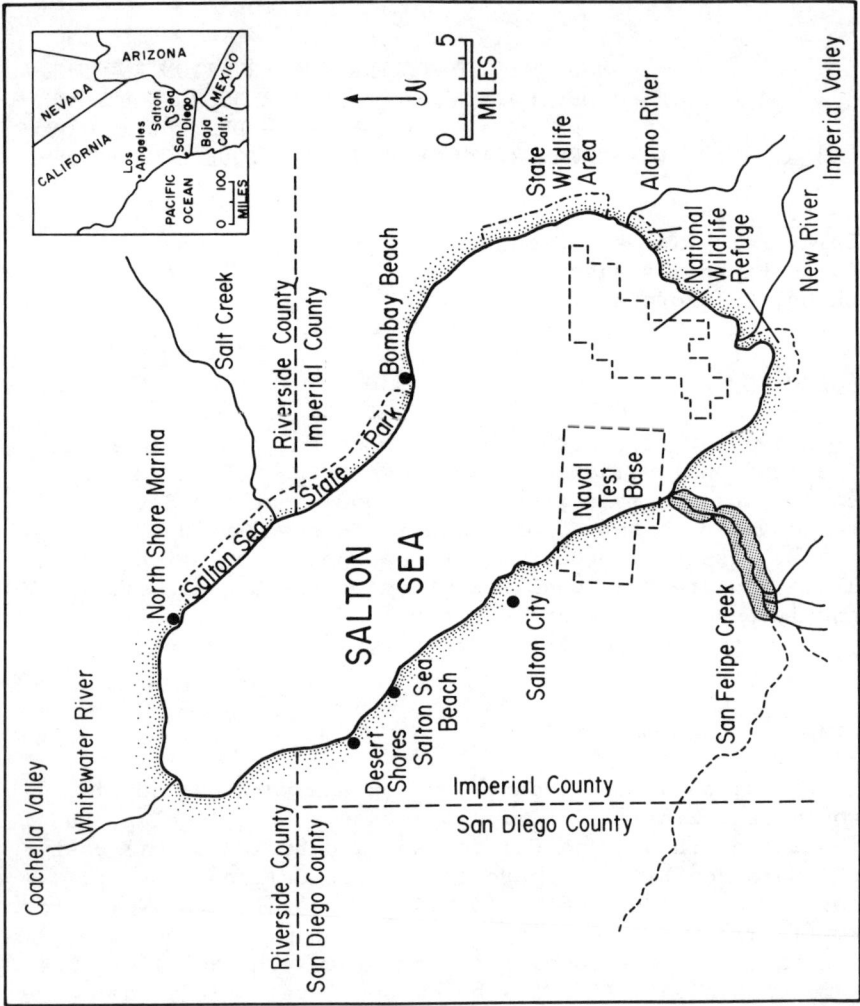

Figure 1. Salton Sea.

the Sea's depth had receded to approximately 9 m (30 ft) by
evaporation, and its surface area was further reduced by
construction of levees and other flood control measures. The
trend reversed, and by 1925 the level had begun to slowly
rise due to increased irrigation of surrounding desert lands
and subsequent drainage of return flows into the Sea [2].
In 1910, the Sea was set aside by a Presidential Act (Public
Water Reserve 90) as a repository for agricultural drainage
water.

Rainfall and natural runoff contributes an average of
43,000 cubic decameters [dam^3 (35,000 acre-ft)] or 2.5 percent
of the total annual inflow while agricultural runoff provides
a mean of 1.7 million dam^3 (1.4 million acre-ft) or 97.5
percent of the total annual inflow to the Sea [1]. Major
agricultural areas border the north (Coachella Valley) and
the south (Imperial Valley) ends of the Sea. The Colorado
River is the ultimate source of irrigation water via the
All American and the Coachella canals. Salt-laden agricultural wastewater is collected and carried to the Sea by an
elaborate network of drains which either flow directly into
it or empty into one of three major rivers (Alamo, New, and
Whitewater, Figure 1) that in turn flow to the Sea. It is
important to note that the Salton Sea has no outlet.

RECREATION

The Salton Sea is important not only as a repository
for agricultural drainage, but also for its recreational
values. It is estimated that the Sea provides approximately
1.5 million recreation days annually [3]. Activities
include fishing, hunting, boating, camping, nature study,
and sightseeing.

Sport Fishery

In 1916, the sport fish of the Salton Sea consisted of
five species that entered from the Colorado River (i.e.,
carp, bonytail chub, razorback sucker, rainbow trout, and
striped mullet) [4], via the irrigation system in Imperial
Valley. By 1929, due to increasing salinities, only the
razorback sucker and striped mullet remained [5]. Striped
mullet provided a limited sport and commercial fishery [6]
until the late 1940s when their access to the Sea was blocked
by a series of dams built along the lower Colorado River.
There was never any evidence that they spawned in the Sea,
and they eventually became extinct [7].

Between 1929 and 1948, the California Department of
Fish and Game attempted to establish a sportfishery at the

365

Sea by planting striped bass, silver salmon, anchovies, and anchovetas, but none of the plants were successful [7]. In a series of plants between 1950-56, involving 30 species acquired from the Gulf of California at San Felipe, the Department was finally successful in establishing a self-sustaining sportfishery [7]. The three species that survived these plants were the orangemouth corvina, bairdiella, and sargo [7].

These fish provide one of the best self-sustaining partyboat sportfisheries in California as evidenced by comparing the average catch per angler of orangemouth corvina in the fishery at the Sea [8] with that at similar fisheries elsewhere in the State (Table I). Harvest estimates from angler interviews conducted by the Department of Fish and Game from 1964-69 indicated that between 557,000 and 1,100,000 of the three sportfish species were caught; estimates of annual angler use from these same interviews ranged from 246,000 to 277,000 [9].

In the last four years another species of sportfish has somehow become permanently established within the Sea - and unidentified species of Tilapia, probably a hybrid. This fish has become the most numerous species caught by anglers and in biological surveys. It is not known what impact it is having on the three previously established sportfish populations, but it has been widely accepted by shore anglers who commonly catch 50 or more a day and by corvina fishermen who use them for bait.

Waterfowl Hunting

The Salton Sea provides important wintering habitat for migratory waterfowl along the Pacific flyway. Mid-winter waterfowl censuses of ducks, geese, and coots have averaged 116,000 birds annually at the Sea for the past 12 years [10].

Approximately 25 km^2 (6,200 acres) of wetlands are managed by the State and Federal governments for hunting, control of waterfowl crop depradation by enticing them off farmers' lands, and protection of waterfowl species; in addition, another 14 km^2 (3,400 acres) of wetlands are managed by 55 private duck clubs [11]. It is estimated that 64,000 people hunt waterfowl on the public and private lands at the Salton Sea each year [10,11].

Nongame Fish

Shoreline pools, portions of several irrigation drains, and some tributaries are occupied by the only native nongame

Table I. Comparison of Catch Per Angler Trip for Selected
Species from Various Partyboat Fisheries in California -
California Fish and Game Published and Unpublished Records.

Location	Species	No. Caught/ Per Angler Trip	Years
Pacific Ocean - So. Calif.	White seabass	0.01	1975-77
Pacific Ocean - So. Calif.	Pacific bonito	0.4	1975-77
Pacific Ocean - So. Calif.	Kelp bass	0.8	1975-77
Pacific Ocean - So. Calif.	Pacific mackerel	1.1	1975-77
San Francisco Bay - Delta	Striped bass	1.6	1938-77
Pacific Ocean - So. Calif.	Rockfish (30 species)	1.8	1975-77
Salton Sea	Orangemouth corvina	5.3	1962-72

species of fish, the desert pupfish. As recently as 1961,
this fish was considered to be "abundant" at the Sea [12].
Since then, several non-native fish, which may compete for
food, prey on it, and even interfere with its spawning
activities, have displaced pupfish in these habitats [13,14].
The desert pupfish has been listed by the State of California
as an endangered species [15] and is currently under "status
review" by the U.S. Fish and Wildlife Service for proposed
federal listing [16].

Nongame Wildlife

Thirty-five species of shorebirds, 47 species of water
birds, and four species of rails have been reported near the
Sea and its mudflats, marshes, and other riparian areas [17,
18]. A few of the many nesting species include the: white
pelican, great blue heron, least bittern, white-faced ibis,
black rail, and black skimmer [17,18]. Tens of thousands of
shorebirds and water birds use the Sea annually; Leitner and
Grant [19] censused over 4,000 white pelicans on one
occasion in 1977 at the Sea.

Essential habitat is also provided for such endangered,
threatened, rare, and species of concern as the: brown
pelican, Aleutian Canada goose, southern bald eagle, American
peregrine falcon, Yuma clapper rail, osprey, California rail,
and snowy plover [10,20].

THREATS TO FISH, WILDLIFE AND RECREATION

Two major threats to fish, wildlife, and recreation values at the Salton Sea have been identified in previous years: (i) fluctuating water levels; and (ii) rising salinity due to the accumulation of salts in a closed system [21].

Past and Present Water Levels and Salinities

In 1907, after the formation of the Sea, the surface elevation was approximately 60 m (195 ft) below sea level, and the salinity was approximately 3,000 ppm [2]. Since 1948, the elevation has increased gradually from 74 m (242 ft) to 69 m (227 ft) below sea level ([1], Figure 2). The salinity dropped from 40,000 ppm TDS in 1948 to 33,500 ppm TDS in 1955, rose again to 40,000 ppm TDS in 1969, and has dropped to its present level of around 38,000 ppm TDS ([1], Figure 2). The marked increase in surface elevation and the associated lowering of salinity concentration since 1970 has been due to greater than normal precipitation and increased irrigation [1](Figure 2).

Future Water Levels and Salinities from Proposed Projects

Water Conservation in Imperial Valley

It is anticipated that the total demand for Colorado River water will exceed the average supply in the 1990s due to large diversions in the lower basin by the Central Arizona Project and to increased demands in the upper basin [22]. To help Imperial Valley farmers get the most from their water, the Bureau of Reclamation (BREC) has begun a water conservation opportunities study in the Imperial Irrigation District (IID) to look at such means of annually conserving as much as 430,000 dam^3 (350,000 acre-ft) of Colorado River water through canal and lateral lining, regulating reservoirs, and wastewater reuse [1]. This represents 25 percent of the annual flow to the Sea.

The effects of this level of water conservation on the surface elevation and salinity of the Salton Sea [1] are projected in Figures 3 and 4 to reverse the trends of the past 11 years (Figure 2). By 2012, the Bureau estimates that the surface elevation of the Sea with full water conservation measures will be approximately -74 m (-243 ft) that is 5 m (16 ft) below present levels, without water conservation the elevation is projected to be -70 m (-230 ft) less than one meter (2 ft) below present day (Figure 3). Salinity levels with full water conservation measures are predicted to rise by the year 2007 to 101,000 ppm TDS and

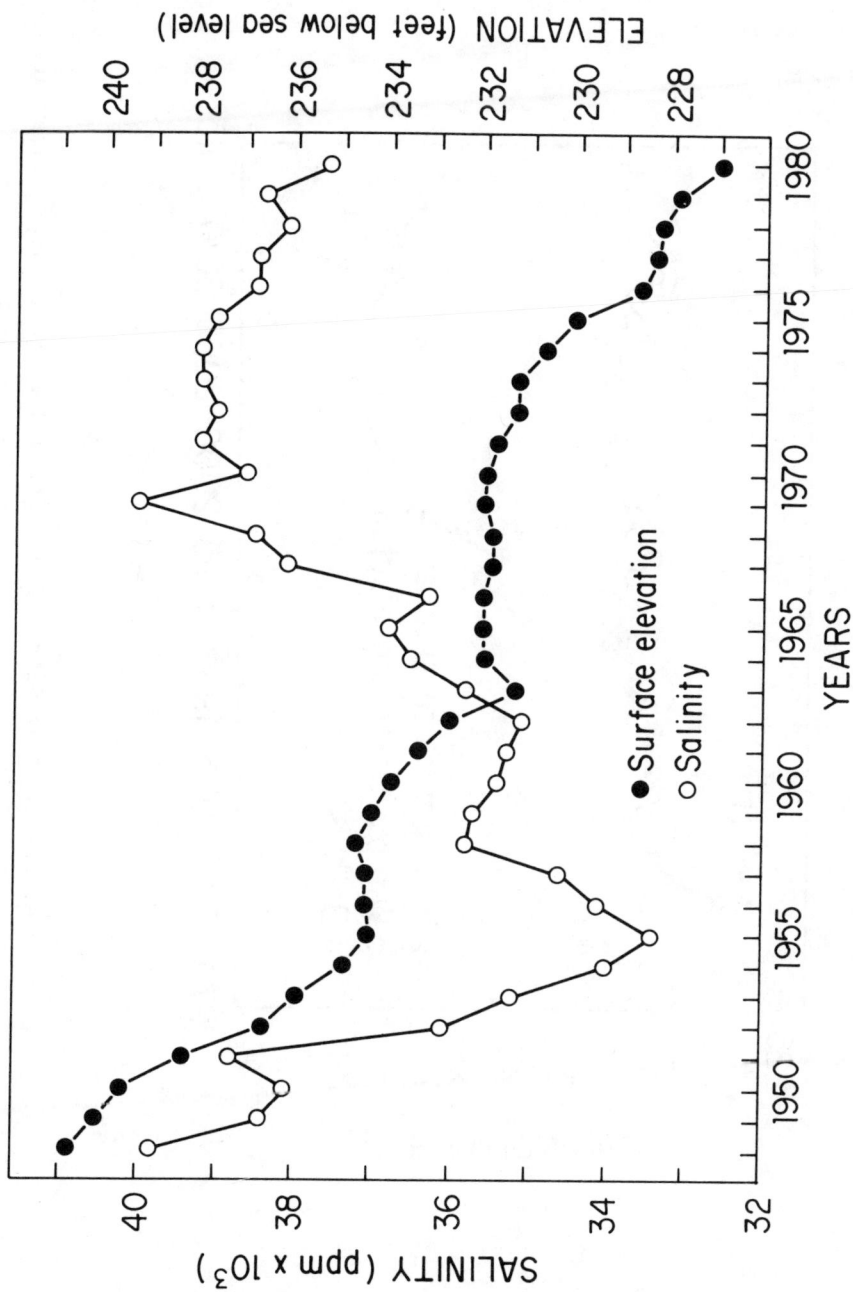

Figure 2. Historic surface elevations and salinities at the Salton Sea 1948–1980.

369

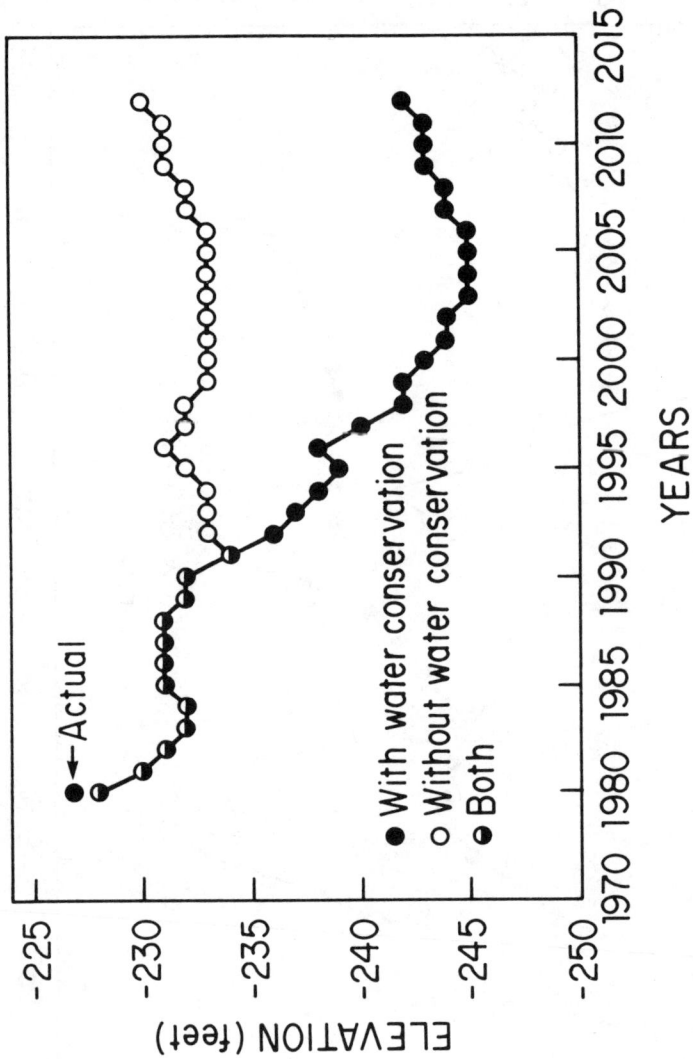

Figure 3. Projected elevations for the Salton Sea, with and without water conservation.

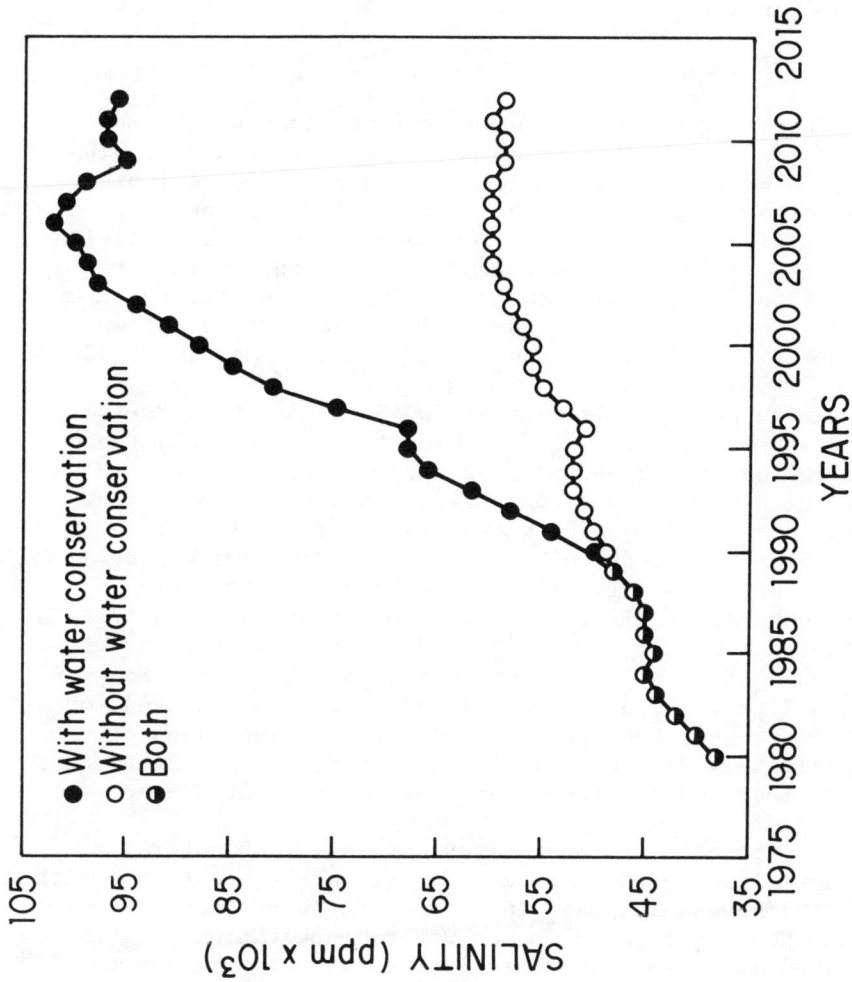

Figure 4. Projected salinities for the Salton Sea, with and without water conservation.

drop to 96,000 ppm TDS by 2012 (Figure 4). This is 2.5 times the salinity of the present-day Sea. It is important to note that by 2012, without water conservation, the Sea's salinity will still rise to approximately 59,000 ppm TDS, or 1.5 times present levels due to the continued accumulation of salts from agricultural runoff. Special attention should also be given to the fact that it will not be known for several years what level, if any, of water conservation will be achieved by the Bureau.

Geothermal Use of Water in Imperial Valley

The Imperial Valley contains four known geothermal resource areas (KGRAs) which may be able to generate 3,000 to 5,000 MW of electrical energy annually [23]. (Figure 5). An important requirement for geothermal energy development is sufficient cooling water for power plants. In addition to water for cooling, geothermal power plants within the Salton Sea, Heber, and Brawley KGRAs are required by Imperial County to reinject a minimum of 80 percent of withdrawn geothermal fluids back into the ground to control subsidence.

To date, approximately 61,500 dam^3 (50,000 acre-ft) of New River water have been approved for use in cooling and reinjection operations of 10 geothermal power plants to be built within the Heber KGRA [24]. An application is now being considered for the withdrawal of a total of 111,000 dam^3 (90,000 acre-ft) of New and Alamo River water for cooling purposes only, within the Brawley KGRA; this represents development of only one-half of the KGRA. It is projected that 102,000 dam^3 (83,000 acre-ft) of water will be needed for commercial development of the Salton Sea KGRA [25]. If all the above identified needs are approved for geothermal use, then this will represent diversions of approximately 274,000 dam^3 (223,000 acre-ft), or 16 percent of the mean annual freshwater flow to the Salton Sea.

The impacts of these diversions of fresh water for geothermal purposes will be to accelerate the rate at which salinity levels increase and to lower the surface elevation of the Sea. Should the major water conservation program outlined in the previous section, along with the known needs of fresh water for geothermal development come about, this would amount to diverting approximately 705,000 dam^3 (573,000 acre-ft) of fresh water annually before it reaches the Sea; this represents 41 percent of the mean annual flow. Overall, the Sea will probably shrink to a little more than half its present size and the salinity will be much higher than predicted in Figures 3 and 4.

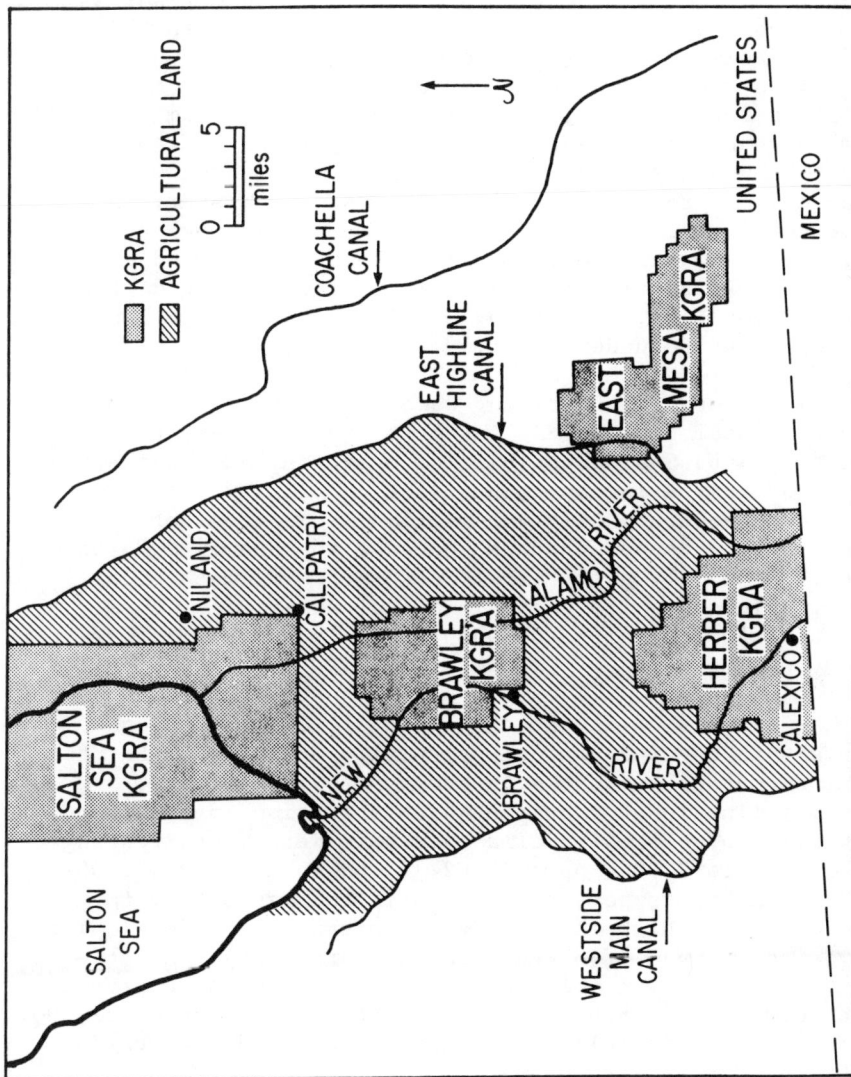

Figure 5. The Imperial Valley, California, and its Known Geothermal Resource Areas (KGRAs).

Known and Predicted Impacts of Surface Level Fluctuation and
Salinities on Recreation, Fish, and Wildlife

The recent increase in surface elevation of the Sea due
to greater than normal inflows of fresh water has caused
flooding and has been instrumental in reducing the number of
recreational facilities from more than a dozen marinas in
the mid-1960s to four at present. Concomitant with this has
been a decrease in the amount of boating activity on the
Sea. Since the establishment of the Salt Sea National
Wildlife Refuge in 1930, all of the original 130 km^2 (32,000
acres) have been inundated by the rising level of the Sea
[25]. The Sea has also flooded approximately 324,000 m^2
(80 acres) of state waterfowl land leased at Wister, and
only extensive diking has kept it from encroaching on more
[26]. Future water declines will reverse this trend,
present-day recreational facilities will be a long distance
from the shoreline, and the original federal refuge lands
will no longer be under water.

The effects of the rising water level have not been
all detrimental because, as was mentioned earlier, the
salinity of the Sea has stabilized over the last 12 years
instead of increasing as it had done previously. The
addition of approximately 4.5 million metric tons of salt
[2] annually presents problems because the Sea is a "closed
system." The only natural salt removal is from chemical,
precipitation of salts, and biological activities; these
processes are insignificant when compared to the large salt
inflow.

The effects of elevated salinity levels on the aquatic
organisms in the Sea are only partially understood, but the
best information available is summarized in Table II.
Optimum salinities for growth and respiration in orangemouth
covina, sargo, and bairdiella have been determined to range
from 33,000 to 37,000 ppm TDS [27]. Salinities of 40,000
ppm TDS and higher have been shown in the laboratory to
cause extremely high and even total mortality of developing
embryos and larvae of bairdiella and sargo [28]. Ninety-six
hr shock bioassay treatments indicate that juvenile orangemouth corvina and sargo cannot tolerate salinities of 62,500
ppm TDS while bairdiella can; bairdiella could not survive
75,000 ppm TDS levels [29]. The identity of the fourth
sportfish species in the Salton Sea, the tilapia, is not
precisely known; however, the highest tolerable salinity
that has ever been reported for any species from this genus
is 150 percent of sea water or approximately 50,000 ppm TDS
[30]. Scanty information on the endangered desert pupfish
indicates that it has been found in salinities up to 90,000

Table II. The Effects of Salinity on Fish and Invertebrate Reproduction, Survival, and Growth at the Salton Sea.

Effects of Different Salinity Levels (ppm TDS)

Species	40,000	45,000	50,000	62,500	70,000	75,000	80,000	90,000
*Orangemouth corvina	mortality of larvae?	growth in adults ceases		mortality of adults				
*Bairdiella	mortality of larvae	growth in adults hampered				mortality of adults		
*Sargo	mortality of larvae	growth in adults hampered		mortality of adults				
Tilapia sp.			mortality of adults?					
**Desert pupfish					mortality of eggs			mortality of juveniles
Pile worm			mortality of early life stages				mortality of adults?	
Barnacles			can tolerate salinities at least this high					

* 33,000 to 37,000 ppm TDS considered to be optimum salinity for growth
** 35,000 ppm TDS considered to be optimum salinity for growth

Not only will the fishes of the Sea be adversely affected by high salinities, but so will at least one of the major invertebrate species, the pileworm, which is considered to be a major forage item for the sportfish [7]. Adult pileworms may not be able to survive Salton Sea salinities higher than 80,000 ppm TDS [32], and total mortality of the young may take place at 50,000 ppm TDS and above [33]. Another important invertebrate present in the Sea, the barnacle, can tolerate salinities at least as high as 50,000 ppm TDS [3]; it is not known what the maximum salinity tolerance is for this species.

Avian diversity could be adversely affected by the loss of fish and major invertebrates. It is not known, however, what magnitude of impact the loss of any or all of these fish and invertebrates will have on the multitude of aquatic-dependent waterbirds and shorebirds, since almost no information is available on their food habits in the Sea [10].

Based on what is currently known about the salinity tolerances of Salton Sea aquatic organisms and the possibility of salinity levels at the Sea going far above those limits, it is fairly safe to suggest that all present fish species and several important invertebrates will totally disappear from the main body of the Sea. Some may continue to survive in the deltas and upstream in the major tributaries.

It is possible that more salt-tolerant fish species could be found that would reproduce and survive in the Sea but is is highly unlikely. The probability is greater that invertebrates, such as brine shrimp, could be introduced that would survive, reproduce, and provide forage for avian species. The question to ask is whether the benefits to be gained from major fresh water diversions are worth the loss of one of the State's most valuable sportfisheries, especially if there are means to preserve the present ecosystem yet still promote water conservation and energy developments.

POSSIBLE SOLUTIONS FOR PERPETUATION OF PRESENT ECOSYSTEM

In 1974, a joint federal-state feasibility report recommended diking or impounding 104 km^2 (40 mi^2) of the Salton Sea to act as an evaporation pond for the removal of salts in order to permanently limit salinity levels to approximately 35,000 ppm TDS [35]. The cost of the desalinization pond was estimated at $58 million, and the time that it would require to lower salinity to optimum levels was estimated at 18 years [35]. At the time, neither state nor federal funding could be found for this project, even though the economic benefits far out-weighed the costs [35].

Salton Sea Solar Pond - The Long-Term Solution?

The Southern California Edison Company is currently considering the building of a pilot solar pond which would use saline water from the Sea to aid in producing 5 MW of electrical power [36]. This project could begin operation by the end of 1985 [37]. If the pilot project is successful, a 600 MW commercial solar pond that would impound approximately 120 km² (46 mi²) of the Sea could be in full operation by the mid-1990s [37]. This solar pond could serve the dual purposes of producing much needed energy and desalinizing the Salton Sea due to its need for large quantities of saturated brine during the pond's start-up period. Subsequently, further salt removal from the Sea could be accomplished by creation of adjacent highly saline brine ponds operated by the state and/or federal governments.

It is important to note that the commercial plant would not be in full operation for another 15 years, lagging behind the time frame for geothermal uses of fresh water and water conservation measures that may be instituted. By the mid-1990s, diversions of fresh water for these purposes will have lowered the Sea's surface elevation significantly and raised the salinity to levels that the fish could not tolerate. Thus, interim measures would need to be taken so that the valuable fishery is not destroyed prior to the operation of the solar pond or any other major desalinization project.

Use of Excess Colorado River Water - A Short-Term Solution

A report by the State of California's Department of Water Resources states that there will be flows in the Colorado River, downstream from Lake Mead, in excess of commitments until 1985 when the Central Arizona Project is scheduled to begin diversions [22]. California is entitled to 50 percent of the surplus flows, as determined by the U.S. Supreme Court, and can use them for wildlife refuges on an interim basis [22].

Since both state and national wildlife refuges are present at the Salton Sea, one possible beneficial use of these excess Colorado River flows would be to flood additional refuge lands to create new habitat. Some of the runoff would eventually drain to the Sea, thus adding more fresh water. Excess river water could also be shunted directly into the Sea in which case it would have even more of a benefit on salinity levels; Colorado River water is much lower in salts than runoff-water from agricultural lands in the Imperial Valley [38]. Although the majority of Salton Sea National Wildlife Refuge lands are under water, the aquatic organisms using this area would directly benefit from keeping salinity levels in check.

None of the bird refuges in or around the Sea have been identified in the above-mentioned report as being considered as recipients of any excess Colorado River flows [22]. This situation needs to be re-evaluated.

Use of Salton Sea Water for Geothermal Reinjection - A Short-Term Solution

The 1974 report on the feasibility of different salinity control methods for the Sea considered the need by geothermal developers for water to reinject into the ground to control subsidence, and it recommended that Salton Sea water could be used [35]. Geothermal development in the Imperial Valley was still considered to be very speculative at that time. Not enough was known concerning the amounts of water that would be needed or even if this development would occur so it was not considered a viable method of salinity control [35]. The report did recommend that the situation be re-evaluated in eight to ten years if a major desalinization project had not yet been funded and geothermal development looked more eminent in the Imperial Valley.

Geothermal development has progressed a long way since then, and there are now several proposals for development of three of the four KGRAs. Layton [23] has suggested that Salton Sea water could suitably be used as injection fluid for subsidence control. It would benefit not only the developer, because it would free steam condensate for cooling, but it would have the added benefit of lowering the Sea's salinity. Goldsmith [39] concluded that 148,000 dam^3 (120,000 acre-ft) of Salton Sea water would have to be removed annually for 40 years to lower the salinity to 35,000 ppm TDS. It is not known what the reinjection requirements are for all four KGRAs; however, the Heber KGRA is going to require the use of 61,500 dam^3 (50,000 acre-ft) of water annually for cooling and reinjection [24]. Thus, it is not possible to say whether the use of Sea water for reinjection by all geothermal developers would be enough to lower salinities to the above levels. Due to the limited 30-year [25] life span of a geothermal plant, this is not the long-term answer to stabilizing salinity levels; but it is very important over the short term.

CONCLUSIONS

This paper has shown the Salton Sea to possess a unique ecosystem that supports an irreplaceable fish and wildlife resource that is utilized by many people. In fact, it has been estimated that the Sea could provide approximately 4.3 million recreation days annually if its problems of rising salinity and fluctuating water levels could be solved [3].

The means for accomplishing this on the short-term are at hand in the forms of reinjecting Sea water as part of geothermal energy development and diverting surplus Colorado River flows into the Sea. A long-term answer is a little farther away (solar ponds). However, geothermal energy development reinjecting river water together with a water conservation program could destroy the aquatic-dependent resources of the Sea before a permanent solution can be implemented.

The fauna of the Salton Sea and the recreational, scientific, and aesthetic values that they provide can only be maintained for the enjoyment of future generations through the interest and cooperation of local, state, and federal agencies, as well as a concerned public. Because of the large number of decision-making entities (i.e., state, federal, and local) that influence water use, water quality, and energy development, it seems imperative that a multi-agency committee be formed that has as its goals the perpetuation of the fish and wildlife resources as well as the development of energy at the Sea. The committee can accomplish these goals through the implementation of a plan designed to permanently stabilize salinity and water elevation by using energy development to meet the energy needs of this state as well as the recreational needs. Without this type of approach, the outlook for the Sea's fish, wildlife, and recreational resources is indeed gloomy.

LITERATURE CITED

1. U. S. Bureau of Reclamation, Lower Colorado Region, September 1981. Water conservation opportunities Imperial Irrigation District, California. Salton Sea Operation Study, Draft Report, 58 p.

2. Hely, A. G., Hughes, G. H., and Irelan, B. 1966. Hydrologic regimen of Salton Sea, California. U. S. Geological Survey, Geological Survey Professional Paper 486-C, 32 p.

3. California State Water Resources Control Board 1969.

Economic benefits derived from the waters of and lands
surrounding the Salton Sea. Prepared by Development
Research Associates, Los Angeles, Calif., 105 p.

4.   Evermann, B. W. 1916.   Fishes of the Salton Sea.
     Copeia, 34:61-63.

5.   Coleman, G. A. 1929.   A biological survey of the Salton
     Sea.   Calif. Fish and Game, 15(3):218-227.

6.   Thompson, W. F., and Bryant, H. C. 1920.   The mullet
     fisheries of Salton Sea.   Calif. Fish and Game, 6(2):
     60-63.

7.   Walker, B. W. (editor) 1961.   The ecology of the Salton
     Sea, California, in relation to the sportfishery.
     Calif. Dept. Fish and Game, Fish Bull. 113, 204 p.

8.   Black, G. F. 1974.   The partyboat fishery of the Salton
     Sea and the apparent effect of temperature and salinity
     on the catch of orangemouth corvina, *Cynoscion xanthu-
     lus*. Calif. Fish and Game, Inland Fisheries, Admin.
     Rept. 74-5, 14 p.

9.   Hulquist, R. G. 1981.   A summary of Salton Sea creel
     censuses, 1958, 1963 through 1967, and 1969.   Calif.
     Fish and Game, Inland Fisheries, Region 5 Infor. Bull.
     0004-3-1981, 46 p.

10. Garcia, J. 1981.   Wildlife input related to the Salton
     Sea.   Calif. Dept. Fish and Game, Region 5, Niland,
     Calif. Memorandum.   November 1981, 2 p. + attachments.

11. Frederickson, L. H. 1980.   An evaluation of the role of
     feeding in waterfowl management in Southern California.
     (final rept.) U. S. Dept. Interior, Office of Migratory
     Bird Management, Wash. D. C., 173 p.

12. Barlow, G. W. 1961.   Social behavior of the desert
     pupfish, *Cyprinodon macularius*, in the field and in
     the aquarium.   Amer. Midl. Nat., 65(2):339-359.

13. Black, G. F. 1980.   Status of the desert pupfish,
     Cyprinodon macularius, (Baird and Girard), in
     California.   Calif. Fish and Game, Inland Fish. Endang.
     Species Program, Spec. Publ. 80-1, 42 p.

14. Matsui, M. 1981.   The effects of introduced teleost
     species on social behavior of *Cyprinodon macularius
     californiensis*.   M. S. Thesis, Occidental College,

Glendale, California, 61 p.

15. California Fish and Game 1980. At the crossroads, 1980. A report on California's endangered and rare fish and wildlife. Calif. Fish and Game, December 1980, 147 p.

16. Kobetich, G. C. 1981. Desert pupfish meeting in Parker, Arizona. U. S. Fish and Wildlife Service, Office of Endangered Species, Sacramento, Calif. Memorandum. July 1981, 1 p.

17. McCaskie, G. 1970. Shorebird and waterbird use of the Salton Sea. Calif. Fish and Game, 56(2):87-95.

18. U. S. Fish and Wildlife Service 1970. Birds of the Salton Sea National Wildlife refuge. U. S. Dept. of Interior, Refuge Leaflet 186-R3.

19. Leitner, P., and Grant, G. S. 1978. Observations on waterbird flight patterns at the Salton Sea, California, October 1976 - February 1977. Rept. Prepared for Lawrence Livermore Laboratory.

20. U. S. Bureau of Land Management 1980. The California desert conservation area, final environmental impact statement, and proposed plan. U. S. Dept. Interior: pp. 28-31.

21. U. S. Dept. Interior and the Resources Agency of California 1969. Salton Sea Project California, federal-state reconnaissance report. October 1969, 160 p.

22. State of California, The Resources Agency, Dept. of Water Resources 1980. Stretching California's water supplies: increased use of Colorado River water in California. Southern District Report. August 1980, 28 p.

23. Layton, D. W. 1978. Water for long-term geothermal energy production in the Imperial Valley. Univ. California Lawrence Livermore Laboratory Dept., VERL-52576, Sept. 1978, 48 p.

24. California State Water Resources Control Board 1980. Decision and Order 1559 in the matter of water. Applications 25794 and 25818, 16 p.

25. County of Imperial, California, Prepared by Westec Services, San Diego, Calif. 1981. Salton Sea anomaly,

master environmental impact report. April 1981, 458 p.

26. Gonzales, C. 1981. Telephone conversation on December 2, 1981. Calif. Dept. Fish and Game, Region 5, Imperial Wildlife Area, Manager.

27. Brocksen, R. W., and Cole, R. E. 1972. Physiological responses of three species of fish to various salinities. Jour. Fish. Res. Bd., Canada, 29(4):399-405.

28. Lasker, R., Tenaza, R. H., and Chamberlain, L. L. 1972. The response of Salton Sea fish eggs and larvae to salinity stress. Calif. Fish and Game, 58(1):58-66.

29. Hanson, J. A. 1970. Salinity tolerances for Salton Sea fishes. Calif. Fish and Game, Inland Fish Admin. Rept., 70-2, 8 p.

30. Kuhl, D. L., and Oglesby, L. C. 1979. Reproduction and survival of the pileworm, *Nereis succinea* in higher Salton Sea salinities. Biol. Bull., 157:153-165.

31. Vittor, B. A. 1968. The effects of oxygen tension, salinity, temperature and crowding on the distribution, growth and survival of *Balanus amphitrite* Darwin in the Salton Sea, California. M. A. Thesis, San Diego State College, California, 151 p.

32. U. S. Dept. of Interior and the Resources Agency 1974. Salton Sea Project California, federal-state feasibility report. April 1974, 139 p.

33. Southern California Edison 1981. Solar pond project advances to preliminary engineering stage. Research and Development Newsletter, 10(3): 8 p.

34. Stolpe, J. 1981. 5 MW salt pond pilot plant and future 600 MW commercialization plans. Paper presented to California Energy Commission, August 12, 1981, 11 p.

35. U. S. Dept. Interior, Water and Power Resources Service, 1980. Potential consequences of reject stream replacement projects on aquatic, terrestrial, and recreation resources. Prepared by Engineering Science, Arcadia, Calif., Vol. 2, Aquatic Resources, 105 p. + appendices.

36. Goldsmith, M. 1976. Geothermal development and the Salton Sea. Calif. Inst. Techn., Pasadena, Calif., Memorandum 17.

CHAPTER 23

ASSESSING THE EFFECTS OF COAL MINING
AND RELATED ENERGY DEVELOPMENT ON AQUATIC
ENVIRONMENTAL QUALITY IN THE UPPER COLORADO BASIN

Kurt Gernerd
Jay J. Messer*
Fredrick J. Post
Vincent A. Lamarra
    Utah Water Research Labora-
    tory, Utah State University
    Logan, Utah

INTRODUCTION

The Upper Colorado River Basin contains vast deposits of coal, oil shale, and tar sands, and recent economic incentives have greatly increased the rate or probability of development of these fossil fuel resources [1]. There can be little doubt that development of the anticipated magnitude cannot avoid impacting the quality of adjacent surface and groundwaters, as well as their associated aquatic ecosystems. Intelligent benefit–cost or risk analyses must be predicated on a firm data base regarding the magnitude and probability of such impacts. As a tool for use in conducting such analyses in the Upper Colorado Basin, a technical information matrix has been constructed relating impacts on aquatic environmental variables to activities associated with coal, oil shale, and tar sands development [2]. Here the form and construction of the matrix are presented and the results are briefly summarized with respect to coal development. Oil shale development impacts are discussed elsewhere in this volume [3, 4, 5], and at the time of writing, tar sands development in the basin is only in the feasibility and planning stages [6].

TECHNICAL INFORMATION MATRIX

Basic Format

Although the formation of the technical information matrix relies heavily on the impact matrix design popularized

---

*To whom **correspondence** should be addressed.

by Leopold and others [7], the objective is somewhat different. Most impact matrices contain, at the intersection of an
action with a corresponding impact, some ordinal value judgment regarding the magnitude and importance of the relationship [7, 8, 9]. The purpose of the information matrix presented here is to present reference citations, including a
brief abstract, at each matrix address. No attempt has been
made to assess a magnitude or relative importance for the
impacts.

Figure 1 displays the low-resolution form of the matrix,
categorizing the most general activities and impacts. Activities are first grouped into three fossil fuel types (coal,
oil shale, and tar sands) and second into major activities
associated with development of each fossil fuel type. Impact
categories are surface and groundwater quality, aquatic
ecology, and aquifer modification, which might be expected
to affect the first three through alteration of streamflow
and geochemically important flow paths.

Activity categories are expanded (Figures 2 and 3) to
cover various aspects of each general activity category that
might be expected either to generate its peculiar suite of
impacts or to be subject to common control measures. Power
production (IE, Figure 3) relates to mine mouth power plants
whose operations cannot be divorced from the associated
mining activities. Waste streams associated with coal conversion (IF2 and IG2) and oil shale or tar sands retorting
(IIE3 and IIIE3, respectively) are grouped because they may
all be treatable at a common point, rather than for their
impact-related similarity.

Impact categories are expanded (Figure 4) to include
water quality constituents, ecological constructs, and
factors important in aquifer modification. Water quality
parameters were grouped according to criteria relating to
the relative amounts and resolution of available information
on sources and impacts. Aquatic ecological impacts are
grouped so that both species specific (species lists) and
community structural (biomass, diversity, indicator assemblage) and functional (P:R ratios, metabolic rate) investigations can be accommodated.

Impacts resulting from a particular activity are located
through an alphanumeric address label. For example, the
impact of overburden disposal from coal mining on surface
water salinity would be found at IB7.1b. A sample listing at
that matrix address is shown in Figure 5. The other letters
indicate that the reference also includes information regarding salinity, nutrients, pH, dissolved solids, suspended
solids, heavy metals, and sulfates. The parenthetical CO and

384

Activities / Impacts	I Coal							II Oil Shale					III Tar Sands				
	A Exploration	B Extraction	C Ancillary Activities	D Reclamation	E Power Production	F Gasification	G Liquefaction	A Exploration	B Extraction	C Ancillary Activities	D Reclamation	E Retorting Operations	A Exploration	B Extraction	C Ancillary Activities	D Reclamation	E Retorting Operations
1. Surface Water Quality																	
2. Subsurface Water Quality																	
3. Aquatic Ecology																	
4. Aquifer Modification																	

Figure 1. Level I matrix showing general activity and impact categories.

385

I.A. Exploration

1	Surveying
2	Test Drilling
3	Test Pit Mining

I.B. Extraction Activities

1	Operation of Equipment
2	Removal of Surface Features
3	Topsoiling Storage
4	Dewatering
5	Blasting
6	Overburden Disposal
7	Mineral Extraction

I.C. Ancillary Activities

Beneficiation	1	Mineral Storage
	2	Mineral Cleaning
	3	Mineral Loading
Transportation	4	Railroad
	5	Access Roads
	6	Haul Roads
	7	Conveyor Pipelines
Support	8	Fuel & Chemical Storage
	9	Maintenance Yards & Parking Lots
	10	Sewage Treatment Plant
Utilities	11	Electric Transmission
	12	Water Supply
Disposal	13	Septic Tank
	14	Sediment Pond/Strip Lakes
	15	Runoff Controls
	16	Waste Rock (Refuse)

I.D. Reclamation Activities

1	Operation of Equipment
2	Backfilling & Grading
3	Topsoiling & Revegetation
4	Irrigation
5	Treatment Facilities

Figure 2. Expanded activity categories (Level II matrix) for coal mining (I) categories A-D. Oil shale and tar sands will be identical for these categories.

Table 1

I.E. Power Production						I.F. Gasification							I.G. Liquefaction			
						Surface				In Situ						
1	2	3	4	5	6	1	2	3	4	5	6	7	1	2	3	4
Cooling Systems	Scrubber Effluent	Scrubber Sludge	Ash Disposal	Ash Ponds	Cooling Lakes	Gas Quench and Cooling Waters	Waste Streams	Ash/Slag Disposal	Ash Ponds	Explosive Fracturing	In-situ Gasification	Product Waters	Blowdown and Cooling Waters	Waste Streams	Solid Wastes (Residues)	

Table 2

II.E Retorting Operations									III.E Retorting Operations								
Surface						In Situ			Surface						In Situ		
1	2	3	4	5	6	7	8	9	1	2	3	4	5	6	7	8	9
Cooling Systems	Tankage	Processing Waste Streams	Retort Water	Retention Ponds	Spent Shale Disposal	Explosive Fracturing	Residual Matter (Spent Shale Oil Products)	Retort & Condensate Water	Cooling Tower & Boiler Blowdown	Tankage	Processing Waste Streams	Retort Water	Plant Site Retention Pond	Spent Sand Disposal	Explosive Fracturing	Residual Matter (Spent Sand Sand Oil Products)	Retort & Condensate Water

Figure 3. Expanded activity categories (Level II matrix) for coal mining (I) categories E-G, and oil shale (II) and tar sands (III) category E.

387

1.SURFACE WATER QUALITY	a	Temperature
	b	Salinity
	c	Nutrients
	d	pH
	e	Dissolved Solids
	f	Suspended Solids
	g	Toxicants
	h	Carcinogens
	i	Heavy Metals
	j	Sulfates
	k	Radionuclides
	l	Pathogens
	m	Organics

2.SUBSURFACE WATER QUALITY	a	Temperature
	b	Salinity
	c	Nutrients
	d	pH
	e	Dissolved Solids
	f	Suspended Solids
	g	Toxicants
	h	Carcinogens
	i	Heavy Metals
	j	Sulfates
	k	Radionuclides
	l	Pathogens
	m	Organics

3.AQUATIC ECOLOGY	a	Algae
	b	Micro-Invertebrates
	c	Macro-Invertebrates
	d	Vertebrates
	e	Aerobic Decomposers
	f	Anaerobic Decomposers
	g	Macrophytes
	h	Biomass
	i	Species Diversity
	j	Key Species
	k	Population Size
	l	Ecosystem Functions

4.AQUIFER MODIFICATION	a	Water Table Alteration
	b	Transmissivity
	c	Recharge & Discharge
	d	Hydrologic Regime
	e	Storage Coefficient

Figure 4.   Expanded impact categories (Level II matrix) for all activities, categories 1-4.

b,c,d,e f,i,j	Average ion concentrations, discharge volumes, and salt loads are tabulated for Oct. 1973–Nov. 1975. (CO & MT)	(McWhorter, Rowe, Van Liew, Chandler, Skogerboe, Sanada, & Skogerboe 1979)

Figure 5. Typical Reference Citation from Address IB7.1b in Technical Information Matrix.

MT indicates that the studies were performed in Colorado and Montana. The reference at the right can be found in the bibliography. If the reference also dealt with impacts of other activities, it would also appear (perhaps with a different abstract) at another matrix address with a different initial three digit alphanumeric label.

Information Sources

Gathering the information contained in the matrix has thus far consumed approximately two man-years. Much of the research has not yet overcome the two year time lag that is often characteristic of publication in the refereed environmental literature, and conventional (e.g. Chemical and Biological Abstracts) searching techniques are not highly effective. Indeed, the best information was found in a variety of reports by federal and state agencies, symposium and workshop proceedings, graduate student theses, and reports by private industry and consulting firms. Many of the government documents are open file reports, and are not routinely available in libraries containing Federal Repositories.

In order to reference the most recent research results, numerous bibliographies, workshop and conference proceedings were consulted, personal telephone calls were made, and extensive computer data searching was done. Table I lists the sources consulted. There are undoubtedly exclusions, particularly in the area of consulting reports, open-file reports, and masters theses. Perhaps some consolation can be gained from the fact that the most significant of this information, and that with the highest quality control, is probably also found in conference proceedings or government documents with wider circulation.

Several different kinds of information appeared during construction of the matrix. The listing in Figure 5 represents the highest quality of information with respect to observed site specific impacts of a particular activity in the Upper Colorado Basin. Of somewhat lesser utility are empirical data collected elsewhere (e.g., acid mine drainage in Kentucky), although many of the impacts are not

Table I. Information Sources

1. Computer Databases Accessed
   a. NTIS-National Technical Information Service
   b. Water Resources Abstracts
   c. Enviroline
   d. Energyline
   e. Compendex
   f. Agricola
   g. Seaminfo
   h. Scisearch
   i. Pollution Abstracts
   j. Conference Papers Index
   k. Dissertation Abstracts
2. Bibliographies and Published Searches
   a. U.S. Fish & Wildlife Service
   b. National Technical Information Service
   c. Western Energy & Land Use Team
   d. National Science Foundation
   e. Bituminous Coal Research, Inc.
   f. U.S. Dept. of Agriculture
   g. National Coal Association
   h. U.S. Geological Survey
   i. Energy & Mineral Resources Res. Inst.
   j. Bureau of Mines
   k. Oil Shale Task Force
   l. Environmental Protection Agency
   m. Various Colleges & Universities
3. Department of Energy – Technical Information Center
   a. Fossil Energy Update (monthly publication)
   b. Laramie Energy Technology Center.
   c. Environmental Development Plans, Workshops, and
      Environmental Impact Statements
4. Hard Bound Abstract Series
   a. ERDA – Energy Research & Development Administration
   b. Biological Abstracts
   c. Chemical Abstracts
   d. Pollution Abstracts
5. Environmental Protection Agency Research Report Series
   a. Environmental Health Effects Research
   b. Environmental Protection Technology
   c. Ecological Research
   d. Environmental Monitoring
   e. Scientific and Technical Assessment Reports
   f. Interagency Energy – Environment Research and
      Development
   g. Special and Miscellaneous Reports
6. Conference and Symposium Proceedings
   a. >100 referenced in the technical matrix
7. National Research Laboratories
   a. Battelle Columbus Laboratories

Table I. Continued

b.	Brookhaven National Laboratory
c.	Los Alamos Scientific Laboratory
d.	Argonne National Laboratory
e.	Oak Ridge National Laboratory

8. Private Corporations and Consulting Firms

geographically site specific. Of still less immediate value
are speculations in the absence of in situ studies regarding
the probability of an impact, e.g., through knowledge of
waste stream constituents on organisms in bioassays. All of
the above types of references are included in the matrix,
(speculative references are so indicated), on the assumption
that researchable problems may be identified. Also discovered were general summaries ranging in quality from invited reviews in refereed journals to luncheon addresses to
local civic organizations. This last group is collated into
a separate bibliography, with the most valuable and comprehensive reviews indicated.

IMPACTS OF COAL DEVELOPMENT

    In order to illustrate the application of the technical
information matrix, the impact addresses corresponding to coal
development were surveyed. These are summarized briefly
below, according to the major impact categories in Figure 1.
In some cases, no impacts appear to be documented, but
probable impacts are apparent from other studies.

A. Exploration

    1. Surface Water Quality - Most premining operations are
        not large enough to create measurable surface water
        quality alterations. However, if a shot hole pierces
        a saline, artesian aquifer, the resultant spring may
        contribute to salinity in surface streams.

    2. Subsurface Water Quality - Premining drilling per-
        forates both shallow and deep groundwater systems, and
        avenues allowing increased rates of groundwater migra-
        tion and mixing are created. Large areas may be ex-
        plored which will not be mined for various reasons,
        but as a result of exploratory drilling, measurable
        groundwater impacts may occur due to aquifer modifi-
        cation.

    3. Aquatic Ecology - Impacts to aquatic ecosystems from
        premining operations in terms of biotic response have
        not been documented.

391

4. Aquifer Modification – Water table alterations allowing mixing of different quality waters across aquitards may occur as a result of exploratory drilling. Impacts to the hydrologic regime from test pit mining should be minimal and confined to the site.

B. Extraction

1. Surface Water Quality – In the West, carbonates in overburden material often result in mine drainage that is alkaline, rather than acidic, with high TDS values. Salinity is the most significant water quality problem in the Upper Colorado Basin. While alkaline conditions decrease the mobility of most metal species by increasing pH values, it also enhances the transport of other potential contaminants such as Mo, F, B, As, Se, $SO_4^=$, Cd, Hg. However, acid mine drainage may occur in areas low in carbonates. Also, sediment transport in mining areas can be as much as 1000 times greater than in undisturbed areas.

2. Subsurface Water Quality – Soluble salts, principally Ca, Na and $SO_4^=$ are major contaminants. TDS values of as much as 5 times as great as prior to mining are produced at most sites. Disruption from explosives creates fissures and fractures in the rock strata which may introduce acid waters from the surface or saline pollution from below.

3. Aquatic Ecology – Alkaline and acid mine drainage along with heavy silt loads will impact aquatic ecosystems if high concentrations persist over a sufficiently long period. At present there is very limited knowledge of dose–response relationships between the amount of contaminant to which organisms or communities are exposed, and a quantitative measure of its effects on the organisms or communities.

4. Aquifer Modification – The most serious water related mining problem associated with development of western coal fields is disruption of aquifers resulting in lowered water tables and well levels. Premining groundwater flow patterns will frequently reestablish themselves in reclaimed areas and will not differ greatly from premining patterns, and storage capacity may actually increase.

C. Ancillary Activities

1. Surface Water Quality – Runoff from coal piles may be acidic if acid production exceeds the buffering

392

capacity of the ash, in which case pH, $SO_4^=$, alkalinity, Fe, Mn, and heavy metals may violate water quality standards. Contaminants in the process water from wet coal cleaning consist of suspended solids, which are chiefly fine clay and coal, and dissolved solids. Water effluents may also contain surface active organic compounds such as alcohols or kerosene, which are added in some coal cleaning plants to enhance frothing. Silt and coal fines often constitute the major water pollution hazards of coal refuse. Sediment yield from improperly designed and constructed coal-haul roads can be as great as that from the stripping phase.

2. Subsurface Water Quality - Analyses indicated that sulfur and the heavy metals: Sb, Ar, Be, Cd, Hg, Ni and Se are "markedly enriched" at the base of the coal seams and coal storage piles. Contamination of groundwater will occur from refuse piles by acid drainage which contains high concentrations of TDS, $SO_4^=$, Fe, and other metals.

3. Aquatic Ecology - Handling of coal near surface water will create siltation impacts by altering benthic communities, fish spawning areas, food web interrelationships, and diversity and/or stability of the biota. Zooplankton and benthos diversity is lower along with fewer wetland plant communities and fewer plant species within each community in strip-mine ponds.

4. Aquifer Modification - Ancillary activities impacting hydrologic parameters have not been documented.

D. Reclamation Activities

1. Surface Water Quality - Reclaimed mine land, even when graded to acceptable specifications, is left with an extremely high potential for erosion and sediment yield.

2. Subsurface Water Quality - Groundwater in replaced oxidized spoil contains significantly higher sulfate, dissolved metal, and cation concentrations.

3. Aquatic Ecology - Floc formed as a product lime neutralization of acid mine drainage could destroy the normal habitat of bottom fauna, and also reduce organism numbers and taxa, as it has in some eastern streams.

393

4. Aquifer Modification – Replaced mine overburden has greater transmissivity and water storage ability than do original premining conditions.

E. Power Production

1. Surface Water Quality – Thermal pollution has been recognized as an environmental problem for cooling waters for some time. Chlorine is often used for biofouling control in cooling water and has been demonstrated under some conditions to cause formation of chlorinated hydrocarbons. The potential for damage to aquatic ecosystems of trace elements identified in power plant effluents are: Cd, Hg > As, Cr, Pb, Se, Z > Ba, Cu, Mn > Co, Mo, Ni.

2. Subsurface Water Quality – Contamination by flyash landfills is highly variable, but has been shown to be generally high in Si, Al, Fe, Ca, Mg, and S among the major constituents, and in such trace elements as Ti, Mn, K, Cu, Zn, Cr, Mo, Ni, and Pb. Se, B, and Cr pose the greatest threat to groundwater from scrubber sludge leachates.

3. Aquatic Ecology – Discharges at temperatures higher than those of the receiving waters will create alterations in the entire biotic community. Biocides, (both chlorine and chloramines) principally used to control fouling on inside surfaces, are toxic to aquatic organisms frequently at extremely low concentrations.

4. Aquifer Modification – The creation of cooling lakes can alter the configuration of local flow systems and increase the discharge of groundwater to local wetlands.

F. Gasification

1. Surface Water Quality – The particulate removal and gas cooling steps produce ash and water contaminated with condensable hydrocarbons, many of which are toxic. The major contaminants in primary coal gasification streams are: tars, oils, flyash, sulfur compounds, ammonia, hydrogen cyanide, thiocyanide, and phenols. Some form of treatment (biological, chemical precipitation, or adsorption) must be applied to coal gasification plant waste waters before they are released from the site.

394

2. Subsurface Water Quality — Leachate of ash and slag disposal represent a potential problem in violating secondary drinking water standards (Cu, Zn, Fe, Mn, SO_4, and TDS). The reaction products, such as ash and tars, that remain underground following gasification, are a potential source of localized groundwater contamination. However, the concentration of important contaminants such as phenols show a significant decrease due to adsorption by the surrounding coal. All detectable contaminants have been shown to rapidly decrease in concentration with distance.

3. Aquatic Ecology — Ammonia and phenol, especially acting synergistically, are major components affecting the acute toxicity of condenser water to Daphnia, minnow, and trout populations.

4. Aquifer Modification — Water influx from natural aquifers is a major underground problem following coal gasification.

G. Liquefaction

1. Surface Water Quality — The foul water condensate is viewed as the process waste stream containing the heaviest organic loadings and potentially the greatest environmental hazard. The unstripped foul water contains high concentrations of BOD, COD, phenols, NH_3-N, and sulfate. Typically, heavy metals concentrations were low, although high concentrations of Pb, Ni, and Al may be observed periodically.

2. Subsurface Water Quality — Leachate of trace elements Mn, Ag, and Ni from solid residues have been observed to exceed recommended potable water standards.

3. Aquatic Ecology — Tests indicate that untreated liquid effluents are highly toxic to aquatic life.

4. Aquifer Modification — Not applicable.

DISCUSSION

Benefit-cost and benefit-risk analyses of energy production and aquatic environmental quality require reliable quantitative information upon which to base management decisions. The information explosion presently occurring on such development activities as coal mining in the Rosebud Creek drainage in Wyoming and oil shale development in the White River drainage in Colorado and Utah requires some sort of data

395

organization which we have endeavored to provide through the
technical information matrix presented here.

Several caveats have become apparent during construction
of the matrix. First, there is no doubt information that was
missed, either through unawareness of the availability of the
information or through having only an abstract to use. Computerized searching is highly sensitive to the choice of key-
words [10], and one cannot always optimize tradeoffs between
efficiency (volume searched) and thoroughness. To expedite
use of the matrix the document will be bound looseleaf, and
it is hoped that individual users will expand the matrix
citations from their own libraries. Communications with users,
together with additional on-line searches, will be used to
provide annual updates to the document.

Finally, we note that in ecology, the whole often appears
greater than the sum of its parts. Combinations of independent water quality impacts may result in synergistic effects
that go far beyond an additive response. Impacts still must
be judged by specialists and modelers in environmental constraints. It is our hope that the technical information
matrix described here will contribute valuable information to
such analysts.

LITERATURE CITED

1. EPA. 1979. Energy from the West: Energy resource
   development systems report. EPA-600/7-79-060a-f. U.S.
   Environmental Protection Agency, Washington.

2. Messer, J. and F. Post [eds.]. 1982. Impacts of western
   coal, oil shale, and tar sands development on aquatic
   environmental quality: A technical information matrix.
   UWRL/P-82/04. Utah Water Research Lab., Utah State
   University, Logan

3. Bulot, M., and M. Donovan (this volume).

4. Dietz, D., and R. Tucker (this volume).

5. Grant, V. (this volume).

6. Nielsen, D. (personal communication). Salt Lake City, Ut.

7. Leopold, L., F. Clarke, B. Hanshaw, and J. Balsley.
   1971. A procedure for evaluating environmental impact.
   Circular 645. U.S. Geol. Surv., Washington, D.C.

8. Canter, L. 1977. Environmental impact assessment.
   McGraw-Hill, New York.

9. Munn, R. E. [ed.]. 1975. Environmental Impact Assessment. SCOPE 5. Internat. Council of Scientific Unions, Toronto, Canada.

10. Garfield, E. 1981. Automatic indexing and the linguistics connection. Current Contents 8 (Feb. 23):8-11.

CHAPTER 24

POTENTIAL IMPACTS OF ENERGY DEVELOPMENT
UPON WATER QUALITY OF LAKE POWELL
AND THE UPPER COLORADO RIVER

Allen J. Medine
   Civil, Environmental and Architectural Engineering
   University of Colorado - Boulder
   Boulder, Colorado 80309

INTRODUCTION

The upper Colorado River system, herein defined as the major tributaries upstream from Lake Powell, includes the Colorado River, Green River, San Juan River, Gunnison River, Dolores River, White River and other tributaries (Figure 1). Lake Powell, created by impoundment of the Colorado River system, by Glen Canyon Dam, has been described as a warm monomictic reservoir [1] although it shows meromictic properties [2] with a distinct chemocline between 50 and 75 meters. The annual average flows ($13.8 \times 10^9 m^3/yr$) from major tributaries result in an average detention period of 2.5 yr. including evaporation losses of $570 \times 10^6 m^3/yr$. There has been substantial research conducted on the Lake Powell system including a major previously funded NSF effort, "Lake Powell Research Project," to examine the ramifications of increased water demands, energy development and population growth.

Development of the oil shale industry in western Colorado, eastern Utah and southwest Wyoming will have significant impacts affecting the land, environmental quality, the economy as well as the general population of the upper Colorado River system. The total magnitude of these impacts is not yet known. Various wastes associated with oil shale processing will cause major problems for the emerging industry.

The Green River Formation (Figure 1) is the world's largest oil shale resource covering a $42,700 \text{ km}^2$ area in the Colorado, Utah and Wyoming [3]. This formation is divided into primary formations including the Piceance Creek Basin (Colo.), Uinta Basin (Utah) and the Green River

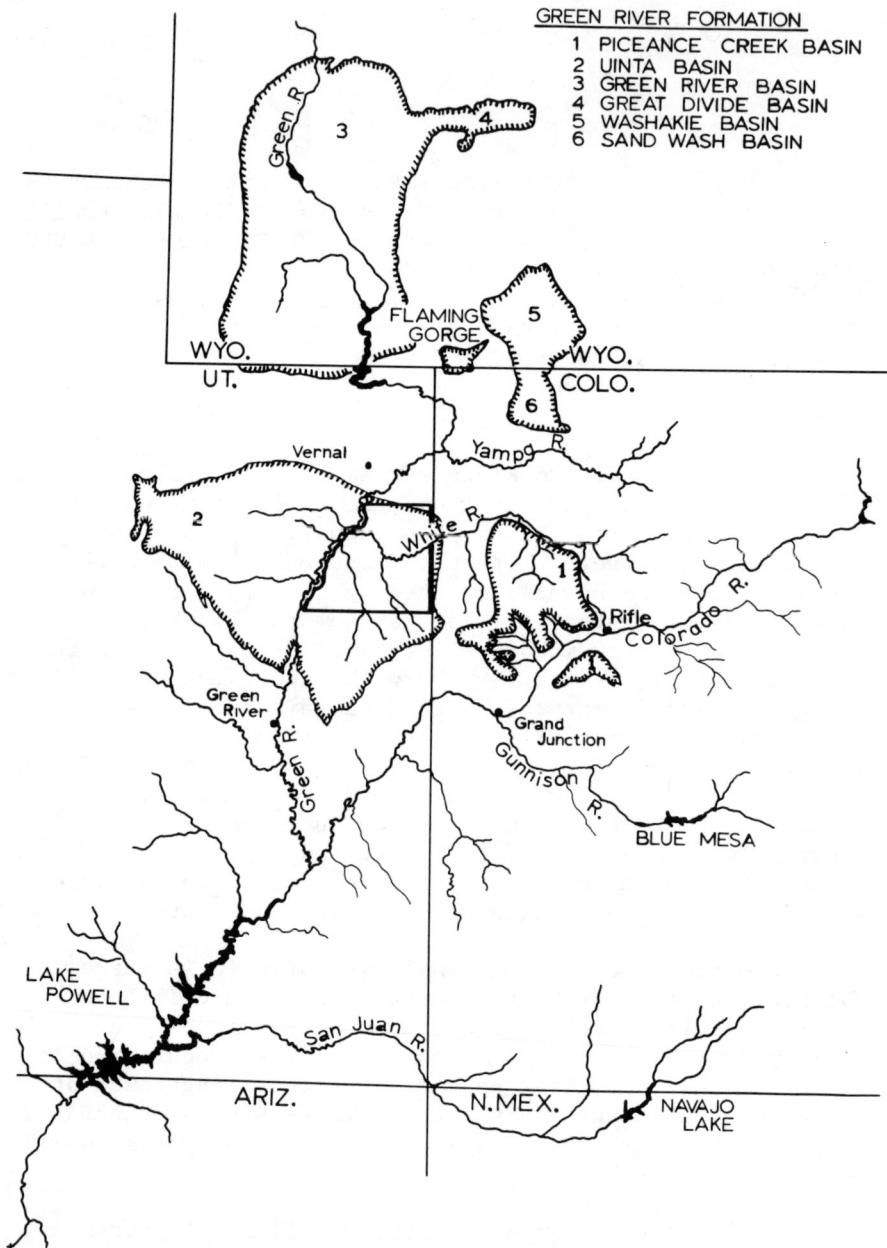

Figure 1. Location Map for the Upper Colorado System

400

and Washakie Basins (Wyoming) with combined total in-place resources of 217 x 10^9 tons (1,500 x 10^9 equivalent barrels of oil) of potentially recoverable oil. Approximately 69% of this estimate lies in the Piceance Basin of Colorado [3]. Addition of oil shale in the Green River Formation above, below and geographically outside the main resources would boost total resources to about 1,800 x 10^9 barrels of oil. If only oil shale deposits with a production capacity of 105 L/tonne are included, the total resource is about 87 x 10^9 tons (600 x 10^9 barrels) [3]. The state of Colorado is at the same time fortunate and unfortunate to possess this major resource within its boundaries.

The objective here was to examine the projected oil shale scenarios with respect to environmental concerns, particularly the impact of wastes to water quality in the upper Colorado River and Lake Powell. It is beyond the scope of this chapter to examine potential impacts of development of the entire oil shale resource. Consequently, this discussion is directed at impacts associated with resource development in the Piceance Basin and development of ancillary support facilities (power production, municipal sources, etc.). The Piceance Basin drains to two major rivers (Figure 1), the White River (via Piceance Creek and Yellow Creek) and the Colorado River (via Roan Creek and Parachute Creek) although most drainage will be toward the White River. Groundwater flows are estimated to contribute 80% of the total annual water discharge from the basin. It should be noted that development of Utah's prime reserve (shown within the heavy lines in the Uinta Basin on Figure 1) would exert additional impacts. The major water pollutants (contained in groundwater flow, runoff, etc.) would impact the White River and the Green River downstream to the Colorado River.

Following a discussion of oil shale scenarios projected for the Piceance Basin including population and energy needs, the major environmental concerns (atmospheric pollutant emissions, water pollutants and solid wastes) are addressed to provide backgound data. Current water quality of the Colorado River system will be considered with respect to heavy metals (primarily Fe, Cu, Cd, Zn, Ni, Hg, Pb, Cr, As, and Se) and projected water quality impacts are estimated using mass emissions and flow characteristics of the major tributaries. Our current knowledge and understanding is insufficient to determine absolute impacts on such a dynamic system and it is hoped that this chapter may provide additional insight for future research.

OIL SHALE DEVELOPMENT SCENARIOS

Current estimates for oil shale production fall somewhere between 200,000 and 400,000 bbl/day by 1990. Maximum synfuels production has been suggested to be 15 x 10^6 bbl/day in 2010 with about 50% of that from the Piceance Basin in Colorado. A reasonable production level estimate, under an extreme development, would not exceed 8 x 10^6 bbl/day in 2000 (in Colorado).

Based upon these figures and industry projections, the Colorado Energy Research Institute [4] conducted a study on possible oil shale development of the Piceance Basin in northwestern Colorado. Four major scenarios were developed to provide a preliminary assessment of the potential problems which may result from a massive influx of people to the Rocky Mountain West. The scenarios are based upon oil shale production in the year 2000 and are shown in Table I along with energy, water, and population projections. Assuming the Colorado production was 50% of the total synfuels industry, an equivalent production might be envisioned for Utah's shale development program, along with corresponding energy, water and population, and wastewater increments.

ENVIRONMENTAL CONCERNS

The environmental pollutant concerns associated with oil shale development may be classified as atmospheric, water quality and solid waste. Wastes may originate from the oil shale mining, processing, retorting and disposal operations or from the ancillary development including energy production, municipal wastewaters, industrial wastewaters, etc. Of primary concern here is the potential atmospheric pollutants generated at the oil shale sites, the discharge of various wastewaters (oil shale retorting, mine dewatering, oil shale upgrading, municipal), surface runoff and leachates from spent and raw shale.

Atmospheric Emissions

Atmospheric emissions of particulates, sulfur oxides, nitrogen oxides, hydrocarbons and carbon monoxide originating from mining, processing, retorting and disposal are shown in Table II for various oil shale scenarios. The data have been extrapolated from emissions of a 50,000 bbl/day plant provided in a report by the EPA Oil Shale Research Group [5] and represent preliminary estimates of potential emissions. It should be noted that the range of values is considerable and that the average may not reflect actual commercial scale emissions.

Table I. Oil Shale Scenarios for Colorado in the Year 2000 [4].

Scenario	Oil Shale Production 10^6 bbl/day	Population Growth[1] in Millions	New Electrical Power Requirement[2] MW	Increased Water Consumption[3] 10^8 m^3/yr	Municipal Wastewater Flow[4] 10^8 m^3/yr
Extreme	4.000	1.248	8,000	10.7	2.58 (187 mgd)
High	1.800	0.735	4,000	4.3	1.52 (110 mgd)
Medium	0.900	0.554	1,600	2.28	1.15 (83 mgd)
Low	0.400	0.456	800	1.00	0.945 (68 mgd)

1 Baseline population in year 2000 is 310,000 (Current = 175,000 in 1981)

2 Colorado's current generating capacity is about 4500 MW

3 The Extreme Scenario represents 18% of the Colorado River flow at Grand Junction

4 Baseline wastewater flows are estimated at 0.725×10^8 m^3/yr (52.5 mgd) using 568 lpcd (150 gal/capita-day)

Table II. Mass Emissions of Pollutants by Source
from a 50,000 bbl/day Oil Shale Facility

Source	Mass Emission (tonnes/day)				
	Part.	SO_x	NO_x	HC	CO
Mining					
Surface	22.5	1.28	20.1	2.18	11.4
Underground	0.56	0.03	1.76	0.22	1.94
Processing					
Surface	7.49	--	--	--	--
Underground	0.63	--	--	--	--
Retorting					
Surface	3.69	2.08	12.5	2.93	0.75
In-Situ	0.20	0.20	--	--	--

Mining and Processing

Surface mining emissions include contributions by
excavation, blasting, crushing, transportation and
equipment emissions. Underground mining includes emission
due to excavation, blasting and ground vehicles while
processing of raw shale prior to retorting produces
emissions from primary and secondary crushing, storage and
transportation. As evident from Table II, the control of
fugitive dust in surface mining operations (data reflect a
30-50% control of fugitive dust) will be critical for the
control of these emissions.

Retorting

Various concentrations of pollutants will be generated
during the retorting process. The data given in the EPA
report [5] have been carefully selected to represent recent
emission estimates by the industry and have excluded the
initial higher emission levels. Sulfur is released
primarily from organic sulfur compounds (1/3 total S in oil
shale) in the form of SO_2, H_2S, carbon disulfide, carbonyl
sulfide, and thiocyanates. The data (Table II) show
current estimates for particulates, SO_x, NO_x,
hydrocarbons and carbon monoxide. Trace metal emissions
need further study but emissions of As, Hg, Pb, Fe, Cr and
Zn may be of concern in off-gas and particulate fractions
[5]. The SO_x emission represents an emission control of
greater than 95%.

Emission from oil shale facilities are significant in terms of potential air quality degradation and, following precipitation, water quality impacts. SO_x and NO_x emissions are particularly important due to formation of acid rain through atmospheric reactions as follows:

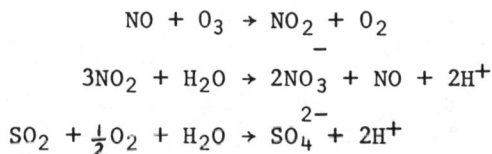

$$NO + O_3 \rightarrow NO_2 + O_2$$

$$3NO_2 + H_2O \rightarrow 2NO_3^- + NO + 2H^+$$

$$SO_2 + \tfrac{1}{2}O_2 + H_2O \rightarrow SO_4^{2-} + 2H^+$$

Acid rain may be significant depending upon diffusion of pollutants in the atmosphere and is greatly affected by terrain and local meteorological conditions. Acid rain impacts upon water quality may result in an overall lower pH in streams and lakes with a corresponding increase in heavy metals from increased weathering and acid redissolution of sedimented metals. A comparison of Adirondak Mountain Lakes (pH=4.7, pH=6.7) showed a tenfold increase in Al, Fe, Mn and Zn and indicated a release of previously precipitated metals [6]. While the main rivers of the Colorado River system have a high buffer capacity and could tolerate a significant amount of acid rain, the small streams and lakes in the White River headwaters could be impacted through alteration of soil and water chemistry.

Atmospheric modeling of emissions has been performed by the EPA using the EPA Valley Model to predict ambient air quality resulting from the previous emission data. The application of the model may not be entirely valid due to rough terrain, but nonetheless, it does provide a "first-cut analysis" of potential air quality degradation. Using the results of the model for various emission levels and the scenarios previously discussed, calculations of the predicted ambient air quality in Piceance Valley have been summarized (Table III) assuming that, at the 400,000 bbl/day production level (low scenario), the plants are separated sufficiently so that ambient air concentrations are the same as that from a 50,000 bbl/day facility. Increases in production above 400,000 bbl/day are compared directly to this output level. This rough analysis indicates that an output level of between 900,000 and 1,800,000 bbl/day would result in ambient air quality levels approaching State and Federal standards. Air quality degradation in areas east will be particularly

Table III. Estimated Ambient Air Quality
Under Various Scenarios

Pollutant Process	Ambient Air Quality ($\mu g/m^3$) Scenario (bbl/day)			
	Low 400,000	Medium 900,000	High 1,800,000	Extreme 4,000,000
SO_x				
Retorting				
Surface	8	18	36	80
In Situ	1	2.3	4.5	10
Surface Mining	4.7	10.6	21.2	47
Underground Mining	<1	<2.3	<4.5	<1.0
NO_x				
Retorting Surface	11	24.8	49.5	111
Surface Mining	18.7	42.1	84.2	187
Underground Mining	1.75	3.9	7.9	17.5

Standards: Federal SO_2=80 mg/m^3, NO_2=100 $\mu g/m^3$
State (Colo.)-SO_2= 20 $\mu g/m^3$

important, i.e. in the Flat Tops Wilderness area and the
White River National Forest, due to current high quality.
Further refinement of rough terrain modelpredictions,
improved understanding of the dispersion characteristics of
the Piceance Valley and the interactions of point source
emissions, and the actual mass emissions may significantly
alter the data in the table. This analysis does not
include degradation of air quality resulting from
development of equal size facilities in Utah's prime
resource area which would undoubtedly result in a mass
influx of SO_x and NO_x into the region (Utah's
current air quality standard for SO_x is 60/$\mu g/m^3$).

Water Quality Impacts

Assessment of potential water quality impacts
attributed to commercial oil shale production is extremely
difficult due to the variability of oil shale processing
technology, the experimental nature of retorting processes,

site geology, wastewater treatment options, spent shale
disposal, as well as production scenarios. The waste
effluents generated by retorting possess an extremely
complex matrix of inorganic and organic contaminants.
Currently, most oil shale developers are planning a zero
discharge of wastewaters from their facilities [5] which
dictates a very complicated water management and reuse
system. Discharges of process waters to surface waters
will be only accidental through improper design or failure
of holding ponds.

To assess potential impacts, the waste streams from
retorting must be considered along with other potential
liquid waste streams. The following appear to be of most
concern for the assessment of water quality impacts:

- Wastewater from retorting
- Wastewater from oil shale upgrading
- Water treatment plant waste streams (on site)
- Municipal wastewater treatment effluents
- Mine drainage
- Mine dewatering waters
- Leachate from spent shale or raw shale
- Leachate from in situ retorting facilities
- Runoff from disturbed areas

If a zero discharge scheme is maintained, then the first
three items may not be of primary concern. According to
Stollenwerk and Runnells [7], the disposal of retorted
waste shale poses a major problem for the industry due to
the potential contamination of groundwater and surface
water through leaching. Leaching studies on spent and raw
shale are receiving considerable attention [7, 8, 9, 10,
11, 12, 13, 14] to provide better understanding of the
potential for increased loading of toxic pollutants in the
Colorado River System. Data related to potential
contamination of surface runoff and mine drainage are
lacking. The important questions to answer relate to

a) the types and concentrations of contaminants which
   may enter surface waters (and groundwaters),
b) the extent of release (continuous, temporary) from
   spent and raw shale,
c) the potential impacts upon stream and reservoir
   ecology,
d) the increment of pollutants relative to natural or
   background levels, and
e) natural mechanisms for the attenuation of pollutant
   concentrations in surface waters.

407

The last question is of major importance in assessing
overall impacts of oil shale activities on the upper
Colorado River and Lake Powell. Natural phenomena which
affect the transport, transformation and accumulation of
organic chemicals and heavy metals in natural environments
include chemical precipitation, adsorption, volatilization,
photolysis, oxidation-reduction, hydrolysis, bioaccumulation, biodegradation, and biotransformation. To determine the extent to which these phenomena exert their
influence on toxic substances is a very complicated
problem. The most critical problems facing researchers
investigating the dynamics of pollutants in the Colorado
River system and Lake Powell relates to kinetic
descriptions of the above phenomena, in addition to
characterizing mass flow of pollutants as a function of
hydrologic characteristics. While a substantial data base
exists for heavy metals in the Colorado River system
upstream from Lake Powell, it is very difficult to evaluate
long term trends in the data as a function of stream flow.
Further research is necessary in this regard. In this
chapter, impacts upon water quality will be examined as a
function of leachates from spent and raw shale, potential
acid precipitation impacts (pH perturbations) and increased
organic,nutrient and heavy metal loading from municipal
wastewaters). These aspects are discussed below.

Leachates from Raw and Spent Shale

     Development of the oil shale industry will most likely
involve both surface and "modified in situ" retorting. In
either method, leaching of natural water (surface water,
groundwater, or precipitation) or applied waters through
spentshale and retorted shale is an important concern as
evidenced by current research [5, 7, 8, 15]. Contaminants
which may enter groundwater or surface waters (i.e.
principal materials in leachates) include soluble salts,
reduced sulfur species, soluble organics (variable
composition/distribution), Mo, B, F, trace metals (As, Se,
Pb, Zn, Cu, Hg, Fe), NO_3 and other undetermined
contaminants [8, 16]. The potential impact to surface and
groundwaters depends upon physical-chemical interactions,
the total leachate quantity [5], net water movement,
containment and/or treatment, and ultimate water quality of
leachates. Owing to the variations reported for
contaminants in various shales and identical shales under
different processes, the overall impacts are as yet
unknown. This, coupled with the inability to confidently
describe groundwater flows and leachate dynamics in natural
soils or aquifers, provides a difficult task for the
environmental engineer/scientist. The complexity of the

408

natural environment and problems associated with
determining, quantitatively, the biological
physical-chemical factors, reactions and interactions
affecting the distribution and role of contaminants in
natural systems preclude a straightforward analysis of
leachate impacts. It appears that the current state of
technology with respect to understanding leachate dynamics
[16, 17] is not sufficient to accurately assess leachate
importance in surface water quality. Groundwater
contamination by leaching through "in situ" spent shale
(and raw shale) has major importance considering an
estimated 80% of the stream flow in Piceance Creek is
groundwater [5, 18], thus increasing the potential surface
water contamination.

The range of values reported for various leachates
from spent and raw shales is shown in Table IV. The range
of reported values reflects different shales as well as
different leaching experiments. Under saturated
conditions, it has been shown that contaminant levels
decrease substantially [8] but are still of concern. The
overall time period for pollutant leaching to become
insignificant is also unknown [5] as well as the effects of
alternating wet/dry cycles. The EPA [5] has summarized
potential pollutants and possible concentrations in waste
shale leachates (Table V). Furthermore, they have ranked
potential pollution from spent shale leachates in the
"Highest Source Priority" and classified pollutants as
follows:

   Highest Pollutant Ranking: TDS,Na,SO_4,As,B,Se,F, organics
   Intermediate Pollutant Ranking: Ca,Mg,Zn,Cd,Hg, organics
   Lowest Pollutant Ranking: Pb,Cu,Fe, etc.

Water quality impacts from leachates will probably have
their greatest effects in upstream areas, i.e. Piceance
Creek, Yellow Creek and Roan Creek, with the impact
diminishing as pollutants move downstream and are
attenuated through dilution and natural physical-chemical-
biological phenomena. It has been estimated that leaching
of "modified in situ" retorts in the Piceance Creek Basin
[18] could result in significant TDS increases (Table VI).

Municipal Wastewater Impacts

Increased municipal and industrial wastewater discharges
can be expected to contribute significant amounts of
pollutants to the river system. Current industry plans
and other estimates place a majority of the population
increases in the area between Grand Junction and Eagle,

Table IV. Spent and Raw Shale Leachate Water Quality [5,7,9]

Parameter	Spent Shales (ppm)	Raw Shales
Fe	<.01 – 16	0.01 – 1.3
Cu	.02 – 16.6	<0.025 – 0.69
Cd	<.025 – .080	--
Zn	.04 – 3.4	<0.01 – 6.8
Ni	.01 – .2	<0.025 – 0.60
Hg	0.001 – 1.0	<.0001 – .0035
Pb	<.005 – 0.31	<.04 – 1.9
Cr	.01 – .15	<.025 – 0.68
As	<.001 – 1.03	<.005
Se	<.001 – 13.4	<.005 – <.01
TDS	970 – 42,000	70 – 30,130
Mo	0.44 – 4.3	0.09 – 9.0
B	0.73 – 4.6	<.025 – 43
F	1.31 – 13	0.8 – 75

Table V. Summary of Pollutants in Waste Shale Leachates

MAJOR INORGANICS (mg/ℓ)		TRACE ELEMENTS (mg/ℓ)	
TDS	140,000	Hg	0.005
Na	35,000	Pb	0.004
Ca	3,000	Cd	0.006
Mg	4,700	As	0.2
K	600	Cu	0.2
SO$_4$	90,000	Zn	3.0
Cℓ	3,000	Se	2.0
F	17	Fe	2.0
		B	10.0

ORGANICS

Oil & Grease	Unknown
Phenols	Unknown
TOC	3-5% by weight
Benzene Extracts	2,500 ppm (?)
Carcinogens	Unknown

Table VI. Potential TDS Increase Due to
Spent Shale Leachates [18]

Watercourse	Average Annual Discharge acre-ft/yr	Maximum Potential TDS Increase Due to Leachate Discharge in Yellow Creek and Piceance Creek (mg/ℓ)
Piceance Creek at White River	14,500	700 - 42,000
White River near Green River, Utah	532,000	20 - 1,270
Green River near Green River, Utah	4,427,000	3 - 150
Colorado River at Lees Ferry, Ariz.	12,426,000	1 - 50

Colorado (including Battlement Mesa) and north of Rifle to
Meeker. Scenarios for increased municipal wastewater flows
are given in Table I and indicate significant increases
above baseline flows for Medium to High Production
scenarios. Potential contaminants include salts (Ca, Mg,
Cl, SO_4, etc.), cyanide, BOD, TOC, detergent builders,
phenols, PCB's, nitrogen species, phosphorus, phthalate
esters along with increased loadings of trace metals and
other toxicants. Possible impacts upon water quality will
range from toxic impacts and biostimulatory responses to
impacts upon dissolved oxygen. Estimated concentrations of
selected metals in secondary effluents are shown in Table
VII and have been used to estimate impacts upon the metals
balance in the following section. Increased nutrient loads
(15 mgN/ℓ, 5-10 mgP/ℓ) will increase the overall nutrient
loading in Lake Powell and could increase productivity of
the upper reservoir significantly [2].

WATER QUALITY OF LAKE POWELL AND THE UPPER COLORADO

Water quality data for the Colorado River System and
Lake Powell indicates variable concentrations of heavy
metals with respect to time and sampling location. A recent

411

Table VII. Average Total Metal Concentrations in Selected Secondary Effluents

Metal	Urban Area Average[1] mg/ℓ	Average Low Industry[1] mg/ℓ	Moab WTP Effluent[2] mg/ℓ	Denver Effluent[3] mg/ℓ	Data Used in Analysis mg/ℓ
Fe	.700	--	.164	--	0.250
Mn	--	--	.083	--	.083
Cu	.224	.086	.018	.12	.075
Ba	--	--	.060	--	.060
Cd	.021	.017	.00043	.024	.020
Zn	.584	.160	.0634	.26	.160
Ni	.274	.276	.0225	.06	.060
Hg	.0072	.0009	.00005	--	.0009
Pb	.148	.054	.0103	.07	.045
Ag	.013	.015	.0075	.00003	.0075
Cr	.303	.028	.0213	.04	.030
Al	--	--	--	--	.100
Se	.0135	--	.00617	--	.0060
As	.009	.014	.0291	--	.017

1 Data obtained from Morel and Schiff (1980) [19]
2 Data from EPA STORET system
3 Data from Denver WTP personnel.

study on the metallic cations in Lake Powell and
tributaries [20] revealed that concentrations in Lake
Powell are generally below limits which may adversely
impact aquatic life for high quality waters (Table VIII).
However, major tributaries showed concentrations of Pb, Zn
and Cu which exceeded limits, while Cd, Cr and Se
approached 50% of the lower limits in Table VIII. Current
data in the EPA data retrieval system (STORET) were
examined for heavy metal concentrations (years 1976-1981)
for total metals (TOT) and dissolved metals (DISS) in
tributaries (Table IX) and verifies similar values
determined by Kidd and Potter [20]. The data presented
represents average values. The fluctuations of values as a
function of streamflow, time and location is largely
unknown due to the system complexity and size. Levels of
metals have not shown consistent increases or decreases
with time.

Compared to previous data [20], levels of Pb and Cd
increased at all tributaries except for a decrease in Cd

Table VIII. Water Quality Standards (in mg/l)

| Dissolved Constituent | Domestic Supply | Aquatic Life Protection | | Colorado[2] |
| | | Utah[1] | | |
		3A/3B/3C	3D	
As	0.05	*[3]	*	0.05
Ba	1.00	*	*	*
Cd	0.01	0.004[4]	*	0.004 - 0.005
Cr(VI)	0.05	0.100	.10	0.025
Cu	1.00	0.01		0.005 - 0.01
Fe	0.30	1.00	1.00	1.00
Pb	0.05	0.05		0.004 - 0.050
Hg(total)	0.002	.00005	.00005	0.00005
Se	0.01	.05	*	0.05
Ag	0.05	.01	*	0.00001-0.00015
Zn	5.00	.05	*	0.05 - 0.10
B	--	*	*	*
F(Utah)	1.4-2.4	*	*	*
Ni	--	*	*	0.05 - 0.20

1) Standards for Aquatic Wildlife as of October 23, 1978
2) Standards for Aquatic Wildlife as of May 16, 1981
3) Insufficient Evidence to Warrant Standard
4) A Standard of 0.0004 mg/l was indicated for cold water
   game fish, i.e., Class 3A

in the Dolores River. Fe and Cu decreased or stayed the same in all tributaries, while Cr and Zn increased in the Colorado River (Moab) and Dolores River and decreased in the Gunnison and Green Rivers. The data are summarized in Table IX and indicates general increases for Cd, Cr, Pb and Zn and decreases for Fe, As, Se and Cu.

Using average long-term flowrates for the major tributaries, mass loadings were computed for dissolved (Table X) and total (Table XI) metals. The total loading to Lake Powell was computed by summation of loadings due to the Colorado (Moab), San Juan and Green Rivers. By input-output comparison to the mass flow out of Lake Powell (STORET station below Glen Canyon Dam), it appears that the reservoir is a sink for Fe, Cu, Zn, Pb and As with removals greater than 70%. Se, Cr and Ni are removed by lesser amounts (40%, 15% and 8%, respectively), while Hg was released from the reservoir (or anthropogenic loading increased the mass flux into the system). Similar conclusions were reached by Kidd and Potter [20] but a comparison of the data in Tables X and XI to their data [20] shows a net increase in Cd, Pb, and Zn(DISS) loading and a decrease in Zn(TOT), Fe and Cu(DISS) loading to Lake Powell. Cu(TOT) loading remained about the same and a slight increase was noted for Ca and Mg. The relative importance of tributaries (Table XII) indicated that the Colorado (Moab), Green and Gunnison rivers contribute the majority of the mass loading to Lake Powell. The current water quality analysis presented here suggests that the primary metals of concern for protecting aquatic life in Lake Powell might be Hg, Cu, Cd, Zn, Pb and Se.

METAL SPECIATION AND DYNAMICS IN NATURAL WATERS

Knowledge concerning the dynamics and speciation of heavy metals in the Colorado River System is necessary to evaluate potential impacts of increased energy development. Understanding the natural mechanisms controlling metal distribution is hampered due to the complexity and size of the system, the difficulty in obtaining necessary data and the incompleteness of current data. Metals in natural waters may undergo numerous reactions with other dissolved and particulate species (or materials). A simplified scheme for these reactions (Figure 2) reveals the possibilities, including the following:

Inorganic Complexes $Zn(OH)_3^-$, $CdCl^-$, $FeOH^+$

Inorganic Ion Pairs $PbCO_3^\circ$, $CuCO_3^\circ$

Table IX. Metal Concentrations in Tributaries to Lake Powell (Average Values)

Metal	Colorado River Moab DISS	Moab TOT	Grand Jct DISS	Grand Jct TOT	Gunnison River DISS	Gunnison TOT	Dolores River DISS	Dolores TOT	San Juan River DISS	San Juan TOT	Green River DISS	Green TOT	White River DISS	White TOT
					CONCENTRATION IN MG/L									
Ca	89.9	--	90.0	--	133.6	--	121.7	--	85.5	--	69.3	--	66.1	--
Mg	29.3	--	31.7	--	50.6	--	49.5	--	25.3	--	28.3	--	23.5	--
Fe	0.017	2.05	0.041	1.41	0.070	0.79	0.079	1.58	0.008	2.64	0.036	6.89	0.026	2.60
					CONCENTRATION IN µG/L									
Cu	3.16	19.4	4.0	11.0	2.5	12.1	2.6	7.9	7.1	32.0	3.6	13.4	5.0	10.8
Cd	1.56	6.8	3.7	3.3	1.65	5.3	1.2	0.6	2.1	2.0	0.7	2.0	0.8	2.3
Zn	5.0	56.0	14.0	30.0	8.6	26.8	20.8	175.0	6.4	95.0	38.7	33.2	19.8	50.7
Ni	14.1	27.5	3.3	4.3	1.0	7.2	3.8	7.6	15.0	21.3	2.4	29.0	1.7	23.8
Hg	--	0.123	--	0.086	--	0.037	0.08	0.10	--	0.16	0.058	0.04	0.102	0.12
Pb	4.0	42.5	18.3	15.9	8.4	52.5	13.8	16.7	9.5	11.4	7.4	81.4	2.9	61.1
Cr	8.3	14.2	5.7	8.6	2.7	4.6	5.0	10.4	4.4	6.5	1.77	6.0	2.2	6.7
As	1.57	4.56	1.14	2.00	1.63	1.91	1.60	2.00	1.20	1.50	1.90	3.88	1.76	4.00
Se	0.94	1.17	6.86	7.00	8.91	11.1	0.70	2.10	1.07	1.08	1.71	1.87	1.18	1.00

*Data obtained from the EPA STORET system and includes data from 1976-1981

Table X. Mass Loading of Dissolved Metals to Lake Powell from Major Tributaries (kg/yr)

Metal	Colorado River at Moab	Colorado River below Grand Jct	Colorado River above Grand Jct	San Juan River	Green River	Gunnison River	Dolores River	White River	Total
Ca	6.01×10^8	5.37×10^3	2.60×10^8	1.65×10^8	3.62×10^8	2.77×10^8	8.79×10^7	3.95×10^7	1.13×10^9
Mg	1.96×10^8	1.89×10^8	8.40×10^7	4.88×10^7	1.48×10^8	1.05×10^8	3.57×10^7	1.40×10^7	3.93×10^8
Fe	1.16×10^5	2.43×10^5	9.8×10^4	1.61×10^4	1.88×10^5	1.45×10^5	5.70×10^4	1.57×10^4	3.20×10^5
Cu	2.11×10^4	2.39×10^4	1.87×10^4	1.38×10^4	1.87×10^4	5.23×10^3	1.88×10^3	2.99×10^3	5.36×10^4
Cd	1.04×10^4	2.21×10^4	1.87×10^4	4.13×10^3	3.71×10^3	3.43×10^3	8.66×10^2	4.60×10^2	1.82×10^4
Zn	3.34×10^4	8.35×10^4	6.56×10^4	1.24×10^4	2.02×10^5	1.79×10^4	1.50×10^4	1.18×10^4	4.38×10^5
Ni	9.43×10^4	1.96×10^4	1.75×10^4	2.89×10^4	1.27×10^4	2.08×10^3	2.71×10^3	1.01×10^3	1.36×10^5
Pb	2.67×10^4	1.09×10^5	9.17×10^4	1.83×10^4	3.84×10^4	1.73×10^4	9.93×10^3	1.76×10^3	8.34×10^4
Cr	5.57×10^4	3.41×10^4	2.85×10^4	8.48×10^3	9.24×10^3	5.63×10^3	3.61×10^3	1.29×10^3	7.34×10^4
As	1.05×10^4	6.80×10^3	3.41×10^3	2.31×10^3	9.92×10^3	3.39×10^3	1.16×10^3	1.05×10^3	2.27×10^4
Se	6.29×10^3	4.09×10^4	2.24×10^4	2.06×10^3	8.93×10^3	1.85×10^4	5.05×10^2	7.04×10^2	1.73×10^4

Table XI. Mass Loadings of Total Metals to Lake Powell from Major Tributaries (kg/yr)*

Metal	Colorado River at Moab	Colorado River below Grand Jct	Colorado River above Grand Jct	San Juan River	Green River	Gunnison River	Dolores River	White River	Total Lake Powell Loading
Fe	1.37×10^7	8.43×10^6	6.77×10^6	5.09×10^6	3.60×10^7	1.65×10^6	1.14×10^6	1.55×10^6	5.48×10^7
Cu	1.30×10^5	6.55×10^4	4.05×10^4	6.17×10^4	6.99×10^4	2.51×10^4	5.70×10^3	6.45×10^3	2.62×10^5
Cd	4.55×10^4	1.96×10^4	8.70×10^3	3.86×10^3	1.04×10^4	1.09×10^4	4.33×10^2	1.39×10^3	5.98×10^4
Zn	3.74×10^5	1.79×10^5	1.23×10^5	1.83×10^5	1.73×10^5	5.57×10^4	1.26×10^5	3.03×10^4	7.30×10^5
Ni	1.84×10^5	2.56×10^4	1.07×10^4	4.11×10^4	1.51×10^5	1.49×10^4	5.51×10^3	1.42×10^4	3.76×10^5
Hg	8.23×10^2	5.13×10^2	4.36×10^2	3.08×10^1	2.09×10^2	7.68×10^1	7.22×10^1	7.16×10^1	1.06×10^3
Pb	2.84×10^5	1.09×10^5	2.00×10^4	2.20×10^4	4.25×10^5	1.09×10^5	1.21×10^4	3.65×10^4	7.31×10^5
Cr	9.50×10^4	5.11×10^4	4.17×10^4	1.25×10^4	3.13×10^4	9.45×10^3	7.51×10^3	3.98×10^3	1.39×10^5
As	3.05×10^4	1.19×10^4	7.93×10^3	2.89×10^3	2.03×10^4	3.97×10^3	1.44×10^3	2.39×10^3	5.37×10^4
Se	7.82×10^3	4.18×10^4	1.87×10^4	2.08×10^3	9.76×10^3	2.31×10^4	1.52×10^3	5.97×10^2	1.97×10^4

*Calculated by summation of mass loadings from the San Juan, Green and the Colorado (at Moab) Rivers.

Table XII. Relative Importance of Major Tributaries in
the Metal Input to Lake Powell

Metal	River Rank (Highest Importance to Lowest)
Ca, Mg, As (DISS/TOT) Ni (TOT)	Colorado (Moab)>Green>Gunnison >San Juan>Dolores>White
Cu(DISS/TOT) CR(DISS/TOT)	Colorado (Moab)>Green>San Juan >Gunnison White>Dolores
Ni(DISS),Zn(TOT)	Colorado (Moab)>San Juan>Green>Dolores >Gunnison>White
Pb(DISS),Cd(DISS)	Colorado (Moab)>San Juan>Green >Gunnison>Dolores>White
Hg(TOT)	Colorado (Moab)>Green>Gunnison >Dolores>White>San Juan
Cd(DISS)	Colorado (Moab)>Gunnison>Green >San Juan>White>Dolores
Pb(TOT)	Green>Colorado (Moab)>Gunnison>White >San Juan>Dolores
Fe(DISS)	Green>Gunnison>Colorado (Moab)>Dolores >White>San Juan
Fe(TOT)	Green>Colorado (Moab)>San Juan >Gunnison>White>Dolores
Zn(DISS)	Green>Colorado (Moab)>Gunnison>Dolores >San Juan>White
Se(DISS	Gunnison>Green>Colorado (Moab) >San Juan>Dolores>White
Se(TOT)	Green>Colorado (Moab)>Gunnison >San Juan>Dolores>White

Precipitation	Zn_2SiO_4, $Pb(OH)_2$, $Fe(OH)_3$
Redox Reactions	$Cr(VI) \rightarrow Cr(III)$, $Fe(II) \rightarrow Fe(III)$
Metal organic Complex	Humics (Hg ,Cu ,Pb ,Zn)
	Oil Shale Organics
	Wastewater Organics
	NTA, EDTA
Cation Exchange	Clay Minerals
	Organic Substances
	$Fe(OH)_3$ ----FeOH groups
Adsorption	Clays (Pb ,Ni ,Cu ,and Zn)
	Fe/Mn Oxides
	Natural water organics(humics)
Coprecipitation	With FeS, $CaCO_3$ and $Fe(OH)_3$
Coagulation-Flocculation	Organic Acids (colloids) + M^+
	Metal organic coplexes + M^+
	Metal humates (precipitates or neutral colloids)

To understand the impacts resulting from increased metal loading and increased emissions (or discharges) of BOD TOC, complexing organics, nutrients, it is necessary to develop chemical models for Lake Powell and the Colorado

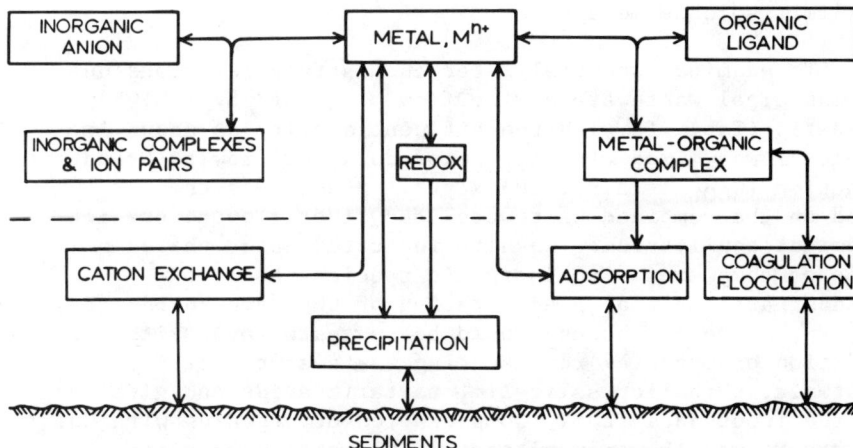

Figure 2. Heavy Metal Dynamics in Natural Waters [6,24].

River System and, then, successfully couple the models to hydrologic and biologic models. To examine the speciation of metals in the system, the potential for chemical precipitation, impacts of wastewater organics and pH perturbations, a computer chemical program, MINEQL, [21] was utilized. The results, using average current water quality for major tributaries and Lake Powell as input data, provide useful estimates of dominant species of various metals. Calculations performed with MINEQL indicate the importance of inorganic complexation and ion pair formation in the Colorado River System (Table XIII) and that the important solid phases for regulating aqueous soluble metal may be $CaCO_3$, $Ca_5(OH)(PO_4)_3$, $ZnSiO_3$ (Zn_2SiO_4), $Fe(OH)_3$, $Ba_3(AsO_4)_2$, $Pb(OH)_2$ and $Cu_2(OH)_2CO_3$. This is especially important since the previous input-output calculations for Lake Powell indicate the reservoir is a Fe, Cu, Zn, Pb and As sink and that $CaCO_3$ precipitation has been indicated.

Additional reactions (Figure 2) may also be significant (cation exchange, absorption, etc.) and further research will be necessary to improve chemical modeling in the Colorado River System. Silica equilibria appears to be important in regulating soluble Zn [2] and calcium carbonate precipitation [22,2] appears to be a significant sink implicating possible coprecipitation of other heavy metals in Lake Powell. The impacts of metal speciation upon phytoplankton dynamics will be important in the overall impact from energy development. Increased organics, which may complex heavy metals will be significant in potential increases in heavy metals and may counteract the solid phase reactions which may be tending to limit soluble metal.

To examine potential water chemistry alterations due to municipal wastewaters, the flow projected by a high scenario (Table I) with the effluent quality as shown in Table VI was mixed with the Colorado River flow at Grand Junction ($Q_{COLO} = 5.965 \times 10^{12} \ell/yr$). MINEQL was then used on the combined wastewater and river flow to predict chemical equilibrium. Results indicated about the same type of complexes with inorganic species should be predominant and that precipitation of the same solids (Fe, Ca, Zn, As solid phases) would be suggested even with the addition of organics representing wastewater (acetic, glutamic, phthalic, salicylic, tartaric acids and glycine) to the input data [23]. Complexation of organics with Cu, Ca and Mg was the only alteration to metal speciation. Insufficient data concerning complexation with wastewater

Table XIII. Probable Metal Species in Various Surface Waters of the Colorado River System

Metal	Lake Powell (pH=8.0)	Colorado River (Hite) pH=8.1	Green River (pH=8.0)	White River (pH=8.0)
Fe	$Fe(OH)_3(s)$, $FeOH^+$	$Fe(OH)_3(s)$, $FeOH^+$	$Fe(OH)_3(s)$, $FeOH^+$	$Fe(OH)_3(s)$, $FeOH^+$
Cu	$Cu(NH_3)_2^{2+}$, $CuNH_3^{2+}$, $CuCO_3^o$	$CuCO_3^o$, $CuOH^{2+}$, $CuB(OH)_4$	$CuOH^+$, $CuB(OH)_4$, Cu^{2+}	$CuCO_3^o$, $CuB(OH)_4$ $Cu_2(OH)_2CO_3(s)$
Cd	Cd^{2+}, $CdSO_4^o$, $CdOH^+$, $CdNH_3^{2+}$	Cd^{2+} $CdSO_4^o$, $CdCl^+$, $CdOH^+$	Cd^{2+}, $CdSO_4^o$, $CdOH^+$	Cd^{2+}, $CdOH^+$, $CdSO_4^o$
Zn	$Zn_2SiO_4(s)$, Zn^{2+}, $ZnSO_4^o$	$Zn_2SiO_4(s)$, Zn^{2+}, $ZnSO_4^o$	$Zn_2SiO_4(s)$ Zn^{2+}	$Zn_2SiO_4(s)$, Zn^{2+}, $ZnSO_4$
Ni	Ni^{2+}, $NiOH^+$, $NiSO_4^o$, $NiNH_3^{2+}$	Ni^{2+}, $NiOH^+$, $NiSO_4^o$	Ni^{2+}, $NiOH^+$, $NiSO_4^o$	Ni^{2+}, $NiOH^+$, $NiSO_4^o$
Hg	$Hg(NH_3)_2^{2+}$, $Hg(OH)_2^o$	$Hg(OH)_2^o$	$Hg(OH)_2^o$	$Hg(OH)_2^o$
Pb	$PbCO_3^o$, $Pb(CO_3)_2^{2-}$, $PbOH^+$	$PbCO_3^o$, $Pb(CO_3)_2^{2-}$, $PbOH^+$	$PbOH^+$, $PbCO_3^o$, Pb^{2+} $Pb(OH)_2(s)$	$PbCO_3^o$, $PbOH^+$, $Pb(CO_3)_2^{2-}$
Cr	CrO_4^{2-}	CrO_4^{2-}	CrO_4^{2-}	CrO_4^{2-}
As	$Ba_3(AsO_4)_2(s)$	$Ba_3(AsO_4)_2(s)$	$Ba_3(AsO_4)_2(s)$	$H_2AsO_4^-$, $HAsO_4^{2-}$
Se	$HSeO_4^-$, SeO_4^{2-}	$HSeO_4^-$, SeO_4^{2-}	$HSeO_4^-$, SeO_4^{2-}	$HSeO_4^{2-}$, SeO_4^{2-}

organics and potential oil shale organics prevent a
detailed examination of complexation reactions.

CONCLUSIONS

Potential impacts of extensive energy development with
respect to heavy metals in the Colorado River System and
Lake Powell need to be further clarified. Levels of metals
in sediments and aqueous phase of the tributaries are high
enough for concern over future increments. Continued
research on metal dynamics in the System will provide
necessary data and knowledge concerning speciation, natural
removal mechanisms and potential toxicity of heavy metals
in the Colorado River System. Mass loadings of pollutants
from various tributaries should be continuously monitored
to assess the magnitude of the increases due to energy
development. Of utmost concern is the development of a
scientifically sound monitoring program to permit the
development of statistical and simulation models to examine
the fate and effects of pollutants.

ACKNOWLEDGEMENTS

The support of the Department of Civil, Environmental
and Architectural Engineering at the University of
Colorado-Boulder is gratefully acknowledged. In addition,
I would like to acknowledge the assistance of graduate
students, Sharon Bryan, Glenn Friedman and Joe
Dollerschell, for data analysis and computer programming.

REFERENCES

1. Standiford D.R., L.D. Potter and D.E. Kidd. 1973. Mercury in the Lake Powell Ecosystem. Lake Powell Research Project, Bulletin No. 1, National Science Foundation.
2. Medine, A., D.B. Porcella. 1980. "Heavy Metal Effects on Photosynthesis Respiration of Microecosystems Simulating Lake Powell, Utah/Arizona, In R.A. Baker, Ed., Contaminants and Sediments. Ann Arbor Science, Ann Arbor, MI.
3. Smith, J.W. 1981. Oil Shale Resources of the United States Mineral and Energy Resources. Colorado School of Mines (0192-6179/81/2306-0001).
4. Colorado Energy Research Institute. 1981. Colorado Oil Development Scenarios, 1981-2000, prepared for Governor's Blue Ribbon Panel, Colorado School of Mines.
5. Bates, E.R. and T.L. Thoem. 1980. Environmental Perspective on the Emerging Oil Shale Industry. EPA Oil Shale Research Group. EPA-600/2-80-205a.
6. Forstner, U. and G.T.W. Wittman. 1981. Metal Pollution in the Aquatic Environment. Springer-Verlag Berlin.
7. Stollenwerk, K.G. and D.D. Runnells. 1981. Composition of Leachate from Surface-Retorted and Unretorted Colorado Oil Shale. Environ. Sci. and Tech., 14:11.
8. McWhorter, D.B. 1980. Reconnaissance Study of Leachate Quality from Raw Mined Oil Shale-Laboratory Columns. U.S. EPA, Grant No. R806278. Cincinnati, Ohio.
9. Skogerboe, R.K., D.F.S. Natusch, D.R. Taylor and D.L. Dick. 1978. Potential Toxic Effects on Aquatic Biota from Oil Shale Development. Proc. of Oil Shale Symposium #11, Colorado School of Mines.
10. Cleave, M.L., D.B. Porcella, and V. Dean Adams. 1980. Potential for Changing Phytoplankton Growth in Lake Powell Due to Oil Shale Development. Environ. Sci. and Tech., 14:6, 683-690.
11. Jackson, K. and L. Jackson. 1980. Heavy Metals Pollution Potential From Oil Shale Leachates as Determined by EPA Proposed Extraction Procedures. Proc. of Oil Shale Symposium #13, Colorado School of Mines.
12. Leenheer, J.A., H.A. Stuber, and T.I. Noyes. 1981. Chemical and Physical Interactions of an In Situ Oil-Shale Process Water with a Surface Soil. Proc. of Oil Shale Symposium #14. Colorado School of Mines.
13. Runnels, D.D. and E. Esmaili. 1980-1981. Release Transport, and Fate of Potential Pollutants in Trace

423

Elements in Oil Shale, Progress Report DOE-10298-2. Center for Environmental Studies, University of Colorado, Denver.

14. Sievers, R.E., J. Stanley, and M. Conditt. 1980-1981. Trace Elements and Organic Ligands in Oil Shale Wastes. In Trade Elements in Oil Shale. Progress Report, DOE-10293-2. Center for Environmental Studies, University of Colorado Denver.

15. Amy, G.L., A.L. Hines, J.F. Thomas, and R.E. Selleck. 1980. Groundwater Leaching of Organic Pollutants from In Situ Retorted Oil Shale-- A Mass Transfer Analysis. Environ. Sci. and Tech., 14:7, 831-835.

16. Wildung, R.E. and J.M. Zachara. 1981. Effects of Oil Shale Solid Waste Disposal on Water Quality: Current Knowledge, Information Requirements and Research Strategy. Presented at symposium, Oil Shale--The Environmental Challenges II, Vail, Colorado, August 10-13.

17. Ramirez, W.F. 1980-1981. Mathematical Modeling and Transport Mechanisms for Leaching of Spent Oil Shale. In Trace Elements in Oil Shale. Progress Report DOE-10298-2. Center for Environmental Studies, University of Colorado, Denver.

18. Persoff, P. and J.P. Fox. 1979. Control Strategies For Abandoned In Situ Oil Shale Retorts," Proc. of Oil Shale Symposium #16, Colorado School of Mines.

19. Morel, F.M.M. and S.L. Schiff. 1980. Geochemistry of Municipal Waste in Coastal Waters. Report No. 259, Ralph M. Parsons Laboratory for Water Resources and Hydrodynamics, Mass. Inst. Tech.

20. Kidd, D.E. and L.D. Potter. 1978. Analysis of Metallic Cations in the Lake Powell Ecosystem and Tributaries. Lake Powell Research Project. Bulletin No. 63. National Science Foundation.

21. Westal, J.C., J.L. Zachary, and F.M.M. Morel. 1976. Mineql-A computer Program for the Calculation of Chemical Equilibrium Composition of Aqueous Systems. Technical Note No. 18, Ralph M. Parsons Laboratory for Water Resources and Environmental Engineering. Mass. Inst. Tech.

22. Reynolds, R.C., Jr. and N.M. Johnson. 1974. Major Element Geochemistry of Lake Powell. Lake Powell Research Project. Bulletin No. 5. National Science Foundation.

23. Sposito, G. 1981. Trace Metals in Contaminated Waters. Environ. Sci. Tech., 15:4, 396-403.

24. Stumm, W. and J.J. Morgan. 1981. Aquatic Chemistry. John Wiley and Sons, New York.

PART 7

SALINITY

CHAPTER 25

EROSION AND SALINITY
PROBLEMS IN ARID REGIONS

Richard H. French
    Water Resources Center
    Desert Research Institute
    Las Vegas, Nevada

William W. Woessner
    University of Montana
    Department of Geology
    Missoula, Montana

INTRODUCTION

The mineral quality problem in southwestern rivers is a
complex problem which is critically important not only on a
regional basis but also on national and international levels.
Mineral quality, commonly termed salinity or total dissolved
solids (TDS), is a particularly serious water quality problem
on the main stem of the Colorado River whose drainage basin
covers one-twelfth of the continental United States and
serves as the water supply for a population exceeding ten million people in the Lower Colorado Basin alone (5). If salinity levels continue to rise in this river, then by the year
2010 damages due to salinity may exceed 1.24 billion dollars
(6). The predicted adverse salinity impacts include: 1) reduced agricultural productivity, 2) reduced suitability of
Colorado River water for municipal and industrial use, and 3)
salinity concentrations in the water reaching Mexico which
will exceed internationally established standards. The elimination of salinity increases in the Colorado River requires
an accurate understanding of both the man-made and natural
sources of salinity.

Man-made sources of salinity include municipal and industrial consumptive use of water, irrigation, and evaporation from reservoirs. In the Colorado River above Hoover Dam
man-made sources of salinity account for approximately 34% of
the total salinity load (5). It should be noted that Blackman et al. (7) claim that evaporation from Lakes Powell and

Mead alone cause an increase in salinity of 100 milligrams per liter (mg/l) at Hoover Dam.

Natural sources of salinity include both point and non-point or diffuse sources. Point sources such as springs and seeps account for approximately 12% of the total salinity load above Hoover Dam (5), non-point sources include the dry-fall of salinity into reservoirs, (14), and the interaction of surface water with natural salt bearing geologic formations (13). Above Hoover Dam, these non-point sources account for approximately 54% of the total salinity load (5).

In Southern Nevada, Las Vegas Wash has been identified as one of the primary sources of salinity to Lake Mead. Although previous to the development of the Las Vegas metropolitan area Las Vegas Wash was an ephemeral stream, it is now a perennial stream fed by sewage effluent and runoff from the urban area. This change in flow regimes led to the development of extensive marsh areas in the Lower Las Vegas Valley and also serious and extensive erosion. It was the contention of the authors that the erosion of highly saline soils in the Lower Las Vegas Valley could result in a significant contribution to the salinity of the Lower Colorado River. In 1980 the U.S. Bureau of Reclamation (USBR) authorized a reconnaissance level survey of salt storage in the Lower Las Vegas Valley. In this context, salt storage is defined to be the salinity associated with the soil above the water table. Although this research is continuing, a number of preliminary results demonstrating the significance of salt storage and erosion to the Colorado River salinity problem are available.

SALT STORAGE IN THE LOWER LAS VEGAS VALLEY

The Lower Las Vegas Valley lies within the Basin and Range Province, and the topography is characterized by sub-parallel mountain ranges with a central basin modified by encroaching alluvial fans. The total relief in this valley is 3260 m (10,700 feet). The surrounding mountains are composed of Paleozoic carbonates and Tertiary volcanics.

Las Vegas Wash is the remnant of a perennial flow referred to as the pluvial Las Vegas River which was active 30,000 years before present (B.P.) (1). The drainage extended from Indian Springs, Nevada, to the Colorado River which is a distance of 113 km (70 miles). Evidence suggests either marshes or shallow lakes occurred in the basin from 30,000 to 15,000 years B.P. and stream flow was again active until 6,000 years B.P. Subsequently, decreased precipitation and spring flow and increasing aridity resulted in an ephemeral drainage (1, 11). In the Lower Las Vegas Valley, the result of these geologic processes was alluvial pediments of Quaternary age which in general have a thin cover of relatively permeable sand and gravel ranging from less than 6 m (20 feet) to greater than 30 m (98 feet) in thickness. The deeper alluvial fill in this region is believed to be composed of the Muddy Creek formation which is a Tertiary sequence of unconsolidated and semiconsolidated sediments dominated by sands, silts and clay but locally displaying caliches, gypsum beds, halites, and lenses of coarse sand and gravel.

The hydrogeologic data available strongly suggest that: 1) long term natural processes have concentrated the most soluble minerals in the lower end of the Las Vegas Valley, and 2) the natural flow of the groundwater in the area has mobilized the soluble minerals and transported them to areas of groundwater discharge. Four short term or historic period mechanisms of salt storage can also be hypothesized. First, the discharge of wastewater from the Las Vegas Metropolitan area has encouraged the development of significant stands of phreatophytes along Las Vegas Wash. This vegetation along Las Vegas Wash and its tributary Duck Creek consume an estimated 1.7×10^7 m^3 (13,800 acre feet) of water per year. This consumptive use of water concentrates salts in the vadose zone. Second, direct evaporation must also be considered a primary mechanism of salt storage. Evaporation of water from the fine grained soils in the study area consumes water and leaves the salt behind. Third, direct evaporation coupled with infiltration may be a significant salt storage mechanism even in areas where the groundwater table is relatively deep. Fourth, natural erosional processes continue to transport salt into the study area.

Because only limited resources could be dedicated to assessing the new and controversial concept of salt storage, a small study area approximately 39 km^2 (15 square miles) in extent was defined in the Lower Las Vegas Valley near Henderson (Figure 1). Within this area, 78 sample sites were defined along lines which were selected to provide a maximum amount of information at a minimum cost. At each sample site, a hole was augered from the ground surface to the water table, and soil samples were taken at the surface, 15 cm, 30 cm, and then at 30 cm intervals to 3 m. Below 3 m, samples were taken

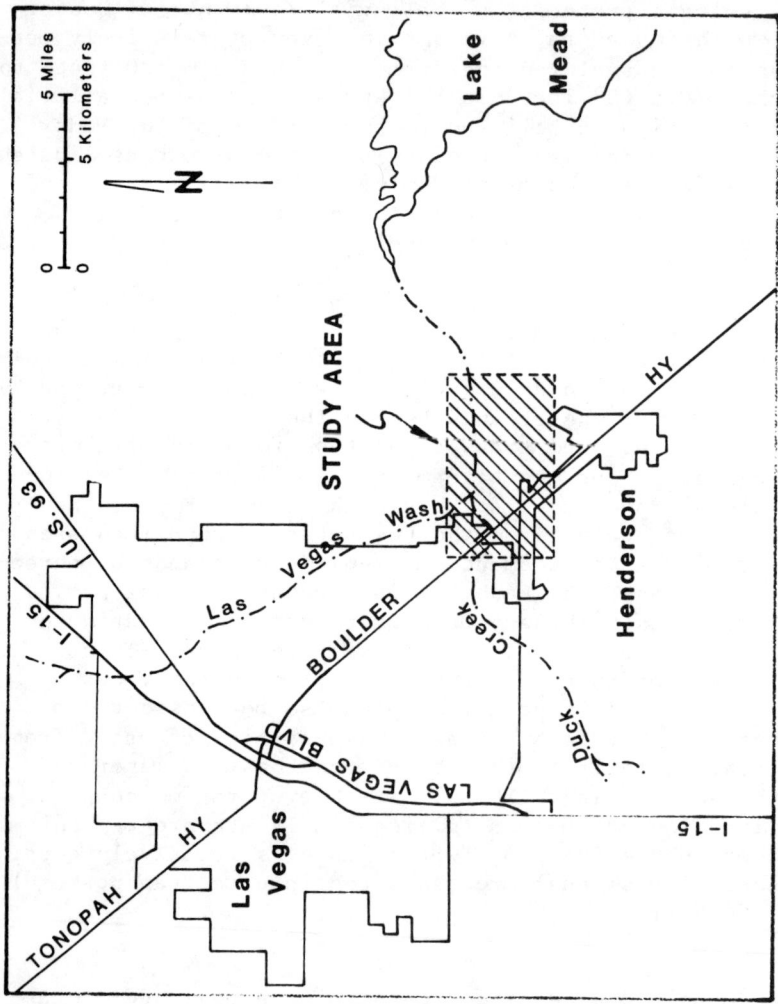

Figure 1. The lower Las Vegas Valley showing the location of Las Vegas Wash and the salt storage study area relative to Las Vegas and Lake Mead.

at 60 cm intervals until the water table was reached.

The soil samples were analyzed by standard ASTM methods to determine the soil moisture, and the mass of <u>readily</u> soluble salts associated with the soil was determined by a procedure developed by the Desert Research Institute. The methodology used was:
1. Thirty grams of oven dried soil was placed in 1,500 ml of distilled water - a 50:1 dilution ratio - and the container was tightly capped to prevent evaporation.
2. Each sample bottle was shaken for 30 sec. at 30 min. intervals, a minimum of four times.
3. After 24 hr., the electrical conductivity of the supernatent sample liquid was measured, and the total dissolved solids present were calculated from the electrical conductivity using a calibration curve developed for the study area.

This laboratory procedure determined the salt content of the soil in terms of the (mass of salt) per (mass of dry soil). It is noted that this procedure yields estimates of soil salinity which are slightly higher than the estimates which result from the method recommended by USBR (4). However, both of these procedures are based on the same principles and an empirical relationship between the methods can be defined for the study area.

The sampling and laboratory programs resulted in a three dimensional array of salt storage concentration values for the study area. Since the soil samples were taken at definite depths below the ground surface, salt storage is actually defined on a set of planes parallel to the ground surface. For numerical convenience, salt storage is by definition zero at all sample locations which are below the water table. It is noted that this convention does not contradict the definition of salt storage and results in 78 values of salt storage being defined on every plane.

The salt in storage between any two adjacent sampling planes can then be determined by numerical integration. If salt storage was defined on a regular cartesian grid with a common origin in each plane, then a very simple integration scheme could be used. However, as noted previously, the sample site locations were chosen to provide information along "arbitrary" lines rather than to provide numerical data which could be easily integrated. Therefore, it was necessary to use a bicubic spline interpolating method to interpolate values of salt storage concentrations onto a regular cartesian grid, (8). This method of analysis used the given field data to estimate salt storage concentration values onto a 33 x 33 cartesian grid in each plane. Then, the salt in storage between any two adjacent planes can be found by:

$$\overline{S}^K_{i,j} = \frac{S^K_{i,j} + S^K_{i,j+1} + S^K_{i+1,j} + S^K_{i+1,j+1}}{4}$$

where $\overline{S}^K_{i,j}$ = average salt concentration in cell i,j of sampling plane K in (mg of salt)/(kg of soil), $S^K_{i,j}$ = salt concentration at node i,j of sampling plane K in (mg of salt) /(kg of soil). Then

$$W^{K+\frac{1}{2}} = \sum_{j=1}^{N_C-1} \sum_{j=1}^{N_R-1} \frac{\overline{S}^K_{i,j} + \overline{S}^K_{i,1}}{2} \quad G\gamma\Delta x\Delta y\Delta h\phi$$

where $W^{K+\frac{1}{2}}$ = salt in storage between sampling planes K and K+1 in kilograms, G = specific gravity of dry soil, γ = specific weight of water, Δx and Δy = incremental distances on sampling planes between nodes, Δh = vertical distance between sampling planes, N_C = number of columns in salt storage concentration matrix, and N_R = number of rows in salt storage concentration matrix. It is noted the that specific gravity of the soil was measured by standard USBR (4) methods at a number of representative sites throughout the study area. The average specific gravity was 1.71 with a standard deviation of 0.127. In the computations defined above, G was treated as a normal random variable.

The results of this analysis are summarized in Figures 2 and 3. With regard to these results, the following should be noted. First, there are 3.9 x 10^6 Mg (42.6 x 10^5 tons) of salt in storage in the study area between the ground surface and 3 m (ten feet). Thus, on the average there are 33.7 kg (2.1 pounds) of readily soluble salt per m^3 (cubic foot) of material in this area. Second, there are large quantities of salt stored in the first 15 cm (half foot) of soil. In this first 15 cm (half foot) there is on the average 51.4 kg (3.2 pounds) of readily soluble salt per m^3 (cubic foot). Third, in some areas in the vicinity of Las Vegas Wash field and laboratory measurements demonstrate that there is in excess of 802 kg (50 pounds) of readily soluble salt per m^3 (cubic foot). Fourth, Figure 3 demonstrates that there is nearly a linear relationship between the cumulative weight of

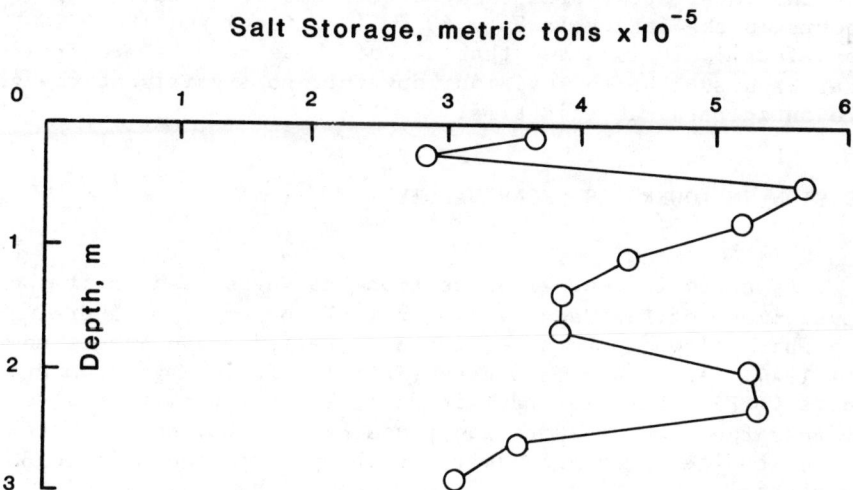

Figure 2. Salt storage as a function of depth below the ground surface.

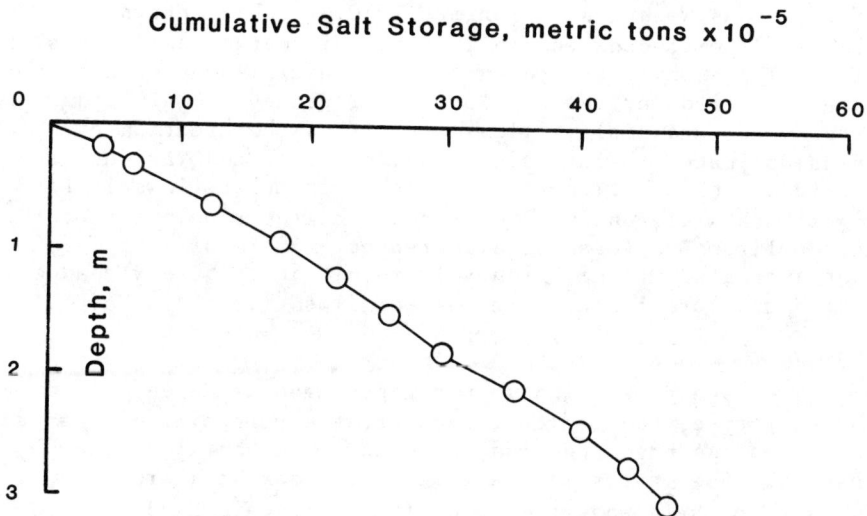

Figure 3. Cumulative weight of salt as a function of depth below the ground surface for the salt storage study area.

salt in storage and depth. Fifth, Figure 2 demonstrates that for the total study area the greatest amount of salt storage occurs in the intervals 30 - 60 cm (1-2 feet) and 180 - 240 cm (6-8 feet). It is noted that all of these results are preliminary and subject to revision; however, no significant changes are anticipated at this time.

EROSION IN LOWER LAS VEGAS VALLEY

As noted in a previous section Las Vegas Wash until the development of Las Vegas was a typical ephemeral arid drainage which flowed only after intense precipitation events until 1944. At this time a ditch from the Basic Magnesium Products (BMP) industrial park in Henderson was constructed to by-pass the plant evaporation ponds, (12). Spring discharge which resulted from groundwater recharge from the unlined BMP industrial waste ponds was also noted in the early 1940's, (10). Perennial flow above Henderson began in 1956 when treated municipal sewage effluent from Las Vegas was routed to Las Vegas Wash for disposal, (12).

Since the late 1950's, perennial flow has gradually increased due to increased discharges of industrial cooling water and sewage effluent. In 1957 the mean annual discharge of Las Vegas Wash was approximately 0.6 cubic meters (21 cubic feet) per second and in 1978 the discharge was 2.3 cms (82 cfs). This slow increase in the perennial flow provided the water required for the development of stands of salt cedar, arrow weed, and small willows as well as the creation of marsh areas dominated by rushes, sedges, cattails and various grasses (3). It is anticipated that the perennial flows in Las Vegas Wash will continue to increase as the population of the metropolitan Las Vegas area increases. It is also anticipated that increased urbanization will result in tributary channelization and more frequent and higher flash flood discharge.

The rate of erosion in Las Vegas Wash is dependent on the soil type, the vegetative cover, the perennial discharge rate, and the magnitude and frequency of flash flood discharges. Soil types in the Lower Las Vegas Wash are alluvial deposits of silt and fine sand of the Glendale-Land Association which are particularly susceptible to erosion, (2). The unconsolidated material is from 6 to 15 m (20 to 50 feet) in thickness in general, but thicknesses of greater than 15 m (50 feet) have been encountered. Although the development of stands of vegetation have aided in retaining the soil, in-

432

creased perennial flow and channel gradients which average
0.0093 are conducive to erosion in the form of stream bed degradation and head cutting. USGS suspended sediment data indicate that, at an average daily discharge of 2.32 cms (82 cfs)
per day, 218 Mg (240 tons) per day of suspended sediment were
being removed in 1978. Thus, on the average, 80,000 Mg
(88,000 tons) of sediment are removed from the Lower Las Vegas
Valley in a year - an estimate based on suspended sediment
data.

Flash flood events also remove significant amounts of
soil. In February, 1976 an instantaneous suspended sediment
load of 25,000 Mg (28,000 tons) per day was measured at a
flow rate of 18 cms (620 cfs). A large flow event in July,
1975 (Q = 68 cms (2,400 cfs)) eroded 3.7×10^4 m^3 (1.3
million cubic feet) of material or approximately 63,000 Mg
(69,000 tons) of soil in the Lower Las Vegas Valley.

Between August, 1975 and April, 1979 flood events and
the increased perennial flow in Las Vegas Wash have eroded
4.5×10^5 m^3 (16 million cubic feet) of channel material
or approximately 777,000 Mg (854,000 tons) of soil. Between
August, 1975 and April, 1979 the Las Vegas Wash headcut had
advanced at an average rate of 1.2 m (four feet) per day and
eroded channel material at an average rate of 364 m^3
(12,860 cubic feet) (628 Mg (690 tons of soil)) per day. It
is noted that this last erosion rate is significantly larger
than the estimate derived from the suspended sediment data.
Although future rates of erosion will be partially determined
by the geology of the area, there is no reason to expect that
the actual amount of material eroded will decrease as the
main stem erosion spreads to the tributary channels; in fact,
the amount of erosion may increase.

EROSION - SALT STORAGE INTERACTIONS

The foregoing material has defined both salt storage in
the Lower Las Vegas Valley and the erosion which is occurring
along Las Vegas Wash in this area. Although it has always
been accepted that the erosion of saline soils contributes to
the Colorado River salinity problem, this research is to the
authors' knowledge the first instance in which accurate data
regarding this situation are available for the Lower Colorado
area. USGS records for Las Vegas Wash at North Shore Road indicate that during water year 1978 the average flow was 2.32
cms (82 cfs), the average salinity concentration was 2,574

mg/l and the average sediment concentration was 1,086 mg/l. Thus during this time period approximately 191,100 Mg (210,000 tons) of salt and 80,000 Mg (88,000) tons of sediment were exported from the Las Vegas Valley to Lake Mead by Las Vegas Wash. The interrelationship between the salt storage concept and erosion is summarized in Table I under the assumption that the 39 km^2 (15 square miles) of study area soils are representative of the sediment being transported in Las Vegas Wash. It is realized that if unsaturated soil becomes saturated during precipitation and flood events prior to physical erosion, salts may be removed by leaching. The physical displacement and transport of a given volume of soil were utilized in this study to calculate the salinity loading by the Las Vegas Wash to the Colorado River.

Table I: Computation of Salt Eroded from the Las Vegas Valley

Depth Meters D, m (1)	Volume Meters^3 to D, m (2)	Metric Tons M_{sa}, Tons (3)	Metric Tons M_{sq}, Tons (4)	Ratio M_{sa}/M_{so} (5)	Average Salt Transported To Colorado River via Erosion Metric Tons/Year (6)
0-0.15	5.8 x 10^6	0.301 x 10^6	9.9 x 10^6	0.030	2,400.
0-0.30	11.6 x 10^6	0.536 x 10^6	20.0 x 10^6	0.027	2,160.
0-1.52	58.0 x 10^6	2.13 x 10^6	99. x 10^6	0.022	1,760.
0-3.05	116 x 10^6	3.88 x 10^6	200. x 10^6	0.020	1,600.

Table I shows the computation of a range of different estimates in column (6) that would result from the erosion of different total depths of soil as shown in column (1). Column (2) shows the volume of the study area for each specified total depth, and column (4) shows the estimated total weight of the soil in the study area based on an assumed specific weight of 1720 Kg per m^3 (107 pounds per cubic foot). Column (3) is the estimated total weight of the salt in the study area based on the conductivity measurements and analysis described in the foregoing material. Column (6) is obtained by multiplying the annual suspended sediment load, 80,000 Mg (88,000 tons), by the salt to soil ratio, column (5). Although not all of the sediment passing the USGS gage used in this analysis comes from the Lower Las Vegas Valley, 96% of the sediment passing this gage does come from the salt storage area. The estimates in Table I are extremely conser-

vative since in the vicinity of Las Vegas Wash where most of
the soil is eroded, the ratio M_{sa}/M_{so} is approximately 0.10
(9). This would indicate that erosion in the Lower Las Vegas
Wash may contribute 4,000 kg (8,800 pounds) of salt on an annual basis to the Colorado River. Thus, it is concluded that
depending on where the sediment being eroded is located as
much as 10% to as little as 2% of the total salinity entering
the Colorado River system from Las Vegas Wash is attributable
to erosion. USGS records also demonstrate that in the reach
of Las Vegas Wash which passes through the study area the annual increase in salinity transport is 67,000 Mg (74,000
tons). If only this reach is considered, then erosion may
account for as much as 12% of the salinity increase.

In the case of extreme flow events, the contribution of
erosion to the salinity problem is even more significant.
For example in Las Vegas Wash the July 1975 flow event which
removed 63,000 Mg (69,000 tons) of soil also removed between
1270 - 6280 Mg (1,400 to 6,900 tons) of salt. Also, between
August 1975 and April 1979 between 15,560 - 77,700 Mg (17,100
and 85,400 tons) of salt was removed. In comparison it is
noted that Blue Springs near the mouth of the Little Colorado
River and considered the largest point source of salinity in
the entire Colorado River Basin (5) contributes 471,000 Mg
(518,000 tons) of salt per year to the Colorado River System.

In addition to the salt storage study, described here in
some detail, the Water Resources Center has also analyzed the
solution-salt loading effect of flash floods on the salinity
problem in four arid, ephemeral, undeveloped watersheds tributary to Lake Mead (14, 15). This study area covered 497
km^2 (192 square miles) in which eight intermittent flow
events were recorded and sampled in 1978 and 1979. The estimated discharge of flash flood water was 2.1×10^6 m^3
(1,700 acre feet) in 1978 and 9.62×10^5 m^3 (780 acre
feet) in 1979. The average TDS of these waters ranged from
1,270 to 2,000 mg/l. Based on these data, it is concluded
that 2730 and 1183 Mg (3,000 and 1,300 tons) of salt entered
the Colorado River System from the 497 km^2 (192 square
mile) study area at rates of 5.6 Mg (16 tons) per square kilometer (square mile) per year and 2.4 Mg (6.9 tons) per square
kilometer (square mile) per year in 1978 and 1979, respectively. These data lend additional support to the authors' contention that erosion is a contributor of salinity to the
Colorado River system.

435

CONCLUSION

Based on the preliminary results of salt storage-erosion analyses and flash flood salinity loading work, it is concluded that erosion contributes to the salinity problem in the Colorado River System. Analyses of salt storage data revealed that 2% to 10% of the total salt balance for the Las Vegas Wash can be attributed solely to erosion. Analyses of data indicate that erosion may account for 12% of the salinity increase recorded for the reach of Las Vegas Wash which passes through the study area. These results highlight the significance of assessing potential national salinity control measures with complete salt balance information. Additional erosion-salt loading evaluations are necessary.

ACKNOWLEDGMENTS

The research described in this paper was supported by the Water and Power Resources Service (0-07-30-V0126) and the Bureau of Land Management (YA-512-CT-200). The authors also wish to acknowledge the aid and advice of Messrs. Robert Barton and D.A. Tuma of the Water and Power Resources Service, Boulder City, Nevada.

REFERENCES

1. _____, "Las Vegas Wash Interim Report No. 2, Water Quality Series: Clark County 208 Water Quality Management Plan, Clark County, Nevada, Las Vegas Nevada," URS Co., Las Vegas, Nevada, 1977.

2. _____, "Soil Survey: Las Vegas and Eldorado Valley Area, Nevada," U.S. Department of Agriculture, Soil Conservation Service, Washington, D.C., 1967.

3. _____, "Comprehensive Plan: Task 1 - Existing Conditions," Clark County Department of Comprehensive Planning, Las Vegas, Nevada, 1980.

4. _____, Earth Manual, U.S. Department of the Interior, Bureau of Reclamation, Second edition, Washington D.C., 1974, pp. 448-450.

5. _____, "The Mineral Quality Problem in the Colorado River Basin: Summary Report," U.S. Environmental Protec-

tion Agency, Washington, D.C., 1971.

6. _____, "River Water Quality Improvement Program," U.S. Department of the Interior, Bureau of Reclamation, Washington, D.C., 1974.

7. Blackman Jr., W.C., Rouse, J.V., Schillinger, G.R., and Shafer Jr., W.H., "Mineral Pollution in the Colorado Resources Center, Reno, Nevada, in press.

8. Foley, T.A., "Computer-Aided Surface Interpolation and Graphical Display." Desert Research Institute, Water Resources Center, Reno, Nevada, in press.

9. French, R.H., Mifflin, M.D., Edkins, J., "Salt Storage in the Lower Las Vegas Valley," Water Resources Center, Desert Research Institute, Las Vegas, Nevada, 1982.

10. Maxey, G.B. and Jameson, C.H., "Geology and Water Resources of Las Vegas and Pahrump and Indian Springs Valleys, Clark and Nye Counties, Nevada," State of Nevada Water Resources Bulletin, No. 5, 1948.

11. Mifflin, M.D., personal communication, February, 1981.

12. Patt, R.O., "Las Vegas Valley Water Budget: Relationship of Distribution, Consumptive Use, and Recharge to Shallow Groundwater," EPA-600/2-78-159, U.S. Environmental Protection Agency, Washington, D.C., 1978.

13. Riley, J.P., Bowles, D.S., Chadwick, D.G., and Grenney, W.J., "Preliminary Identification of Price River Basin Salt Pick-up and Transport Processes," Water Resources Bulletin, American Water Resources Association, Vol. 15, No. 4, August, 1979.

14. Woessner, W.W., "Reconnaissance Evaluation of Water Quality – Salinity Loading Relationships of Intermittent Flow Events in a Desert Environment, Las Vegas, Nevada," Water Resources Center, Desert Research Institute, Publication 44021, September, 1980.

15. Woessner, W.W., "Intermittent Flow Events – Salinity Loading Relationships in the Lower Colorado River Basin, Southern Nevada," Hydrology and Water Resources in Arizona and the Southwest, Arizona Sec. AWRA, Tucson, Arizona, Vol. 10, 1980, pp. 109-119.

CHAPTER 26

USE OF HYDROELECTRIC DAMS TO CONTROL EVAPORATION
AND SALINITY IN THE COLORADO RIVER SYSTEM

L. J. Paulson
    Lake Mead Limnological Research Center
    University of Nevada, Las Vegas

INTRODUCTION

   The main stem reservoirs on the Colorado River comprise
one of the largest and most heavily used freshwater bodies
in the nation. These reservoirs (Lake Powell, Lake Mead,
Lake Mohave and Lake Havasu) can store up to 53,590,400
acre-feet ($66 \times 10^9 \, m^3$) of water at their maximum capacities.
Nonetheless, local water shortages still exist in some areas
of the Colorado River Basin. There is also concern that salt
concentrations are approaching levels that could severely
affect municipal and agricultural uses [1]. Water shortages
will become even more acute as demands for water increase
with continued urban and agricultural development in the
basin.
   Water conservation and salinity control programs have
already been adopted, or are under investigation, in most
states using Colorado River water. Reductions in consumptive
water uses through more efficient irrigation practices, power plant cooling and wastewater reuse will, to some extent,
help alleviate future water shortages. However, this will
not offset the rising demands, and basin-wide shortages
could occur by the year 2000 [2]. Similarly, recent estimates indicate that salt concentrations in the river at
Imperial Dam will rise to 1150 mg/1 as a result of flow depletions projected to occur during this century [1,3]. Construction of salinity control projects approved by Congress
under PL 93-320 will significantly reduce salinity, but implementation of these projects will be costly and time consuming [3].
   Water shortages and salinity control in the Colorado
River system have thus far been addressed from the standpoint of reducing water uses and controlling point source
salt inputs. Little attention has been given to investigating methods of reducing evaporation from the reservoirs, but
studies conducted in 1952 and 1953 [4] showed that it was a
major water loss from the Colorado River system. Moreover,

high evaporation directly influences salinity because it increases the concentration of salts in the reservoirs. Although various schemes have been offered for reducing evaporation from Lake Mead, it has usually been viewed as an uncontrollable water loss. However, during the mid-1960s, U.S. Geological Survey and Bureau of Reclamation scientists estimated that cold-water discharges from Glen Canyon Dam would reduce evaporation in Lake Mead. The estimates were never published in report form but did appear in internal government memoranda and newspaper articles (Arizona Republic, May 19, 1966; Phoenix Gazette, July 28, 1966). Our analysis of historical evaporation data, and recent investigations in Lake Mead [5] indicate that evaporation did indeed decrease after Lake Powell was formed in 1963.

Advective energy (heat) inputs (Colorado River inflow) and outputs (Hoover Dam discharge) have a significant influence on the heat budget of Lake Mead [4,6]. Historically, the Colorado River inflow contributed large quantities of heat to the reservoir during the spring and early summer. However, the construction of Glen Canyon Dam and formation of Lake Powell in 1963 altered the natural temperature and flow cycles of the river [7]. Discharges of cold water from the hypolimnion (230 ft, 70 m) of Lake Powell have significantly reduced energy inputs to Lake Mead. Similarly, it appears that heat losses from the reservoir could be increased if Hoover Dam were operated from a surface, rather than deep-water, discharge. The combined effects of a cold-water discharge from Glen Canyon Dam and a surface discharge from Hoover Dam could reduce evaporation from Lake Mead by over 200,000 acre-feet ($2.47 \times 10^8 m^3$)/yr and result in considerable decreases in salinity. The purpose of this paper is to present data in support of these conclusions and to describe how the hydroelectric dams can be operated to minimize evaporative water losses from Lake Mead and reduce salinity in the Colorado River.

STUDY AREA

Lake Mead was formed in 1935 by construction of Hoover Dam. It extends 114 miles (183 km) from the mouth of Grand Canyon to Black Canyon, the site of Hoover Dam (Figure 1). Lake Mead is one of the largest reservoirs in the country with a surface area of 163,088 acres (660 km^2) and a volume of 29,185,245 acre-feet ($36 \times 10^9 m^3$), at the maximum operating level of 1227 ft (374 m) [8]. It is separated into two large basins by Boulder Canyon, located midway through the reservoir (Figure 1). The area above Boulder Canyon is referred to as the Upper Basin and that below as the Lower Basin. Hoover Dam is equipped with intake gates at 1045 ft (319 m) and 895 ft (273 m) elevations. The dam has been operated from the lower gates since 1954.

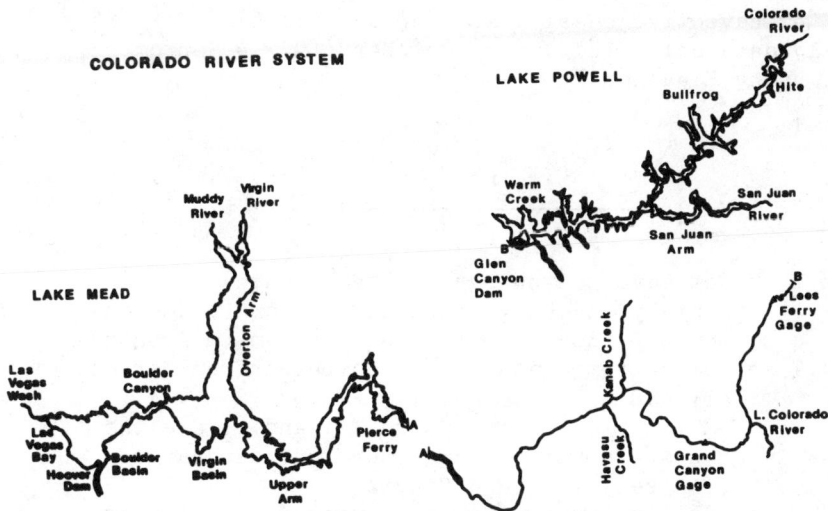

Figure 1. Map of the Colorado River System (Lake Mead and Lake Powell).

The Colorado River inflow to Lake Mead was unregulated prior to 1963 when Glen Canyon Dam was constructed 280 miles (451 km) upstream (Figure 1). Annual discharges are high [8,354,000 acre-feet (10.3 x $10^9 m^3$) in 1978], and seasonal discharge peaks usually occur during winter and summer. Discharges from Glen Canyon Dam are withdrawn from the hypolimnion (230 ft, 70 m) of Lake Powell and temperatures range from 7.5-13.5°C. The Colorado River inflow, via discharges from Lake Powell, comprises 98% of the inflow to Lake Mead. The remainder is derived from the Virgin and Muddy Rivers, which discharge into the Overton Arm, and Las Vegas Wash, which discharges secondary-treated sewage effluents into Las Vegas Bay (Figure 1).

DATA SOURCES AND METHODS

Historical evaporation data for Lake Mead and discharge data for Grand Canyon were obtained from "Surface Waters of the United States," U.S. Geological Survey Water-Supply Papers, Part 9, Colorado River Basin, until 1967. Grand Canyon temperature data and salinity data for Lake Mead were derived from the "Quality of Surface Waters in the United

441

States," U.S. Geological Survey Water-Supply Paper, Part 9,
Colorado River Basin. After 1967, these data were obtained
from "Water Resources Data for Nevada" or "Water Resources
Data for Arizona" of the U.S. Geological Survey Water-Data
annual reports.

Net advective energy was computed for Lake Mead from
monthly data collected during October, 1977 - September,
1978, using Equation 1.

$$Q_v = \frac{q_i\,(T_i - T_r) - q_o(T_o - T_r)}{A} k_i - - - k_n \qquad (1)$$

where Q_v = net advected energy (cal/cm^2·month)

q_i = monthly discharge in Grand Canyon (m^3/month)

q_o = monthly discharge from Hoover Dam (m^3/month)

T_i = inflow temperature (°C) computed from Harbeck et
al. [4] equation of $(T_{gc} + 2.6°C) - (.04\ T_{gc})$
$- 2.1 \times 10^{-5} \times q_i$, where T_{gc} and q_i are the
average monthly temperature and discharge
(ft^3/sec) in Grand Canyon

T_o = outflow temperature (°C) measured at the lower
intake gates 295 ft (90 m) near Hoover Dam from
Paulson et al. [5]

T_r = reference temperature of 4.4°C

A = average monthly surface area in Lake Mead
(cm^2) from Lara and Sanders [8]

k_i = unit conversion factors

Estimates of net advective energy for a surface discharge at Hoover Dam were computed by Equation 1 with T_o =
monthly surface temperature (°C) near Hoover Dam from
Paulson et al. [5].

Differences in evaporation rates from Lake Mead for a
surface and hypolimnion discharge on Hoover Dam were computed from Equation 2 [6].

$$E = \frac{Q_{vh} - Q_{vs}}{L(1+R)} k_i - - - k_n \qquad (2)$$

where E = annual evaporation rate (cm/yr)

Q_{vh} = average net advective energy for the
hypolimnion discharge (cal/cm^2·day)

Q_{vs} = average net advective energy for a surface
discharge (cal/cm^2·day)

L = latent heat of vaporization (585 cal/cm^3)

R = average Bowen Ratio as estimated for Lake

Mead by Anderson and Pritchard [6] and
Harbeck et al. [4] using monthly data
from October 1977 - September 1978

k_i = unit conversion factors

Total evaporative water loss from Lake Mead was then
determined by extrapolation from volume curves [8].

Using USGS data collected during the period from Octo-
ber 1977-September 1978 as initial model conditions (Table
I), the effects of decreased evaporation on salinity in Lake
Mead were determined by Equation 3.

$$Sc_t = \frac{Ss_{t-1} + (Si_t\alpha - So_t)}{V_t + Ev_t} \quad k_i \text{---} k_n \quad (3)$$

where Sc = salt concentration in Lake Mead (mg/l)
 Ss = salt storage in Lake Mead (kg)
 Si = salt inputs to Lake Mead (kg)
 α = salt retention coefficient
 So = salt output at Hoover Dam (kg)
 V = Lake Mead volume (m^3)
 Ev = evaporation reductions (m^3)
 t = time interval (yr)
 k_i = unit conversion factors

Table I.   Parameters and Data Used in the Salinity Model for
Lake Mead. Data Collected Oct. 1977 - Sept. 1978.

Parameter	Symbol	Average	Units
Lake Mead Volume	V	25.48	m^3 x 10^9
Colorado River Inflow	I	10.43	m^3 x 10^9
Hoover Dam Discharge	O	9.48	m^3 x 10^9
Salt Input	Si	73.714	kg x 10^8
Salt Retention	α	0.8693	-
Salt Storage	Ss	172.76	kg x 10^8
Salt Output	So	Variable	kg x 10^8
Evaporation	Ev		
Reduction Minimum		1.48	m^3 x 10^8
Maximum		2.63	m^3 x 10^8

Evaporation reductions of 120,000 acre-feet (1.48 x
$10^8 m^3$) and 213,000 acre-feet (2.63 x $10^8 m^3$) were used in the
salinity model. These evaporation reductions were added to
the 1978 water year average volume in Lake Mead (25.48 x
$10^9 m^3$) during the first year of modeling. In subsequent
years, these evaporation reductions were added to the annual
discharges from Hoover Dam, using the 1978 water year (9.48

x $10^9 m^3$) as the initial discharge rate. Salinity decreases
projected to occur from the Las Vegas Wash, Nevada and Grand
Valley, Colorado Salinity Control Projects [3] were also incorporated in the salinity model.

RESULTS AND DISCUSSION

Temperature and Discharge Cycles

The construction of Glen Canyon Dam in 1963 drastically
altered the seasonal temperature and discharge cycles in the
Colorado River (Figure 2). River temperatures have increased
by nearly 5°C during the late fall and winter but decreased
by 10°C during the rest of the year. These temperature
changes were caused by cold-water releases from Glen Canyon
Dam. Water is withdrawn from the hypolimnion of Lake Powell,
and discharge temperatures average about 8°C throughout the
year. In the summer, river temperatures increase to 10-11°C
at Grand Canyon and 15-16°C at Pierce Ferry, where the river
enters Lake Mead. However, river temperatures are still
nearly 10°C colder than for comparable spring and summer
periods prior to 1963.

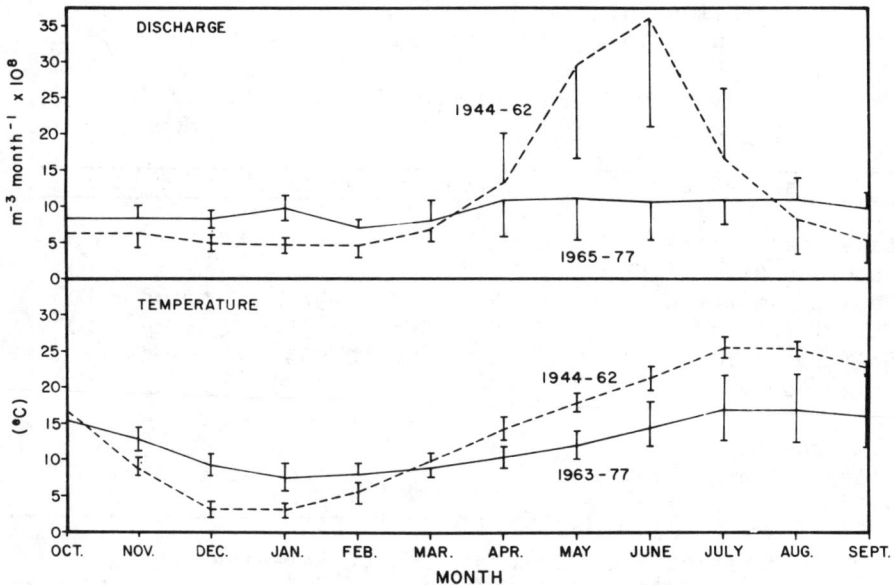

Figure 2. Historical Temperature and Discharge Data (±SD)
           in Grand Canyon for Pre-and Post-Lake Powell
           Periods [USGS Data]. From [7].

Discharges from Glen Canyon Dam are regulated for power generation and flood control purposes. This has eliminated the spring discharge pulse that occurred historically due to runoff from the upper Colorado River drainage system (Figure 2). Monthly discharges are now subject to much less variation and peak discharges usually occur in summer when power demands are greatest.

Energy Advection

The alterations in temperature and discharge cycles in the Colorado River have had a significant influence on energy advection into Lake Mead. Investigations conducted in 1948 by Anderson and Pritchard [6] and in 1952-53 by Harbeck et al. [4] showed that large quantities of energy were advected into Lake Mead during spring and early summer (Figure 3). Advection contributed 300-400 cal/cm^2·day of heat to the reservoir during these periods. This was nearly half that derived from solar radiation. In contrast to pre-Lake Powell periods, advection now contributes minimal heat to Lake Mead (Figure 3). Cold-water discharges from Glen Canyon Dam resulted in a net heat gain of only 9.04 cal/cm^2·day during 1977-1978. This has had a marked influence on evaporation rates from the reservoir.

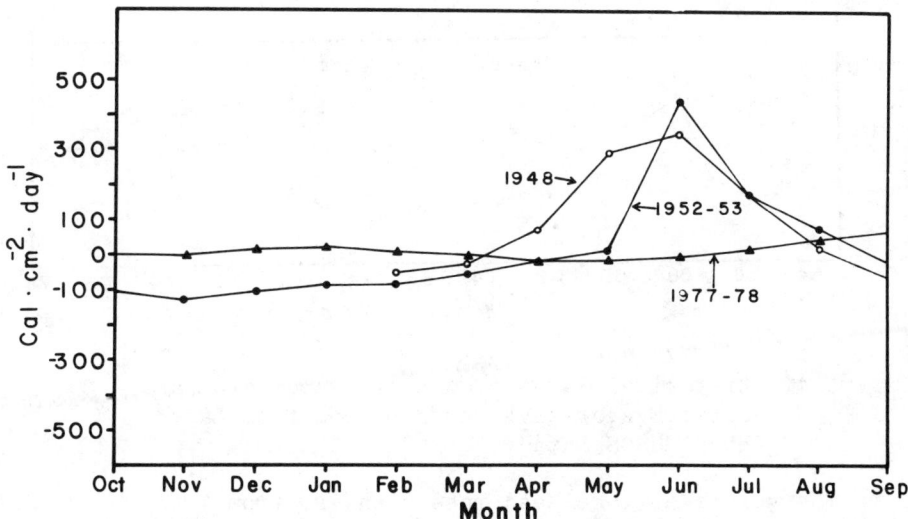

Figure 3.   Net Advective Energy in Lake Mead During 1948 [6], 1952-53 [4] and in 1977-78 [This Study].

Annual evaporation rates from Lake Mead, as reported by the U.S. Geological Survey, averaged 85.2 inches (216 cm)/yr prior to the construction of Glen Canyon Dam (Figure 4). Evaporation rates decreased significantly after 1964 when Lake Powell was filled to operating levels and discharges were increased to normal. In the period from 1965-1970, evaporation rates decreased to about 74 inches (188 cm)/yr which reflects the changes in energy advection caused by cold-water discharges from Glen Canyon Dam. Advection was especially pronounced during this period because of low lake elevations in Lake Mead and relatively high discharges from Glen Canyon Dam (Figure 5). Annual discharges were 65% of the Lake Mead volume in 1965 and averaged nearly 50% throughout the period.

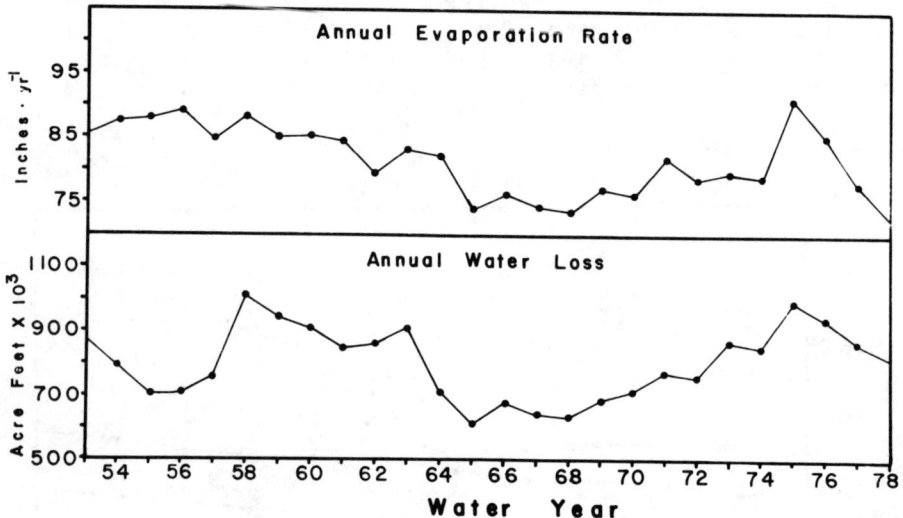

Figure 4.   Historical Rates of Annual Evaporation and
            Total Evaporative Water Losses from Lake
            Mead [USGS Data].

The volume of Lake Mead rose steadily from 1964 through 1974, but river discharges remained fairly constant after 1965 (Figure 5). This lessened the influence of advection on the reservoir heat budget, and evaporation rates increased somewhat during 1970-1974. The abrupt increases in evaporation rates in 1975-76 and subsequent decreases in 1977-78

(Figure 4) were not related to changes in river discharges or reservoir volumes (Figure 5). Rather, it appears these variations were caused by changes in methods of estimating evaporation.

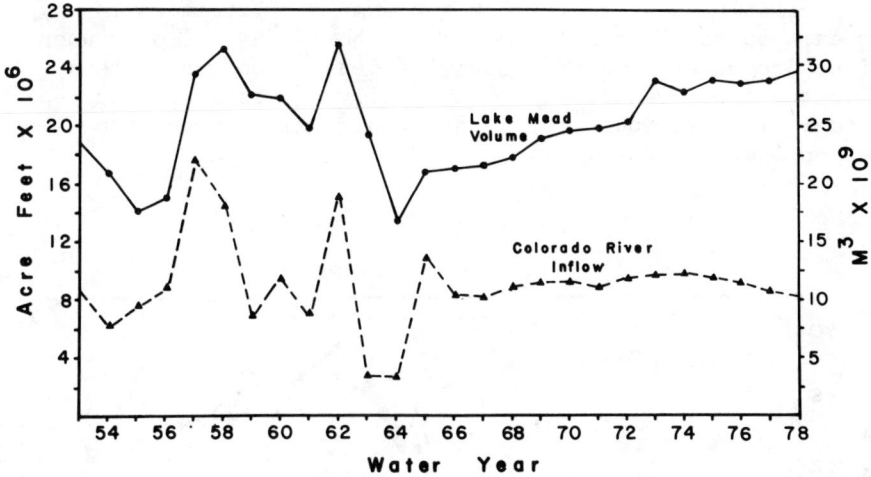

Figure 5. Average Annual Volumes for Lake Mead and
Inflows from Grand Canyon [USGS Data].

Evaporation rates in Lake Mead have historically been estimated with the mass transfer method using equations developed by Harbeck et al. [4]. Evaporation rates were routinely adjusted for changes in energy advection and storage. This was discontinued in October, 1974, and evaporation rates rose sharply in 1975 and 1976. The mass transfer method was still used to estimate evaporation, but, in February of 1976, the coefficient in the equation was changed, and evaporation rates immediately decreased. This indicates that the abnormally high evaporation rates for 1975 and 1976 were caused by failures to adequately compensate for advection. Although evaporation rates for 1977 and 1978 appear reasonable in comparison to other post-Lake Powell years, recent data collected in limnological studies of Lake Mead [5] indicate that evaporation rates are still being overestimated.
Temperatures in the Upper Basin of Lake Mead are generally colder than in the Lower Basin [5]. In 1980, surface temperatures in Virgin Basin were often 1-2°C colder than in

Boulder Basin (Figure 6). This was especially evident during
the spring and early summer, and only on a few occasions did
surface temperatures in Virgin Basin exceed those in Boulder
Basin. Although these temperature differences could reflect
regional variations in climatology over the reservoir, they
are most likely due to advection from the Colorado River
inflow. The circulation patterns in Lake Mead are such that
the Colorado River inflow is confined primarily to the Upper
Basin [5]. The river forms a density current that extends to
Virgin Basin and into the Overton Arm. The Virgin Basin
appears to act like a large "mixing bowl" [6] and only when
river discharges are high does the density current extend
into the Lower Basin [5]. This usually occurs during late-
summer after periods of prolonged, high discharges from Glen
Canyon Dam.

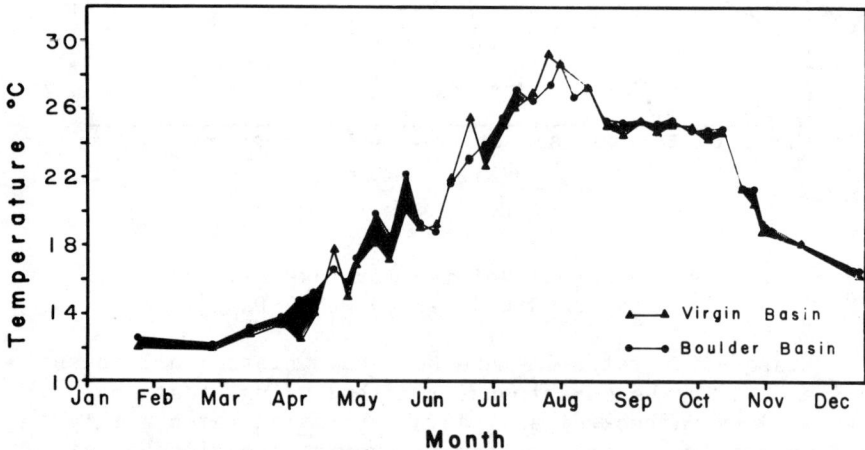

Figure 6.   Surface Temperatures in Boulder Basin and
             Virgin Basin During 1980 [Lake Mead
             Limnological Research Center Data].

        Historically, adjustments to evaporation rates for
changes in energy storage in Lake Mead have been based on
temperature measurements made at Hoover Dam intake towers
[4]. This decision was reached on the basis of data collected in 1952-1953, which showed that temperature differences
between the Upper and Lower Basins were minimal. This is no
longer the case with cold-water discharges from Glen Canyon

Dam. The Hoover Dam intake towers, being located at the distant end of the reservoir, would be one of the last areas in Lake Mead to be influenced by cold-water discharges. Temperatures at the dam could be considerably higher than elsewhere in the reservoir, particularly in comparison to the Upper Basin. Estimates of reservoir-wide evaporation based on data from the Hoover Dam intake towers could, therefore, also be higher than actual evaporation. Hydrologists at the Bureau of Reclamation have consistently observed an overall gain of water in Lake Mead. Based on a ten-year average during 1960-1970, the measured Lake Mead contents exceeded water budget estimates by approximately 230,000 acre-feet $(2.84 \times 10^8 m^3)$/yr (USBR data). This could, in part, be due to an overestimate of evaporation from the reservoir since 1963 when advection was altered by construction of Glen Canyon Dam.

Although measured evaporation rates may be somewhat too high in the period after 1963, it is still evident that cold-water discharges from Glen Canyon Dam have significantly reduced evaporation from Lake Mead. If we exclude the 1975 and 1976 values, which are clearly too high, pre- and post-Lake Powell evaporation rates average 85.2 inches (216 cm)/yr and 76.8 inches (195 cm)/yr. This is equivalent to a reduction in annual water loss of at least 93,376 acre-feet $(1.2 \times 10^8 m^3)$, which is very similar to predictions made during the mid-1960s. Government scientists reported that cold-water discharges would reduce evaporation by about 100,000 acre-feet $(1.23 \times 10^8 m^3)$/yr. Operation of Glen Canyon Dam from a deep discharge is thus an extremely effective method of reducing evaporation from Lake Mead.

Manipulation of Evaporation Rates

It has long been known that reservoirs operated from a deep discharge store heat, whereas, those operated from a surface discharge dissipate heat [9]. The principle here is quite simple and depends only on the formation of thermal gradients in the reservoir. In Lake Mead, surface temperatures exceed hypolimnion temperatures during all periods of the year, except winter when the reservoir is completely mixed and isothermal. The temperature gradient is particularly sharp during summer when surface temperatures reach 27-30°C, compared to 11-12°C in the hypolimnion. In the period from October, 1977 - September, 1978, operation of Hoover Dam from the deep discharge resulted in an average, net advective heat gain of 9.04 cal/cm^2·day (Table II). However, this would have decreased to -29.55 cal/cm^2·day if the dam had been operated from a surface discharge over this period. The net difference in advection between surface and deep discharge would be -38.59 cal/cm^2·day (Table II). Using

Equation 2, with L = 585 cal/cm^3 and a Bowen Ratio (R) of -0.108, this would be equivalent to a decrease in reservoir evaporation rates of -0.07395 cm/day or -26.99 cm/yr (-10.6 inches/yr). At the average lake elevations for 1977-78 (1186 ft), this would result in an annual reduction in water loss of 119,779 acre-feet (1.48 x 10^8 m^3). The approach used to derive this estimate is very simplified in that other variables in the heat budget were not included in the calculations. It was assumed that solar radiation, net radiation and change in energy storage would be similar regardless of discharge depth. As was pointed out by U.S. Bureau of Reclamation scientists in their review of a previous report [10], these assumptions may not be entirely valid. Extensive studies will be required to determine how other variables in the heat budget will change with discharge depth. Nonetheless, the estimate appears to be a reasonable approximation of water loss savings based on conclusions from earlier studies on Lake Mead. Harbeck et al. [4] made similar estimates with data collected in 1952-53 and concluded that a surface discharge would reduce evaporation in Lake Mead by 72,000 acre-feet (8.9 x 10^7 m^3)/yr at lake elevations of 1174 ft (358 m). This is similar to the present estimate if differences in lake elevations are taken into consideration.

Table II. Net Advective Energy Estimates in Lake Mead For a Surface and Hypolimnion Discharge at Hoover Dam [5].

Month	Colorado River			Hoover Dam					Lake Mead		
	Inflow[1] Temp.	Discharge	Energy[4]	Hypolimn.[2] Discharge Temp.	Epilimn.[2] Discharge Temp.	Discharge	Hypolimn.[4] Discharge Energy	Surface[4] Discharge Energy	Surface Area	Hypolimn. Discharge Net Advect. Energy	Epilimn. Discharge Net Advect. Energy
	(°C)	(m^{3}x10^8)	(calx10^{16})	(°C)	(°C)	(m^{3}x10^8)	(calx10^{16})	(calx10^{16})	(cm^{2}x10^{12})	(cal·cm^{-2}·day^{-1})	(cal·cm^{-2}·day^{-1})
Oct 77	13.0	5.2	0.447	12.0	22.3	5.3	0.403	0.949	5.34	+2.66	-30.30
Nov 77	12.4[3]	4.8	0.384	12.0	20.5	5.6	0.426	0.902	5.31	-2.64	-32.52
Dec 77	11.5	9.5	0.675	12.0[3]	16.6[3]	5.9	0.448	0.720	5.35	+13.69	-2.71
Jan 78	10.9	10.9	0.709	12.0	12.7	2.9	0.220	0.241	5.46	+28.89	+2.65
Feb 78	11.0	7.1	0.469	12.0	12.5	5.3	0.403	0.429	5.50	+4.29	+2.60
Mar 78	12.0[3]	9.3	0.707	12.1	16.4	9.1	0.701	1.092	5.55	+0.35	-22.38
Apr 78	13.0	5.9	0.507	11.7	15.1	11.1	0.810	1.188	5.46	-18.50	-41.58
May 78	12.0[3]	6.9	0.524	12.0	19.3	10.6	0.806	1.579	5.46	-16.66	-62.33
Jun 78	11.0	9.3	0.614	12.3	22.0	8.1	0.640	1.426	5.46	-1.59	-49.57
Jul 78	15.8	9.4	1.072	12.0	27.0	10.3	0.783	2.328	5.36	+17.39	-75.59
Aug 78	14.7	13.?	1.360	12.0	24.8	11.0	0.836	2.244	5.37	+31.48	-53.10
Sep 78	14.7	12.5	1.288	12.3	22.5	6.2	0.490	1.122	5.41	+49.1?	+10.2?
										+9.04[a]	-29.55[a]
											[a]Average

[1]Inflow temperature = (T + 2.6°C) - 0.04 (T) - (2.1 x 10^{-5} x q)
Where T and q are Grand Canyon temperature in °C and mean discharge in CFS (Harbeck et al. 1958). Grand Canyon temperature from USGS data.

[2]Discharge temperatures are the Lake Mead temperatures at each depth of discharge near the Hoover Dam intake towers. Lake Mead surface and hypolimnion (90 m) temperatures from UNLV data.

[3]Average of preceding and following months.

[4]Advected energy computed reference 4.4°C.

450

It thus appears that the ideal strategy for reducing evaporation from Lake Mead would be to continue operating Glen Canyon Dam from a deep discharge and shift Hoover Dam to a surface discharge. This could result in a combined reduction in water loss of 213,155 acre-feet (2.63 x $10^8 m^3$)/yr, at the 1977-78 lake levels. Such reductions in water loss would constitute one of the best water conservation programs available for the Colorado River.

Influences on Reservoir Salinity

Reductions in evaporative water losses from Lake Mead would result in significant decreases in salinity of the reservoir. Inflows of cold water from Glen Canyon Dam are probably already causing reductions in salt concentrations in the Upper Basin of Lake Mead, although data are not available to estimate the magnitude. Water loss reductions derived from operating Hoover Dam with a surface discharge would act to further decrease salinity, especially in the Lower Basin.

Dissolved solids concentrations at the Hoover Dam intake towers in Lake Mead averaged 676 mg/l during water year 1978. Evaporation reductions of 120,000 acre-feet (1.48 x $10^8 m^3$), achieved with a surface discharge at Hoover Dam; or 213,000 acre-feet (2.63 x $10^8 m^3$), achieved with a cold-water discharge on Glen Canyon Dam and a surface discharge on Hoover Dam, would reduce salinity in Lake Mead by 9 mg/l and 16 mg/l, respectively (Figure 7). These salinity reductions would occur within a five-year period and are comparable to those which will be achieved by the Las Vegas Wash, Nevada (8 mg/l) and Grand Valley, Colorado (19 mg/l) Salinity Control Projects (Figure 7) [3]. This would serve to effectively augment salinity control projects on the Colorado River.

Feasibility of Operating Hoover Dam from a Surface Discharge

There are several potential problems associated with operation of Hoover Dam from a surface discharge [4]. First, this would require modifying the intake structures. Hoover Dam is currently equipped with intake gates at 895 ft (273 m) (lower gates) and 1045 ft (319 m) (upper gates) elevations. At the 1978 lake elevations of 1186 ft (361 m), operation from the upper gates would still result in withdrawal of cold, hypolimnion waters [12]. Intake gates would have to be installed at higher elevations to permit withdrawal of warm water. Engineering studies would have to be done to evaluate the feasibility and cost-effectiveness of such modifications. Hydraulic studies should also be conducted to insure that the intake structures would indeed withdraw sur-

451

face waters and not pull cold water from deeper strata in
the reservoir.

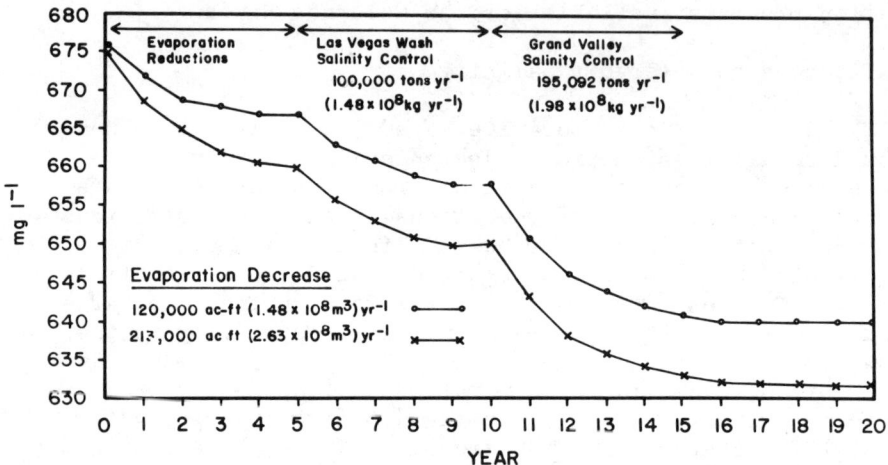

Figure 7. Salinity Model for Lake Mead Projecting
Average Reservoir Salt Concentrations for
Evaporation Reductions and Various Salinity
Control Projects.

A second problem that needs to be considered relates to
the impacts of warm-water discharges on downstream uses. The
Black Canyon area below Hoover Dam supports a popular cold-
water trout fishery that could be adversely influenced by
warm-water discharges from Hoover Dam. Recent studies, however, have shown that warm-water discharges could benefit
reproduction of aquatic insect populations that comprise an
important food resource for trout [13]. Aquatic insects require seasonal temperature cycles, like those that existed
historically, to complete their life cycle [14]. Discharge
temperatures from Hoover Dam are now virtually constant at
12-13°C throughout the year and appear to be the cause for
declines in aquatic insect populations in Black Canyon. Operation of Hoover Dam from a surface discharge would restore seasonal temperature cycles in the river and perhaps
enhance production of aquatic insects. This, combined with
stocking of warm-water tolerant rainbow trout, could insure
that a viable trout fishery was still preserved in Black
Canyon.

452

Warm-water discharges from Hoover Dam could also result in increased evaporation from Lake Mohave and Lake Havasu, the downstream reservoirs. The temperature structure in upper Lake Mohave is currently influenced by cold-water discharges from Hoover Dam [5]. However, the river forms an underflow in Lake Mohave and mixing is not sufficient to advectively cool the entire reservoir. Surface temperatures in Lake Mohave frequently exceed those in Lake Mead, and discharges of warm water from Hoover Dam will probably not contribute more heat to the reservoir than it currently assimilates from solar radiation. Temperature data are too limited to allow for speculations on how evaporation could be altered in Lake Havasu. However, the surface area of Lake Mohave and Lake Havasu are each roughly one-third that of Lake Mead. Net water losses and salinity in the river would probably still be reduced, even if warm-water releases did increase evaporation rates in these reservoirs.

Finally, operation of Hoover Dam from a surface discharge would alter the nutrient budget for Lake Mead. Total nitrogen retention in Lake Mead would increase by 66% and total phosphorus by 60% with a surface discharge on Hoover Dam [11]. This, in turn, would elevate productivity in the reservoir, particularly in the Lower Basin where wastewater inflows from Las Vegas Wash contribute large amounts of phosphorus to the reservoir [15]. However, this could be beneficial to the largemouth bass population which has undergone a serious decline in Lake Mead. This decline appears to be related to a decrease in reservoir fertility that occurred after Glen Canyon Dam was constructed in 1963 [16,17]. High nutrient losses from the deep-water discharge at Hoover Dam have further contributed to this decline in fertility. A surface discharge could help sustain greater fertility in Lake Mead, and perhaps provide a better food base for the bass populations [11].

The environmental and engineering problems associated with operation of Hoover Dam from a surface discharge do not appear to be insurmountable. Some of the environmental questions are being addressed in limnological studies currently being conducted for the Office of Water Research and Technology, or in fisheries investigations being conducted by the regional fisheries biologists and the U.S. Fish and Wildlife Service. The engineering problems, however, will clearly require additional investigations to determine whether the existing intake structures can be modified to cost-effectively withdraw surface waters from Lake Mead.

In addition, further limnological studies should be done in Lake Mead to determine if the present methods of estimating evaporation are accurate, evaluate estimates of water loss savings made in this paper, and better assess the relationship of salinity to evaporation in the reservoir.

This should be accompanied by similar investigations in Lake Mohave and Lake Havasu to evaluate how evaporation rates and salinity in those reservoirs would change with a surface discharge on Hoover Dam. The operation of the proposed pump-storage units (Spring Canyon and Rifle Range sites) should be included in these investigations since it is likely they can be used to further reduce evaporation in the reservoirs. If water were withdrawn from the hypolimnion of the reservoir with the pump-storage units and released via a diffuser into the epilimnion, it could result in significant cooling of the surface waters.

There are, therefore, a number of possible ways to operate hydroelectric facilities to reduce evaporative water losses and salinity in the Colorado River system. Cold-water discharges from Glen Canyon are already operating to reduce evaporation and salinity in Lake Mead. Operation of Hoover Dam from a surface discharge or use of pump-storage systems to further cool surface waters could result in greater reductions in evaporative water losses and salinity. These methods would certainly help preserve precious water supplies and water quality in the Colorado River.

ACKNOWLEDGEMENTS

I am especially grateful to Mr. Gary Bryant, U.S. Bureau of Reclamation, for his continual assistance and support of our research program. Mr. Robert Barton, Mr. David Overbolt, Mr. Gordon Mueller, Mr. David Solbeck and Mr. Art Tuma of the Bureau provided reviews of the report. I also wish to thank Mr. John R. Baker and James E. Deacon for their suggestions and assistance; Ms. Sherrell A. Paulson, Penelope E. Naegle for drawing the illustrations and editing; Jim Williams for the photographing and Laurie Vincent for typing the report.

REFERENCES

1. U.S. Dept. of Interior (USDI). 1981. Quality of Water, Colorado River Basin. Prog. Rept. 10. 190 pp.

2. GAO. 1980. Water supply should not be an obstacle to meeting energy development goals. Govt. Acct. Off. Rept. No. CED-80-30. 79 pp.

3. U.S. Dept. of Interior (USDI). 1977. Colorado River Water Quality Improvement Program. Vol. I. n.p.

4. Harbeck, G.E. Jr., M.A. Kohler and G.E. Koberg. 1958. Water-loss investigations: Lake Mead studies. U.S. Geol. Surv. Prof. Paper 298. 100 pp.

5. Paulson, L.J., J.R. Baker and J.E. Deacon. 1980. The limnological status of Lake Mead and Lake Mohave under present and future powerplant operations of Hoover Dam. Lake Mead Limnological Res. Ctr. Tech. Rept. No. 1. Univ. Nev., Las Vegas. 229 pp.

6. Anderson, E.R. and D.W. Pritchard. 1951. Physical limnology of Lake Mead. Lake Mead sedimentation survey. U.S. Navy Electronic Lab. San Diego, California. Rept. No. 258. 153 pp.

7. Paulson, L.J. and J.R. Baker. 1980. Nutrient interactions among reservoirs on the Colorado River. Pages 1647-1656 in H.G. Stefan, ed. Symposium on surface water impoundments. June 2-5, 1980. Minneapolis, MN.

8. Lara, J.M. and J.I. Sanders. 1970. The 1963-64 Lake Mead survey. U.S. Bur. Rec. Rept. No. REC-OCE-20-21. 169 pp.

9. Wright, J.C. 1967. Effect of impoundments on productivity, water chemistry and heat budgets of rivers. Pages 188-199 in C.E. Lane, ed. Reservoir fisheries resources. Symp. Amer. Fish. Soc. Spec. Publ. No. 6.

10. Paulson, L.J. 1981. Use of hydroelectric dams to control evaporation from Lake Mead. Lake Mead Limnological Res. Ctr. Tech. Rept. No. 9. Univ. Nev., Las Vegas. 28 pp.

11. Paulson, L.J. 1981. Nutrient management with hydroelectric dams on the Colorado River system. Lake Mead Limnological Res. Ctr. Tech. Rept. No. 8. Univ. Nev., Las Vegas. 39 pp.

12. Baker, J.R. and L.J. Paulson. 1980. Evaluation of possible temperature fluctuations from proposed power modifications at Hoover Dam. Lake Mead Limnological Res. Ctr. Tech. Rept. No. 3. Univ. Nev., Las Vegas. 23 pp.

13. Paulson, L.J., T.G. Miller and J.R. Baker. 1980. Influence of dredging and high discharge on the ecology of Black Canyon. Lake Mead Limnological Res. Ctr. Tech. Rept. No. 2. Univ. Nev., Las Vegas. 58 pp.

14. Lemkuhl, D.M. 1972. Change in thermal regime as a cause of reduction of benthic fauna downstream of a reservoir. J. Fish. Res. Board Can. 29:1329-1332.

15. Baker, J.R. and L.J. Paulson. 1980. Influence of Las Vegas Wash density current on nutrient availability and phytoplankton growth in Lake Mead. Pages 1638-1646 in H.G. Stefan, ed. Symposium on surface water impoundments. June 2-5, 1980. Minneapolis, MN.

16. Paulson, L.J., J.R. Baker and J.E. Deacon. 1979. Potential use of hydroelectric facilities for manipulating the fertility of Lake Mead. Pages 269-300 in G.A. Swanson, Tech. Coord. The mitigation symposium: A national workshop on mitigating losses of fish and wildlife habitats. July 16-20, 1979. Fort Collins, Colo. Gen. Tech. Rept. No. RM-65, Rocky Mt. Forest and Range Exp. Sta.

17. Prentki, R.T., L.J. Paulson and J.R. Baker. 1981. Chemical and biological structure of Lake Mead sediments. Lake Mead Limnological Res. Ctr. Tech. Rept. No. 6. Univ. Nev., Las Vegas. 89 pp.

CHAPTER 27

THE EFFECTS OF IMPOUNDMENTS ON SALINITY IN THE COLORADO RIVER

L.J. Paulson
J.R. Baker
    Lake Mead Limnological Research Center
    University of Nevada, Las Vegas

INTRODUCTION

The increase in salinity of our western rivers has been identified as one of the most serious water quality problems in the nation [1]. This is of special concern in the Colorado River where salinity has increased from pristine levels estimated at 380 mg/l [2] to present-day levels of 825 mg/l at Imperial Dam [3,4]. Flow depletions, associated with decreased runoff and increased evaporation and diversions, coupled with high salt loading from natural and man-created sources are considered the primary causes for rising salinity in the river [5]. The urban and agricultural development projected to occur in the basin through this century could deplete flows by an additional 2 million acre-feet $(2.5 \times 10^9 m^3)$/yr [4]. Salinity models indicate that depletions of this magnitude will elevate total dissolved solids concentrations (TDS) to 1150 mg/l at Imperial Dam. Since this would have an enormous economic impact on municipal and agricultural water uses [6], salinity control programs are being implemented in the basin to maintain TDS at or below the 1972 levels.

Historical data for the Colorado River, however, indicate that TDS concentrations are not increasing as rapidly as the models predict. Despite the extensive development and large flow depletions that have already occurred in the basin, TDS concentrations in Grand Canyon and below Hoover Dam have not changed appreciably since monitoring began [4]. Water quality monitoring has recently shown that TDS concentrations throughout the Lower Colorado River Basin have been decreasing since 1972. This is thought to be a transient phenomenon caused by changes in flow patterns, salt routing or possibly inundation of saline sources in the Upper Colorado River Basin following completion of Lake Powell and other impounds during the 1960s [7]. This might also reflect more permanent reductions in TDS due to changes in chemical processes operating in the impoundments.

The U.S. Geological Survey (USGS) has monitored ion and TDS concentrations in the inflows and outflow of Lake Mead and Lake Powell since early impoundment. The purpose here is to present results of our analysis of the USGS salinity data and describe how these large impoundments have historically influenced ion and TDS concentrations in the Colorado River. The implications of these findings are discussed relative to current efforts to control salinity in the Colorado River Basin.

SALINITY STUDIES DURING EARLY IMPOUNDMENT

Large impoundments, like Lake Mead and Lake Powell, are generally thought to have a detrimental effect on salinity. This view stems from the observation that TDS concentrations in the impoundment outflow exceed those in the inflows. Concentration of salt by evaporation is considered a primary cause for this increase in TDS [1]. Evaporation in Lake Mead ranged as high as 900,000 acre-feet $(1.1 \times 10^9 m^3)$/yr during early impoundment [8,9]. Howard [10] noted that this caused a slight increase in TDS below Hoover Dam. Evaporation in Lake Powell has been estimated at about 500,000 acre-feet $(6.2 \times 10^8 m^3)$/yr [11] and causes a 16 mg/l increase in TDS below Glen Canyon Dam [12].

Evaporation clearly has an effect on TDS, but it appears to be relatively small by comparison to that caused by salt dissolution processes occurring in the impoundments. The chemical composition of the Colorado River is strongly influenced by the regional geology. Calcium, sulfate and carbonate have historically comprised 60-70% of the TDS. The impoundment of Lake Mead in 1935 further increased calcium and sulfate concentrations in the river [10,13]. This was derived primarily from dissolution of gypsum deposits which were prevalent in the Muddy Creek geologic formations in Las Vegas Bay and Virgin Basin [14]. The U.S. Bureau of Reclamation [15] estimated that there were only 22 acres $(8.9 \times 10^4 m^2)$ of exposed salt outcroppings in the reservoir floor prior to inundation. They predicted that rates of dissolution would be high during early impoundment but then diminish as the outcrops dissolved or became silted over. Subsequent studies have not been conducted to evaluate this prediction, or to determine if similar dissolution processes occur in Lake Powell. However, Gloss et al. [12] noted that there was a slight increase in sulfate concentrations at Lees Ferry after the formation of Lake Powell. They suspected that this was caused by dissolution of gypsum and also predicted that rates would diminish as the impoundment aged.

The increases in TDS caused by evaporation and dissolution appear to be offset to some extent by precipitation of

458

calcium carbonate (calcite) that occurs in the impoundments. Large quantities of calcite were precipitated in Lake Mead during early impoundment [10]. The formation of Lake Powell substantially reduced calcite precipitation in the Upper Basin of Lake Mead, but rates are still high in the Lower Basin [16]. Appreciable quantities of calcite also precipitate in Lake Powell [12,17]. The combined impoundment system may therefore precipitate more calcite than what historically occurred just in Lake Mead. An increase in overall rates of calcite precipitation and/or a decrease in the rates of gypsum dissolution could be possible reasons for recent decreases in TDS observed in the Lower Colorado River Basin.

DATA SOURCES AND METHODS

The U.S. Geological Survey has monitored flow rates, ion and TDS concentrations in the Colorado River and storage in the impoundments for several years. Prior to 1970, the flow rates and storage data were compiled in "Surface Waters of the United States" and ion concentrations in "Quality of Surface Waters of the United States," both of which were published annually in the U.S. Geological Survey Water-Supply Papers, Part 9, Colorado River Basin (1926-1970). These records have since been published in the "Water Resources Data" series of the U.S. Geological Survey Water-Data annual reports for individual states in the Colorado River Basin. In our analyses, we used data from the "Water Resources Data for Arizona", "Water Resources Data for Nevada" and "Water Resources Data for Utah" reports for water years 1970-79.

Flow vs. TDS concentration relationships were evaluated from data collected for various time periods at Lees Ferry, Grand Canyon and below Hoover Dam (Figure 1). Sufficient data were available to evaluate these during pre-impoundment periods of 1926-42 and 1951-60 in Grand Canyon and 1951-60 at Lees Ferry. Data were also available to assess these relationships at Grand Canyon and Lees Ferry during the post-Lake Powell period of 1970-79 and below Hoover Dam during the post-Lake Mead periods of 1935-43, 1951-60 and 1970-79. Flow-weighted average TDS concentrations were computed for individual years within each time period. Changes in TDS concentrations vs. flows were computed between each year in these time periods and used to construct statistical relationships for each location.

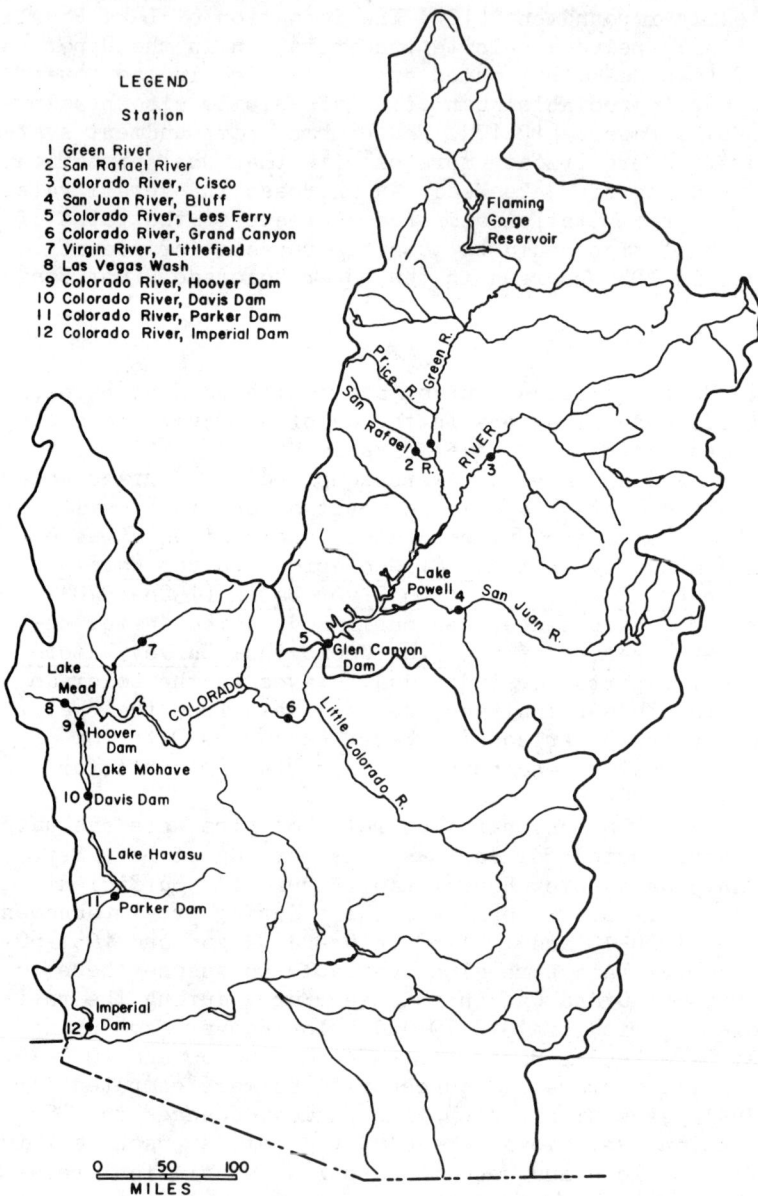

LEGEND

Station

1 Green River
2 San Rafael River
3 Colorado River, Cisco
4 San Juan River, Bluff
5 Colorado River, Lees Ferry
6 Colorado River, Grand Canyon
7 Virgin River, Littlefield
8 Las Vegas Wash
9 Colorado River, Hoover Dam
10 Colorado River, Davis Dam
11 Colorado River, Parker Dam
12 Colorado River, Imperial Dam

Figure 1. Map of Colorado River System.

Ion and TDS budgets were also constructed for Lake Mead during the 1951-60 and 1970-79 periods and for Lake Powell during the 1970-79 period. Annual ion and TDS loads were computed from monthly flow and concentration measurements at

460

the principal inflows and outflow of each impoundment (Figure 1). Individual ion and TDS data were not available for the Las Vegas Wash inflow to Lake Mead during the 1951-60 period or for the Muddy River inflow during either period. The annual data were used to calculate flow-weighted average concentrations over the 10 yr periods on the combined inflows and outflow of each impoundment.

The effects of evaporation on ion and TDS concentrations could not be evaluated directly. Gross evaporation rates have been measured annually in Lake Mead since 1952, but water inputs from precipitation and ungaged inflows, which are necessary to compute net evaporation, have only been measured during special studies [8,9]. Gross evaporation rates and other variables of the water budget were only measured in Lake Powell during 1973-74 [11]. Water losses from the impoundments, nonetheless, exceed water inputs. In order to assess the effects of this on ion and TDS concentrations, we computed average annual, net rates of water loss for each impoundment and time period using Equation 1.

$$R = Ig - Og - Od - \Delta S \qquad (1)$$

where R $\;=\;$ net water loss (precipitation + ungaged inflows) - (evaporation + Δbank storage)
Ig $\;=\;$ gaged inflows (as in Figure 1)
Og $\;=\;$ gaged outflows (as in Figure 1)
Od $\;=\;$ diversions
ΔS $\;=\;$ change in impoundment storage

The ion and TDS loads computed for the inflows to each impoundment were then divided by the term (Ig-R) to estimate ion and TDS concentrations expected in the outflow as a result of the net water losses. The net water losses were assumed to be due primarily to evaporation, and we acknowledge that factors like bank storage will have some influence on estimated changes in ion and TDS concentrations for years when lake fluctuations are more severe. However, averages over long term periods should not be largely affected.

Ion budgets, adjusted for net water losses, were used to estimate chemical precipitation and dissolution rates in the impoundments. Calcite precipitation was estimated from molar changes in carbonate (= bicarbonate/2.03) [18]. Gypsum (CaSO$_4$) and halite (NaCl) dissolution were estimated from molar changes in sulfate and chloride.

461

RESULTS

The various model predictions regarding future TDS
levels in the Colorado River are based on the assumption
that TDS concentrations will vary inversely with flows.
Sufficient data were available to statistically evaluate
these relationships at Lees Ferry and Grand Canyon during
pre- and post-Lake Powell periods and below Hoover Dam during three post-Lake Mead periods.

The flow vs. TDS concentration relationships developed
for these locations are presented in Table I. It was evident
that TDS concentrations were inversely related to flows during pre-Lake Powell periods at Lees Ferry and Grand Canyon.
The correlation coefficients for these relationships were
highly significant at each location (Table I). There was no
apparent relationship between TDS concentrations and flows
at Lees Ferry or Grand Canyon during the post-Lake Powell
period (Table I). A similar situation existed below Hoover
Dam (Table I), even though annual flow variations were similar in magnitude to those at Lees Ferry and Grand Canyon
during the pre-Lake Powell period. The lack of a relationship, or poor relationship, between TDS concentrations and
flows during the post-impoundment periods indicated that
other factors were operating in the impoundments to influence TDS.

Table I. Regression Equations and Correlation Coefficients
for Relationships Between Changes in Flows (X) and
TDS Concentrations (Y) for Various Time Periods and
Locations in the Colorado River, (ΔFlow Units as
$m^3 \times 10^{10}$ and ΔTDS Concentrations as mg/l) [USGS
Data].

Location	Time Periods	N	Regression Equations a b $\times 10^{-10}$	r
Lees Ferry	1951–60	9	Y= 0.53–151.712(X)	−.852*
Lees Ferry	1970–79	9	Y= 4.40–110.844(X)	−.384
Grand Canyon	1926–42	16	Y= 4.66–211.686(X)	−.786*
Grand Canyon	1951–60	9	Y= −2.84–172.953(X)	−.894*
Grand Canyon	1970–79	9	Y= −4.92– 42.164(X)	−.208
Hoover Dam	1935–43	8	Y= −1.28– 11.391(X)	−.079
Hoover Dam	1951–60	9	Y= −2.01–121.480(X)	−.568
Hoover Dam	1970–79	9	Y= 5.18– 6.297(X)	−.011

*Significant at $P_{(.01)}$

The ion and TDS budgets for Lake Mead and Lake Powell revealed that the concentrations of certain ions were drastically altered by the impoundments (Figure 2). Lake Mead increased average sulfate concentrations by 63 mg/l during the 1951-60 period which was over 50 mg/l higher than that expected due to net water losses (Figure 2). There was less increase in sulfate during the 1970-79 period in Lake Mead, but concentrations were still 26 mg/l higher than expected (Figure 2). Lake Powell elevated average sulfate concentrations by 24 mg/l in the 1970-79 period which was about 10 mg/l higher than expected (Figure 2).

The increases in sulfate were offset to varying degrees by reductions in carbonate concentrations (Figure 2). In Lake Mead, carbonate was reduced by 29 mg/l below the expected value during the 1951-60 period (Figure 2). Silica and calcium were also slightly lower than expected. These reductions were not sufficient to offset increases in sulfate, and TDS concentrations rose by 42 mg/l. [TDS concentrations expressed as the sum of constituents do not always agree with TDS measured as residue. Both are reported in Figure 2, but we only refer to sum of constituents.] In the 1970-79 period, average carbonate concentrations were only 9 mg/l lower than expected for Lake Mead (Figure 2). TDS concentrations therefore increased by 57 mg/l which was 15 mg/l higher than the previous period and 12 mg/l higher than expected. Lake Powell reduced carbonate by 14 mg/l which offset nearly one-half the increase in sulfate (Figure 2). TDS concentrations increased by 34 mg/l and were only slightly higher than expected during the 1970-79 period.

The concentrations of other ions were not altered appreciably by the impoundments (Figure 2). There was no measureable change in magnesium and potassium concentrations in Lake Powell during 1970-79 or in Lake Mead during 1951-60. These ions only increased by 4 mg/l in Lake Mead during the 1970-79 period. Sodium and chloride were slightly higher than expected in Lake Powell during 1970-79 and in Lake Mead during the 1951-60 period. However, during the 1970-79 period, sodium and chloride concentrations were lower than expected for Lake Mead (Figure 2), indicating that these ions were being retained in the impoundment. This seemed unlikely because of the conservative nature of both ions. Rather, it appears that sodium and chloride loading to Lake Mead are being underestimated because of a sampling error that developed in Grand Canyon after flows were regulated by construction of Glen Canyon Dam in 1963. We explain this in greater detail in the discussion section of the paper.

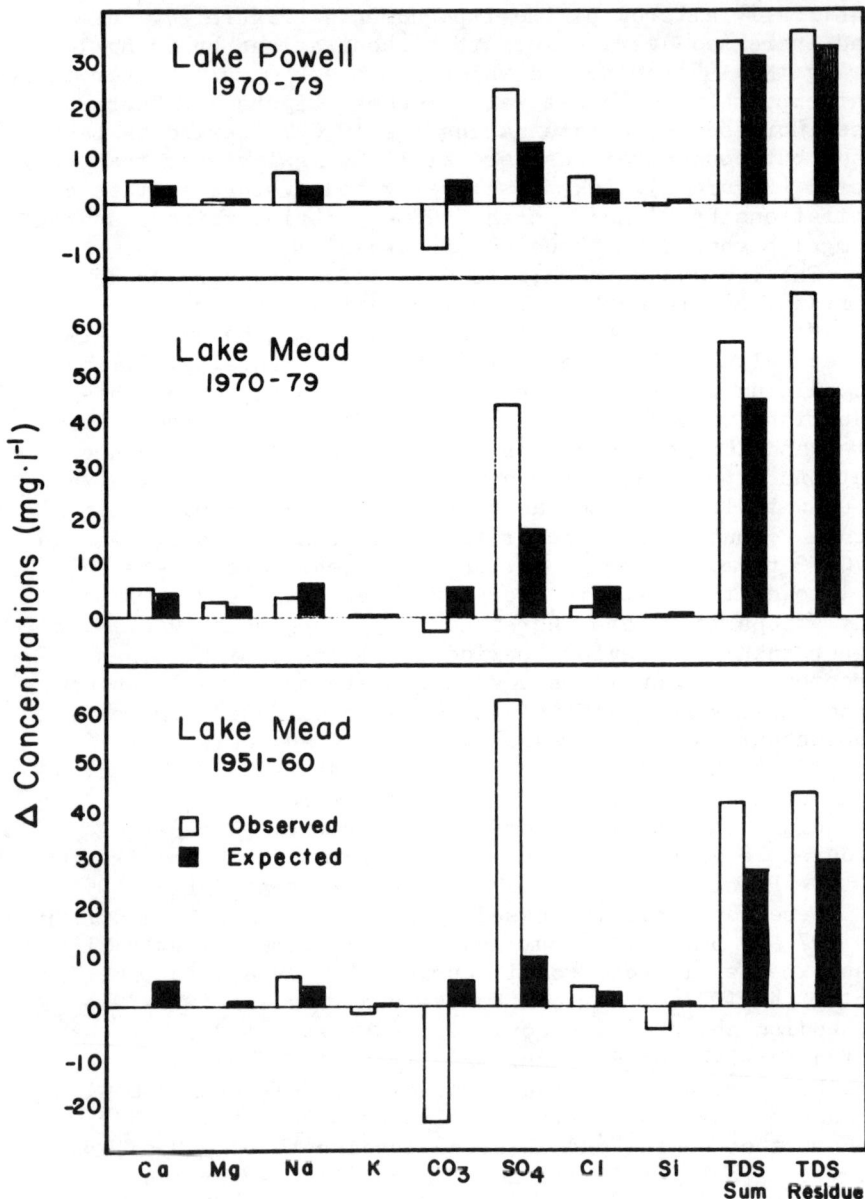

Figure 2. Observed and Expected Changes in Ion and TDS Concentrations in Lake Mead During the 1951-60 and 1970-79 Periods and in Lake Powell During the 1970-79 Periods. (Carbonate Estimated From Bicarbonate/2.03) [18]. [USGS Data].

464

Net water losses estimated for Lake Mead during the 1970-79 period were higher than the 1951-60 period. This was also unexpected because evaporation rates in Lake Mead have decreased significantly due to cold-water discharges from Glen Canyon Dam [19,20]. However, water levels in Lake Mead have risen steadily since Lake Powell was formed in 1963. Significant quantities of water are retained in bank storage in Lake Mead when levels increase after extended periods of draw down [9]. This appears to be the principal reason for higher net water losses during the 1970-79 period.

DISCUSSION

The large impoundments on the Colorado River clearly have a significant effect on salinity. The relationships of flow and TDS differed markedly between pre- and post-impoundment periods. The reason for this is that concentrations of sulfate, carbonate and calcium, the principal ions in the river, were altered by dissolution and precipitation processes occurring in the impoundments (Table II). Dissolution of gypsum significantly elevated concentrations of sulfate during all time periods. Halite dissolution also caused slight increases in concentrations of sodium and chloride. Conversely, calcite precipitation caused marked reductions in concentrations of calcium and carbonate. However, this was not sufficient to offset increases caused by dissolution, and TDS concentrations were elevated above expected levels in all time periods.

Table II. Estimated Average Annual Rates of Calcite ($CaCO_3$) Precipitation and Gypsum ($CaSO_4$) and Halite (NaCl) Dissolution for Various Time Periods in Lake Mead and Lake Powell. Calcite Precipitation Was Estimated from Molar Changes in Carbonate Ion and Gypsum Dissolution from Molar Changes in Sulfate Ion. Calcium Concentrations Given in Parentheses. [USGS Data].

Locations/Time Periods	$CaCO_3$ (mg/l)	$CaSO_4$ (mg/l)	NaCl (mg/l)
Lake Powell/1970-79	23 (9)	16 (5)	5
Lake Mead/1970-79	15 (6)	37 (11)	-
Lake Mead/1951-60	48 (19)	75 (22)	3
Lake Mead/1935-48*	47	123	19**

*From Howard [10], Gould [13].
**Includes potassium.

It appears that significant changes have occurred in the rates of dissolution and precipitation in Lake Mead since early impoundment. Rates of gypsum dissolution decreased from an average of 123 mg/l during 1935-48, to 75 mg/l during 1951-60 and 37 mg/l during 1970-79 (Table II). Average rates of halite dissolution decreased from 19 mg/l during 1935-48 to 3 mg/l during 1951-60. It was not possible to compute rates for 1970-79. The changes observed in the rates of dissolution in Lake Mead seem to confirm predictions made by the USDI [15] that rates would decrease as the salt outcroppings dissolved or became silted over.

The changes in sulfate concentrations in Lake Mead may also have been influenced by the activity of sulfate reducing bacteria. These bacteria convert sulfate ion to hydrogen sulfide under anaerobic conditions. The hydrogen sulfide often combines with iron to form insoluble iron sulfide precipitates that are retained in the sediments [21]. Howard [10] noted that substantial populations of sulfate reducing bacteria were present in Lake Mead sediments. Rates of sulfate reduction have never been measured directly, but sulfate diffusion coefficients were determined for sediments in Las Vegas Bay and Bonelli Bay [22]. It is possible to estimate rates of sulfate reduction from sulfate diffusion coefficients if we assume estimates for Las Vegas Bay are representative of the Lower Basin of Lake Mead and those for Bonelli Bay are representative of the Upper Basin. These calculations indicate that sulfate reduction would decrease sulfate concentrations in the outflow from Lake Mead by 8 mg/l·yr. This functions to offset some of the increase in sulfate concentrations caused by dissolution of gypsum.

Rates of calcite precipitation estimated by Howard [10] indicated that 47 mg/l were precipitated annually in Lake Mead during 1935-48 (Table II). Those measured by Prentki et al. [16] were considerably lower during this period. Rates of calcite precipitation in Lake Mead decreased from an average of 48 mg/l during the 1951-60 period to 15 mg/l during the 1970-79 period (Table II) which does agree with changes measured in Lake Mead sediments and those expected as a result of decreased phytoplankton productivity [16]. The reductions in phosphorus loading that occurred after Lake Powell was formed [23] caused productivity in the Upper Basin of Lake Mead to decrease from an average of 4612 mg C/m^2·day to 503 mg C/m^2·day [16,24]. Increased phosphorus loading from sewage effluent discharges into Las Vegas Bay increased productivity in the Lower Basin from 937 mg C/m^2·day to 1582 mg C/m^2·day [24]. However, this was not sufficient to offset the decreases that occurred in the Upper Basin, and reservoir-wide productivity in Lake Mead decreased by 78% after Lake Powell was formed in 1963. Reservoir-wide calcite precipitation decreased from an

466

average of 397 x 10^3 t/yr to 180 x 10^3 t/yr over the same period [16]. The greatest decrease occurred in the Upper Basin where rates dropped from 240 x 10^3 t/yr to 28 x 10^3 t/yr. The changes in calcite precipitation were therefore closely related to changes in productivity.

Calcite precipitation in Lake Powell averaged 23 mg/l during the 1970-79 period (Table II). Reynolds [17] demonstrated that polyphenols in the Colorado River inflow to Lake Powell significantly inhibited calcite precipitation in the upper end of the impoundment. The polyphenols are derived from forested regions of the Upper Colorado River Basin, and concentrations vary directly with seasonal flow patterns [12]. During spring, when the river forms an overflow in Lake Powell, polyphenol concentrations are sufficient to inhibit calcite precipitation in the upper one-third of the impoundment. Calcite precipitation is limited primarily to summer months and only occurs in the lower end of Lake Powell where dilution reduces polyphenol concentrations [12,17].

Rates of calcite precipitation in Lake Powell during the 1970-79 period were roughly one-half as high as those estimated for Lake Mead during the 1935-48 and 1951-60 periods (Table II). This could reflect differences in factors influencing solubility (temperature) or possibly indicate that polyphenol inhibition was not as high in Lake Mead when it received runoff directly from the Upper Colorado River Basin. Ratios of autochthonous organic carbon sedimentation to autochthonous calcite precipitation indicate that inhibition was, and still is occurring in the Upper Basin of Lake Mead [16]. These ratios did not change appreciably after Lake Powell was formed indicating that polyphenols are still being supplied to Lake Mead either via export from Lake Powell or possibly from inputs in the Grand Canyon. Thus, even though rates of calcite precipitation were relatively high in Lake Mead during early impoundment, it is likely that they would have been even higher were it not for the inhibition that appears to be caused by polyphenols.

The inhibition of calcite precipitation that occurs in both impoundments, and decreases that occurred in Lake Mead after Lake Powell was formed, reduce the combined effectiveness of the impoundments for calcite removal. However, the two impoundments still removed an average of 38 mg/l of calcium carbonate during the 1970-79 period (Table II) which is extremely significant from the standpoint of salinity control. Lake Powell increased sulfate concentrations by 16 mg/l, but rates of gypsum dissolution have decreased considerably in Lake Mead. The combined impoundment system now contributes 53 mg/l calcium sulfate to the river, but this is still considerably lower than what occurred in Lake Mead

467

during the 1935-48 and 1951-60 periods (Table II). This has a pronounced effect on TDS because sulfate alone comprises nearly one-half the TDS in the Colorado River. The changes in rates of gypsum dissolution may, therefore, be an important factor in causing the decrease in TDS observed in the Lower Colorado River Basin during recent years.

Accuracy of Ion Budgets

Precipitation and dissolution estimates based on mass balance calculations have been questioned in a recent salinity study [25]. Messer et al. [25] contend that salinity decreases attributed to calcite precipitation [10] can often be accounted for by salt storage in the impoundments. Sufficient data are rarely available to accurately estimate salt storage, and it was excluded from our calculations. It is unlikely, however, that this introduced errors in the ion budgets.

Salt storage is a function of salt loads in the inflows and outflows (discharge and diversions). Salt concentrations, however, can only vary with changes in inflow salt concentrations or evaporation, if precipitation or dissolution processes are not occurring in the impoundment. The expected outflow concentrations, estimated from 10 yr flow-weighted average inflow concentrations, adjusted for net water losses, differed significantly from measured concentrations for several ions (Figure 2). Outflow concentrations of carbonate were consistently lower and sulfate consistently higher than expected. These differences simply cannot be explained by salt storage since the estimates encompass roughly three flushings of the impoundments. The stoichiometry was not exact in that calcium lost by calcite precipitation did not balance that derived from gypsum dissolution (Table II). This probably reflects the influence of sediment diagenesis processes (e.g. sulfate reduction) [22], but the net affects of such processes on ion concentrations are unknown. There is, nonetheless, little doubt that calcite precipitation is a major loss, and gypsum dissolution a major source, of salinity in the Colorado River.

It was not possible to estimate halite dissolution in Lake Mead during the 1970-79 period. The observed decreases in sodium and chloride concentrations were nearly 5 mg/l lower for each ion than those expected due to net water losses (Figure 2). This did not occur in Lake Mead during the 1951-60 period or in Lake Powell during the 1970-79 period, indicating that it was not caused by retention of sodium chloride in the impoundment. Rather, this appears to be caused by sampling problems that developed in Grand Canyon after flows were regulated by Glen Canyon Dam.

The Little Colorado River enters the main stem Colorado

468

about 25 miles (40 km) above the Grand Canyon gaging station (Figure 1). Flows in the Little Colorado are derived from surface runoff, which is highly variable seasonally, and a nearly constant base flow of 223 CFS (6.3 m³/sec) from several springs and seeps, collectively referred to as Blue Springs [26]. Blue Springs contribute about 550,000 tons of salt per year to the Colorado River [4]. Comparison of flow-weighted average ion concentrations during the 1970-79 period at Lees Ferry and Grand Canyon (Table III) indicate that the Blue Springs input is comprised primarily of sodium and chloride.

Table III. Flow-weighted Average Ion and TDS Concentration at Lees Ferry and Grand Canyon During the 1970-79 Period. [USGS data].

Location	Constituents (mg/l)								
	Ca	Mg	Na	K	CO₃	SO₄	Cl	Si	*TDS
Lees Ferry	72	25	76	3.9	79	238	52	8.2	554
Grand Canyon	75	26	96	4.1	84	243	81	8.5	618
Δ	3	1	20	0.2	5	5	29	0.3	64

*Sum of constituents

Concentrations of these ions in Grand Canyon will vary inversely with river flows. Glen Canyon Dam stabilized seasonal flows but resulted in extreme variations in hourly flows. It is common for daily flows to vary from 2000 CFS (57 m³/sec) to over 20,000 CFS (566 m³/sec). USGS sampling in Grand Canyon has almost always been conducted at the beginning of each month, but the actual time of sampling varies from month to month. Sampling conducted when flows are high will cause more dilution of sodium and chloride concentrations and result in underestimates of loading to Lake Mead. This appears to be the case during the 1970-79 period which accounts for the unusual discrepancy observed in the sodium and chloride budgets.

This illustrates the kind of problems that can develop where sampling is conducted below a point source tributary in a regulated river. Grand Canyon is an extreme case due to the unique nature of the Little Colorado River, but a similar situation could exist in the Green River, which is regulated by Flaming Gorge Dam. The USGS gaging station in the Green River at Green River, Utah is located below the Price River which is a significant TDS point source [4]. At these locations, it seems that composite, rather than grab, sampling should be conducted to insure that flow-induced variations in ion concentrations are adequately represented by the sample.

Implications for Salinity Control

Congress authorized the U.S. Department of Interior
(Bureau of Reclamation) to proceed with construction of four
salinity control projects under Title II of the "Colorado
River Basin Salinity Control Act" (PL 93-320) of 1974. These
included: the Grand Valley and Paradox Valley units in Colorado, the Las Vegas Wash unit in Nevada and the Crystal Geyser unit in Utah [3]. The Crystal Geyser unit has since been
dropped from further consideration, but implementation of
the others is proceeding on schedule. Collectively, these
salinity control projects will decrease TDS concentrations
at Imperial Dam by 65 mg/l [4]. Numerous other projects are
in the planning or feasibility stage, and, if all are implemented, these would reduce TDS by 130 mg/l at Imperial Dam
[4].

Control of point sources is obviously an effective
method of reducing TDS concentrations, and this approach
appears to be warranted in view of predictions that TDS
could increase to 1150 mg/l by year 2000 [4]. These predictions, however, are based on models that assume TDS concentrations are inversely related to flows. This was the case
during pre-impoundment periods, and the assumption is perhaps still valid for extreme variations in flow. However,
flow vs. TDS relationships have been highly modified by the
large impoundments. The concentrations of principal ions in
the Colorado River are now altered significantly by mineral
dissolution and precipitation and evaporation processes
occurring in Lake Powell and Lake Mead. Rates of calcite
precipitation are closely linked to rates of phytoplankton
productivity. Inhibition of calcite precipitation occurs to
varying degrees in both impoundments. Rates of gypsum and
halite dissolution have been decreasing in Lake Mead since
early impoundment and will probably decline even more in the
future. A similar trend has been predicted to occur in Lake
Powell [12]. Although net water losses in Lake Mead were
higher, apparently due to increased bank storage during the
1970-79 period, evaporation rates appear to have been reduced considerably by cold-water discharges from Glen Canyon
Dam [19,20]. It is estimated that this has reduced average
TDS concentrations by 9 mg/l [20]. The impoundments have
thus caused numerous changes in ion and TDS concentrations
that cannot be modeled by simple flow vs. TDS relationships.
Rates of calcite precipitation, mineral dissolution, and
evaporation must be incorporated into the models, if they
are expected to have any predictive value.

Moreover, the whole concept of controlling TDS point
sources seems illogical in view of the natural ion composition of the river and the effect that these have on various
beneficial uses. Sulfate comprises nearly one-half the TDS

470

in the Colorado River. Studies conducted on a variety of
irrigation waters by the FAO [27] show that sulfate has no
appreciable effect on agricultural uses. In fact, gypsum is
often applied to agricultural soils to maintain calcium at
levels sufficient to avoid permeability or toxicity problems
that develop where irrigation waters are high in sodium
[27]. Sulfate also appears to have little effect on municipal uses, even though concentrations in the Lower Colorado
River Basin are slightly higher than the first tier drinking
water standards of 250 mg/l.

There are no drinking water standards on calcium and
carbonate, but it is well known that these are the principal
hardness agents responsible for severe scaling problems in
municipal water systems of Colorado River water users.
Kleinman and Brown [6] estimated that $240,500 in economic
damages were incurred per mg/l TDS by municipal users in the
Lower Colorado River Basin. These estimates would probably
be considerably higher if they were expressed per mg/l calcium carbonate. Lake Powell and Lake Mead collectively removed an average of 38 mg/l of calcium carbonate per year
over the 1970-79 period. This has greatly reduced the economic damages to municipal water systems. Moreover, rates of
calcite precipitation in the impoundments would be even
higher were it not for polyphenol inhibition. Similarly,
calcite precipitation in Lake Mead would probably increase
significantly if phytoplankton productivity could be
restored to pre-Lake Powell levels.

Sodium and chloride are the only other ions that make
up a significant fraction of the TDS in the Colorado River.
Dissolution of halite and evaporation in the impoundments
caused a slight increase in concentrations of these ions.
Sodium, via the effect it has on sodium adsorption ratios,
is especially harmful to agricultural crops [27]. Based on
the FAO recommended guidelines, sodium adsorption ratios are
approaching levels at Parker and Imperial Dams that could
present a problem for agricultural users (Paulson unpubl.
data). This may warrant some form of control in the future,
which could probably best be achieved by controlling point
source sodium and chloride inputs, rather than TDS.

Based on information presented by USDI [4], and USGS
data collected in rivers near salinity control projects, it
appears that the Las Vegas Wash and Grand Valley sources are
primarily sulfate salts. Implementation of these projects
would decrease TDS by 4 mg/l and 43 mg/l, respectively, but
this would probably have little effect on beneficial uses
because the salts are primarily sulfate. However, the salts
originating from the Paradox Valley area appear to be comprised primarily of sodium and chloride. Similarly, Glen-
wood and Dotsero Springs in Colorado, which are being considered for salinity control, are comprised primarily of

sodium and chloride [4]. Implementation of these projects could be extremely effective in reducing impacts on agricultural uses.

Selective control of specific ions, coupled with enhancement of natural salinity control processes, like calcite precipitation, seem to constitute a more cost effective approach to salinity management than indiscriminate control of TDS point sources. We are hopeful that results of this paper will stimulate a move in that direction.

ACKNOWLEDGEMENTS

We wish to thank Penelope Naegle, Jamie Meyer and Brian Klenk for assistance with data reduction; Laurie Vincent and Thom Hardy for typing the manuscript and Jim Williams and Sherrell Paulson for drawing and photographing the illustrations.

REFERENCES

1. Pillsbury, A.F. 1981. The salinity of rivers. Scientific American. 15:54-65.

2. Gardner, B.D. and C.E. Stewart. 1975. Agriculture and salinity control in the Colorado River Basin. Pages 63-82 in International Symposium on salinity of the Colorado River. Nat. Resour. Jour. Vol. 15.

3. U.S. Dept. of Interior (USDI). 1977. Colorado River Water Quality improvement program. Vol. I. n.p.

4. U.S. Dept. of Interior (USDI). 1981. Quality of water Colorado River Basin Progress Report No. 10. 190 pp.

5. Evans, N.A. 1975. Salt problem in the Colorado River. Pages 55-62 in International symposium on salinity of the Colorado River. Nat. Resour. Jour. Vol. 15.

6. Kleinman, A.P. and F.B. Brown. 1980. Colorado River salinity economic impacts on agricultural, municipal, and industrial uses. U.S. Dept. Int., Water and Power Resour. Ser., Eng. and Res. Ctr. Denver, CO. 19 pp.

7. C.R.B.S.C.F. 1981. Water quality standards for salinity, Colorado River system. Proposed report on the 1981 review, Colorado River Basin Salinity Control Forum (CRBSCF). 116 pp.

8. Harbeck, G.E. Jr., M.A. Kohler and G.E. Koberg. 1958. Water-loss investigations: Lake Mead studies. U.S. Geol. Surv. Prof. Paper No. 298. 100 pp.

9. Langbein, W.B. 1960. Water budget. Pages 95-102 in W.O. Smith, C.P. Vetter, G.B. Cummings, eds. Comprehensive survey of sedimentation in Lake Mead, 1948-49. U.S. Geol. Surv. Prof. Paper 295.

10. Howard, C.S. 1960. Chemistry of water. Pages 115-124 in W.O. Smith, C.P. Vetter, G.B. Cummings, eds. Comprehensive survey of sedimentation in Lake Mead, 1948-49. U.S. Geol. Surv. Prof. Paper 295.

11. Jacoby, G.C. Jr., R. Nelson, S. Patch and O.L. Anderson. 1977. Evaporation, bank storage and water budget at Lake Powell. Lake Powell Res. Proj. Bull. 48. 98 pp.

12. Gloss, S.P., R.C. Reynolds, L.M. Mayer and D.E. Kidd. 1980. Reservoir influences on salinity and nutrient fluxes in the arid Colorado River Basin. Pages 1618-1629 in H.G. Stefan, ed. Symposium on surface water impoundments. June 2-5, 1980. Minneapolis, MN.

13. Gould, H.R. 1960. Erosion in the reservoir. Pages 209-213 in W.O. Smith, C.P. Vetter, G.B. Cummings, eds. Comprehensive survey of sedimentation in Lake Mead, 1948-49. U.S. Geol. Surv. Prof. Paper 295.

14. Longwell, C.R. 1936. Geology of the Boulder Reservoir floor, Arizona-Nevada. Geol. Soc. Amer. Bull. 47: 1429-1432.

15. U.S. Dept. Interior (USDI). 1950. Geological investigations. Boulder Canyon Project Final Reports, Pt. 3, Bull. 1. 232 pp.

16. Prentki, R.T., L.J. Paulson and J.R. Baker. 1981. Chemical and biological structure of Lake Mead sediment. Lake Mead Limnological Res. Ctr. Tech. Rept. No. 6. Univ. Nev., Las Vegas. 89 pp.

17. Reynolds, R.C. 1978. Polyphenol inhibition of calcite precipitation in Lake Powell. Limnol. Oceanogr. 23: 585-597.

18. Brown, E., M.W. Skougstad, and M.J. Fishman. 1970. Techniques of water-resources investigations of the United States Geological Survey. Book 5, Chapter A1.

Methods for collection and analysis of water samples for dissolved minerals and gases. U.S. Geol. Surv. 84 pp.

19. Paulson, L.J. 1981. Use of hydroelectric dams to control evaporation from Lake Mead. Lake Mead Limnological Res. Ctr. Tech. Rept. No. 9. Univ. Nev., Las Vegas. 28 pp.

20. Paulson, L.J. (this volume).

21. Golterman, H.L. 1975. Physiological limnology, an approach to the physiology of lake ecosystems. Elsevier Scientific Pub. Co., New York. 489 pp.

22. Murray, J.W., C.J. Jones, K. Kuivila and J. Sawlan. 1981. Diagenesis of organic matter in Las Vegas Bay and Bonelli Bay, Lake Mead. Univ. Washington Spec. Rept. No. 96. 55 pp.

23. Evans, T.D. and L.J. Paulson. (this volume).

24. Prentki, R.T. and L.J. Paulson. (this volume).

25. Messer, J., E.K. Israelson and V. Adams. 1981. Natural salinity removal processes in reservoirs. UWRL/Q-81/03. Utah Water Research Laboratory. 84 pp.

26. Cole, G. and D.M. Kubly. 1976. Limnologic studies on the Colorado River from Lees Ferry to Diamond Creek. Colorado River Res. Program Tech. Rept. No. 8. Grand Canyon National Park Rept. Series. 88 pp.

27. Ayers, R.S. and D.W. Westcot. 1976. Water quality for agriculture. Irrigation and Drainage paper 29. FAO Rome. 97 pp.

CHAPTER 28

CALCIUM CARBONATE PRECIPITATION
IN RESERVOIRS

Gordon R. Dutt
David M. Hendricks
    Soils, Water & Engineering Department
    The University of Arizona
    Tucson, AZ

INTRODUCTION

Over the last decade, there has been considerable interest in salinity in the Colorado River. It has been pointed out (1) that development of water resources along the river has increased salinity. Other researchers (2) have found that $CaCO_3$ is or should be precipitating in the reservoirs along the river which reduces salinity by five percent or 19-29 ppm (3) in Lake Powell (2). If precipitation does occur, and the water does not redissolve calcite upon leaving the reservoir, the dams along the Colorado are, indeed, improving water quality with respect to total salinity, rather than causing degradation as previously suspected.

The chemical reactions occurring in water involving calcite have received considerable attention by researchers and have been discussed in textbooks (4). Several factors affect the solubility of calcite. These factors include crystal purity and size, temperature, other solutes present in solution and partial pressure of CO_2. In addition, other processes which affect any of the above also affect the solubility. Thus, if two waters, both saturated with calcite but of different chemical composition, are mixed, calcite may precipitate. If a water saturated with calcite is warmed at constant CO_2 partial pressure, precipitation will occur. If the CO_2 partial pressure is reduced by photosynthesis in plant life, again calcite will precipitate. All of the above processes occur in reservoirs. On the other hand, other processes are present which would be expected to increase the solubility of calcite, e.g., respiration production of CO_2 by plant and animal life. The net effect of what occurs is complex and a method to evaluate how much calcite is precipitating and, indeed, prove that calcite precipitation is occurring, is needed.

As indicated above, biological activity controls the concentration of dissolved CO_2 to a large extent (2), and hence the partial pressure, P_{CO_2}. In turn, the solubility of calcite is a function of the P_{CO_2} (5). Biological activity is a function of growth factors, such as nutrients, temperature, and light, which continually change throughout the year in a lake. Thus, evaluating calcite precipitation would be best carried out at a lake, since the conditions would be difficult to reproduce in a laboratory.

An approach which would seem suitable for the purposes of this study would be to place water samples in polyethylene bags in the lake. Since polyethylene transmits heat and CO_2 while being relatively impervious to water and salt, the water samples being investigated would be subjected to the same partial pressure of CO_2 as the surroundings. Thus, water samples could be collected from water entering a lake, placed in the plastic bags, equilibrated, and finally analyzed to determine quantitatively if calcite precipitates.

CO_2 Diffusion Across A Polyethylene Membrane

A system (Figure 1) was designed to test the permeability of polyethylene membranes to dissolved CO_2. One liter of boiled deionized water containing a pH indicator was added to each side of the membrane. This indicator was prepared by dissolving 1 g of solid chlorophenol red in 1 ℓ of boiled deionized water. Twenty mℓ in 1 ℓ of boiled deionized water produced the solution to be used on each side of the membrane.

Both chambers were agitated constantly with magnetic stirrers, while one received, at a constant rate, one percent CO_2 in dry, compressed air. In that chamber, a color change of the indicator could be detected within five m.

Several membrane thicknesses were used in testing this system. The first was a six mil membrane. This membrane allowed diffusion, as shown by the pH change in Table I, but was judged unsatisfactory because of the restrictive time element. The second membrane tested was of two mil thickness. Diffusion proved adequate, but the membrane was too flimsy to withstand even the slightest mechanical disturbances. The final choice of membranes was the four mil thickness. This thickness allowed adequate diffusion, yet was sturdy enough to withstand moderate physical abuse without perforating. This is an important consideration for operating in the field.

MEMBRANE

PRESSURE RELIEF

SOURCE OF 1% CO_2 IN COMPRESSED AIR

CONSTANT TEMP. WATER BOTH 30°C

WATER LINE

CO_2 ⇄ CO_2

MAGNETIC STIRRING

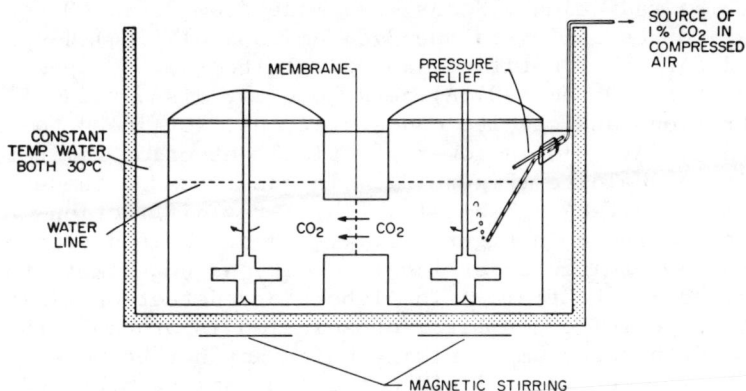

Figure 1. Apparatus for determining CO_2 diffusion across a four mil polyethylene membrane.

Table I. Membrane permeability to CO_2 at 30° and 5°C.

Run Number	Membrane Thickness	°C	Duration of Run	Starting pH	Ending pH Cell 1	Ending pH Cell 2
1	6 mil	5°	21 days	8.30 (both)	4.1	4.9
2	2 mil	30°	4 days	8.30	4.5	7.0
3	4 mil	5°	2 days	8.25	4.6	5.9
4	4 mil	30°	2 days	8.30	4.4	6.0
5	4 mil	30°	2 days	8.20	5.0	5.7
6	4 mil	30°	4 days	8.20	5.2	6.3
7	4 mil	30°	3 days	8.30	5.6	7.3

The four mil membrane was laboratory tested, as above, with the pH's being measured before and after each run with a Beckman model 76 expanded scale pH meter (Table I).

Calcite Detection Limits

Five solutions of 0.005 g to 0.025 g reagent grade calcite, $CaCO_3$ concentration, and a deionized blank were prepared. These amounts of calcite were diluted to one ℓ by adding distilled water. The suspensions were transferred to clean storage bottles and immersed in a constant 35°C water bath.

After three days, these solutions were mixed thoroughly, and filtered through a 0.8 micron millipore filter and dried. The retained calcite was analyzed by X-ray diffraction by mounting the filters directly on a Phillips diffractometer

477

using Cu Kα radiation. Scans were made from 20 to 40 degrees. The peak corresponding to the 3.04 nm (304 Å) (104) d-spacing was used for the identification of calcite (6). Graphs based on the 3.04 nm (304 Å) peak heights, dissolved Ca^{++} concentrations and calcite concentrations, were used to graphically determine a lower practical detection limit for precipitated calcite (Figure 2). The analysis of these solutions is given in Table II. The X-ray diffraction patterns are shown in Figures 3, 4, and 5. This detection limit was determined under ideal conditions not likely to be encountered in the field. The laboratory detection limit of 0.55 ppm calcite per liter of solution is probably the result of thorough sample mixing which may not be possible in the buoy. We expect field detection limits to exceed one ppm calcite/liter of solution filtered.

Calcite Precipitation: A Laboratory Model

Seventy-five liters of synthetic Lake Powell water were prepared after studying the analysis of water samples provided by the U.S. Bureau of Reclamation. With the exception of Ca^{++} and HCO_3^-, the ion concentrations in the synthetic mixture fall within the range demonstrated by analysis of actual water samples from Lake Powell. A solution was prepared from reagent grade NaCl, Na_2SO_4, $MgSO_4 \cdot 7H_2O$ and K_2SO_4, as shown in Table III. It was then brought to proper volume by the addition of sufficient distilled water to produce 75 ℓ of solution. Excess $CaCO_3$ was added to raise the Ca^{++} concentration to saturation. The $CaCO_3$ was dissolved by constant bubbling through the agitated solution of first one percent CO_2, then 100 percent CO_2. As the pH dropped, solutions containing successively higher concentrations of Ca^{++} were prepared.

Four mil polyethylene bags were fitted with a device which allowed them to be filled from the filtered stock solution and allowed samples to be drawn from a separate, smaller tube which was fitted with a millipore filter assembly (Figures 6 and 7). After each bag had been filled from the stock solution, samples were drawn from both inlet and outlet tubes. These samples were then analyzed for Ca^{++} using 0.0111 N CDTA, Calcein indicator, and a 0.5 mℓ 8 N KOH buffer. Sample pH's were also taken using a Beckman model 76b expanded scale pH meter.

After initial sampling, the inlet tube was clamped shut. The entire assembly was then immersed in a water bath at 35°C. To this bath a small circulating pump was attached which provided adequate mixing and good aeration. After immersion, samples were periodically drawn through the 0.8 μm

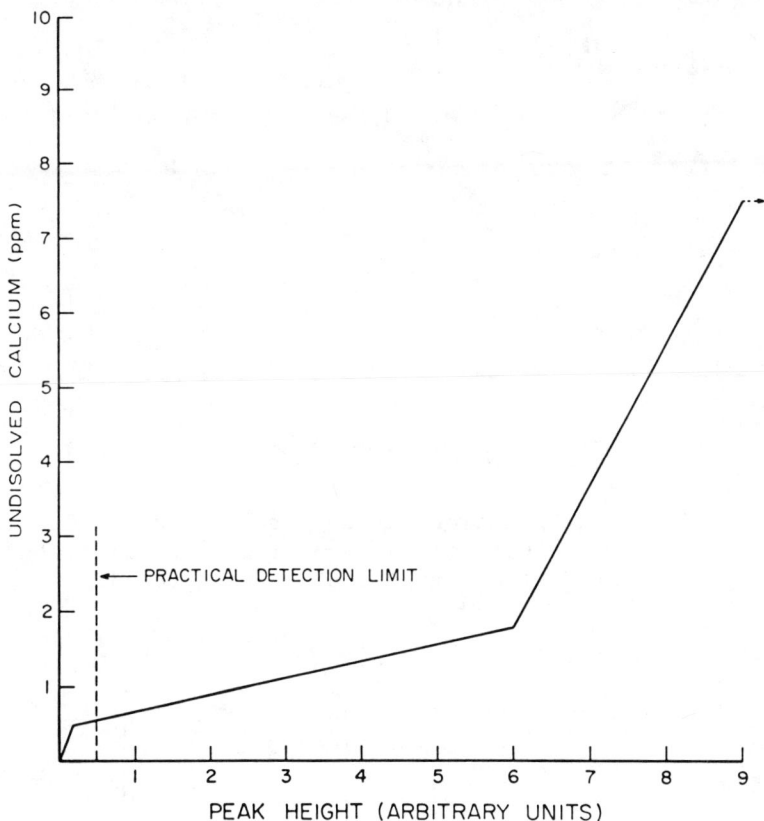

Figure 2.   Undissolved calcium ($CaCO_3$) in ppm vs X-ray diffractometry peak height.

Table II.   Standard calcite ($CaCO_3$) solutions after three days

Initial $CaCO_3$ g/l	pH	Total available Ca^{++} ppm	Ca^{++} determined by titration (ppm)	Precipitated $CaCO_3$ in ppm[*]
Blank	8.78	0.0	0.0	0
.005	8.75	2.0	2.0	0
.010	8.85	4.0	3.8	0.5
.015	8.90	6.0	5.2	1.79
.020	8.85	8.0	5.0	7.45
.025	9.00	10.0	5.2	11.95

[*] $[(Ca^{++}$ available $- Ca^{++}$ actual$)\, 2.5] = CaCO_3$ (solid) ppm

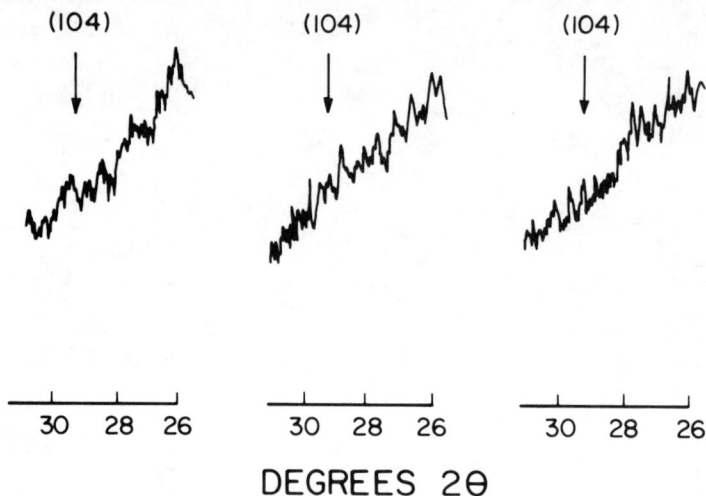

Figure 3. X-ray diffraction patterns from standard CaCO₃ solutions with (104) d-spacing of precipitated calcite not observable above background.

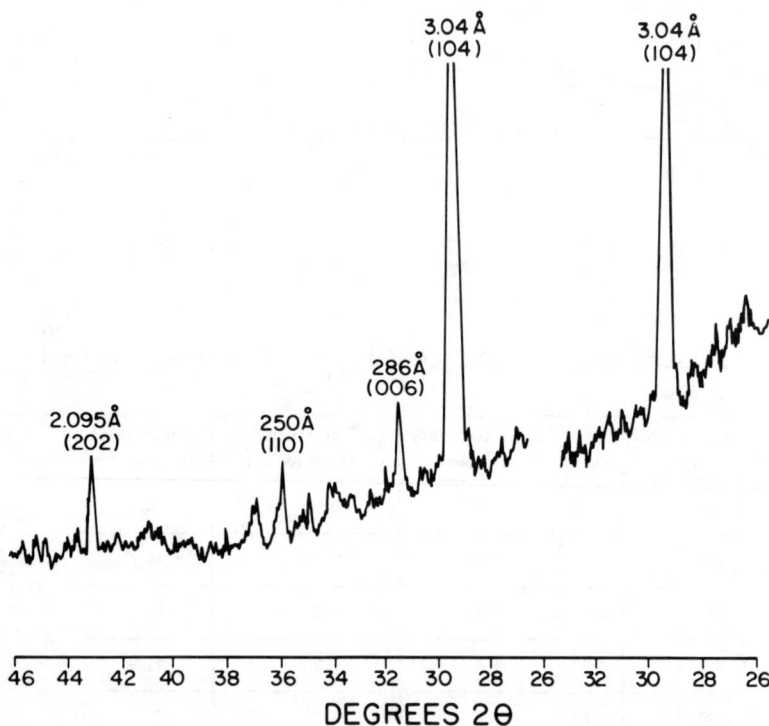

Figure 4. X-ray diffraction pattern from standard CaCO₃ solutions showing moderately strong calcite peaks.

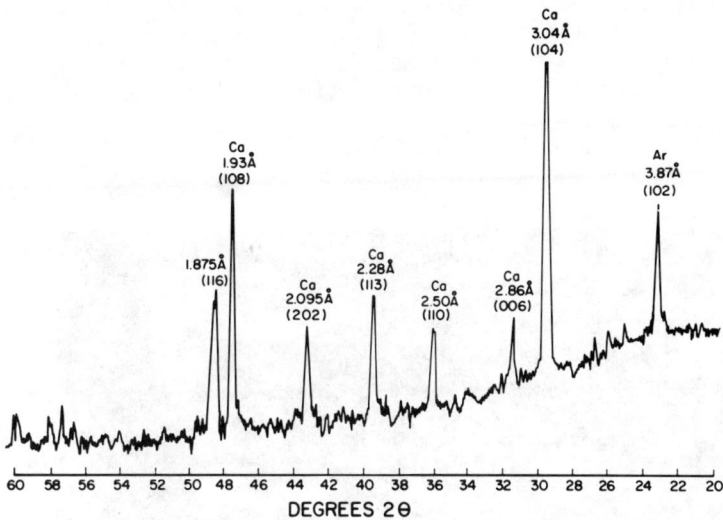

Figure 5. X-ray diffraction pattern from standard $CaCO_3$ solutions showing strong calcite peaks.

Table III. Synthetic Lake Water-Chemical Composition

Compound	g/ℓ
NaCl	0.105
Na_2SO_4	0.134
$MgSO_4$	0.130
K_2SO_4	0.009

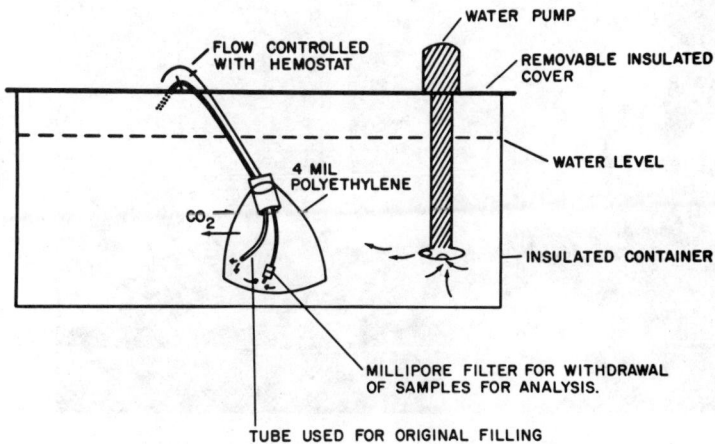

Figure 6. Diagram of laboratory apparatus for $CaCO_3$ precipitation in the laboratory.

Figure 7. Laboratory apparatus for determining $CaCO_3$ precipitation.

millipore filter, their pH's recorded, and their Ca^{++} concentrations determined. A record of these analyses appears in Table IV and in Figure 8. After an adequate number of samples had been drawn and analyzed, the solution remaining in each bag was allowed to pass through the outlet tube and the millipore filter. The apparatus was then disassembled and the millipore filter removed. After drying, the filter was analyzed for the presence of crystalline calcite ($CaCO_3$) by X-ray diffraction.

Bags 3, 5, A, B, and C were analyzed by X-ray for the presence of calcite. Good peaks were recorded on filters from bags B and C (Figure 9). The $CaCO_3$ peaks from bags A and 5 were not apparent on first analysis. In order to retrieve another sample, bags A and 5 were filled with 100 mℓ of deionized water and shaken to retrieve as much $CaCO_3$ as possible. The contents were then poured into a buret and filtered through 0.8 micron millipore filters. The X-ray analyses showed strong calcite peaks in both cases (Figures 10 and 11) and a strong aragonite peak from bag A.

Under some conditions, lake water apparently can become supersaturated to the point that aragonite precipitates and with time the aragonite decays to form calcite.

Water Quality Buoy

A prototype buoy was constructed and field tested in Kennedy Lake, a small recreational lake located in Tucson, Arizona. This lake has a maximum depth of 6.7 m and covers an area of 3460 m^2.

Table IV. Ca^{++} concentration as a function of time in the solutions.

Calcium Precipitation

BAG NUMBER: Ca^{++} ppm vs. time

DATE	2	3	4	5	'A'	'B'	'C'	'1'	BOUY SAMPLES
2-14	33.20	44.50	13.60	129.45					
2-18	30.06	43.88	13.54	124.44					
2-21	38.07	39.07	13.62	79.55					
2-24	38.07	46.09	13.22	73.34					
2-25					182.36	284.56			
2-28							315.22		
3-3		48.29		51.50	120.03	166.33	280.15		
3-5		46.09		47.09	85.37	132.06	170.74		161.12
3-6									124.84
3-7		46.09		45.69	64.32	93.38	91.98		137.87
3-11		47.49		43.88	51.50	66.73	72.74	346.89	
3-12		47.49		45.69	47.09	64.32	57.71	342.48	
3-13								340.27	
3-14					44.48	53.30	52.10	232.46	

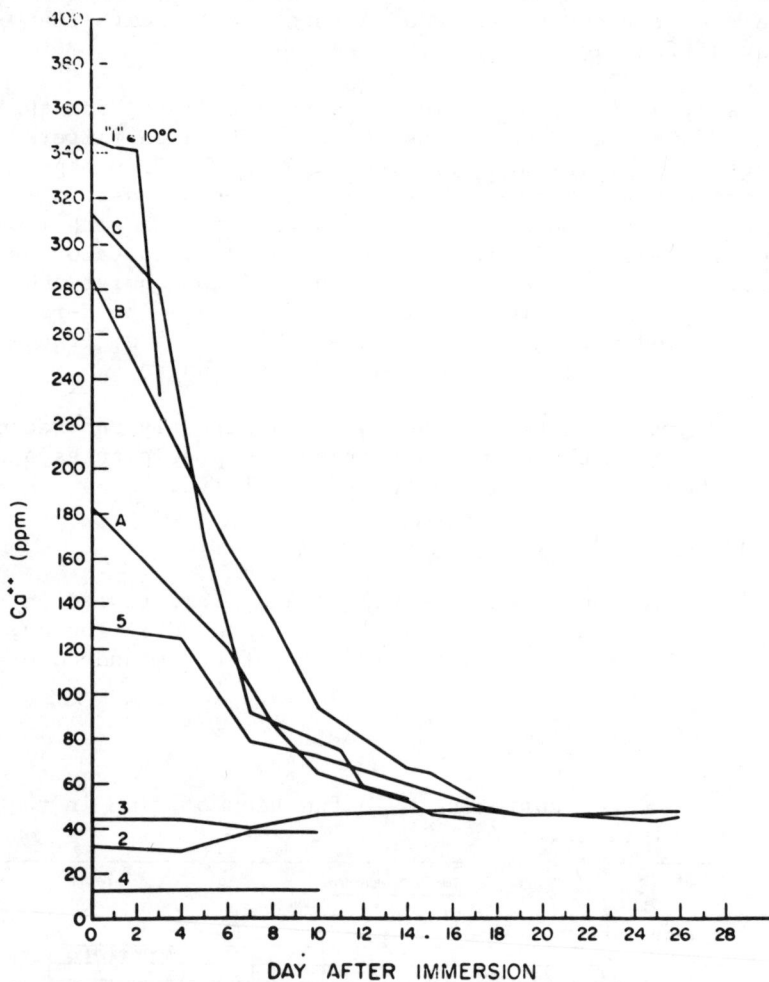

Figure 8.   Ca^{++} concentration as a function of time of immersion.   Labels refer to bag number listed in Table IV.

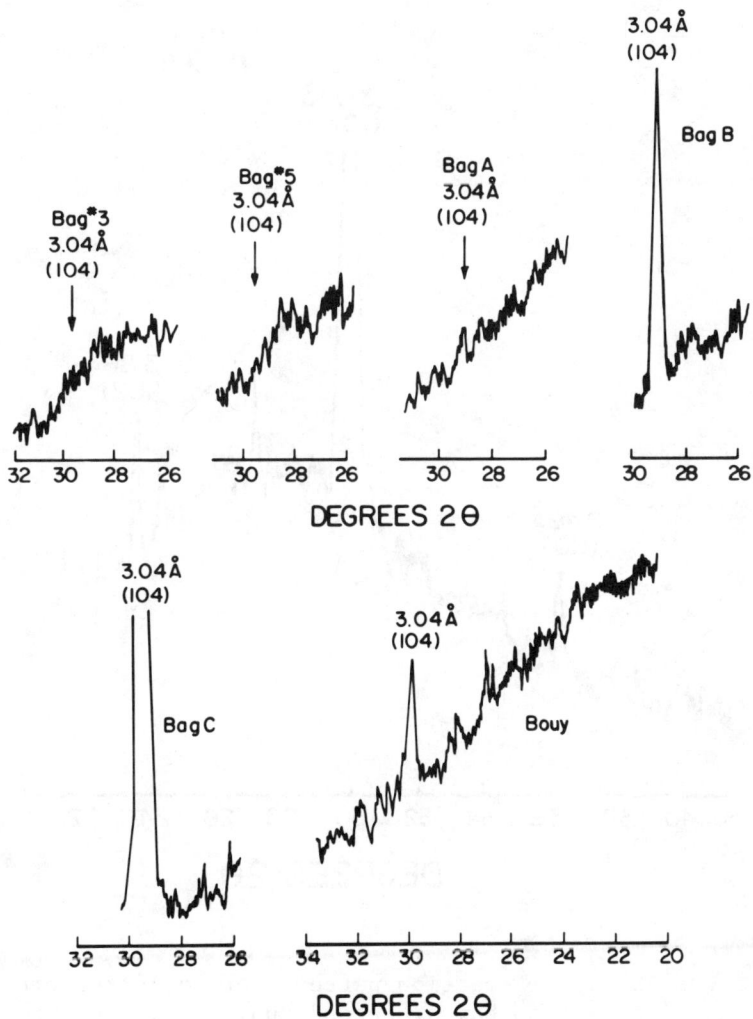

Figure 9. X-ray diffraction patterns for collected precipitate from bags 3, 5, A, B, C, and buoy.

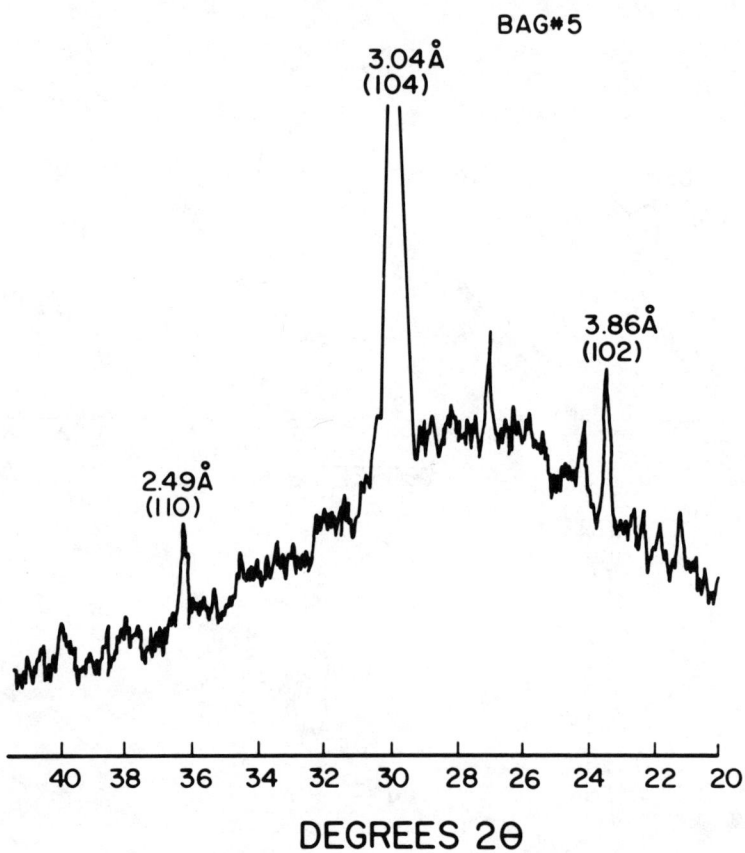

Figure 10.   X-ray diffraction pattern for collected precipitate from bag 5, second run.

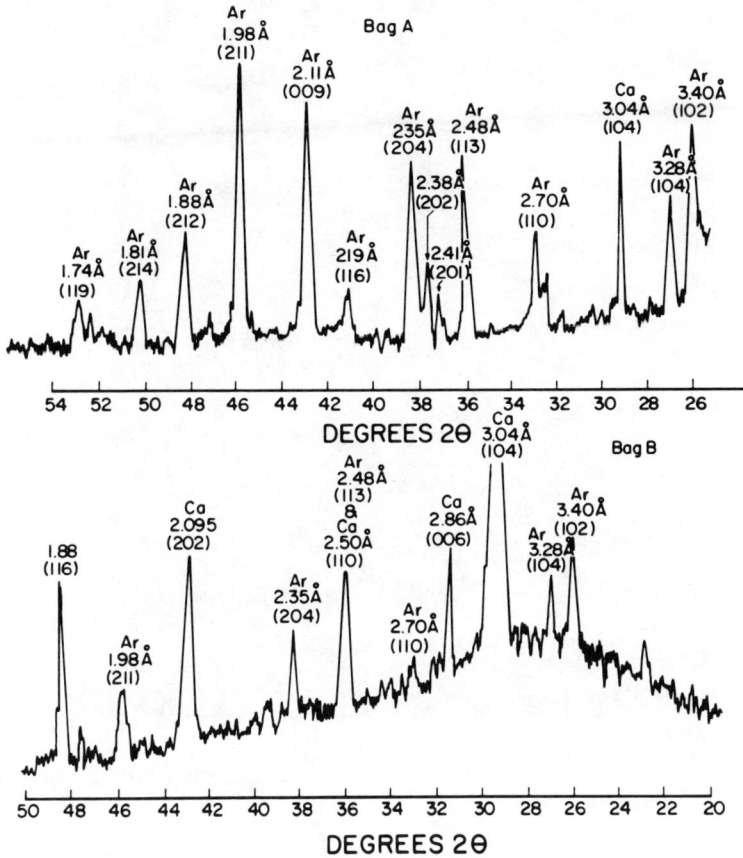

Figure 11.   X-ray diffraction patterns for collected precipitation from bags A and B, second run.

The main body of the buoy was fabricated from 15 cm
diameter polyvinyl chloride (PVC) pipe and standard PVC
fittings. Six kg of lead ballast in the bottom gave stability under windy conditions. The overall height is approximately 3.6 m. PVC pipe of 2.5 cm diameter and fittings
were attached to the side of the buoy to support the polyethylene bag and to serve as a conduit for Tygon tubing for
adding water to the bags (Figure 12). The water is filtered
through a millipore filter before entering the sampling tube.

The main buoy was surrounded by four smaller metal
buoys anchored by 6.4 mm polyethylene rope to concrete
blocks. Additional water and ballasts could be added or
withdrawn. Without water ballast, the buoy floats

487

Figure 12. Apparatus for determining calcite precipitation
in the field.

horizontally (Figure 13). When the water ballast is added,
the buoy floats vertically in the water (Figure 14).

Water treated as described earlier to insure high
calcium content, was placed in the bag. Water added on
February 26, 1975 contained 284.0 ppm Ca^{++} and the pH was
6.52. The sample removed from the bag on March 6, 1975 had
a pH of 7.5 and a concentration of Ca^{++} of 140.00 ppm.
Finally a sample was removed on March 10, 1975. The precipitate on the millipore filter was analyzed for calcite by
X-ray diffraction. The Ca^{++} concentration of the last sample
was 95 ppm and the pH was 7.5. The X-ray diffraction pattern
showed calcite to be present on the millipore filter.

It was concluded that the buoy was functioning as
planned and would be a useful tool in studying calcite
formation in lakes.

Figure 13. "Water quality" buoy. Note bag at far end.

Figure 14. Buoy floating vertically after ballasting.

LITERATURE CITED

1. Jensen, M. E. "Design and operation of farm irrigation
    systems." ASAE Nomograph Number 3. 1980.

2. Anderson, O. L., and Perking, P. C. "Some consequences
    of restricting the maximum elevation of Lake Powell."
    Lake Powell Research Project Progress Report for the
    National Science Foundation, 1973.

3. Reynolds, R. C., Jr., and Johnson, N. M. "Major element
    geochemistry of Lake Powell." Lake Powell Research
    Project Bulletin #5. National Science Foundation, RANN,
    1974.

4. Garrels, R. M., and Christ, C. L. "Solutions, minerals
    and equilibrium." Harper and Row, New York, 1965.

5. Hem, J. D. "Study and interpretation of the chemical
    characteristics of natural water." U.S.G.S. Water-Supply
    Paper #1473, 1959.

6. Doner, H. E., and Lynn, W. C. Carbonate halide, sulfate
    and sulfide minerals. In Minerals in Soil Environments.
    Dixon, J. B., and Weed, S. B. (ed.) 1977.

CHAPTER 29

NATURAL SALINITY REMOVAL IN MAINSTEM RESERVOIR: MECHANISMS, OCCURRENCE, AND WATER RESOURCES IMPACTS

Jay J. Messer
Eugene K. Israelsen
V. Dean Adams
    Utah Water Research
    Laboratory, Utah State
    University, Logan, Utah

INTRODUCTION

Interest in natural salinity removal processes in western reservoirs probably stems from the calculations that led [1] to propose that calcium carbonate precipitation was partially ameliorating the salinity leaching from the drowned sediments of Lake Mead on the Lower Colorado River. Since that time, similar mechanisms have been proposed for other reservoirs, including Lake Powell [2, 3] and Flaming Gorge [4] in the Upper Colorado River Basin, and Lake Mead in the Lower Basin [5]. Such salinity removal processes have engendered considerable excitement because of projections that each mg/l increment of salinity removed from the river will result in an annual savings downstream of $448 million in 1980 dollars [6].

These analyses led Messer et al. [7] to investigate alternative mechanisms by which natural salinity removal in reservoirs may occur, along with the principal controlling factors. Because of its proximity to the laboratory, these studies were conducted on Oneida Narrows Reservoir, a small hydroelectric reservoir on the Bear River in southeastern Idaho, that a previous study [8] had shown to act as a salinity sink. Because the results of this study may have some applicability to Colorado River resources, they will be summarized briefly below. However, the principal objective in this chapter will be to discuss shortcomings of previous salinity removal case studies and their implications for reservoir management.

We hasten to indicate at the outset that it is not our intention to discredit the work of previous investigators in

491

this area of research. Salinity mass balance modeling, pre-
and postimpoundment, usually necessitates working with data
from preexisting monitoring programs that seldom were
designed to provide the most useful information. Furthermore, the large volumes and high shoreline development
indices of Colorado mainstem reservoirs require that even
cursory sampling trips to determine accurate ion storages
be measured in days, rather than hours. Similarly, salinity
impact studies on crops and plumbing fixtures have focused
(understandably) on differential effects of highly saline
water, rather than the more subtle nuances of mg/1 increments
of sodium or calcium ions. What we wish to demonstrate here
is that previous salinity budget calculations have contained
sufficient model error and data uncertainty to put the true
water quality significance of this mechanism into considerable
doubt. Furthermore we suggest that, although calcium
carbonate precipitation undoubtedly occurs in reservoirs,
its importance as a management technique for controlling
downstream salinity may be limited.

POSSIBLE MECHANISMS FOR NATURAL SALINITY
REMOVAL IN RESERVOIRS

Messer et al. [7] have proposed four biogeochemical
process categories that potentially could remove salinity
from surface waters during transit through reservoirs.
These mechanisms (shown schematically in Figure 1) include
precipitation, coprecipitation, coagulation, and bioassimilation. Precipitation involves congruent processes in which
two or more dissolved species react to form a more or less
crystalline precipitate. Coprecipitation involves incongruent
processes in which solid phases participate as reactants and
products, including ion exchange. Coagulation may be purely
physico-chemical, in response to increasing ionic strength,
or may involve bridging by biopolymers secreted by aquatic
bacteria [9]. Biochemical assimilation includes incorporation of dissolved ions into shells, bones, tests, and other
biogenic structures with long lifetimes relative to sedimentation rates.

Results of an extensive literature review, thermodynamic
modeling, in situ and laboratory studies of the four biogeochemical categories led Messer et al. [7, 10] to conclude
that calcite precipitation, together with some bioassimilation of silica by diatoms, were the only significant salinity
removal mechanisms operating in the reservoirs studied.
Although one laboratory experiment with Oneida Narrows
Reservoir water suggested some magnesian calcite precipitation, no evidence was found for participation of sodium,
potassium, or silica ions in congruent reactions. These
observations are consistent with those of Jones and Bowser

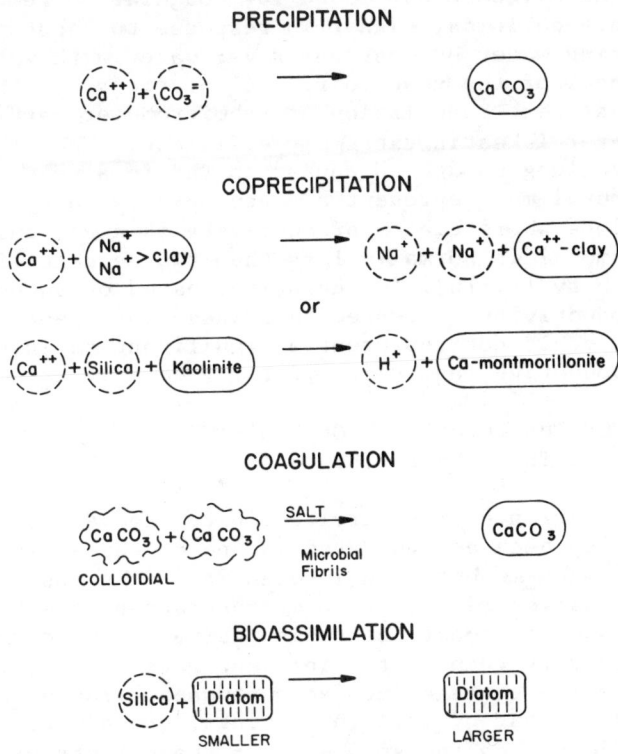

PRECIPITATION

$Ca^{++} + CO_3^{=} \longrightarrow Ca\,CO_3$

COPRECIPITATION

$Ca^{++} + \left(\begin{matrix}Na^+\\Na\end{matrix}\right. > clay \longrightarrow Na^+ + Na^+ + (Ca^{++}\text{-clay})$

or

$Ca^{++} + Silica + Kaolinite \longrightarrow H^+ + (Ca\text{-montmorillonite})$

COAGULATION

$Ca\,CO_3 + Ca\,CO_3 \xrightarrow[\text{Microbial Fibrils}]{\text{SALT}} CaCO_3$

COLLOIDIAL

BIOASSIMILATION

$Silica + \boxed{Diatom} \longrightarrow \boxed{Diatom}$

SMALLER   LARGER

Figure 1. Example of Biogeochemical Processes That Could
           Potentially Transfer Dissolved Solids (Dashed
           Boundaries) to a Particulate Phase (Solid
           Boundaries) in Natural Waters.

[11], who found no evidence for clay mineral authigenesis
occurring in the water column of freshwater lakes. Both
laboratory experiments and thermodynamic calculations
indicated the absence of significant coprecipitation salinity
sinks, including ion exchange, in the Oneida ecosystem.
Indeed ion exchange was shown to act as a small salinity
source under some conditions.

     Natural waters contain many types of particulates held
in dispersion, which have diameters small enough to pass
through a standard glass fiber filter [12]. Glass fiber
filters typically pass 50% of the particles as large as
0.7-0.9 μm [13] which thus appear as total dissolved solids
[$\equiv$ filterable residue, 14] in the analysis. Any mechanism
which results in the coagulation of these particles would
thus result in the apparent removal of salinity, even though

there would be no actual reduction in the truly dissolved species. No evidence was found for coagulation removing nonfiltrable colloids, either in response to inputs of saline spring water or upstream river water with water from the main pool of the reservoir. Calculations by O'Melia [15] suggest that coagulation of submicrometer particles in freshwater lakes is extremely efficient. Thus they may not survive long enough to appear as TDS in a reservoir. Silica removal may represent a measurable salinity sink in situations where diatom productivity is high, but the overall removal is normally less than two percent of the total salinity [7, 16]. Calculations based on annual ecosystem productivity suggested that insect emergence and fish emigration could not represent a significant salinity sink, even in productive reservoirs [7].

EVIDENCE FOR THE OCCURRENCE OF SALINITY REMOVAL IN WESTERN RESERVOIRS

The occurrence of natural salinity removal in reservoirs has generally been demonstrated in one of three ways. The two most common methods are inferential. The easiest (and least conclusive) method is to demonstrate a lower concentration of dissolved constituents in the tailwater downstream from a reservoir than in the influent water. This method generally fails to take into account concentration or dilution brought about by influent springs, groundwater seepage, local precipitation and snowmelt, or the effects of evaporative loss from the lake surface.

A more conclusive demonstration of salinity removal can be obtained through the use of a mass balance on the constituents of interest. Such a balance employs a continuity equation of the form

$$\Delta m_{i_R} = \Delta \ \Sigma c_{i_R} D_{i_R} = \Sigma c_{i_{IN}} D_{i_{IN}} - \Sigma c_{i_{OUT}} D_{i_{OUT}} - \Delta c_S D_S \qquad (1)$$

where m_i = the mass of the i'th constituent; c_i = the concentration of the i'th constituent over some time period, t; and D_i equals the volume of water represented by c_i. The subscripts R, IN, OUT, and S refer to removal, input, output, and storage, respectively. The input terms should include all inflow from the main stem, precipitation, snowmelt, seepage, and dry deposition ($D_i=0$). Output terms include tailwater and evaporation losses ($c_i=0$). The storage term should be measured accurately at the beginning and end of the period in question, and should include bank storage that can actively exchange ions with the water column.

In practice, it is seldom practical to measure these terms accurately within the budget of most projects.

Typically only the major tributaries and reservoir tailwater
are monitored for chemical fluxes with sufficient accuracy
and frequency to give reliable discharge values. Even in
these cases major ions are seldom sampled more frequently
than monthly. Chemical sampling stations usually are
located at USGS stream gages, and are often upstream or
downstream of the ideal locations for establishing a mass
balance [5]. Because such stations are also located above
river reaches which are not part of the reservoir, riverine
processes may be confused with reservoir processes. Precipitation and evaporation are seldom gaged with precision
and analyses of dry deposition are seldom available, although
such sources can contribute significantly to lake mineral
budgets in arid areas [15]. Groundwater movements and
chemistry are seldom known with accuracy, and minor tributaries are seldom gaged. Although in western states, the
proportional contributions of such sources to lowland reservoirs probably is small, they may be highly mineralized and
contribute disproportionately to the salt load. The converse
may be true in upland reservoirs. Hydraulic outputs are
often better known, because gaging stations are conveniently
located below dams.

    The final method for demonstrating salinity removal is
deductive in nature: to observe the occurrence of the
suspected process itself. Project budget limitations and the
desire to manipulate individual variables have restricted
most of this type of work to laboratory studies. These may
be restricted to water column models [18] or may include
sediment phases [7]. In situ studies include observations
of calcite formation [19], silica precipitation in the form
of diatom frustules, and sediment trap studies in lakes [20],
but such investigations have been relatively rare in western
reservoirs. A variant on this process is the analysis of
depositional history through sediment core analysis [21].

    In the following sections, we shall review some of the
previous studies conducted on western reservoirs from which
evidence of natural salinity removal was inferred. In some
cases, the primary intent of the writers was not to demonstrate salinity removal, but their data were used to draw
tentative conclusions toward that end. The locations of these
reservoirs are shown in Figure 2. They include: Bighorn
Lake on the Bighorn River on the Wyoming-Montana border;
Canyon Reservoir on the Guadalupe River in Texas; Flaming
Gorge Reservoir on the Green River on the Utah-Wyoming
border; Lake Powell on the Upper Colorado River in southern
Utah; and Lake Mead on the Lower Colorado River on the
Nevada-Arizona border. Also shown in Figure 2 is the location
of Oneida Reservoir on the Bear River in southern Idaho.
The salinity budgets of Gloss et al. [3] for Lake Powell and

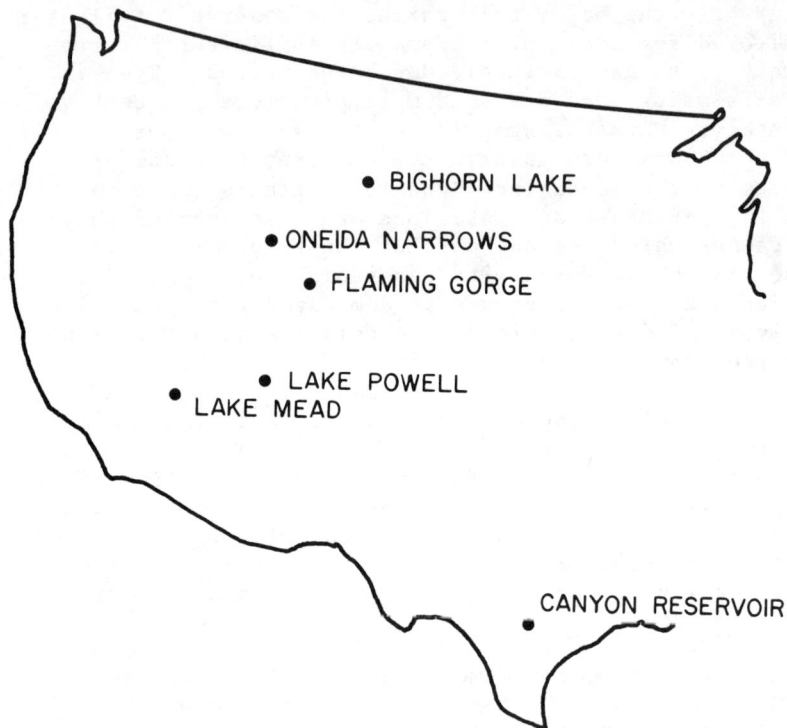

Figure 2. Location Map of Reservoir Studies.

Paulson and Baker [5] for Lake Mead became available too
late for a thorough evaluation.

Lake Mead

As pointed out previously, Howard [1] was the first to
note the apparent importance of salinity removal mechanisms
in a western reservoir. Howard used both concentration
data and an incomplete mass balance. He noted that in the
summer following closure of Hoover Dam in February 1935,
total dissolved solids (TDS) concentrations became lower at
the Willow Beach gage, 10 miles below the dam, than at Lee's
Ferry, Arizona, approximately 355 miles above the dam. He
calculated that during the water years 1935-1948, 127 Gg
(140 x 10^6 tons) of TDS were carried past the Lee's Ferry
station, and that another 7 Gg (8 x 10^6 tons) were added
by unmeasured tributaries above Lake Mead. Output was
measured at the Willow Beach gage between 1935 and 1939, and
immediately below the dam thereafter. The TDS in the
outflow was 123 Gg (136 x 10^6 tons), resulting in a net
decrease of 11 Gg (12 x 10^6 tons). Storage was calculated

to be 20 Gg (22 x 10[6] tons), indicating a net increase in TDS
of 9 Gg (10 x 10[6] tons) of TDS.

Although Howard does not describe his calculations, he
estimated that 8 Gg (9 x 10[6] tons) of calcium carbonate
and 1 Gg (1 x 10[6] tons) of silica precipitated in the lake,
citing deposits of calcium and silica around the lake littoral
as supporting evidence. From increases in sulfate and
chloride output below the dam, however, Howard calculated
that this removal of calcium and silica was overshadowed by
the dissolution of gypsum, together with a smaller amount of
halite, totaling 18 Gg (20 x 10[6] tons). Slight increases
in magnesium, potassium, and nitrate concentrations were
consistent with evaporative concentration, which accounted
for 5% of the average annual inflow.

Howard thus suggested that approximately 6% (9 ÷ 152)
of the total input of TDS to Lake Mead, including internal
loading, is removed by calcium carbonate and less than 1% by
silica precipitation. Data on inflows and outflows were
taken from weighted annual average concentrations published
by USGS. Neither the number of data points used to calculate
fluxes or storage, nor the method of calculating calcium or
silica precipitation or dissolution is presented. It is
worthy of note that the calculated removal rate is similar
in magnitude to the estimated value for input below Harper's
Ferry, and that salinity may vary by a factor of two over
space and depth in Lake Mead over a 4-day sampling cruise
[22]. Either of these data uncertainties could substantially
affect the calculated salinity removal.

Bighorn Lake

Studies by Soltero et al. [23] on Bighorn Lake, a
multipurpose reservoir on the Bighorn River, Montana, are
illustrative of the potential errors inherent in flow
weighted salinity calculations. Soltero et al. [23] published discharge-weighted major ion concentrations of the
inputs and outputs streams for the reservoir during 1968 and
1969, that indicated removal rate of 5-15% for the major
cations and anions (Table I). However, hydrologic budgets
were also given by Soltero et al. [23], from which mass
balances could be calculated. The results of these calculations, also shown in Table I, indicate slight increases in
most major ions as a result of passage through the reservoir
except for sodium and chloride which showed small negative
changes. These calculations do not include a storage terms,
and thus the contributions are in reality even more postive.

In a subsequent paper Soltero et al. [24] discovered
that ungaged snowmelt along the sides of the reservoir was

Table I. Major Ion Budgets for Bighorn Reservoir, Montana, during 1968 and 1969 [based on data from Soltero et al. 23]

Constituent	Flow Weighted Average Basis		Mass Balance Basis		Difference (%)
	Changes (%)		Input	Output	
	1968	1969	10^9 Kg		
Sodium	−10	0	33.0	32.4	−2
Calcium	−1	−3	30.5	31.1	+2
Magnesium	−12	+4	9.55	9.63	+1
Chloride	−9	−14	4.7	4.4	−6
Bicarbonate	−1	−1	75.9	78.5	+3
Sulfate	0	+2	111	116	+4

diluting the lake water, which resulted in the original apparent decrease in salinity. Indeed, Soltero et al. [23] had noted in the first paper that calcium carbonate precipitation was unlikely to be important in the reservoir because the Ca^{++}/Mg^{++} ratio did not change between the influent and effluent waters. We emphasize that the purpose of the comparison in Table I is for heuristic purposes, and not to cast aspersions on the cited work. The original purpose was to describe reservoir biology and chemistry, and not to prove that salinity removal was or was not occurring.

Canyon Reservoir

Studies by Hannan [25] on Canyon Reservoir, an oligo-mesotrophic deep storage reservoir on the Guadelupe River in central Texas, are interesting inasmuch as they present circumstantial evidence for calcite precipitation as a natural salinity removal process. Hannan and Young [26] found that bicarbonate ion concentrations decreased near the center of the reservoir, which the writers associate with the areas of highest chlorophyll a concentrations. Examination of their data (p. 189), however, does not seem to justify such a simple relationship. Specific conductance also decreased between an upstream station and the dam, mostly in the upstream, riverine reach of the reservoir. The writers again attributed this effect to biogenic carbonate precipitation, but without offering either proof or calculations.

Hannan and Broz [27] extended the earlier work, and reported the following average changes in concentrations between a station upstream for the reservoir and the tail-race: alkalinity, −20 to 30%, specific conductance, −34%,

calcium, −25%, magnesium, −23%, sodium, −20%, potassium, +35%.
No chloride or sulfate data were presented. There are no
seasonal trends apparent in any of these concentration
changes, which might be expected if the principal removal
mechanism were calcium carbonate precipitation. Hannan et al.
[28] presented more detailed isopleths, and noted that
bicarbonate reduction was positively correlated with primary
productivity rates, although reference was made to an unpublished thesis, and no data were presented.

In summary, these studies gave circumstantial evidence
of calcium carbonate acting to remove salinity from Canyon
Reservoir. No mass balance was performed, however, and
seepage of less saline water through an alluvial formation
crossed by the reservoir could provide some dilution. Normal
precipitation is low (< 16 cm/yr) in the area, which minimizes the impact of rainfall as a diluent. No mechanisms
for sodium or magnesium removal were suggested.

Flaming Gorge Reservoir

Closing of the Flaming Gorge Dam on the Green River in
northern Utah in 1962 led to an increase in the flow-weighted
average TDS concentration in the outfall, which led Bolke
[4] to construct a salt balance for the reservoir. The
TDS load immediately below the dam site was related to the
load from three gaged upstream tributaries (Green River,
Black's Fork, and Henry's Fork) during 1957-1962 prior to
closure of the dam, the difference being ascribed to unmeasured salt sources (nonpoint flows, minor tributaries,
springs, etc.). The precipitated/leached salt load was
calculated using the mass balance:

$$LL = DSL + OTL + DLF - 1.20 \ MINL \tag{2}$$

in which LL is the net gain or loss from biogeochemical
reactions, DSL is the change in water column storage, OTL is
the measured outflow load, DLF is a storage term for the
smaller Fontanelle Reservoir, downstream from the gaging
station on the Green River, and MINL is the measured inflow
load. The nature of the calculation of the DLF term is not
clear from Bolke's discussion, nor is the reliability of the
data used to calculate DSL apparent.

The net biogeochemical loading rate, LL, was calculated
for eight periods of various lengths, based (presumably) on
the availability of data with which to calculate DSL [29].
The results indicated that, between the closing of the
reservoir in 1962 and the end of the study in 1975, a net
release of 1.95×10^6 metric tons of TDS to the Green River
resulted following construction of the reservoir. The

pattern of TDS release was 1) high during initial filling (3 yr) of the reservoir, 2) low during a following period of declining water levels (3 yr), and 3) a return to high rates during subsequent filling to design capacity. Bolke suggested that the increases were primarily due to leaching of salts from newly inundated rocks, and that loadings would probably decrease in the future.

Bolke [4] also calculated a dissolved ion budget for the period September 1972 to September 1975. The net changes in ion loads (LL) are shown in Table II. Modest increases in sodium, magnesium, and chloride were accompanied by a large increase in sulfate, a large decrease in bicarbonate, and no change in calcium concentration. Bolke [4] reasoned that if the increase in sulfate load were caused by gypsum dissolution, then 190×10^6 kg of calcium would be brought into solution, which would in turn precipitate 280×10^6 kg of carbonate in order to realize the zero balance of calcium inputs and outputs noted in Table II. Bolke then noted that the dissolved bicarbonate calculated in the load budget was 390×10^6 kg, thus leaving 110×10^6 kg unaccounted for. Bolke's calculations indicate that as much as 10% of the TDS (external plus internal) load to the reservoir may be lost by calcium carbonate precipitation.

While such calculations are worthwhile, reference to ion budget data (Table II) reveals some disturbing (and perhaps

Table II. Accumulation or Release of Ion Loads in Flaming Gorge Reservoir between September 1972 and September 1974 (modified from [4]).

	Accumulation (−) or release (+)		
	$\times 10^6$ Kg	$\times 10^9$ equivalents	
Calcium	0	0	
Magnesium	70	6	+8
Sodium	40	2	
Bicarbonate	−390[a]	−6	
Sulfate	460	10	+5
Chloride	30	1	
Total (sum of constituents)	+220		
TDS	+564		

[a]This value is misprinted as -0.49×10^6 metric tons in Bolke (p. 32).

typical) problems. The calculated output of cations is
almost twice the calculated output of anions, which violates
charge balance conditions. Furthermore, the twofold discrepancy between TDS (residue) and sum of constituents
salinity budgets reveals the necessity for basing calculations on the most reliable data. Which data these are in
this case is difficult to guess, inasmuch as TDS by the
residue method is usually smaller than sum of constituents
due to loss of volatile constituents, especially in alkaline
waters [30]. This example particularly points to a need for
clear display of data and calculations, so that such discrepancies can be noted and explained. Inasmuch as USGS
exercises careful analytical quality control, the discrepancy
here is likely to result from inadequate models relating
input or output loadings to flow.

Lake Powell

The intensive NSF-RANN study of Lake Powell during the
1970's led to accumulation of a considerable biogeochemical
data base for this large reservoir on the Utah-Arizona border.
This data base allows for a demonstration of the possible
contributions of model error and data error to the results
of mass balance studies. In an early study, Reynolds and
Johnson [2] determined that between 8 and 12% of the bicarbonate load of the influent water was being removed in Lake
Powell, which would correspond to a salinity reduction of 19
to 29 mg/ℓ, or approximately 5% of the 1965-1970 average
salinity of 500 mg/ℓ. The methodology involved assumptions
similar to those used for our Lake Powell salinity budget
that will be described at the end of this section. Corrobera-
tive evidence was gained from the fact that the surface water
was everywhere supersaturated with respect to calcite, and
subsequent laboratory experiments by Reynolds [18] tended
to support calcite precipitation as an important salinity
removal mechanism. Mayer [16] calculated a net loss (input
minus output) of 1.3 to 2.9 mg/ℓ of silica annually, with
the higher value necessitating considerable bank storage of
dissolved Si. Thus silica removal does not contribute
substantially to salinity removal in Lake Powell.

Because no detailed calculations were published in the
earlier study, we set out to calculate a salinity balance for
Lake Powell using an approach similar to that of Bolke [4] on
Flaming Gorge. Inputs were calculated using a linear regression model that relates the total annual salinity load
from five streams tributary to the reservoir (Figure 3) to
the total salt load passing a downstream station (Lee's Ferry,
Figure 3) prior to closing of the Glen Canyon Dam in 1963.
According to Iorns et al. [31], these influent stations
contribute 96.8% of the salt load to the reservoir. Data

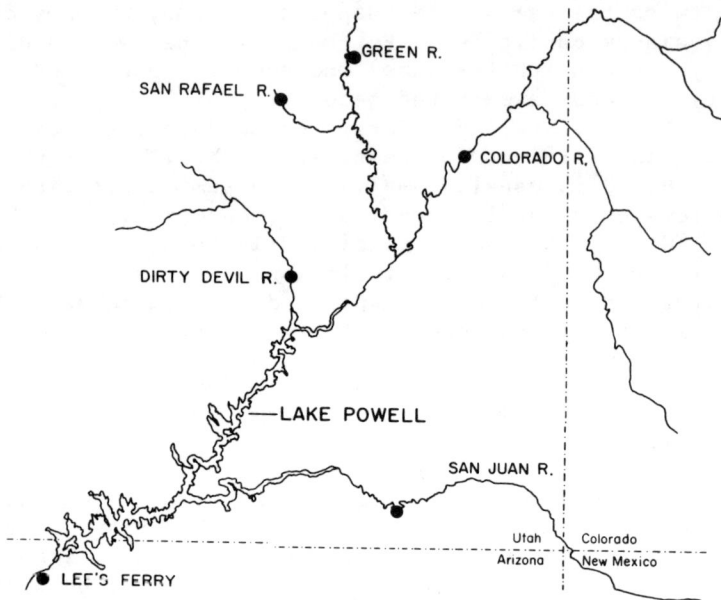

Figure 3. USGS Gaging Stations Used to Construct the
Lake Powell TDS Mass Balance

for salt loads were taken from Bureau of Reclamation records
[6] for the water years 1941-1963 and are shown in Table III.

The linear model took the form

$$TDS_{OUT} = 0.446 + 1.02\ TDS_{IN} \qquad\qquad (3)$$

where TDS loads are expressed in 10^6 metric tons/yr. The
regression accounted for 91% of the variation in annual
salinity output at Lee's Ferry. Assuming that closure of
the dam had no effect on salinity levels in the river,
Equation 3 predicts that total salt output for the 13 water
years 1963 through 1975 at Lee's Ferry should have been 99.1
x 10^6 metric tons, whereas the observed salt load passing
the station totaled only 76.2 x 10^6 metric tons, leaving
22.9 x 10^6 metric tons apparently stored in the reservoir
(Table IV).

The actual storage of salt in the water column can be
estimated with fair accuracy. Total dissolved solids concentrations were calculated by integrating constituent ion
isopleths measured by Merritt and Johnson on a sampling trip
at the end of September in 1975 [32]. The volume weighted

502

Table III. Salinity Inputs and Outputs to Lake Powell During
1941-1975 Water Years (Based on [7])

Year	Input 1000 metric tons	Output 1000 metric tons
1941	10720	11649
1942	9020	8510
1943	7471	7598
1944	7513	7734
1945	7182	7712
1946	6102	6664
1947	8121	8630
1948	7300	7739
1949	8404	9030
1950	7073	7347
1951	6672	7106
1952	9849	10338
1953	6446	6790
1954	5463	5793
1955	5372	5940
1956	5529	5909
1957	9517	11472
1958	7382	8419
1959	5391	6138
1960	5537	6434
1961	5441	6409
1962	7835	9361
1963	4920	1595
1964	6007	3246
1965	9083	8172
1966	6223	4934
1967	7126	5794
1968	7427	7000
1969	7969	7173
1970	6967	6314
1971	6653	6573
1972	6472	6539
1973	8967	6584
1974	6506	5978
1975	7098	6399

Table IV. Total Dissolved Solids (TDS) Balance for Lake
           Powell, Colorado River (all values in 10^6 metric
           tons)

---

1. TDS output at Lee's Ferry predicted based on a
    model for water years 1941-1962 ($\pm$ ISD,   = 0.1)

    $TDS_{out} = 0.446 + 1.02\ TDS_{in}$ 99.1 $\pm$ 3.2

    $r^2 = 0.91$, n = 22

2. TDS output observed at Lee's Ferry   -76.2
    (USGS surface water records)

3. TDS in water column
    (Based on integration of sum of   -14.8
    constituent ion isopelths,
    September 1975 [32])

4. TDS in bank storage
    (560 mg/1 historical flow-weighted   -6.2
    TDS, 1963-1975, x 11 X 10^9 m^3)

5. Discrepancy between predicted output and $\overline{+1.9}$ (2%)
    observed output plus storage

---

mean TDS concentration for the reservoir at this time was 547
mg/1. Multiplied by the storage volume of 2.76 x 10^{10} m^3,
this leads to a total water column storage of 14.8 x 10^6
metric tons corrected for a 1.5% decrease in water volume
due to sediment storage. This leaves 8.1 x 10^6 metric tons
of salt, or 8% of the calculated input, to have been removed
either in bank storage or by some natural removal process.

     Bank storage of salinity, the most frequency neglected
term in salinity budgets, depends on the movement of dissolved
salts relative to stored water, in cracks, crevices, and
permeable formations inundated by the reservoir water. Bank
storage also includes interstitial water associated with
sediments deposited in the reservoir since its filling.
Bank storage cumulatively increases over the life of a
reservoir, although bank storage may decrease during years
of reservoir drawdown [33].

     Qualitatively, one may envision that, in a filling
reservoir, if the hydraulic migration rate of water away
from the reservoir exceeds the molecular diffusion rate
back toward the reservoir of salts released to the water

from the flooded minerals, bank storage may act as a sink for
salinity. This is most likely to be the case during initial
filling of the reservoir. As the reservoir reaches steady
state, or during drawdown, salts are free to leach out (or
are washed out) of interstices in the rocks and sediments,
and bank storage may act as a salinity source [33]. If we
assume that transported sediment is near equilibrium with
respect to the reservoir water at the time of deposition,
less leaching might be expected from these sources than from
newly inundated strata in the banks.

Although the magnitude of bank storage and the concom-
mitant transport of salinity cannot be resolved with information presently available, a crude calculation offers some
perspective. Bank storage in Lake Powell in September 1975
was approximately 11×10^9 m^3 [34]. If we assume an average
historical flow weighted TDS concentration of 560 mg/l was
stored with this water (i.e., no biogeochemical interactions
occurred), the storage of salt would be 6.2 metric tons,
which would account for most of the remaining discrepancy
between input and output, leaving only 1.9% of the input
unaccounted for (Table IV). Although this calculation is
likely to represent an absolute maximum storage because it
neglects biogeochemical salt loading from drowned rock and
soil formations and stored sediments, it suggests that massive
biogeochemical salinity removal is in no way conclusively
demonstrated by the salinity mass balance.

The mass balance displayed in Table IV also contains
important model and data error terms. For example the error
in the estimate of the 1963-1975 salinity export at Lee's
Ferry is $\pm$ 3.2 x 10^6 MT (α = 0.10, 2 tailed test), a value
much larger than the remaining discrepancy in the salt
balance. Additionally, there is data error associated with
the model tabulation converting daily EC measurements to
salt loads [6], which are not corrected for seasonal changes
in water chemistry which affect the EC:TDS relationship
[32]. Also, salinity records for the San Rafael station
were estimated during 1941-45, and for the Lee's Ferry
station during 1941-42 and 1946-47 because continuous EC
readings were not available before 1970, and loads in these
years were based on composited USGS monthly samples [6].
With respect to water column storage, no lateral transects
appear to have been run, which may somewhat alter the
storage estimate, especially if horizontal variations are
similar to those encountered in Lake Mead [22].

Summary of Reservoir Studies

Although the authors of reservoir studies in the semi-
arid western states have frequently suggested that salinity

505

removal may occur, the magnitudes are usually small and within the range of error of the calculations. High estimates are 6% for Lake Mead, 3 to 8% for Lake Powell, 10% for Flaming Gorge, and a net gain of 3% in Bighorn Reservoir. Decreasing ion concentrations were observed in Canyon Reservoir, Texas, but no mass balance was performed. Most published mass balance calculations are plagued by poor cation-anion balances, hidden calculations, and unexpressed or unsupported assumptions. In the case of Lake Powell model error, sampling error in determining salt storage in the water column, and failure to account for bank storage are sufficient to account for the apparent salinity removal.

MANAGEMENT IMPLICATIONS FOR NATURAL SALINITY REMOVAL PROCESS

In addition to being a limnological curiosity, natural salinity removal processes in reservoirs have been of particular interest to water resources managers because of an estimate by Kleinman and Brown [35] that salinity removal in the Colorado Basin is worth $448 million (1980 dollars) per mg/l increment removed at Imperial Dam. This calculation was based on a regression of anecdotal accounts of replacement rates of plumbing fixtures and appliances against TDS of water supplies, or upon LP modeling of effects of increasing salinity from 900-1400 mg/l upon management actions or damage functions in the agricultural sector. All calculations assume that 1) without mitigation the salinity at Imperial Dam will reach 1225 mg/l; 2) municipal costs exhibit linear increments of $240,500 per mg/l (1976 dollars) over the 800-1400 mg/l range of salinity; and 3) that agricultural costs can be directly related to TDS and exhibit an indirect cost multiplier of 5.32 [35]. While the foregoing analysis of economic impacts of salinity in the Colorado is likely to be the best possible given the available data, the savings resulting from the differential removal of calcium carbonate must be interpreted with some caution.

Impacts of Calcium Carbonate Removal

As to damage to materials, corrosion is the normal response to increasing concentrations of monovalent cations and the nonalkaline anions, while carbonate hardness is associated with scaling of pipes, boilers, water mains, and so on. Because municipal costs are based on replacement of corroded or scaled fixtures, the removal of calcium (relative to sodium) may increase or decrease the $240,500 per mg/l cost, depending on the relative damage of scaling versus corrosion, which was not considered in the analysis.

If water treatment were upgraded to avoid damages, however, natural calcium removal would be much less valuable than sodium removal if damages from the two effects are otherwise assumed to be equal. While hardness and carbonate alkalinity can be removed through conventional lime-soda softening, sodium, chloride, and sulfate must be removed through more expensive desalination procedures such as reverse osmosis. Thus although reduction of salinity involves the same thermodynamic cost per mole of solute removed (usually measured using the corresponding change in vapor pressure of the solution), some removal processes require higher quality energy and more intensive (and thus expensive) technology.

Agricultural damages present thornier problems. The value of irrigation water depends not only on its salinity, but also on the ratio of sodium to divalent cations. Because sodium causes alkalization and loss of tilth of clayey soils, reduction of salinity through removal of divalent ions alone, without a corresponding decrease in sodium, will increase the exchangeable sodium percentage and thus harm the soil structure. The sodium absorption ratio, SAR [36], has long been used in the U.S. to quantify this phenomenon. The ratio:

$$SAR = \frac{meq/l \ Na^+}{\sqrt{\dfrac{meq/l \ Ca^{++} + Mg^{++}}{2}}} \qquad (4)$$

should not exceed 6 in the range of salinity resulting in an electrical conductivity of 750 μmho/cm. Reynolds and Johnson [2] report values of approximately 4 meq/l for sodium and calcium, and 2 meq/l for magnesium in Lake Powell in summer. If half of the calcium were removed by precipitation (representing a decrease in 100 mg/l salinity), the SAR would be increased from 2.3 to 2.8, a value still well within the low salinity range. Nonetheless, removal of calcium relative to magnesium will produce a poorer quality irrigation water, despite its lower salinity, in that more must be applied in order to enjoy the same benefit when leaching to reclaim saline soil. For the sake of argument, if we assume that removing calcium alone results in no benefits, then the savings are reduced 30% (agricultural losses plus indirect effects) to $313,400 (1980) per mg/l (cf. Kleinman and Brown [35, pp. 18–19]. Nonetheless, the savings per mg/l remain substantial.

Management of Calcium Carbonate Removal

Assuming then that calcium carbonate removal is desirable, the method by which it is to be removed requires

507

some thought. Equilibrium modeling of Oneida Narrows Reservoir water [7] indicates that temperature and pH appear to be the factors which most strongly affect the saturation index of calcite. Table V indicates the results of methodically varying temperature, pH, and various ions of interest which either contribute to or compete for calcium and carbonate ions. Temperature acts to force the inorganic carbon equilibrium toward carbonate, and less substantially to decrease the calcite solubility product, while pH also contributes to the former process. Calcium and alkalinity of course provide the constituent ions of calcite, while magnesium competes for carbonate through ion pairing. Sulfate and chloride primarily interfere with precipitation through the diverse ion effect, although their impact is relatively slight. Other considerations are listed in Table VI.

The analyses in Tables V and VI suggest that promising methods of calcite removal include maintaining a high pH through encouraging photosynthesis, and maintaining a high surface temperature. The latter desideratum would suggest hypolimnetic release of cold water. However, this practice may lead to increased salinity concentration through evaporation, and to loss of phosphorus which is apparently lost by advection to the hypolimnion through adsorption onto solids, thus reducing photosynthetic potential in the epilimnion [3]. Conversely, mechanisms that would mix phosphorus rich hypolimnetic water into the epilimnion, thus encouraging photosynthesis, would reduce epilimnetic temperatures, and consequently calcite supersaturation. Raising the pH through photosynthesis also has the likely effect of encouraging dominance by blue-green algae [37]. Indeed, perhaps the best alternative is to seek a method for reducing the concentrations of phenolic organic compounds that inhibit the efficient precipitation of calcite at a given supersaturation ratio [18]. Tradeoffs between salinity reduction thus must be weighed against recreational, aesthetic, and ecological losses resulting from eutrophication and hypolimnetic deoxygenation.

Additionally, one must speculate on the biogeochemical homeostatic mechanisms that may be operating in the Colorado River reservoirs. Gypsum dissolution has been suggested as a major contributor to sulfate loading in Lake Mead [1] and Flaming Gorge [4]. Although kinetic and more complex equilibrium factors undoubtedly make such simple approach naive, precipitation of calcite may increase the rate or extent of gypsum dissolution through removal of the common ion calcium. The apparent decrease in gypsum dissolution accompanying reduction of calcium carbonate precipitation in Lake Mead [5], may be indicative of such a process, although effects of

Table V. Effect of Varying Physical and Chemical Parameters of Oneida Narrows Reservoir Water on the Calcite Saturation Index as Determined by Equilibrium Modeling. The asterisked value was held constant when evaluating the remaining variables

Parameter	Value	Calcite Saturation Index
Temperature	0 C	1.32
	5	1.65
	10	2.05
	15	2.55
	20*	3.22
	25	3.89
pH	7.5	0.52
	7.7	0.82
	7.9	1.29
	8.1	2.04
	8.3*	3.22
	8.5	5.07
Calcium	59 mg Ca/l	2.67
	84*	3.22
	109	3.63
Alkalinity	176 mg/l as $CaCO_3$	2.28
	252*	3.22
	328	4.14
Magnesium	6.6 mg/l	3.63
	16.6*	3.22
	26.6	2.90
Sulfate	16 mg/l	3.25
	40*	3.22
	64	3.19
Chloride	10.2 mg/l	3.23
	25.6*	3.22
	64	3.21

Table VI. Factors Which May Affect Carbonate Precipitation
in Western Reservoirs (greatly modified from [21])

Equilibrium factors
  Saturation Index (IAP/K_{SP})
    Evaporation
    Temperature
    Alkalinity
    Loss of CO_2 from supersatured water
    pH (especially photosynthesis)
  Ionic strength
    Groundwater mixing
    Evaporation
    Mg/Ca Ratio
Kinetic factors
  Availability of seed crystals
  Rate of seeding
  Rate of nucleation
  Rate of supersaturation
  Inhibitors (phosphates, organic films)

---

covering old gypsum beds by allochthonous sediment deposition
also must be considered.

We also note that for a reaction such as:

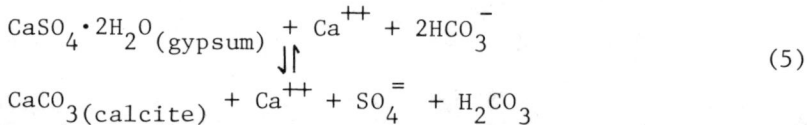

$$CaSO_4 \cdot 2H_2O_{(gypsum)} + Ca^{++} + 2HCO_3^{-} \updownarrow CaCO_{3(calcite)} + Ca^{++} + SO_4^{=} + H_2CO_3 \qquad (5)$$

conductivity would increase 32%, TDS would increase 35%
(assuming 50% loss of HCO_3 and 100% loss of H_2CO_3 at 180°C),
and the true salinity would increase by 22%. Although a
Ca^{++}/SO_4 water may be more desirable than a Ca^{++}/HCO_3 water,
the actual salinity would increase. This points to the need
for more well-defined (i.e., scientifically rather than
legislatively) concepts of salinity management.

As a final comment, the incremental savings per mg/l TDS
removed at lower concentrations ($\cong$800-900 mg/l) are much
lower and more uncertain, than those at high salinities
(1300-1400 mg/l). Prudence dictates that benefit-cost
analyses of management techniques that encourage eutrophication as a positive management goal be geared toward ambient,
rather than future salinity values. It also appears that we
have time to pursue thoughtful studies of the optimum conditions for carbonate precipitation in reservoirs, as well as
investigating beneficial side effects such as coprecipitation

of heavy metals that may migrate downstream from mining
operations, prior to initiating expensive management plans.

SUMMARY

Investigations of chemical changes occurring as a result
of impoundment have been conducted for decades, but salinity
removal has not often been systematically examined in such
studies. Typically such studies have indicated that the most
likely range of gross salinity removal is between 0 and 10%
of the annual input load. They further suggest that salinity
removal mechanisms such as calcium carbonate precipitation
often only partially offset salt loading from the reservoir
sediment, the net result being a gross increase in salinity
downstream. Overall, however, the vagaries of sampling to
determine water column salt storage and the failure to
account for bank storage, together with often necessarily
unfounded assumptions and unnecessarily unclear calculations,
cast doubt as to the validity of these conclusions. There
is little doubt that calcium carbonate precipitation occurs,
but its quantitative effect is not clear from past studies.
If a generalization can be made from the results of the
previous studies, it would be that calcium carbonate precipitation is more important in the newer Colorado River mainstem
reservoirs (Flaming Gorge and Lake Powell) than in the older
Lake Mead. Other ions, except possibly silica, do not appear
to be removed.

Examination of the biogeochemical literature indicates
that the four types of processes most likely to contribute
significantly to natural salinity removal in reservoirs are
homogeneous precipitation; coprecipitation processes that
include clay diagenesis and ion exchange reactions which
replace lighter counter ions with heavier ones; coagulation,
including biocoagulation; and bioassimilation followed by
sedimentation or emigration. Calcite precipitation, driven
either by photosynthetically induced increases in pH or by
increasing temperature, has been amply demonstrated in lakes,
reservoirs, and microcosm experiments, and appears to be the
most important salinity removal mechanism. It is worthy of
note in terms of management models that the expected stoichiometry between photosynthetic carbon removal and carbonate
precipitation is seldom found [7, 38]. The remaining biogeochemical processes appear less promising, although ion
exchange processes may be important.

With respect to management implications, calcium carbonate precipitation may not lead to a significant increase in
downstream water quality, and management techniques aimed at
this process may be neither feasible, cost-effective, nor
environmentally desirable. Above all, a strong orientation

toward competent chemical analyses using specific ion data, such as have been pursued in recent years, should lead to more useful results than classicial conservative species (EC and TDS) constructs.

LITERATURE CITED

1. Howard, C. 1960. Chemistry of the water. In: W. Smith et al. [eds.]. Comprehensive survey of sedimentation in Lake Mead, 1948-1949. U.S. Geol. Surv. Prof. Paper 295. Washington, D.C. pp. 115-124.

2. Reynolds, R., and N. Johnson. 1974. Major element geochemistry of Lake Powell. Lake Powell Res. Proj. Bull. 5. NSF-RANN, Washington, D.C.

3. Gloss, S., R. Reynolds, Jr., L. Mayer, and D. Kid. 1981. Reservoir influences on salinity and nutrient fluxes in the arid Colorado River basin. pp. 1618-1629 In H. G. Stefan [ed.], Surface Water Impoundments. ASCE, New York.

4. Bolke, E. L. 1979. Dissolved oxygen depletion and other effects of storing water in Flaming Gorge Reservoir, Wyoming and Utah. USGS Water Supply Paper 2058. Washington, D.C.

5. Paulson and Baker (this volume).

6. USDI. 1981. Quality of water: Colorado River Basin. Progress Rept. No. 10. U. S. Dept. of Interior. Washington, D.C.

7. Messer, J. E. Israelsen, and V. Adams. 1981. Natural salinity removal processes in reservoirs. UWRL/Q-81/03. Utah Water Research Laboratory.

8. Sorensen, D., T. Hughes, C. Israelsen, A. Huber, E. Israelsen, M. Mandavia, and L. Baker. 1976. Inventory related to water quality objectives. Bear River Basin Type IV study, Idaho-Utah-Wyoming. Utah Water Research Lab., Logan.

9. Leppard, G., A. Massalski, and D. Lean. 1977. Electron-opaque microscopic fibrils in lakes: their demonstration, the biological derivation, and their potential significance in the redistribution of cations. Protoplasma 92:289-309.

10. Messer, J., V. Adams, and E. Israelsen. 1982. Natural salinity removal in mainstem reservoirs mechanisms,

occurrence and water resources impacts. (in preparation).

11. Jones, B. F., and C. J. Bowser. 1978. The minerology
    and related chemistry of lake sediments. In: A. Lerman
    [ed.]. Lakes Chemistry Geology Physics. Springer-
    Verlag. pp. 179-235.

12. Stumm, W. E., and J. J. Morgan. 1981. Aquatic chemis-
    try. 2nd ed. Wiley Interscience.

13. Sheldon, R. 1972. Size separation of marine seston by
    membrane and glass fiber filters. Limnol. Oceanogr.
    17:494-498.

14. APHA. 1975. Standard methods for the analysis of
    water and wastewater, 14th Ed. New York.

15. O'Melia, C. 1980. Small particles in water. Environ.
    Sci. Techol. 14:1052-1060.

16. Mayer, L. 1977. The effect of Lake Powell on dissolved
    silica cycling in the Colorado River. Lake Powell
    Research Project Bull. 42. NSF-RANN. Washington,
    D.C.

17. Paterson, M. P. 1976. Is air chemistry monitoring
    worth its salt? pp. 777-787 In J. Nriagu [ed.].
    Environmental Biogeochemistry. Ann Arbor Science.

18. Reynolds, R. C. 1978. Polyphenol inhibition of calcite
    precipitation in Lake Powell. Limnol. Oceanogr. 23:585-
    597.

19. Dutt (this volume).

20. Kelts, K., and K. J. Hsu. 1978. Freshwater carbonate
    sedimentation. In: A Lerman [ed.]. Lakes Chemistry
    Geology Physics. Springer-Verlac. pp. 295-324.

21. Prentki et al. (this volume).

22. Anderson, E., and A. Pritchard. 1960. Circulation and
    evaporation. In: W. Smith et al. [eds.]. Comprehen-
    sive survey of sedimentation in Lake Mead, 1948-1949.
    U.S. Geol. Surv. Prof. Paper 295, pp. 125-148. Washing-
    ton, D.C.

23. Soltero, R. A., J. C. Wright, and A. A. Harpestad.
    1973. Effects of impoundment on the water quality of
    the Bighorn River. Water Res. 7:343-354.

24. Soltero, R., J. Wright, and A. Harpestad. 1974. The physical limnology of Bighorn Lake-Yellowtail Dam, Montana: Internal density currents. Northwest Sci. 48:107-124.

25. Hannan, H. 1979. Chemical modifications in reservoir-regulated streams. In: R. Ward and S. Stanford [eds.]. The Ecology of Regulated Streams. Plenum, New York. pp. 75-94.

26. Hannan, H., and W. Young. 1974. The influence of a deep-storage reservoir on the physico-chemical limnology of a central Texas river. Hydrobiologia 44:177-207.

27. Hannan, H., and L. Broz. 1976. The influence of a deep storage and an underground reservoir on the physicochemical limnology of a permanent central Texas river. Hydrobiologia 51:43-63.

28. Hannan, H., I. Fuchs, and D. Whitenberg. 1979. Spatial and temporal patterns of temperature, alkalinity, dissolved oxygen, and conductivity in an oligomeso-trophic deep-storage reservoir in Central Texas. Hydrobiologia 66:209-221.

29. Bolke, E., and K. Waddell. 1975. Chemical quality and temperature of water in Flaming Gorge Reservoir, Wyoming and Utah, and the effect of the reservoir on the Green River. U.S. Geol. Surv. Water Supply Paper 2039-A. Washington, D.C.

30. Hem, J. D. 1970. Study and interpretation of the chemical characteristics of natural water. 2nd Ed. U. S. Geol. Surv. Water Supply Pap. 1473. U. S. GPO, Washington, D. C.

31. Iorns, W., C. Hembree, and G. Oakland. 1965. Water resources of the Upper Colorado River Basin. Technical Report, Prof. Paper 441. U. S. Geol. Surv., Washington, D. C.

32. Merritt, David, Bureau of Reclamation, personal communication.

33. Langbein, W. B. 1960. Water budget. pp. 95-102. In W. Smith et al. [eds.]. Comprehensive survey of sedimentation in Lake Mead, 1948-1949. U. S. Geol. Surv. Prof. Paper 295. Washington, D.C.

34. Morrison, Lee, Bureau of Reclamation, personal communication.

35. Kleinman, A., and F. Brown. 1980. Colorado River salinity economic impacts on agricultural, municipal, and industrial users. Bureau of Rec., U. S. Dept. of Interior, Washington, D. C.

36. USDA. 1954. Diagnosis and improvement of saline and alkali soils. Agr. Handbook No. 60. U. S. Salinity Laboratory.

37. Shapiro, J. 1973. Blue-green algae: why they become dominant. Science 179:382-384.

38. Megard, R. 1968. Planktonic photosynthesis and the environment of calcium carbonate deposition in lakes. Interim. Rept. No. 2. Limnological Res. Cent., Univ. of Minnesota, Minneapolis.

PART 8

FISHERIES

CHAPTER 30

NATURAL VS MANMADE BACKWATERS
AS NATIVE FISH HABITAT

Richard A. Valdez
    U.S. Fish and Wildlife Service
        Yellowstone National Park, Wyoming

Edmund J. Wick
        Colorado State University
        Fort Collins, Colorado

ABSTRACT

    Natural backwaters and embayments are important habi-
tats for native fishes of the Upper Colorado River System.
The protected Colorado squawfish, humpback chub and razorback
sucker use these quiet areas during various life stages.
Water developments in the system have prompted mitigation for
habitat enhancement to favor and recover these species. This
paper explores the feasibility of recreating backwaters and
embayments as part of this enhancement. Side channels diked
for flood control in the Yampa River demonstrate the feasibility of recreating backwaters using existing riverine features. These function as natural backwaters which adult
Colorado squawfish occupy during runoff. They drain with
descending flow and do not become refuges for predaceous and
competitive non-native fishes. Natural embayments are
important low-water habitats, especially for young Colorado
squawfish and humpback chub after runoff. These are created
naturally during high flows by eddies around channel obstruc-
tions or stable shoreline features, and can be artificially
created with appropriately placed wing dams and bank re-
inforcements. Abandoned gravel pits flooded by the Colorado
River near Grand Junction, Colorado have long been viewed as
simulating backwaters for native fish habitat. Most contain
water year around and are refuges for large numbers of
competitive non-native fishes and a host of pathogens. They
do not appear to benefit native species and should be sealed
from the river, and managed as warm-water game fisheries and
waterfowl sanctuaries. However, gravel pits constructed in
midchannel islands, graded gently to flood during runoff and

drain with descending flow, may benefit native fishes while preventing establishment of permanent populations of non-natives.

INTRODUCTION

Natural backwaters and embayments in the Upper Colorado River System are important fish habitat, particularly for native fauna like the protected Colorado squawfish (Ptychocheilus lucius), humpback chub (Gila cypha) and razorback sucker (Xyrauchen texanus). Young-of-the-year (yoy) and juvenile Colorado squawfish frequent these habitats in the Colorado, Green and Yampa Rivers [1,2,3]. Adult squawfish and razorback sucker as well as yoy and juvenile humpback chub also use these habitats.

Water uses of the Upper Colorado River in the last two decades have altered flow regimes and eliminated many backwaters and embayments. Reduced flows and channelization have significantly altered the natural processes that created these habitats. Declines in numbers of endangered fishes have accompanied a diminished prevalence of backwaters and embayments.

Reconstructing these habitats seems to be a reasonable mitigation measure for habitat lost to increasing water uses. We address this hypothesis by first characterizing natural and manmade backwaters and embayments in the Yampa and Upper Colorado Rivers, and then by contrasting fish use to determine the most desirable aspects of these habitats. Construction alternatives are examined in response to different fish communities; constructing manmade habitats is greatly complicated by the presence of non-native species. These analyses apply only to the Upper Colorado River System in the interest of enhancing and managing habitat for recovering Colorado squawfish, humpback chub and razorback sucker. A similar study was conducted on four artificial backwaters in the Lower Colorado River for managing game fishes [4].

This study is part of the Colorado River Fishery Project conducted by the U.S. Fish and Wildlife Service and the Colorado Division of Wildlife, and funded by the Bureau of Reclamation, Bureau of Land Management and U.S. Park Service.

NATURAL FEATURES

The term backwater has been used to describe shallow shoreline pockets or indentations in the river channel, regardless of size or origin. A backwater, for the purposes of this discussion, forms either in a side channel at low flow, or in a former river channel or intermittent tributary flooded by high flow. An opening to the river is usually at the downstream end of the flooded channel.

Indentations at upstream and downstream ends of sand bars or small islands, usually formed by eddies, we shall call embayments. This distinction between backwaters and embayments is made because the origin of these features greatly influences measures necessary to recreate them on the river. Backwaters and embayments appear to serve a similar purpose, but their origins are very different. Embayments are commonly called backwaters by many biologists afield since both are low-velocity habitats.

Backwaters

Backwaters in the Upper Colorado River vary in size and depth. Average size of five randomly-selected backwaters was 1.06 ha surface area, and average depth was 1.1 m (Table I). Water velocity in all of these backwaters was undetectable although wind and river circulation sometime occur, especially near the outlet. The substrate of these five backwaters was silt and sand, or silt over gravel. These backwaters may be temporarily eliminated by high spring flows when the side channel or tributary has continuous flow. Others exist only during high flow when flood waters inundate low-lying areas created by former river channels (Figure 1).

Fish samples in these five backwaters from July to November 1980 yielded 5 native (17.4% numerical composition) and 10 non-native species (82.6%) (Table II). Red shiner

Table I. Area, depth and substrate of five randomly-selected backwaters of the Upper Colorado River, located as river miles upstream from the Green River confluence.

River Mile	Area (ha)	Depth (m) Max.	Depth (m) Ave.	Substrate
1.4	1.20	3.6	2.8	silt/sand
48.2	0.98	1.1	0.5	sand/silt
120.9	0.98	1.5	1.0	silt/gravel
125.7	1.01	1.5	1.1	silt/gravel
146.9	1.14	1.2	0.3	silt
Means:	1.06	1.8	1.1	silt

Figure 1. Typical backwater of the Upper Colorado River formed by an inundated side channel, river mile 158.

Table II. Percentage composition of native and non-native
fishes in five randomly-selected backwaters of the
Upper Colorado River (n=1895).

Native Species	Percentage	Non-Native Species	Percentage Composition
flannelmouth sucker	10.6	red shiner	30.3
roundtail chub	5.5	fathead minnow	27.4
Colorado squawfish	0.9	sand shiner	17.9
bluehead sucker	0.2	black bullhead	2.6
razorback sucker	0.2	largemouth bass	1.9
total:	17.4	green sunfish	1.2
		common carp	0.6
		channel catfish	0.3
		black crappie	0.2
		white sucker	0.2
		total:	82.6

(Notropis lutrensis), fathead minnow (Pimephales promelas) and sand shiner (N. stramineus) accounted for 75.6% of the fish composition. Flannelmouth sucker (Catostomus latipinnis) and roundtail chub (G. robusta) were the most common native species. Small numbers of adult and juvenile Colorado squawfish occurred as well as small numbers of adult razor-back sucker.

Backwaters in the Upper Colorado River are very produc-tive communities and support a varied complement of fishes, dominated by non-natives. The rare Colorado squawfish and razorback sucker frequently use these backwaters during runoff for staging, resting and feeding; and in low water for feeding.

Embayments

Embayments are generally smaller than backwaters and often form at upstream and downstream ends of sand bars between the bar and the more stable river bank (Figure 2). Embayments are also found on the periphery of sand islands between sand spits. These habitats are usually inundated and formed by eddies during high flows. Eddy currents dig deep concavities in the channel bed. As flow subsides, fine sediments are deposited in the low-velocity areas to form bars with depressions for embayments in the high-velocity areas. Few embayments exist on the river longer than one normal water year because of the shifting river substrate.

Five randomly-selected embayments from the Upper Colo-rado River had an average surface area of 0.13 ha and an average depth of 0.2 m (Table III). No detectable velocity was recorded, but like backwaters, wind circulation and river exchange does occur. The substrate of embayments is usually silt or sand.

A large complement of non-native fishes (96.0% numerical composition) was represented in embayments (Table IV). Red shiner, sand shiner and fathead minnow made up 91.5% of the fish composition. Native fishes composed only 4.0% of the catch; young bluehead sucker (C. discobolus) and yoy and juvenile Colorado squawfish were the most common species.

Embayments in the Upper Colorado River tend to support a greater percentage of non-native fishes than do backwaters, probably because the smaller, shallower and warmer embayments are more conducive to the small non-natives and their young. Embayments also appear to be the principal nurseries for yoy Colorado squawfish and some humpback chub; these fishes appear subjected to a potentially competitive situation.

523

Figure 2. An embayment at low flow on the Upper Colorado River, river mile -16. Note the stable talus bank and large sand bar formed by a high-water eddy.

Table III. Area, depth, and substrate of five randomly-selected embayments of the Upper Colorado River, located as river miles upstream from the Green River confluence.

River Mile	Area (ha)	Depth (m) Max.	Depth (m) Ave.	Substrate
8.5	0.09	0.2	0.2	silt
16.9	0.11	0.3	0.2	silt
31.6	0.20	0.2	0.2	sand
38.3	0.10	0.2	0.1	silt
67.0	0.16	0.8	0.2	sand/silt
Means:	0.13	0.3	0.2	silt

Table IV. Percentage composition of native and non-native fishes in five randomly-selected embayments of the Upper Colorado River (n=3757).

Native Species	Percentage Composition	Non-Native Species	Percentage Composition
bluehead sucker	1.7	red shiner	56.1
Colorado squawfish	1.4	sand shiner	25.1
roundtail chub	0.6	fathead minnow	10.3
flannelmouth sucker	0.2	channel catfish	3.0
speckled dace	0.1	common carp	0.7
Total:	4.0	black bullhead	0.3
		largemouth bass	0.2
		green sunfish	0.1
		plains killifish	0.1
		white sucker	0.1
		Total:	96.0

Permanent populations of non-natives are precluded by the inundating and flushing effects of spring runoff.

MANMADE FEATURES

Manmade features that resemble backwaters and embayments in the Upper Colorado River System are usually the coincidental result of agricultural or industrial activities in or near the river channel. Flood control dikes on side channels, and gravel pits in islands or adjacent floodplains often provide habitat for large numbers of fishes. Use of these features by native and non-native fishes often varies substantially. The intentional design and construction of backwaters as native fish habitat is known from only the Upper Yampa River.

Flood Control Dikes

Earthen dikes, constructed in side channels to prevent flooding and erosion of agricultural lands, often result in large backwaters during high flows. Four such cases are known from the Upper Yampa River where land owners blocked side channels to prevent erosion and flood damage to adjacent fields, and inadvertently created desirable native fish habitat. These backwaters are usually short-lived; they fill at spring runoff in May and drain with descending flows in late June or early July.

One such backwater is located at river mile (RM) 59, just above Cross Mountain Canyon. The dike prevents a side channel from flowing and a backwater is formed by waters rising from the downstream end (Figure 3). The feature is about 1.5 ha in surface area and is 0.3-1.5 m deep, with a silt substrate. Two radiotagged Colorado squawfish, tagged 46 and 62 km upstream in early May, entered this backwater at peak runoff in early June 1981. Both fish freely swam the length of the backwater and periodically stayed at the outlet near the flooding river. The backwater, at this time, had relatively silt-free water and was warmer than the torrential river. One fish spent nearly 2 weeks in the backwater before moving downstream. One week after the fish left, the water level receded sufficiently to isolate the backwater from the river.

Figure 3. Aerial view of side channel on the Yampa River diked for flood and erosion control at river mile 59. The backwater formed by the large earthen dike is sealed from the main channel at low flow.

A second diked side channel at RM 99, above Juniper Canyon, was also used by two radiotagged squawfish during runoff in early June 1981. This feature was about 1.5 m deep with a silt bottom and a 1-m deep narrow channel leading to the river. The feature was drained by descending flows in late June, about 1 week after the squawfish left.

A side channel at RM 64, diked to protect a low-laying alfalfa field, also developed into a backwater during runoff. Maximum depth was 1.5-2.0 m and the backwater was drained by descending flows in late June. One radiotagged squawfish entered this backwater briefly in mid -June for the high-flow period then proceeded downstream and entered the large backwater previously mentioned at RM 59.

A natural backwater in this same area of the Yampa River was also used by these fish, suggesting that adult radio-tagged Colorado squawfish showed a strong tendency to find and occupy natural and manmade backwaters during runoff 1981 in the Yampa River. The capture of numerous squawfish in large backwaters of the Colorado River during runoff also supports this contention. The flood control structures of the Yampa River demonstrate that usable backwaters can be intentionally created to enhance adult and possibly juvenile squawfish habitat during high flows.

Design Backwaters

Fish habitat enhancement efforts were attempted in the Upper Yampa River by private industry in the late 70's. Side channels were diked to create backwaters like those described in the previous section. But, the dikes were constructed of rock instead of earth and eventually breached to allow either continuous flow or flow only during runoff (Figure 4). The side channels with dikes that breached at high flow but sealed at low flow eventually filled with silt and sand and became unusable as native fish habitat. Those with some continuous flow provided good nurseries for native and non-native fishes (Table V). Non-native fishes made up 72.0% of the species composition in one breached backwater while native species made up 28.0%. Redside shiner (Richardsonius balteatus) and fathead minnow accounted for 68.8% of the composition. The most abundant native species was the roundtail chub with 18.1% of the composition. Most roundtail chub were large yoy while most other species were small yoy, suggesting that the native roundtail reproduces earlier, grows faster and has a size advantage on sympatric species. Colorado squawfish are not known to spawn at this location, but if they did they would likely spawn later and the young

Figure 4. Side channel at low flow on the Yampa River diked for fish habitat enhancement at river mile 121. The rock dike in the foreground is partially beached.

Table V. Percentage composition of native and non-native fishes in a designed backwater created from a partially-blocked side channel in the Yampa River, river mile 121 (n=867).

Native Species	Percentage Composition	Non-Native Species	Percentage Composition
roundtail chub	18.1	redside shiner	45.4
speckled dace	8.3	fathead minnow	23.4
bluehead sucker	1.2	white sucker	2.8
flannelmouth sucker	0.4	common carp	0.2
Total:	28.0	speckled dace x redside shiner	0.1
		bluehead sucker x white sucker	0.1
		Total:	72.0

could be similarly disadvantaged in size and face competition from native and non-native species.

These designed backwaters appear to have been only partially successful because the dikes were breached or partially breached. This suggests that backwaters created from side channels should be formed by high non-porous and stable dikes. Also, none of the endangered fishes have been recorded recently from the vicinity of these backwaters, illustrating the need for proper location and placement of these enhancements.

Gravel Pits

Gravel extraction is especially prevalent in the floodplain of the Colorado River near Grand Junction, Colorado. A nearly continuous series of gravel pits now line the floodplain. These vary in size, depth, shape and orientation to the river channel, and are of three types; those permanently open to the river, those open to the river only during high spring flows, and those permanently isolated from the river. The third type is not considered in this analysis. Most industries prefer to excavate away from the high water line to avoid costly delays from 404 permits required by the Water Quality Act and consultation required by Section 7 of the Endangered Species Act. So, most gravel operations are conducted away from the high water line and are of the third type. Proper structural specifications are still needed for advising industries on appropriate design, and excavation and rehabilitation techniques to benefit especially native riverine fishes.

Gravel pits permanently open to the river are most commonly excavated in large midchannel islands or low floodplains, such as the 30 Road Pit near Clifton, Colorado (Figure 5). The 1.0-ha pit was excavated from a midchannel island in 1978, leaving a horseshoe-shaped gravel berm with a downstream opening. Water circulates into the pit through the opening that was nearly sealed by shifting silt and gravel during runoff 1980. Maximum depth of this pit was 5.5 m and average depth was 1.1 m (Table VI). Silt deposits have nearly filled the rear of the backwater and covered the gravel substrate to a depth of 0.3-0.5 m. Emergent vegetation has become established in the silt deposits, and ducks and geese were seen resting and feeding in the pit in October 1980. The pit was sealed from the river in 1981, except during runoff.

Fish composition when open to the river in 1980 and 1981, was dominated by non-native species (73.3%), predominantly carp, red shiner, fathead minnow, sand shiner and

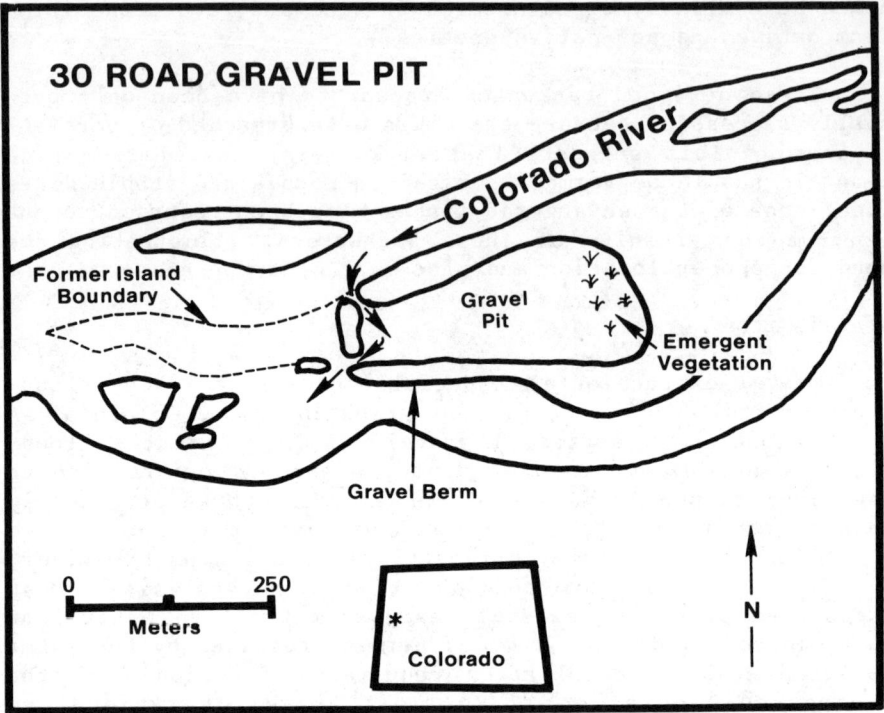

Figure 5. A gravel pit excavated from a midchannel island on the Upper Colorado River, river mile 175.5. The 1978 pit is nearly filled with silt and sediment.

Table VI. Area, depth and substrate of four gravel pits open to the Upper Colorado River near Grand Junction, Colorado.

River Mile	Name	Area (ha)	Depth (m) Max.	Ave.	Substrate
163.6	Walter Walker	18.30	1.9	1.1	silt/gravel
168.0	Connected Lakes	4.00	2.2	1.0	silt/gravel
175.5	30 Road Pit	1.00	5.5	0.6	silt/gravel
177.8	Clifton Ponds	1.10	3.1	0.5	silt/gravel
Means:		6.10	3.2	0.8	silt/gravel

530

black bullhead (Ictalurus melas) (Table VII). The native
fishes were 26.7% of the fish composition in this artificial
feature; roundtail chub, flannelmouth sucker and bluehead
sucker were found primarily at the mouth of the backwater
near the swift main channel current. One adult Colorado
squawfish and one adult razorback sucker were caught in a
trammel net in about 2 m of water within the pit in June
1981. A second squawfish was taken in a fyke net set over-
night to catch fish moving into the pit. Fish composition
when the pit was isolated in summer 1981 was dominated
by green sunfish, fathead minnow, red shiner, and topminnows
(Gambusia affinis).

The second type of gravel pit is open to the river only
during runoff, and is represented by several pits (Table VI),
including Walter Walker Wildlife Area (WWWA). A single outlet, a 5-m cut in a gravel berm, separates this 18.3-ha pit
from the river (Figure 6). The river flows through this
single opening only when flow exceeds about 7,500 cfs. The
river normally flows into the pit from May to mid-July except
for low water years such as 1981, when flow into the pit
occurred for only 2-3 weeks. Much silt is deposited in this
gravel pit during this overflow period, as the energy of the

Table VII. Percentage composition of native and non-native
fishes in the 30 Road Gravel Pit of the Upper
Colorado River (n=569).

Native Species	Percentage Composition	Non-Native Species	Percentage Composition
roundtail chub	15.2	common carp	23.7
flannelmouth sucker	5.4	red shiner	15.6
bluehead sucker	5.0	fathead minnow	11.7
speckled dace	0.7	sand shiner	7.5
Colorado squawfish	0.3	black bullhead	7.2
razorback sucker	0.1	green sunfish	4.5
Total:	26.7	largemouth bass	2.2
		white sucker	0.5
		channel catfish	0.3
		black crappie	0.1
		Total:	73.3

531

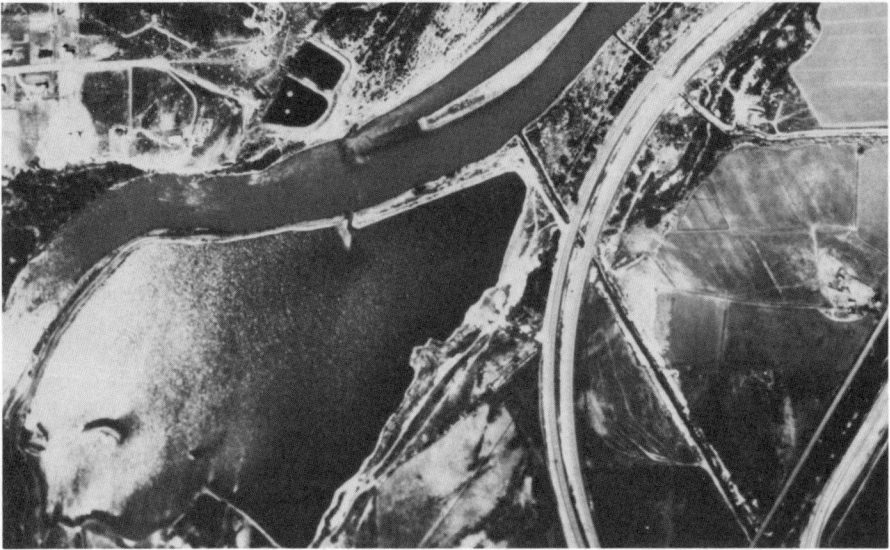

Figure 6. Aerial view of a large abandoned gravel pit at Walter Walker Wildlife Area on the Upper Colorado River, river mile 163.6. Note the narrow gravel berm separating the pit from the river and the single high-water opening.

silt-laden river dissipates in the pond. A layer of silt 0.3-0.6 m deep covers most of the gravel substrate of the pond, except for a narrow band of gravel and cobble along the wind-swept shore.

Colorado squawfish and razorback sucker have been caught in relatively large numbers in this pit [5,6], but these numbers have steadily declined [3,7]. Fish composition, particularly the relative abundance of native and non-native species, changed dramatically with river access in 1980 (Table VIII). Over 96% of the fishes in WWWA during isolation were non-native. That composition changed dramatically to 58.2% non-native and 41.8% native when river flows inundated the pit during runoff.

Semi-open gravel pits such as WWWA are inhabited by large numbers of non-native fishes that flourish in the pit year around. Many native species occupy the pit and similar quiet areas during spring runoff, apparently as refuges from the physical rigors of the river or as staging areas for spawning, or both. Kennedy [4] caught gravid female razor-backs in large backwaters of the Lower Colorado River and suggests that spawning may occur in these habitats. Gravel

Table VIII. Percentage composition of native and non-native fishes in Walter Walker Wildlife Area Gravel Pit when isolated (I, n=734) from and open (O, n=607) to the Upper Colorado River.

Native Species	Percentage Composition		Non-Native Species	Percentage Composition	
	I	O		I	O
flannelmouth sucker	0.5	29.3	black bullhead	32.3	22.1
roundtail chub	–	10.5	common carp	28.5	16.8
Colorado squawfish	0.8	0.8	green sunfish	25.1	6.4
			channel catfish	6.8	5.6
razorback sucker	2.6	0.8	white sucker	2.0	1.5
bluehead sucker	–	0.2	largemouth bass	0.7	0.5
flannelmouth x razorback	–	0.2	bluegill	0.4	–
Totals:	3.9	41.8	black crappie	0.3	0.3
			northern pike	–	0.2
			red shiner	–	4.5
			sand shiner	–	0.3
			Total:	96.1	58.2

pits may also serve as staging and possibly spawning areas for Colorado squawfish and razorback sucker; tuberculate squawfish and razorbacks releasing eggs and milt were caught in WWWA and Clifton Ponds in May and June 1980. The large complement of predaceous fishes in these gravel pits probably precludes any spawning success by these rare species. Native fishes appear extremely sensitive to water levels in these pits and will leave them with descending flow to avoid being stranded in the isolated pits. Razorback sucker were observed moving between Clifton Ponds and the Colorado River on a nearly daily basis with fluctuating water levels.

HABITAT ENHANCEMENT

Habitat requirements of the protected native fishes of the Upper Colorado River System are not completely understood. Those that are known differ by species and life stages; habitat features used by one species or life stage are not likely to satisfy the needs of another. For example, small, low-water embayments and backwaters are used by yoy and some small juvenile Colorado squawfish as well as some yoy and juvenile humpback chub. Medium-sized features are used by juveniles of both species at all flow stages. Large

533

backwaters and embayments are used by adult Colorado squawfish and razorback sucker as refuges and staging areas during
runoff and as foraging areas during low flow. The requirements of these target species cannot be universally satisfied
with construction of a single type of feature. Enhancement
of high-water and low-water habitats is needed.

Interesting comparisons can be made between natural and
manmade backwaters and embayments in the Yampa and Upper
Colorado Rivers that reveal some of the more desirable
aspects of these habitats. Diked side channels, built for
erosion control, appear to offer valuable fish habitat during
high flows, when the fish tend to leave the main channel and
seek refuge in quiet areas. Adult Colorado squawfish use
artificial backwaters as they use naturally-occurring
ones. Diking side channels in areas inhabited by adult
squawfish and razorback sucker warrants consideration for
recreating backwaters lost to reduce flows and channelization. Earthen dikes placed at upstream ends of side channels
to prevent the river from flowing through allow flooding from
the lower end and are a desirable manmade structure. Dikes
should not be allowed to breach, and side channels should
have gentle gradients that allow backwaters to drain with
descending flows and prevent the establishment of potentially
competitive non-native fish populations.

Habitat for juvenile and yoy squawfish and razorback
sucker can be created from partially-breached dikes at heads
of side channels. This allows the channels to flush silts
and particulates during runoff, but to become slow-flowing
backwaters during low flow that support high levels of
macrophytes and insects for young fishes. Care should
be taken before creating these low-water habitats for native
species since the presence of high densities of non-native
species may pose serious competition/predation problems.

Embayments can also be created to provide low-flow nurseries for young native fishes by diverting the river energy
into an eddy effect. Wing dams and reinforced banks can create this digging action, but more study is needed to understand what substrates respond best to various structures.

Abandoned gravel pits constitute unnatural riverine
features that allow competitive and predaceous non-native
fishes to flourish because they remain sheltered from the
river year around. These large, unnatural features should be
permanently sealed from the river and managed as warm-water
sport fisheries. Subterranean flow into these pits is
sufficient to maintain good water levels and quality.
Gently-graded gravel pits that flood during runoff but drain

with descending flow may benefit native fishes by providing staging or resting areas during runoff while precluding permanent populations of non-native fishes.

Habitat enhancement to recover native fishes should utilize natural riverine features, where possible. Existing side channels, tributaries, old river channels and oxbows should be utilized for backwaters to insure a stable, long-lasting feature. The feasibility and expense of diking these channels will depend largely on access and side effects of the procedure. Excavating new channels or backwaters is not advised because the shifting nature of the river is likely to fill these with sand and silt and render the operation a wasted expense.

SUMMARY

1. Backwaters created by diked side channels resemble in appearance and function natural habitats. These benefit protected native fishes by providing resting, staging and feeding areas during high flow, and drain at low flow to prevent establishment of permanent populations of competitive non-native fishes.

2. Backwaters that persist at low flow are used by native fishes for foraging and resting. These are generally inundated and flushed of non-native fishes at high flows.

3. Side channels with breached dikes can provide low-velocity nursery habitats that flush of silts at high flow. These features resemble naturally-occurring habitats that can harbor large numbers of non-natives; so, community structure must be well known before constructing breachable dikes on side channels.

4. Embayments are important low-water nurseries and habitats for young fishes. These can be created with wing dams and reinforced river banks to direct river energy into an eddy effect that digs the embayment and builds protective sand bars. Embayments are flushed of silt by annual runoff and their short-term nature lessens potentially competitive effects of non-native species.

5. Large, deep, open-water gravel pits that are refuges for non-native fishes but attract native fishes during runoff should be sealed from the river and managed as warm-water fisheries and wildlife sanctuaries. This may reduce problems for native fishes associated with competition, predation and pathogens, and reduce the influx of non-native fishes into the river system.

6. Graded gravel pits that flood during runoff and drain with descending flow may benefit native species, by providing feeding, resting and possibly spawning areas while minimizing potential problems with non-native fishes.

7. Backwaters and embayments are only two habitat types that can be enhanced to help recover protected native fishes. No single habitat type will universally benefit all target species.

LITERATURE CITED

1. Holden, P. B. 1978. A study of the habitat use and movement of the rare fishes in the Green River, Utah. Trans. of the Bonneville Chapter of the Amer. Fish. Soc.:64-89.

2. Miller, W. H.; J. Valentine; and D. Archer. 1981. Colorado River Fisheries Investigations. Preliminary Draft Report, U.S. Bureau of Reclamation. Salt Lake City, Utah.

3. Wick, E. J.; T. Lytle; and C. Haynes. 1981. Colorado squawfish and humpback chub population and habitat monitoring. Job Progress Report SE-3-3, Federal Aid to Endangered Wildlife, Colorado Division of Wildlife. Denver, Colorado.

4. Kennedy, D. M. 1979. Ecological investigations of backwaters along the Lower Colorado River. Ph.D. Dissertation. The University of Arizona, Tucson.

5. Kidd, G. T. 1977. An investigation of endangered and threatened fish species in the Upper Colorado River as related to Bureau of Reclamation Projects. U.S. Bureau of Reclamation. Grand Junction, Colorado.

6. McAda, C. W. and R. S. Wydoski. 1980. The razorback sucker, Xyrauchen texanus, in the Upper Colorado River Basin, 1974-76. Tech. Paper 99, U.S. Fish and Wildlife Service. Washington, D.C.

7. Valdez, R. A.; P. G. Mangan; R. P. Smith; and B. C. Nilson. 1981. Upper Colorado River fisheries investigations. Draft Final Report, U.S. Bureau of Reclamation. Salt Lake City, Utah.

CHAPTER 31

EFFECTS OF HABITAT ALTERATION BY ENERGY RESOURCE DEVELOPMENTS IN THE UPPER COLORADO RIVER BASIN ON ENDANGERED FISHES

Terry J. Hickman
U.S. Fish and Wildlife Service
Salt Lake City, Utah

INTRODUCTION

For its size, the Colorado River might be the most used, controlled, and fought-over river in the world. Waters of the Colorado now serve millions of people for municipal and industrial purposes, electric power generation, mining, irrigation, grazing, fish and wildlife, and recreation. Huge volumes of water are exported from the Colorado River system to adjoining areas. Institutions dealing with the water of the Colorado River include the following: resource agencies of seven states, at least 15 federal agencies, several irrigation districts, numerous municipalities, and many energy and environmental agencies.

The Colorado River begins in the Rocky Mountains in Colorado and Wyoming and flows southwesterly for about 2700 km (1700 mi) to the Gulf of California. Its drainage area of 627,000 km^2 (242,000 mi^2) (in the U.S.) represents one-fifteenth of the area of the United States.

The Colorado River drainage is divided into the upper and lower basins, with the demarcation being Lee Ferry, Arizona, about 24 km (15 mi) below Glen Canyon Dam. The Upper Colorado River Basin comprises about 279,000 km^2 (107,900 mi^2) covering portions of five states: Arizona, New Mexico, Utah, Colorado, and Wyoming. This discussion will deal primarily with the Upper Colorado River Basin above Lee Ferry, Arizona.

Except for the mountainous areas, much of the Upper Colorado River Basin is arid or semi-arid. Many of the fishes that evolved in the Colorado River developed unique adaptations to the river's harsh environment of fluctuating flows, temperatures, and turbidities.

Because of these unique adaptations and long isolation from other surrounding river basins (Missouri, Columbia, etc.), the Colorado River, as a whole, has the highest percentage of endemic species of any river basin in North America [1]. There are eight endemic fish found in the upper basin [2], two of which are recognized subspecies of more broadly distributed species. Of these, four are listed under the Endangered Species Act as endangered: Colorado squawfish (Ptychocheilus lucius), bonytail chub (Gila elegans), humpback chub (Gila cypha), and Kendall Warm Springs dace (Rhinichthys osculus yarrowi). Two others, the Colorado River cutthroat trout (Salmo clarki pleuriticus) and the razorback sucker (Xryauchen texanus) are on state lists of threatened and endangered species.

DISCUSSION OF THE ENDANGERED
FISH OF THE UPPER COLORADO
RIVER BASIN

Kendall Warm Springs Dace

The Kendall Warm Springs dace, a subspecies of speckled dace, is found only in the heated outflow of Kendall Warm Springs, Wyoming. Its entire habitat consists of less than 305 m (1000 ft) of a small stream, which flows into the Green River in the most northern portion of the basin, less than 50 km (31 mi) from the river's origin. Because this population of speckled dace occupies a restricted habitat in a restricted location, no further mention of this subspecies will be made. For a more detailed account of the Kendall Warm Springs dace, refer to Binns [3].

Colorado Squawfish

The Colorado squawfish was listed as endangered by the U.S. Fish and Wildlife Service in the Endangered Species List published in the Federal Register on 11 March 1967. It is the largest species of the minnow family (cyprinidae) native to North America. The largest known specimens seen in recent years have been about 0.9 m (3 ft) long and have weighed about 6.8 kg (15 lb) [1]. Maximum weight has been recorded as exceeding 36 kg (80 lb) and lengths recorded of nearly 1.8 m (6 ft).

Early records indicate that the Colorado squawfish was once found throughout the Colorado River system from the

upper Green River in Wyoming to the Gulf of California,
including the Gila River system in Arizona. It was abundant
over all of its range prior to the 1850's [4]. The type
specimen was obtained from the Colorado River in California
[5], where it no longer exists.

The present range of natural populations of the Colorado
squawfish is restricted to the Upper Colorado River Basin.
It is found in the Green River from the confluence of the
Yampa River to its confluence with the Colorado River. It is
also found in the Yampa, lower Duchesne, and White Rivers
tributaries to the Green River. In the mainstem Colorado
River it is found from Lake Powell to above Grand Junction
and in the Gunnison and San Juan Rivers, tributaries to the
mainstem Colorado.

The Colorado squawfish occurs in a variety of habitats,
especially in eddy and pool habitats protected from the main
current. However, because of its feeding and spawning
requirements, it can be found in various habitats throughout
the river [4]. Movement of the Colorado squawfish appears to
be related to flow, temperature, spawning, and feeding.
Recent investigations by the U.S. Fish and Wildlife Service
indicate the Colorado squawfish may require a relatively
unrestricted movement to satisfy all of its life history
requirements. Radio-tagged studies have shown that some
individuals make long upstream and/or downstream migrations
to specific areas [2]. Colorado squawfish have been tracked
over 322 km (200 mi) in just a few months from the area in
which they were first radio-tagged.

Humpback Chub

The humpback chub was also listed as endangered by the
U.S. Fish and Wildlife Service in the Endangered Species
List published in the Federal Register on 11 March 1967. It
was described by R. R. Miller [6] from specimens collected in
the lower basin of the Colorado River (Grand Canyon and
another unknown location). The humpback chub is a medium-
sized [less than 500 mm (20 inches) in total length] freshwater fish of the minnow family. The greatest number of
humpback chubs have been found in deep water canyon areas.
They appear to be adapted to the turbulent waters in these
canyons and are associated with the deep, swift, rocky areas
of the river [7]. The original distribution of the humpback
chub is not fully known; it is assumed to be similar to that
of the squawfish. But because of taxonomic confusion with
related species and a paucity of earlier collections, an accurate assessment of its historic distribution and abundance
is difficult [7].

Present distribution of the humpback chub, as indicated by recent collections include the following upper basin locations: Desolation, Gray, and Labyrinth Canyons in the Green River; Dinosaur National Monument in the Green and Yampa Rivers; Cross Mountain Canyon on the Yampa River; and Black Rocks, Westwater, and Cataract Canyons on the mainstem Colorado River.

In the lower basin, recent collections include the following: Marble and Grand Canyons in the Colorado River and the lower 23 km (14 mi) of the Little Colorado River.

Gray, Westwater, and Black Rocks Canyons appear to support the only major concentrations of humpback chub in the Upper Colorado River Basin [6].

Based on tagged and radio transmitter studies, the humpback chub, unlike the Colorado squawfish, do not typically migrate over large stretches of river. They seldom leave their canyon habitats.

Bonytail Chub

The bonytail chub was listed as endangered by the U.S. Fish and Wildlife Service in the Federal Register on 23 April 1980. It was originally described by Baird and Girard [8] from collections from the Zuni River, a tributary of the Little Colorado River. Bonytails commonly reach 300-350 mm (11-14 inches) in total length in the upper basin. The roundtail chub (Gila robusta) and humpback chub are closely related to the bonytail chub, and the three were often confused in the early literature. Many early references to the bonytail chub refer, in fact, to the roundtail chub [1].

As a result of this confusion, documentation of historic distribution and abundance, as with the other chubs, is difficult [7]. The bonytail chub is believed to originally have ranged throughout the Colorado River system in the main channels and larger tributaries from Mexico to Wyoming.

Presently, small concentrations of bonytail chub are thought to exist in Gray Canyon in the upper basin, and in Lake Mohave [7] and Lake Havasu [1] in the lower basin. No bonytail reproductive success has been identified in any of these areas. The bonytail chub's abundance has declined steadily until it is now the most rare endemic fish in the entire upper basin [2].

Little habitat information is available for the bonytail chub, except that they appear to occupy deep, swift semi-rocky areas in main channels [7].

CAUSES OF DECLINE OF THE
ENDANGERED FISHES

Several papers have recently discussed the causes of decline of the endangered Colorado River fish in the upper basin ([1,4,7], draft bonytail chub recovery plan, draft revised recovery plans for the Colorado squawfish and humpback chub).

The major impacts, according to these papers, are 1) the result of dams and reservoirs, 2) the removal of water from the system, and 3) the introduction of competing non-native fish to the system. Each of these factors has significantly altered the aquatic habitat and community structure.

Dams and Reservoirs

The single most important factor identified by most sources as causing the decline of the endangered Colorado River fishes in the upper basin has been the construction and operation of dams and reservoirs (primary mainstem). Hundreds of miles of flowing river habitat have been converted into great impoundments. Riverine habitat downstream from the dams has been drastically altered in flow, temperature, chemistry, and biota. Migration routes for the larger ranging fish, such as the Colorado squawfish, have been blocked.

Over 20 mainstem and tributary dams (at least half of these in the upper basin) have been constructed on the Colorado River since the first major dam, Hoover Dam, was built in 1935. Over 20 more dams and reservoirs are either authorized, partially completed or are contemplated for the mainstem Colorado and Green Rivers in the upper basin alone.

Dams and reservoirs have affected the endangered fishes of the Colorado River in several ways. They have altered natural flow and temperature regimes and water qualities, reduced total annual discharges through evaporation and diversion (dewatering), converted lotic habitat to lentic, and blocked migration routes. The total effect of these impacts has resulted in a 37% reduction in suitable habitat for the endangered fishes in the upper basin. This does not include the subtle biotic and abiotic impacts which have reduced the suitability and extent of preferred niches in the remaining 63% of the inhabitable river reaches of the upper basin.

Through construction of impoundments across the river, the squawfish's ability for long-range migrations is impeded.

541

This impediment may have a profound effect upon the survival of the Colorado squawfish. Such migrations appear to be related largely to flow, temperature, spawning, and feeding. Dams of various types upstream from Glen Canyon Dam have obstructed access to approximately 396 km (246 mi) of the available habitat in the upper basin. Access to another 216 km (134 mi) of the upper basin could be restricted with the construction of the proposed White River Dam and the proposed Juniper-Cross Mountain Dam on the Yampa River. Approximately 550.3 km (342 mi) of the Upper Colorado River system have been impounded. Another 319.4 km (198.5 mi) in the Upper Colorado River system have been proposed for impoundment.

Molles [9] suggested that the fragmentation of the Colorado River system by dams might isolate subpopulations and restrict gene flow of the endangered fishes, thereby reducing the ability of these subpopulations to adapt to changing environmental conditions.

Subtle impacts, poorly understood and difficult to assess, are those associated with changes in the natural hydraulic cycle with its normal seasonal extremes that were once the pattern of the Colorado River. For example, spring peak flows have been reduced below Flaming Gorge Dam by 50% and baseline flows for the remainder of the year have been increased by 140%. This picture is further complicated by the highly variable daily fluctuations in flow.

This radical alteration in annual flow patterns results in submerging and exposing of the endangered species habitat on an annual basis, which impacts important spawning and foraging areas. Holden [10] suggested that the new flow and temperature regimes below Flaming Gorge Dam were probably the major factors in eliminating the bonytail chub from that area. Suttkus and Clemmer [11] indicated that the future of the humpback chub in the Grand Canyon is questionable due to altered flows and temperature, which fluctuate more than historic conditions.

There are usually significant changes in temperature regimes, turbidity, salinity and other water quality factors below dams. Generally, mean temperatures are often reduced, turbidities decline and salinities increase [10,12]. All of these changes can have a dramatic effect upon the survival of the endangered Colorado River fishes for several miles below the impoundments.

Releases of cold water from Flaming Gorge Dam, after the dam was closed in 1962, effectively eliminated squawfish [1] and bonytail chub [13] from 105 km (65 mi) of the Green River below the dam. Subsequent penstock modifications on Flaming Gorge Dam have resulted in the release of warmer downstream water. This modification may serve to lessen the impact of adverse water qualities below Flaming Gorge.

Much of the Colorado River system has been "tamed" by impoundments, giving a competitive edge to introduce fishes which now predominate in much of the system. Only in the few relatively natural riverine stretches do the native species retain their dominance. Impoundments have created a refuge for exotic fishes that otherwise might not be able to complete their life-cycle in the absence of a lentic or reduced flow environment. The impoundment of additional river reaches will result in a still greater proliferation of exotic species in the upper basin, the impacts of this proliferation on the behavior and survival of the remaining populations of endangered fish species in the Colorado River basin is not easily predicted.

Colorado squawfish, humpback chub, and bonytail chub are not known to reproduce successfully in lentic habitats. Colorado squawfish can live in reservoirs but they have been unable to maintain themselves by natural reproduction [1].

Water Depletion

Water depletions from the Upper Colorado River Basin have drastically altered flow patterns, water quality parameters, and river channel characteristics, and eliminated the quiet backwater nursery areas to a point that much of the essential habitat for endangered fishes is no longer present.

Flow depletion in the upper basin may have immediate and long-term effects on the endangered fish. The immediate effect is loss of flow and reduction of required habitat. The depletion of water during peak runoff periods may lower overall reproductive success of the fishes. Flows below an unknown critical level could result in loss of habitat concentrating the endangered fish populations, thereby increasing the danger of disease and predation by other fish.

Long-term effects of flow reduction change the hydraulic characteristics of the river thus altering stream bank cutting, meander patterns, backwater building, sediment transport capacities, and velocities. With time, eddies, pools, riffles, river banks and beds can be greatly changed along with channel depth, width, and flow patterns. These changes as well as changes in temperatures and turbidities may affect reproduction and other life history stages. The gradual,

cumulative impacts of water diversion or removal on habitat
are much less dramatic and not as obvious as the more sudden
changes created by a large dam and reservoir, but the end
result is similar in relation to the loss of habitat needed
for the continued existence of the listed fish species.

The natural flow of the Upper Colorado River (as recorded at Lee Ferry) has averaged slightly less than 18 x
10^9 m^3 [15 million acre feet (MAF)] annually over the past
80 years. Annual flows have ranged from a low of 7 x 10^9 m^3
(6 MAF) to a high of 30 x 10^9 m^3 (24 MAF). Approximately 25
percent of the natural flow is presently being depleted from
the upper basin. The consumptive use in the upper basin
(including reservoir evaporation) is approximately 5 x 10^9
m^3 (4 MAF) [14]. Table I, from Harris et al. [14], shows
the percentage of estimated future remaining flows as a result of water depletions in the upper basin.

Table I. Estimated future flows as a percentage of present
development in the upper basin [14].

River Basin	Natural Flow	1980	1990	2000	2010	2030
Upper Green River (above Flaming Gorge)	100	78	74	71	67	62
Yampa River	100	99	98	95	93	93
Duchesne River	100	37	37	29	29	29
White River	100	100	79	76	74	74
Total Green River (above Green River, UT)	100	85	79	75	73	71
Upper Mainstem (Colorado River above Cameo)	100	46	42	38	36	36
Gunnison River	100	100	99	99	99	99
Dolores River	100	100	89	86	86	86
Total Colorado River (above Cisco, UT)	100	85	79	75	73	71
Upper San Juan River (above Navajo Dam)	100	89	89	89	89	89
Total San Juan River (above Bluff, UT)	100	82	69	67	70	70
TOTAL COLORADO RIVER (above Lee Ferry, AZ)	100	75	67	64	64	61

It is expected that the projected water use in the upper basin will increase to 6.0×10^9 m^3 (4.9 MAF) by 1990, 6.7×10^9 m^3 (5.4 MAF) by 2000, and to 7.2×10^9 m^3 (5.8 MAF) by 2030, while the average annual flow at Lee Ferry will decrease from 12×10^9 m^3 (10 MAF) at the present to 11×10^9 m^3 (9.1 MAF) in 1990, 11×10^9 m^3 (8.6 MAF) in 2000, and 10×10^9 m^3 (8.2 MAF) by the year 2030 [14]. Table I highlights areas within the upper basin where future demands for water will be the greatest. It may be more prudent to pinpoint areas for recovery of the endangered species based upon the projected outlook for water demands. Certain areas may be more valuable to protect than others, such as those areas which most nearly maintain pristine flows. This would enhance the prospects of the coexistence between energy development and the survival of endangered fish in the upper basin.

Almost 15% [0.64×10^9 m^3 (0.52 MAF)] of the total water depletion [4.8×10^9 m^3 (3.9 MAF)] from the upper basin in 1980 was attributed to reservoir evaporation [15]. As noted earlier, there are over 20 other dams and reservoirs which could be built and be in operation by the year 2000. This could mean that the total water depletion in the upper basin from reservoir evaporation might exceed 1.2×10^9 m^3 (1 MAF) within the next 20 years. The estimated total water depletion from the upper basin by the year 2000 is approximately 6.2×10^9 m^3 (5 MAF) [15], therefore almost 20% of the total water depletion from the upper basin by the year 2000 could be a result of reservoir evaporation.

CHANGES IN FUTURE WATER USE
PATTERNS

Table II shows the future change in water use in the upper basin based upon information presented by Bishop and

Table II. Future change in water use in the Upper Colorado River Basin.

Water Use	1973[1]	1980-2000
Irrigation	73%	34%
Urban	25%	31%
Energy Development	2%	35%

[1]Modified from Bishop and Porcella [16].

Porcella [16] compared with information that was obtained from various sources by the U.S. Fish and Wildlife Service (FWS). Of importance here is not whether the figures are exact, but rather the trend that is depicted. Much of the water being sold to the energy development companies was previously used for agricultural purposes. El-Ashry [17] indicated that the shift of water use from agriculture to energy development would have adverse impacts on surface and ground water and, consequently, on fish and wildlife. This would happen through an increased depletion of available ground water resources which, in some areas, depends on deep percolation of excess water from irrigation runoff. Streams, in many areas, are also sustained by excess water draining from irrigated lands.

It is also possible that another factor could cause a net increase in water use when there is a shift from irrigation to energy development. There are many situations where the full allotment of water owned for the purpose of agricultural use is not completely utilized during a given year; however, it is possible that this may change and energy developers may consistently utilize the majority of the allotted water. This is a result of the different demands for water between the agriculture uses and energy development uses.

One study estimated that only 5% [about 0.1×10^9 m^3 (90,000 AF)] of current agricultural water supplies in Colorado and Utah will have been converted to energy use by the year 2000 [18]. However, several of the energy development companies purchased water rights (many of which had previously been used for agricultural purposes) during the 1950's and 60's, and may not have been included in the above study. For example, a group of energy development companies have proposed removing a maximum of 0.45×10^9 m^3 (362,000 AF) of water annually from the Colorado River near the town of DeBeque, Colorado. Most of these water rights are senior and were purchased 15-30 years ago. This is just one example. The FWS has worked with several energy development companies that have proposed projects that would require significant amounts of water that has already been purchased, some having purchased the water several years ago. It is assumed that much of this water was used for agricultural purposes or is still being used for agricultural purposes (via leases, etc.) until it is needed for energy development.

PROSPECTS FOR THE FUTURE

The continued existence of the endangered Colorado River fishes will require cooperation and communication among diverse interest groups. As demonstrated in the snail darter vs Telico Dam case, all future development will not likely come to a halt because of unique fishes such as the bonytail chub, humpback chub, and Colorado squawfish. Probably several proposed dams and water diversion projects will have to be abandoned. Other projects must expect some delay, compromises, and modifications in projects to maintain certain environmental conditions and avoid the extinction of these rare fishes [1]. However, this rationale is not new, Wheelwright [19] pointed out that the majority of nearly 150 conflicts between the provisions of Section 7 of the 1973 Endangered Species Act and planned, federally involved projects between 1973 and 1977, had been resolved through negotiation and compromise. The majority of Section 7 work in Colorado and Utah, dealing with fish that the FWS has been involved in, has also resulted in some form of negotiation and compromise. In the future we will have to be even increasingly selective with respect to projects allowed to proceed and even more demanding with respect to modifications and constraints to minimize undesirable environmental alterations.

Since the demand for water appears to be the issue concerning the future survival of the native upper basin fish, we must reduce this demand or face the fact that extinction of some unique fish and additional loss and modification of natural riverine ecosystems will occur. Perhaps we should look more to moderating or minimizing our demands for still more water. We cannot afford to be extravagant. The following is a list of some potential means for reducing our future water demands.

1. Use of energy alternatives (solar, etc.).
2. Conservation methods (less water use for domestic purposes, more efficient irrigation systems like those used in some arid countries, etc.).
3. Reduce the level of planned activities.
4. Development of new technologies (where cost/benefit ratios usually prohibit this aspect, the future price tags on water might make them more feasible).
5. Use of ground water that has little or no effect on surface water (Spofford [20], estimated that the recoverable reserve of ground water in the upper 30.5 m (100 ft) of saturated rocks may be as much as 142×10^9 m^3 (115 MAF). However, about 99×10^9 m^3 (80 MAF) (70%) is saline). Care is required if ground water is used because ground water sources can be easily polluted and lost, especially if not readily recharged.
6. Weather modifications (cloud seeding).

With the knowledge that we have about the impacts of dams and reservoirs on the endangered fish species, there is a definite need to consider these species in the planning and operation phases. It will be necessary to have flow and temperature regimes that are much more favorable to these species. Factors, if implemented, that might help to balance the construction of impoundments and at the same time preserve much of the habitat for endangered species include the following:

1. Purchase water rights to replace depletions.
2. Habitat manipulation (construction of nursery areas, backwater habitats, etc.).
3. Construction of fish passage ways (ladders, etc.).
4. Supplemental stocking programs (artificial propagation).

Bishop and Porcella [16] sized up the situation very aptly when they said that the future ability of the Colorado River to sustain its unique fisheries will be dependent on maintaining the quantities and qualities of water required for them.

LITERATURE CITED

1. Behnke, R. J., and D. E. Benson. 1980. Endangered and threatened fishes of the Upper Colorado River Basin. Cooperative Extension Service., Colo. St. Univ., Fort Collins, Colo. Bull. 503A. 34 p.

2. Tyus, H. M., B. D. Burdick, R. A. Valdez, C. M. Haynes, T. A. Lytle, and C. R. Berry. 1981. Fishes of the Upper Colorado River Basin: Distribution, abundance, and status. Prep. for the End. Colo. River Fish. Symp., Am. Fish. Soc., Albuquerque, NM. 18 Sept. 1981.

3. Binns, N. A. 1978. Habitat structure of the Kendall Warm Springs, with reference to the endangered Kendall Warm Springs dace, Rhinichthys osculus thermalis. Fish. Tech. Bull. No. 4, Wyoming Game and Fish Dept., Cheyenne, Wyo.

4. Seethaler, K. 1978. Life history and ecology of the Colorado squawfish (Ptychocheilius lucius) in the Upper Colorado River Basin. M.S. Thesis. Utah State Univ., Logan, UT. 156 p.

5. Girard, C. 1856. Researches upon the cyprinid fishes inhabiting the fresh water of the United States of America, west of the Mississippi Valley, from specimens in the museum of the Smithsonian Institution. Proc. Acad. Nat. Sci., Phila. 8:165-213.

6. Miller, R. R.   1946. The need for ichthyological surveys of the major rivers of western North America. Science 104(2710):517-519.

7. Valdez, R. A., and G. H. Clemmer. 1981. *Gila cypha* and *Gila elegans* life history and prospects for recovery. Prep. for the End. Colo. River Fish. Symp., Am. Fish. Soc., Albuquerque, NM. 18 Sept. 1981.

8. Baird, S. F., and C. Girard. 1853. Fishes. 148-152 pp. *In:* Capt. L. Sitgraves' Rept. of an expedition down the Zuni and Colorado Rivers. U.S. Senate Executive Document 59, 32nd Congress, 2nd Session.

9. Molles, M. 1980. The impacts of habitat alterations and introduced species on the native fishes of the Upper Colorado River Basin. 163-181 pp. *In:* Spofford, W. D., A. L. Parker, and A. V. Kneese (eds.). Energy development in the southwest. Vol. II. Resources for the Future RES paper R-18. Baltimore, MD: Johns Hopkins Univ. Press. 541 p.

10. Holden, P. B. 1980. The relationship between flows in the Yampa River and success of rare fish populations in the Green River System. Bio/West PR-31-1. Logan, UT.

11. Suttkus, R. D., and G. H. Clemmer. 1977. The humpback chub *Gila cypha*, in the Grand Canyon area of the Colorado River. Occas. Pap. Tulane Univ., Mus. Nat. Hist. 1:1-30.

12. Joseph, T. W., J. A. Sinning, R. J. Behnke, and P. B. Holden. 1977. An evaluation of the status, life history, and habitat requirements of endangered and threatened fishes of the Upper Colorado River system. FWS/OBS Rept. No. 24, Ft. Collins, CO. 183 p.

13. Vanicek, C. D., R. H. Kramer, and D. R. Franklin. 1970. Distribution of Green River fishes in Utah and Colorado following closure of Flaming Gorge Dam. Southwest Nat. 14(3):297-315.

14. Harris, R. E., H. N. Sersland, and F. P. Sharpe. 1981. Realities of providing water for endangered fishes in the Upper Colorado River system. Prep. for the End. Colo. River Fish. Symp., Am. Fish. Soc., Albuquerque, NM. 18 Sept. 1981.

15. U.S. Department of the Interior, Bureau of Reclamation. 1981. Upper Colorado Region projected water supply and depletions. Salt Lake City, Utah. Sept. 14 pp.

549

16. Bishop, A. B., and D. B. Porcella. 1980. Physical and ecological aspects of the Upper Colorado River Basin. 17-56 pp. In: Spofford, W. D., A. L. Parker and V. A. Kneese (eds.). Energy development in the southwest. Vol. I. Resources for the Future RES paper R-18. Baltimore, MD: Johns Hopkins Univ. Press. 523 p.

17. El-Ashry, M. T. 1980. Physical and ecological aspects of the Upper Colorado River basin: A discussion. 68-78 pp. In: Spofford, W. D., A. L. Parker and A. V. Kneese (eds.). Energy development in the southwest. Vol. I. Resources for the Future RES paper R-18. Baltimore, MD: Johns Hopkins Univ. Press. 523 p.

18. U.S. Department of the Interior, Water for Energy Management Team. 1974. Water for energy in the Upper Colorado River Basin. Washington, D.C.

19. Wheelwright, J. 1977. The furbish lousewart is no joke. new Repub. 176(20):9-12.

20. Spofford, W. O. 1980. Potential impacts of energy development on stream flows in the Upper Colorado River Basin. 351-429 pp. In: Spofford, W. D., A. L. Parker and A. V. Kneese (eds.). Energy development in the southwest. Vol. I. Resources for the Future RES paper R-18. Baltimore, MD: Johns Hopkins Univ. Press. 523 p.

CHAPTER 32

THE EFFECTS OF LIMITED FOOD AVAILABILITY
ON THE STRIPED BASS FISHERY IN LAKE MEAD

J.R. Baker
L.J. Paulson
    Lake Mead Limnological Research Center
    University of Nevada, Las Vegas

INTRODUCTION

The original range of striped bass (Morone saxatilis) was along the Atlantic Coast. They were introduced into the lower Sacramento River in 1879 and are now also found along the Pacific Coast [1]. A landlocked striped bass fishery was established in Santee-Cooper Reservoir, South Carolina, in 1954, and they have since been introduced into numerous other reservoirs, including Lake Havasu, Lake Mead and Lake Powell on the Colorado River.

Striped bass were introduced into Lake Mead in 1969 in response to declines in the largemouth bass (Micropterus salmoides) fishery that occurred during the 1960s and in order to further utilize the forage base of threadfin shad (Dorosoma petenense). Natural reproduction of striped bass was documented in 1973 [2], and a highly successful fishery developed during the late 1970s. Striped bass comprised 40.1% of the total angler catch in 1979 [3].

The development of the striped bass fishery in Lake Mead was not without cost. A stocking program of rainbow trout (Salmo gairdneri) and other salmonid species was started in 1969. This was also initiated to utilize the surplus threadfin shad production. The trout fishery was considered good from 1970 to 1975, when they comprised 13 to 19% of the total angler catch. This declined to 1% in 1976, despite increased stocking [2]. Food habit studies conducted during this period revealed that rainbow trout occurred in 23% of the striped bass stomachs. The decline in the trout fishery was attributed primarily to predation by striped bass [2]. The occurrence of other gamefish species in striped bass stomachs was low, but threadfin shad comprised 50% of their diet [2]. Striped bass are noted for their voracious appetites and their ability to exploit shad in limnetic areas of reservoirs. This resulted in over exploitation of shad in Santee-Cooper Reservoir, South Carolina [4].

Shad production is closely linked to phytoplankton
productivity because of their planktivorous feeding habits.
Phytoplankton productivity in Lake Mead declined considerably after Lake Powell was formed in 1963 [5], and most of
the reservoir is now oligotrophic-mesotrophic [6]. Shad in
Lake Mead are, therefore, extremely vulnerable to possible
over exploitation by striped bass. The purpose of this paper
is to describe how rapid growth of the striped bass population altered the relative abundance of threadfin shad and
how food limitation may be a factor in limiting future
success of the fishery.

LAKE MEAD DESCRIPTION

Lake Mead was formed in 1935 by the construction of
Hoover Dam and occupies a 183 km reach of the Colorado River
on the Arizona-Nevada border. Morphometric characters of
Lake Mead are given in Table I. Major reaches consist of
Gregg, Temple, and Virgin Basins, collectively referred to
as the Upper Basin, and Boulder Basin referred to as the
Lower Basin (Figure 1). There are also two large embayments,
the Overton Arm of Virgin Basin, which receive discharges
from the Muddy and Virgin Rivers, and Las Vegas Bay, a large
bay of Boulder Basin, which receives secondary-treated sewage effluents from Las Vegas metropolitan area via Las Vegas
Wash. The Upper Basin has been classified as oligotrophic,
Boulder Basin as mesotrophic and Las Vegas Bay as meso-
trophic-eutrophic [6].

Table I. Morphometric Characteristics of Lake Mead from
        Paulson, Baker and Deacon [7].

Parameter	Lake Mead
Maximum operating level (m)	374.0
Maximum depth (m)	180.0
Mean depth (m)	55.0
Surface area (km^2)	660.0
Volume (m^3 x 10^9)	36.0
Maximum length (km)	183.0
Maximum width (km)	28.0
Shoreline development	9.7
Discharge depth (m)	83.0
Annual discharge (1977) (m^3 x 10^9)	9.3
Storage ratio at maximum operating level (years)	3.9

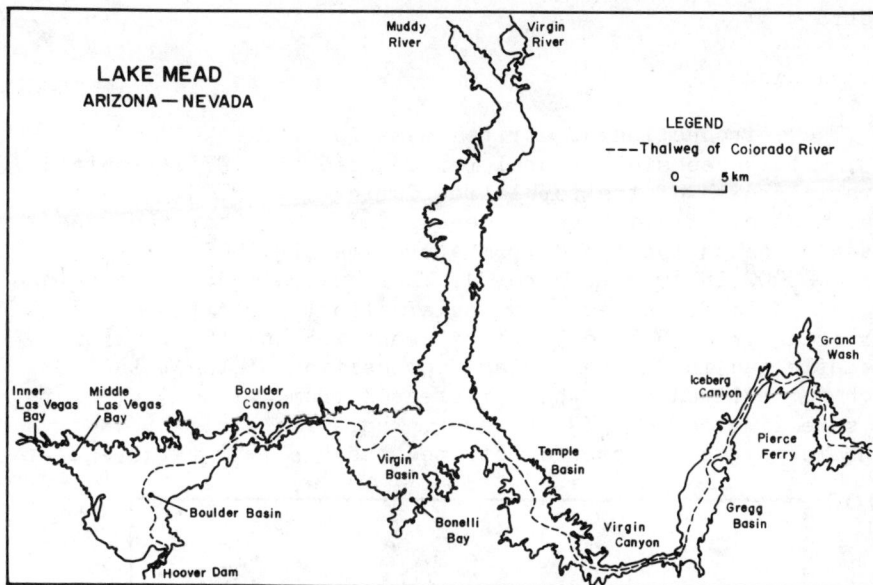

Figure 1. Map of Lake Mead.

The higher trophic state of Boulder Basin and Las Vegas Bay is due to high nutrient loading from Las Vegas Wash. Approximately 60% of the total phosphorus load to Lake Mead in 1977-78 was derived from the Las Vegas Wash inflow [8]. An advanced wastewater treatment plant is being constructed in Las Vegas. Phosphorus loading to Las Vegas Bay will decrease substantially in the future if the plant is operated to specifications. This will probably result in reduced phytoplankton growth in the Las Vegas Bay and Boulder Basin.

METHODS AND DATA SOURCES

Echo-sounding surveys have been made frequently in Lake Mead as part of water quality investigations conducted by Dr. James E. Deacon, University of Nevada, Las Vegas (UNLV) and by the Lake Mead Limnological Research Center (UNLV). Echo-sounding transects were run at various locations in Lake Mead at an approximate speed of 5 mph for 5-15 min at least monthly during 1972, 1974-75, and 1980. A Furuno (FM-22A) recording echo-sounder was used in the surveys. This instrument sounds at a frequency of 50 KHz and the transducer has a 28° beam angle.

Data on striped bass were derived from Nevada Department of Wildlife Job Progress Reports [3,9].

RESULTS

Striped Bass

The introduction of striped bass into Lake Mead in
1969, with successful natural reproduction in 1973, resulted
in rapid growth of the population during the 1970s. Nevada
Department of Wildlife creel census data show that the annual angler catch increased from approximately 1500 in 1973 to
over 400,000 in 1979 (Figure 2). There was a major change in
the Lake Mead fisheries with angler effort for striped bass
increasing from 15.5 to 51.5% between 1978 and 1979. This
was reflected in the percentage composition of the total
catch for striped bass which increased from 4.1 to 40.1% for
the same time period [3]. However, there was a slight decline in the total catch of striped bass in 1980 (Figure 2).

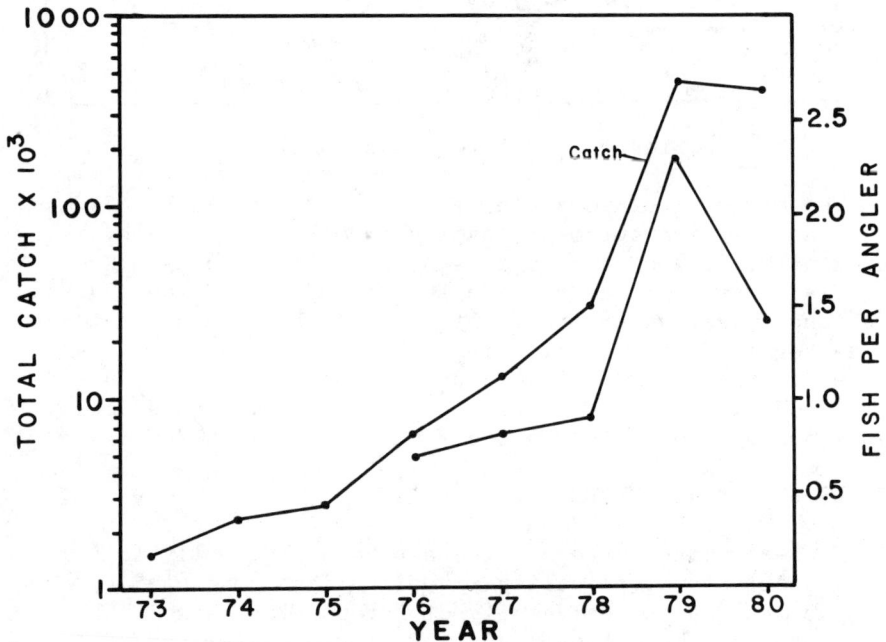

Figure 2. Total Angler Catch and Fish Per Angler for
Striped Bass 1973-80 [3,9].

At this time, fishermen complained that the incidence
of large striped bass in the catch was decreasing and that a
large percentage of the fish caught were emaciated and in
poor condition [10]. There was a marked decline in the overall condition factor (Figure 3) of striped bass in 1980,
substantiating complaints from fishermen. The poor overall
condition factor in 1980 was not related to a shift in the
size range of fish in the catch as the condition factors of

554

most size classes were lower, especially in the smaller fish
15-30 cm [3].

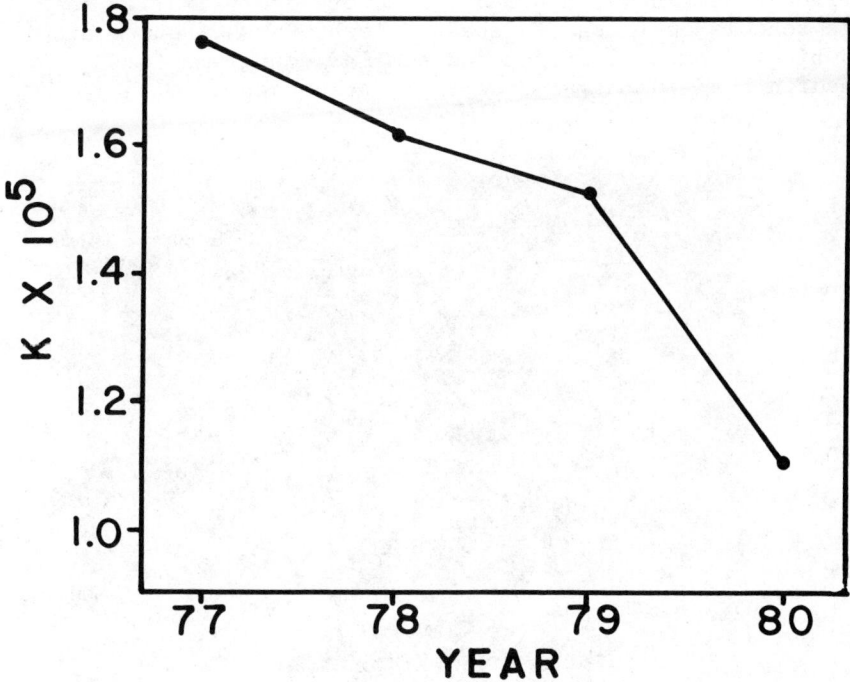

Figure 3.   Striped Bass Condition Factor, 1977-80 [3,9].

        Annual die-offs of striped bass, in the size range of
56-71 cm, occurred regularly in the spring of 1976-79 [11].
These fish were mainly males, and, because the die-offs
occurred after spawning, it was suspected that they resulted
from post-spawning stress. The Nevada Department of Wildlife
initiated an investigation in 1980 to determine the actual
cause for the annual die-offs; however, there was no evidence of a die-off in 1980. Striped bass were nonetheless
collected in July for autopsy. The autopsies revealed that
"...Lake Mead striped bass appear thin, had few parasites,
and few skeletal abnormalities; however, over half (57%) of
the fish had liver abnormalities" [11]. These findings, plus
the poor condition factor, indicated that a nutritional
problem existed and that food may have become limiting.

Threadfin Shad

        Echo-sounding conducted during the early 1970s revealed
that threadfin shad schools were extremely abundant throughout the epilimnetic waters of Lake Mead. This was indicated

by the high density of inverted cones on the echograms
presented in Figure 4. Mid-water trawling conducted by the
Nevada Department of Wildlife [2] and fish trapping [12]
confirmed that these inverted cones were due primarily to
large schools of threadfin shad. Netsch, Kersh, Houser and
Kilambi [13] have also recorded similar echograms for
threadfin shad in Beaver Reservoir, Arkansas.

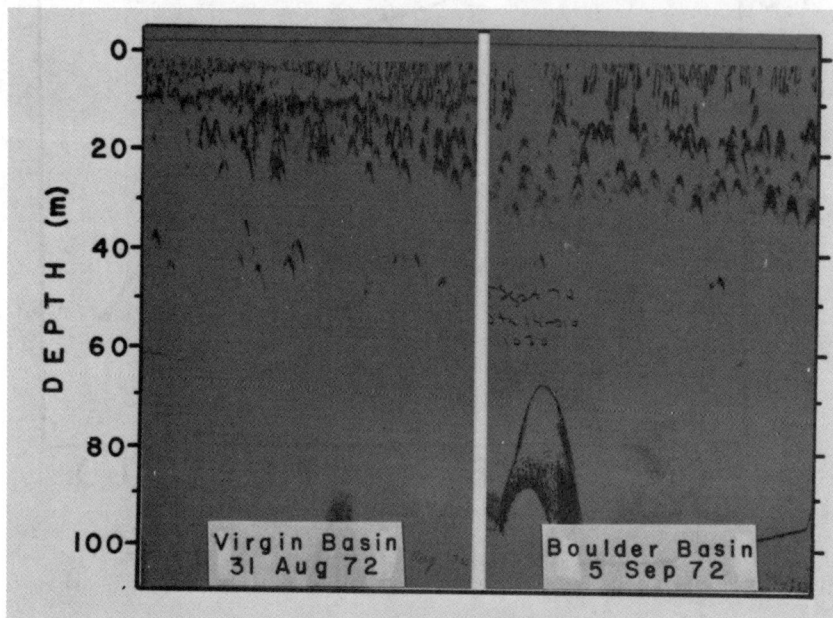

Figure 4.   Echograms from Virgin and Boulder Basins, 1972.

        Although it was not possible to determine absolute
abundance of threadfin shad from the echograms, it was still
evident that shad were numerous in 1975 (Figure 5). Echo-
sounding surveys were not conducted during 1976-79, but
those made during 1980 (Figure 6) showed that threadfin shad
schools were nearly absent in limnetic areas of Lake Mead.
This was surprising since threadfin shad are primarily limnetic in their distribution [13,14]. Scattered schools were
still observed in the littoral areas, and they were fairly
abundant in parts of Las Vegas Bay near the sewage inflow
(Figure 7). Again, no quantitative estimate of the population can be made, but it is apparent that there was a major
decline in threadfin shad abundance in the limnetic areas
sometime during the late 1970s. The decline in threadfin
shad was not due to a winter kill as Lake Mead temperatures
rarely fall below 12°C which is above their critical minimum
temperature of 9°C [15].

556

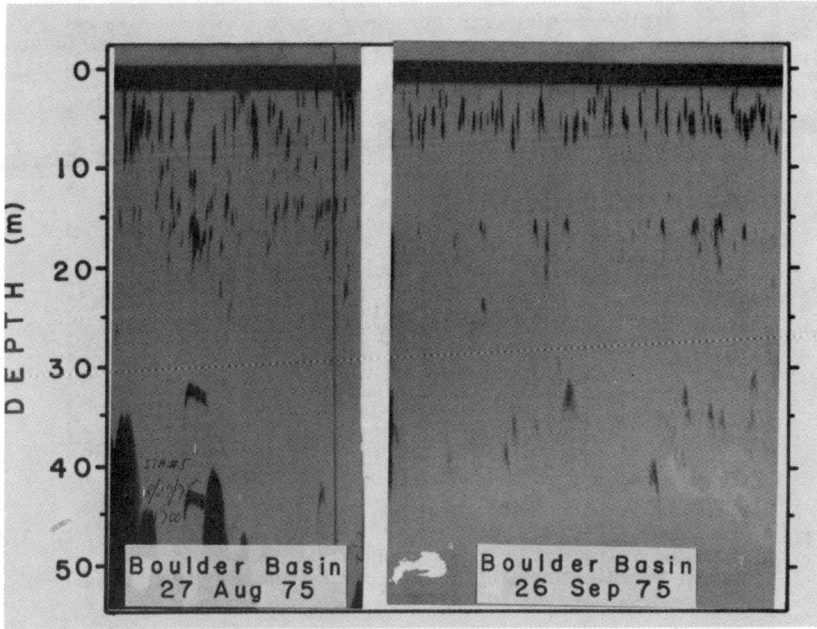

Figure 5. Echograms from Boulder Basin, 1975.

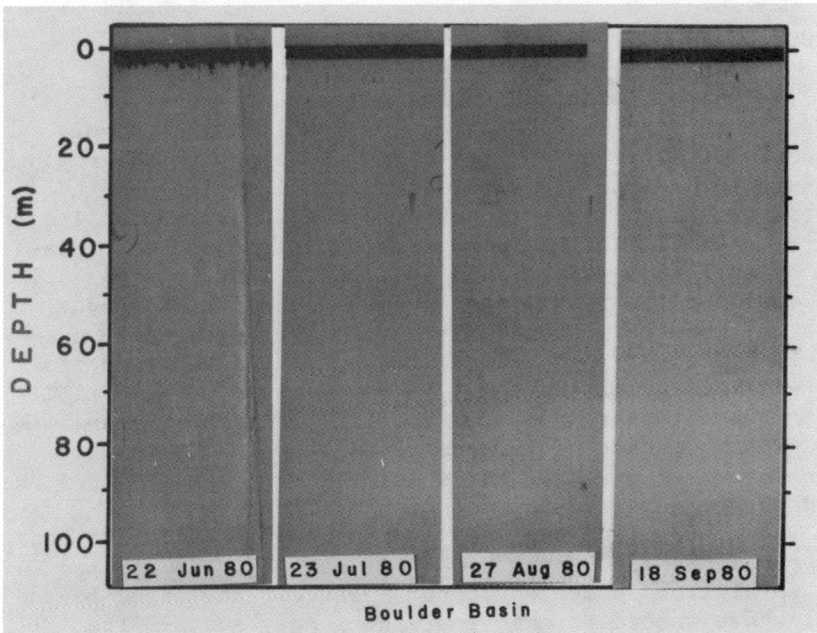

Figure 6. Echograms from Boulder Basin, 1980.

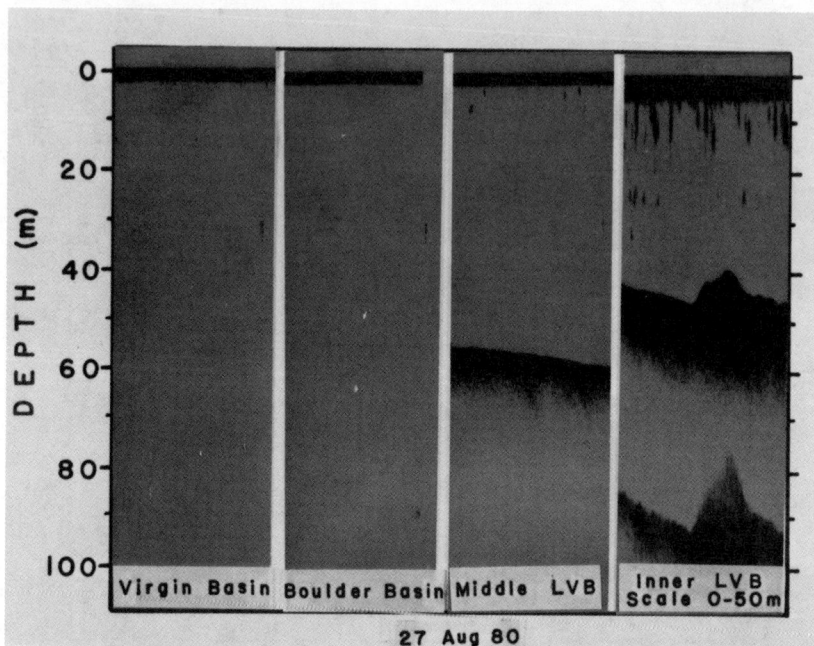

Figure 7.   Echograms from Virgin Basin, Boulder Basin and
             Las Vegas Bay, August, 1980.

DISCUSSION

     The introduction and subsequent establishment of
striped bass into Lake Mead in 1969 resulted in a highly
successful fishery during the 1970s. However, rapid growth
of their population was associated with a decline in the
rainbow trout and threadfin shad populations. The decline in
the trout population has already been attributed to predation by striped bass [2], and it appears that this was also
the cause for the recent decrease in threadfin shad abundance in limnetic areas. Striped bass were apparently extremely efficient in utilizing the surplus shad production
that existed during the early 1970s. However, over exploitation seems to have occurred sometime between 1975 and
1979. The incidence of large striped bass in the angler
catch decreased in 1980, and a large percentage of the fish
taken were emaciated and in poor condition.
     The pattern in Lake Mead is remarkably similar to what
occurred in Santee-Cooper Reservoir in 1960. The striped
bass population in Santee-Cooper Reservoir increased to the
point where threadfin shad could no longer sustain the
predator pressure. This ultimately led to a 30 to 50% decrease in their population. Minckley [16] predicted that the

forage base in Lake Mead and other Colorado River reservoirs was also susceptible to over exploitation by striped bass. This seemed to be compounded in Lake Mead due to low fertility and productivity that developed as a result of decreased nutrient loading from the Colorado River after Lake Powell was formed in 1963 [5]. Chlorophyll-a concentrations in Lake Mead averaged only 1.3 mg/m^3 in the Upper Basin and 3 mg/m^3 in Boulder Basin during 1977-78 [6]. In Las Vegas Bay, near the sewage inflow, chlorophyll-a concentrations averaged 7 mg/m^3 and ranged as high as 23 mg/m^3 during summer [6].

Rinne, Minckley and Bersell [17] found that the horizontal distribution and abundance of threadfin shad was directly related to chlorophyll-a concentrations in the Salt River reservoirs. Zooplankton standing crops were affected both by levels of phytoplankton productivity and threadfin shad predation and were not therefore correlated with fish abundance [17]. Average annual zooplankton standing crops in Lake Mead were generally related to chlorophyll-a concentrations during 1977-78, but there were seasonal variations in this relationship and marked differences in the response of various zooplankton groups [6]. Preliminary experiments conducted during the summer of 1981 indicate that chlorophyll-a concentrations were sufficient to maintain optimal growth and reproduction of Daphnia in Las Vegas Bay, but concentrations in Boulder Basin were not adequate (Baker unpubl. data). Wilde [18] reported that zooplankton abundance in Boulder Basin has decreased considerably since 1971. He attributed this to decreased productivity because it closely paralleled reductions in nitrate loading from the Colorado River after 1970 as well as decreases in chlorophyll-a concentrations in the inner Las Vegas Bay since 1972 [19].

Threadfin shad extensively utilize zooplankton and phytoplankton as food resources in Lake Mead [20], and it is likely that historic reductions in zooplankton abundance have influenced their populations. Echo-sounding surveys revealed that shad were still fairly abundant in Las Vegas Bay during 1980 indicating that perhaps their abundance does increase at higher levels of phytoplankton productivity. The productivity in Las Vegas Bay will probably further decline if phosphorus loading is significantly decreased by operation of the advanced wastewater treatment plant. Studies are presently being conducted to determine what levels of treatment are appropriate to protect beneficial uses. In this regard, the trophic relationships in Lake Mead should be further investigated to determine if present and future phytoplankton production is adequate to maintain shad production at levels necessary to have a productive striped bass fishery.

559

REFERENCES

1. Setzler, E.M., W.R. Boynton, K.V. Wood, H.H. Zion, L.
   Lubbers, N.K. Mountford, P. Frere, L. Tucker and J.A.
   Mihursky. 1980. Synopsis of biological data on striped
   bass, Morone saxatilis Walbaum. NOAA Tech. Rept. NMFS
   Circular 433. FAO Synopsis No. 121. 69 pp.

2. Allen, R.C. and D.L. Roden. 1978. Fish of Lake Mead and
   Lake Mohave. Nev. Dept. of Wildlife, Biological Bull.
   No. 7. 105 pp.

3. Nevada Department of Wildlife. 1980. Job progress
   report for Lake Mead. Proj. No. F-20-17. 209 pp.

4. Goodson, L.F. 1966. Landlocked striped bass. Pages
   407-412 in Al Calhoun, ed. Inland fisheries management.
   Calif. Dept. of Fish and Game.

5. Prentki, R.T., L.J. Paulson and J.R. Baker. 1981.
   Chemical and biological structure of Lake Mead sediments. Lake Mead Limnological Res. Ctr. Tech. Rept. No.
   6. Univ. Nev., Las Vegas. 89 pp.

6. Paulson, L.J., J.R. Baker and J.E. Deacon. 1980. The
   limnological status of Lake Mead and Lake Mohave under
   present and future powerplant operations of Hoover Dam.
   Lake Mead Limnological Res. Ctr. Tech. Rept. No. 1.
   Univ. Nev., Las Vegas. 229 pp.

7. Paulson, L.J., J.R. Baker and J.E. Deacon. 1979.
   Potential use of hydroelectric facilities for manipu-
   lating the fertility of Lake Mead. Pages 269-300 in
   G.H. Swanson, Tech. Coord. The mitigation symposium: A
   national workshop on mitigating losses of fish and
   wildlife habitats. July 16-20, 1979. Fort Collins,
   Colo. Gen. Tech. Dept. No. RM-65, Rocky Mt. Forest and
   Range Exp. Sta.

8. Baker, J.R. and L.J. Paulson. 1980. Influence of Las
   Vegas Wash density current on nutrient availability and
   phytoplankton growth in Lake Mead. Pages 1638-1646 in
   H.G. Stefan, ed. Symposium on surface water impoundments. June 2-5, 1980. Minneapolis, MN.

9. Nevada Department of Wildlife. 1979. Job progress
   report for Lake Mead. Proj. No. F-20-16. 170 pp.

10. Glassburn, Bob. 1980. Mead stripers dying at alarming
    rate. Las Vegas Review Journal, 13 March 1980:130.

11. Sakanari, J. 1981. Lake Mead, Nevada, striped bass collections. Pages 40-41 in Cooperative striped bass study (COSBS), second progress report. Special Prog. Rept. No. 8101-1. Calif. State Water Resources Control Board.

12. Paulson, L.J. and F.A. Espinosa. 1975. Fish trapping: a new method of evaluating fish species composition in limnetic areas of reservoirs. Calif. Fish and Game 61(4):209-214.

13. Netsch, N.F., G.M. Kersh, A. Houser and R.V. Kilambi. 1971. Distribution of young gizzard and threadfin shad in Beaver Reservoir. Pages 95-105 in G.E. Hall, ed. Reservoir Fisheries and Limnology. Special Publ. No. 8. American Fisheries Society, Washington, D.C.

14. von Geldern, C. and D.F. Mitchell. 1975. Largemouth bass and threadfin shad in California. Pages 436-449 in H. Clepper, ed. Black bass biology and management. Sport fishing Institute, Washington, D.C.

15. Hubbs, C. 1951. Minimum temperature tolerance for fishes of the genera Signalosa and Herichthys in Texas. Copeia, 1951:297.

16. Minckley, W.L. 1973. Fishes of Arizona. Sims Printing Co., Inc., Phoenix, Arizona. 293 pp.

17. Rinne, J.N., W.L. Minckley and P.O. Bersell. 1981. Factors influencing fish distribution in two desert reservoirs, central Arizona. Hydrobiologia 80:31-42.

18. Wilde, G.R. 1981. Recent changes in zooplankton standing crops in Lake Mead and their probable effects on largemouth bass fry. Proceedings of the Twenty-fifth Annual Meeting, Arizona-Nevada Acad. Sci., May 1-2, 1981. Tucson, AZ.

19. Paulson, L.J. 1981. Nutrient management with hydro-electric dams on the Colorado River system. Lake Mead Limnological Res. Ctr. Tech. Rept. No. 8. Univ. Nev., Las Vegas. 39 pp.

20. Deacon, J.E., L.J. Paulson and C.O. Minckley. 1972. Effects of Las Vegas Wash effluents upon bass and other game fish reproduction and success. Final rept. Nev. Dept. Fish and Game. 74 pp.

CHAPTER 33

THE EFFECTS OF WATER LEVEL
FLUCTUATIONS ON THE SPAWNING
SUCCESS OF LARGEMOUTH BASS
(*Micropterus salmoides*) IN LAKE MEAD

Sue A. Morgensen
Arizona Game & Fish Department,
Kingman, Arizona

INTRODUCTION

Since the establishment of Lake Powell, the quality
of the largemouth bass fishery in Lake Mead has apparently declined (1,2,3,4,5). In 1977, Arizona Game & Fish
Department (AGFD) and Nevada Department of Wildlife (NDOW)
initiated a study to evaluate factors causing this decline.
The study was funded by the Bureau of Reclamation (BR) for
a 5 year period.

Among the changes that occurred with the establishment
of Lake Powell was a change in operating criteria for
Lake Mead (6). Prior to Powell, Lake Mead water levels
increased through the spring and summer. Currently, the
lake level is being drawn down during the spring/summer
months and increased in the fall/winter period.

Hogue (7) and Summerfelt and Shirley (8) found that
strong year-classes of bass were produced when water
levels were increased during the spring. Poor year-
classes resulted when water levels were decreased in May
and June. Johnson (9) felt that poor bass catch rates in
Lake Carl Blackwell were due to decreased spawning success
as the result of water level fluctuations. Summerfelt (10)
suggests that year-class strength of largemouth bass is
determined by events occurring within the first few weeks
of life. Bass are most susceptible to sharp declines in
temperature and the effect of wave action. Jonez and Wood
(11) expressed concern that spring drawdowns impair bass
production in Lake Mead. As part of the 5 year study on
Lake Mead, an effort was made to evaluate what effect, if
any, spring drawdown has on bass production. Bass nesting,
survival and water level were closely monitored from 1978
through 1981.

STUDY AREA

Lake Mead is a BR impoundment located midway in a chain of reservoirs along the Colorado River. Impounded by Hoover Dam in 1934, Lake Mead was created to provide flood control; water for agricultural, municipal and industrial needs; and hydroelectric power. Lake Powell was created by the construction of Glen Canyon Dam upstream in 1963. This development altered the operating criteria for Lake Mead.

At maximum storage, Lake Mead covers 63,902 ha, has a volume of 32,937 x $10^6 m^3$ and emcompasses 1,363.8 km of shoreline (12). It is located on the Arizona-Nevada border. For purposes of this study, stations sampled by AGFD personnel were located in the upper basin from Temple Basin to Iceberg Canyon (Figure 1). NDOW personnel performed similar analyses at stations located in the lower basin area.

METHODS

Initially, four stations in the upper basin were selected for nesting and survival studies. Stations 4, 11, 14, and 16 (Figure 1) represented different types of cove morphology in the area.

Beginning in March, underwater surveys of these stations were conducted every 3-4 days. When bass nests were located, a numbered, 5 cm, fluorescent orange flag was placed within a 60 cm radius of the nest. Nest status (i.e. eggs, sac-fry, swim-up fry, empty), temperature at the nest and depth of the nest below the surface were recorded for each dive. Nest duration was calculated as the time observed in developing from egg to swim-up fry. The BR provided information on daily lake elevations. The National Park Service (NPS) at Meadview provided weather data.

Initially, nesting success was divided into three categories: minimum success (nests observed in the swim-up fry stage); maximum success (all nests that could have developed to swim-up fry whether or not actually observed in the swim-up fry stage); and unsuccessful (nests known to be destroyed prior to reaching the swim-up fry stage). In 1980, the categories were changed to successful (nests actually observed in swim-up fry stage) and unsuccessful (all other nests) to conform with NDOW reporting procedure. To provide a more accurate assessment of nest success, the interval between dives was decreased to every other day at stations 4 and 11. Stations 14 and 16 were dropped from

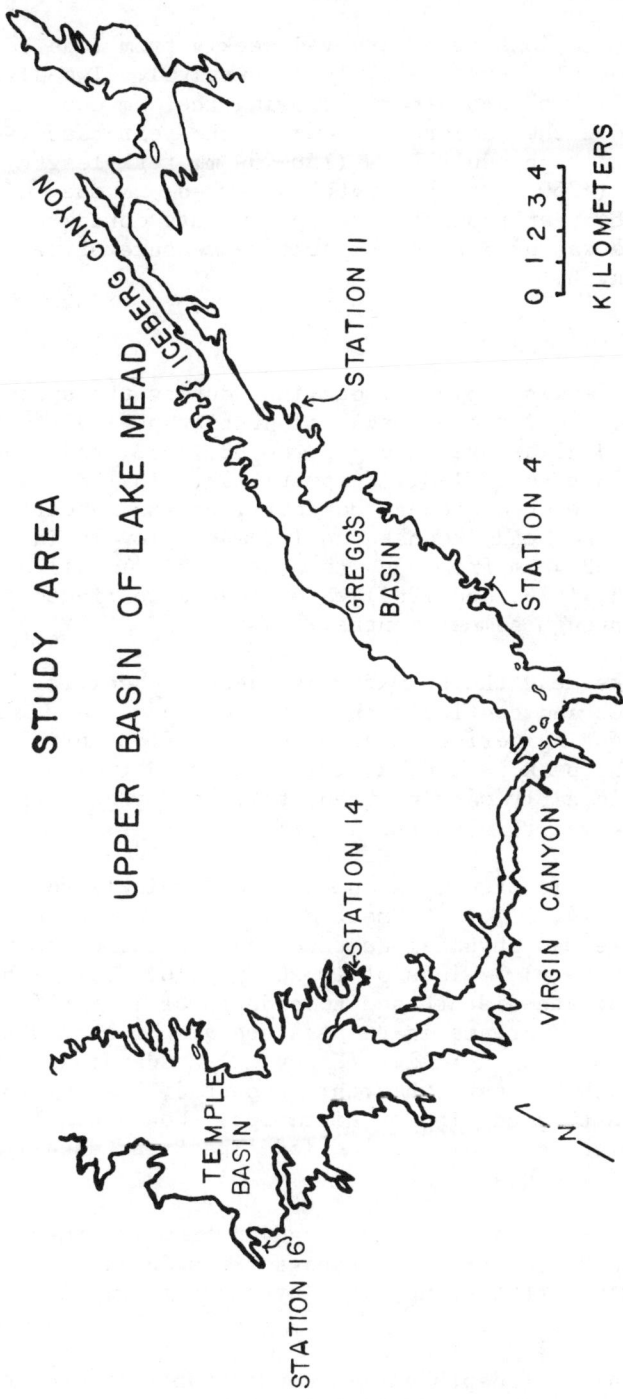

Figure 1. Map of Upper Basin of Lake Mead showing locations (stations 4, 11, 14, and 16) monitored for largemouth bass nesting and survival.

the nesting portion of the study.

All 4 stations were surveyed weekly from June to September of all years to evaluate survival. Two divers swam parallel to each other following the 5 m contour. They recorded the number of young of the year bass (<150 mm fork length), sub-adult bass (150-250 mm fork length) and adult bass (>250 mm fork length) observed. Number of young of the year bass observed per minute during an x minute dive was used as a comparative measure of year-class strength.

RESULTS

Water levels typically declined during the spring (Figure 2). During the peak bass nesting period (April 6-24) this decline averaged 1.38, 0.88, 1.32, and 3.94 cm/day for 1978 through 1981, respectively. In 1980, when Lake Powell reached storage capacity, excess runoff was stored in Lake Mead. Consequently, water levels in Lake Mead rose 152.4 cm from June to September. In all other years (1978, 1979, and 1981) water levels continued to decrease through summer months.

In Lake Mead the primary bass nesting activity began in mid-March and continued through the end of May (Table I). The peak nesting period usually occurred from April 6-24. In 1978, the peak nesting activity occurred from March 20-April 4. An early warming trend followed by cold fronts may have caused this shift.

Spawning is often disjunctive due to inclement weather (13,14,15). The most commonly observed cause of nest failure was physical destruction from wind and wave action (Figure 3). Winds of 10-20 mph (16.1-32.2 kph) caused substrate disturbance to a depth of 5 feet (152.4 cm). A sharp decrease in temperature also caused disruption of nesting activity (Figure 3). Guardians tended to abandon their nests when this occurred, leaving them susceptible to predation or siltation. Consequently, two or three distinct spawning 'periods' per year usually developed (Figure 4).

Temperatures at the nests ranged from $12.5^{o}C$-$25.0^{o}C$. The mean water temperature at nest sites for all years was $17.7^{o}C$. Most nesting activity occurred within $\pm2^{o}C$ of the mean.

Initial nest depth varied considerably from cove to cove and between years (Table I). This was probably due

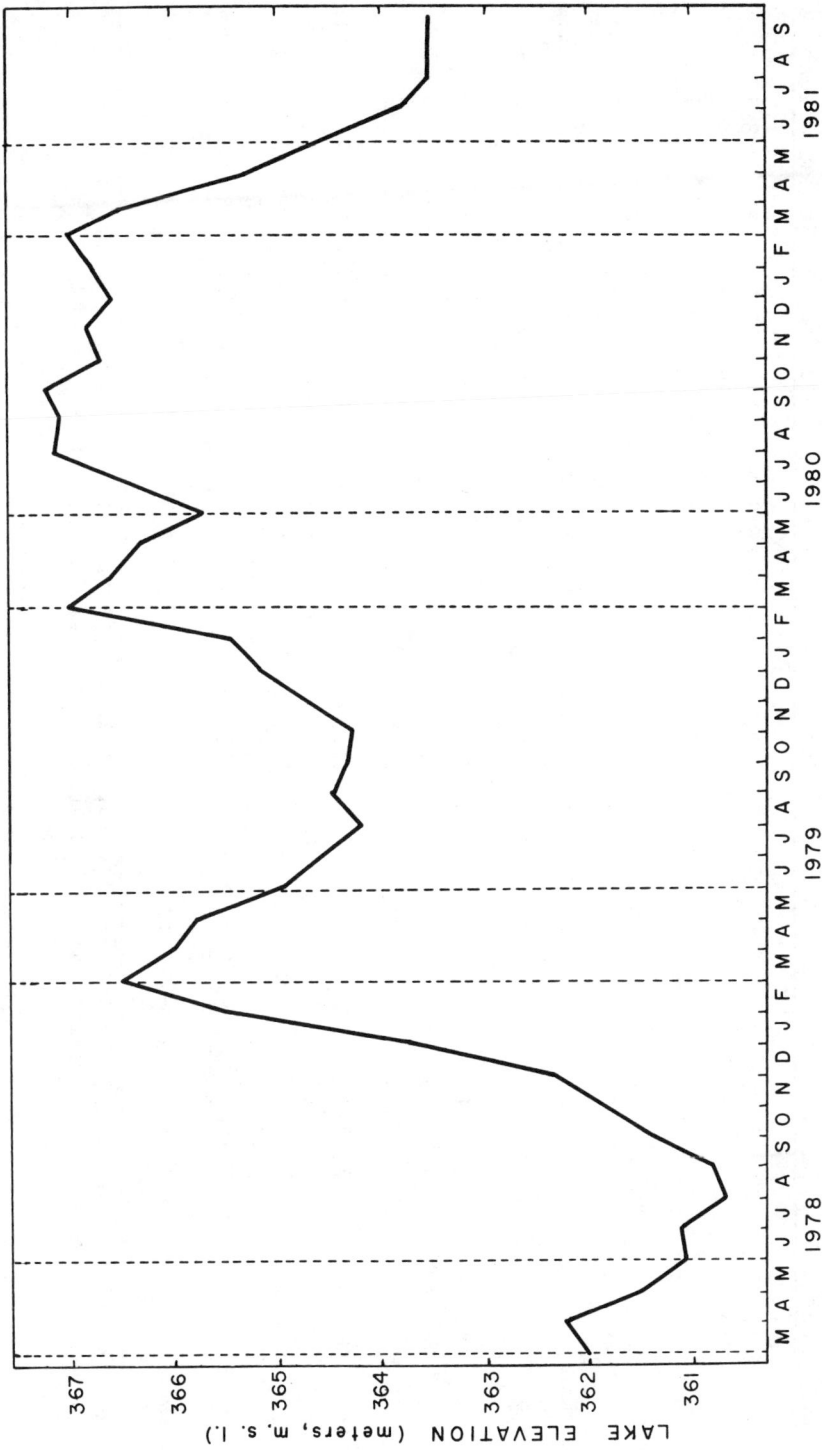

Figure 2. Water level (meters, m.s.l.) of Lake Mead, AZ-NV, March 1978 to Sept. 1981. Bass nesting periods indicated by dotted lines (16).

Table I. Temperature, depth, duration and success of largemouth bass nests on Lake Mead, 1978-1981

Year	Location	Dates	Peak Period	Init. Temp.-(°C) Range	(Mean)	Init. Depth-(cm) Range	(Mean)	Duration-(Days) Range	(Mean)	No. Nests	Percent Successful
1981	Sta 4	3/18-5/18		16.0-23.0	(19.2)	46-305	(122)	4-19	(10.3)	31	58.1
	Sta 11	3/12-5/11		14.6-22.5	(17.4)	30-229	(122)	5-19	(10.5)	29	44.8
TOTAL		3/12-5/18	4/8-24	14.6-23.0	(18.4)	30-305	(122)	4-19	(10.4)	60	51.7
1980	Sta 4	3/23-5/27		14.0-23.0	(17.1)	61-396	(170)	7-19	(13.2)	22	90.9
	Sta 11	3/9-5/27		12.5-25.0	(17.9)	61-274	(140)	4-19	(10.8)	28	32.1
TOTAL		3/9-5/27	4/8-27	12.5-25.0	(17.5)	61-396	(150)	4-19	(12.4)	50	58.0
1979	Sta 4	4/18-5/9		17.0-25.5	(19.5)	60-225	(153)	8-9	(8.5)	6	50.0-83.3
	Sta 11	4/17-5/11		18.0-21.7	(19.9)	60-100	(80)	8	(8.0)	**2	50.0
	Sta 14	3/26-5/1		14.0-22.0	(17.7)	55-150	(92)	7-13	(10.3)	25	32.0-80.0
	Sta 16	4/6-5/3		16.0-22.8	(18.0)	135-240	(182)	10-17	(12.3)	7	42.9-57.1
TOTAL		3/26-5/11	4/6-22	14.0-25.5	(19.1)	55-240	(117)	8-17	(10.0)	40	57.5*
1978	Sta 4	3/21-5/17		16.5-23.0	(18.9)	61-367	(190)	7-15	(11.6)	18	27.8-77.8
	Sta 11	3/21-5/17		16.5-24.0	(19.1)	46-457	(139)	8-12	(10.5)	22	22.7-90.9
	Sta 14	3/18-5/17		16.0-22.0	(17.9)	47-396	(152)	9-15	(12.8)	21	47.1-90.5
	Sta 16	3/21-5/8		14.5-19.0	(16.0)	183-244	(213)	12-15	(13.0)	4	75-100
TOTAL		3/18-5/17	3/20-4/4	14.5-24.0	(18.2)	46-457	(162)	7-15	(12.7)	65	55.4*
OVERALL		3/9-5/27	4/6-24	12.5-25.0	(17.7)	30-457	(135)	4-19	(11.5)	215	55.3

* Best estimate from stage of nest development at last observation
** Several nests were not monitored -- 11 swarms were observed.

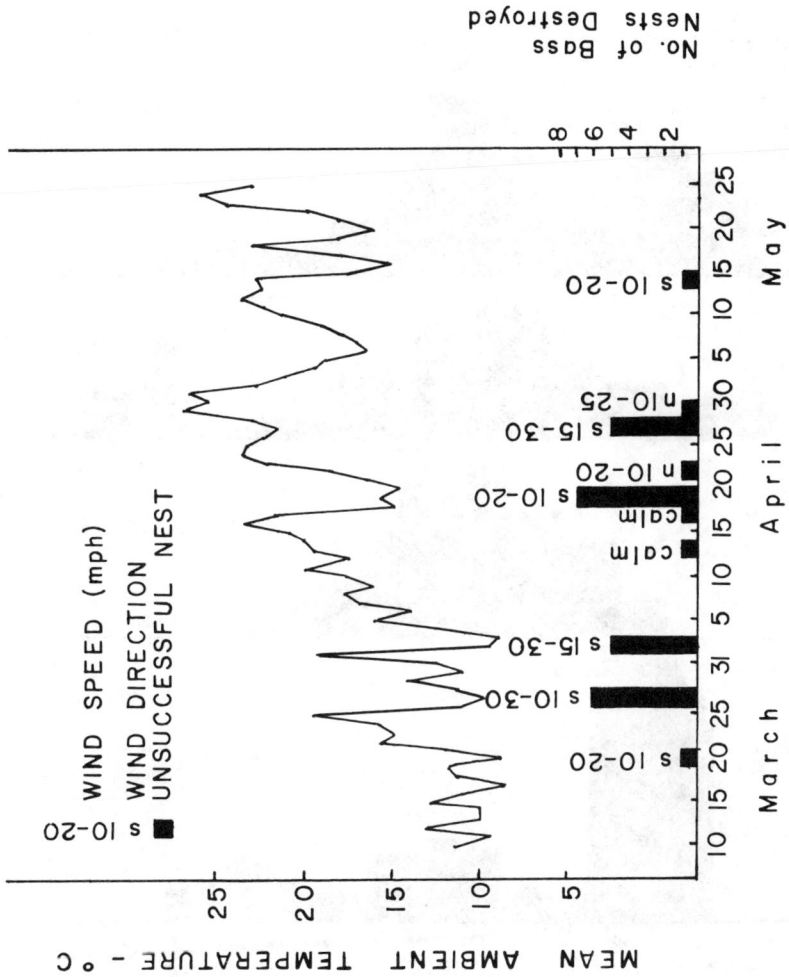

Figure 3. Mean ambient temperature (°C), wind direction, speed (mph), and number of unsuccessful bass nests during a typical (1981) spawning season.

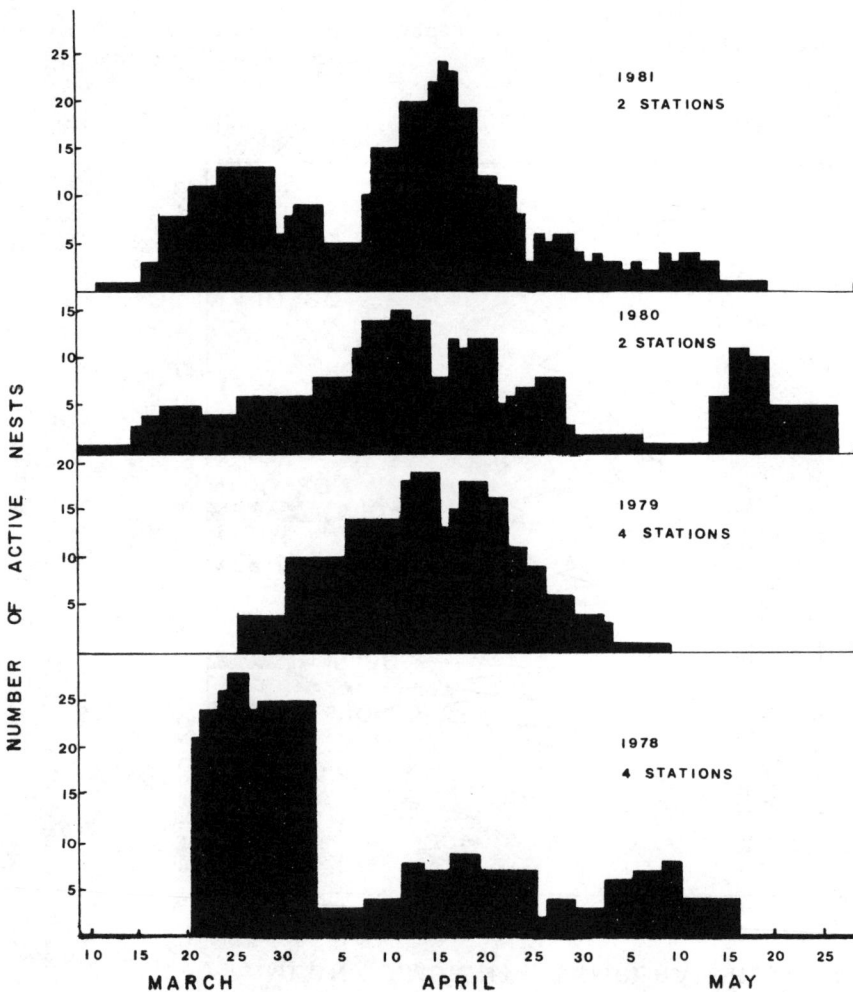

Figure 4.   Number of active largemouth bass nests observed
during March–May, 1978–1981.

to the availability of suitable nesting substrates. Overall, nests ranged from 30 cm to 457 cm deep at construction.
Mean nest depths in 1979 (117 cm) and 1981 (122 cm) were
less than nest depths in 1978 (162 cm) and in 1980 (150 cm).
There did not appear to be a correlation between nest
depth and decline rate of the lake elevation. Water
levels declined most rapidly in 1981 (Figure 2). Nesting
at stations 11 and 14, both steep sided coves, usually
occurred in more shallow water than at stations 4 and 16.
Stations 4 and 16 are shallow and unprotected from prevailing winds.

Duration of nests varied with temperature from 4-19
days (Table II). The mean was 11.5 days at $17.7^{o}C$. This
agrees with the findings of Kramer and Smith (17, 18) and
Laurence (19) .

Table II. Duration of bass nests at various temperatures.

Water Temperature at the Nest (^{o}C)	Mean Duration (days)	Number of Nests
13.5-15.5	14.3	14
16.0-17.5	12.6	36
18.0-19.8	9.4	21
20.0-22.5	7.8	18

Nesting success varied greatly among stations but not
among years (Table I). Annual success rates for the four
years, 1978-1981, were 55.4%, 57.5%, 58.0%, and 51.7%,
respectively. The lowest success rate, 51.7%, was observed in 1981 when water levels dropped faster than in
previous years (Figure 2). Overall success was estimated
at 55.3% for all 4 years.

Survival of young of the year bass was better in 1980,
a year when water levels increased during the summer months.
By September, observations of young of the year bass
averaged 2.1 fish/min. (Figure 5). Survival for 1978 and
1981 was better than 1979. Young of the year bass obser-
vation/min. were 0.9 for September 1978, 0.75 for September 1981 and 0.3 for September 1979. Most mortality
occurred prior to the end of June.

DISCUSSION

Wind and wave action was the most commonly observed
cause of nest failures. Kramer (20) reported destruction

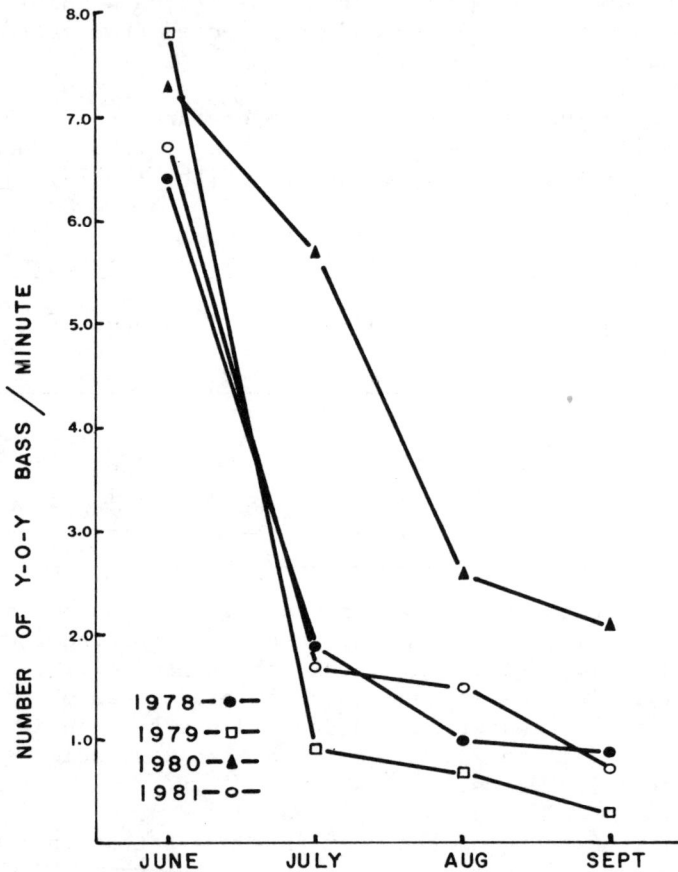

Figure 5. Average number of young of year bass observed per minute snorkel time, June–September, 1978–1981.

of nests located at median depths of 29–30 inches (73.7–76.2 cm) in Lake George Minn. by 17.2–18.6 mph (27.7–29.9 kph) wind. Miller and Kramer (21) found that nests in less than 5 feet (152.4 cm) of water were destroyed by wind and wave action at Lake Powell. Von Geldern (22) observed greater abundance of bass fingerlings in areas protected from prevailing winds. Our findings indicated that southerly winds greater than 15 mph (24.1 kph) over a 2 day period cause nest destruction to a depth of 150 cm. Combined data from the 1980 and 1981 nesting season (Table III) showed that 59.6% of nesting occurs in this depth range. Decreasing water levels resulted in more nests becoming susceptible to wind and wave action. At a drawdown rate of 6.0 cm/day, 88.0% of bass nests would be susceptible to wind and wave destruction. Most successful nests (69%) were located between 30 and 168 cm deep. Kramer (20) found that 63–67% of successful nests in Lake George were located at depths greater than the median.

Table III.   Percent of bass nests located at various depths in Lake Mead and rate of lake level decline at which nests become susceptible to wind and wave destruction.

Depth (cm)	Percent Nests Located	Rate of Lake Level Decline
<30	0.0	Susceptible
<60	2.8	at
<90	20.2	stable
<120	44.0	lake
<150	59.6	levels
<180	77.0	−3.0 cm/day
<210	88.0	−6.0 cm/day
<240	92.6	−9.0 cm/day
<270	95.4	−12.0 cm/day
<300	97.6	−15.0 cm/day
<410	100.0	−27.0 cm/day

Although the mean nesting success on Lake Mead (55.3%) may seem low, it is within published limits. Kramer and Smith (18) reported 72% success in 1957 and 49% success in 1958 in Lake George at relatively stable water levels. Rising water levels (approximately 10 m from April–June 1970) in Lake Powell may have resulted in a nesting success rate as high as 84.3% (23). No nesting information during years of rising water levels is available from Lake Mead.

Hulsey (24) suggested the ideal situation for bass production is maintenance of high water levels in spring and early summer followed by a decrease after forage fish have spawned. Heman et al. (25) found that a midsummer drawdown (June 19-29) improved bass production in Little Dixie Lake. Other investigators have not seen improved bass production when water levels were held constant during nesting (26, 27). Although Kramer and Smith (18) stated that year class strength was set after egg deposition and before fingerlings are 2 weeks old, Benett et al. (26) found that large numbers of fry did not necessarily result in strong year-classes. Aggus and Elliot (28) felt that extreme (4.4-12.6 m/year) water level fluctuation could be an overriding influence on year-class strength in a reservoir with limited habitat and nutrients. High summer water levels were associated with strong year-classes and low water levels with weak year-classes in Bull Shoals Reservoir.

In Lake Mead, increased water levels through the summer of 1980 resulted in better bass survival for this period. Newly inundated vegetation provided cover for young bass as well as increased nutrient input. During the summers of 1978, 1979, and 1981, water levels in Lake Mead declined slowly. Survival for these years was considerably less than 1980. Zweiacker et al. (29) suggest that decreasing water levels have a negative effect on littoral zone invertebrates resulting in decreased growth and survival of young of the year bass.

CONCLUSIONS AND RECOMMENDATIONS

Declining water levels during the springs of 1978-1981 (0.88-3.94 cm/day) apparently had little effect on bass nesting. Success figures ranging from 51.7%-58.0% were sufficient to produce large numbers of bass fry. However, more rapid declines in water level may be harmful to bass nesting, particularly during the peak nesting period in mid-April. Water level decline rates of greater than 6 cm/day will result in more than 85% of nests being susceptible to wind and wave action and should be avoided.

Survival of young of year bass was highest in 1980 when water levels increased 152.4 cm from June to August. This indicated that the bass population in Lake Mead was influenced more by water level fluctuations during their first summer than by conditions prevailing during the nesting period.

ACKNOWLEDGEMENTS

I would like to thank the several employees of the AGFD who have actively participated in the "Black Bass Project". These include Thomas A. Liles, principal investigator, who reviewed this manuscript; W. Bradford Jacobson, project biologist 1977-1979, who conducted preliminary investigation on Lake Mead and established the techniques used; James E. Brooks, project biologist 1979-1980; and Rose A. Lipinski, project assistant, 1981-1982. Funding for this study was provided, in part, by The Bureau of Reclamation under Contract No. 7-07-30-X0028.

LITERATURE CITED

1. AGFD. 1959-1975. Annual statewide survey of aquatic resources reports. Projects F-7-R-2 through F-7-R-17. Phoenix, Arizona.

2. AGFD. 1965-1968. Annual job completion reports. Projects FS-1 through FS-4 Phoenix, Arizona.

3. AGFD. 1973. Annual job completion report. Project FS-7. Phoenix, Arizona.

4. NDOW. 1961-1964. Annual job completion reports, Lake Mead. FA projects F-13-R-4 through F-13-R-7. Las Vegas, Nevada.

5. NDOW. 1965-1975. Annual job completion reports, Lake Mead. FA projects F-20-R-1 through F-20-R-12.

6. USDI. 1975. Coordinated operation of Lake Powell and Lake Mead: Operational Guidelines. U.S. Bureau of Reclamation, Boulder City, Nevada. 26 p.

7. Hogue, J. 1972. An evaluation of high water levels in Bull Shoals Reservoir during spawning season. Admin. Report. Ark. G and F Comm., Mountain Home, Ark. 17 p.

8. Summerfelt, R. C., and K. E. Shirley. 1978. Environmental correlates to year-class strength of largemouth bass in Lake Carl Blackwell. Proc. Okla. Acad. Sci. 58: 54-63.

9. Johnson, J. N. 1974. Effects of water level fluctuations on growth, relative abundance and standing crop of fishes in Lake Carl Blackwell, Oklahoma, Master's Thesis. University of Oklahoma, Stillwater, Okla, 72 p.

10. Summerfelt, R. C. 1975. Relationship between weather and year-class strength of largemouth bass, pages 166-174 in R. H. Stroud and H. Clepper,(eds.) Black Bass Biology and Management. Sport Fishing Institute, Washington, D.C.

11. Jonez, A., and N. Wood. 1964. Tentative findings concerning the drop on Lake Mead in relation to largemouth bass. Special Report, Nev. Dept. of Fish and Game, Las Vegas, Nevada., 4 p.

12. McCall, T. 1980. Fishery investigation of Lake Mead, Arizona-Nevada, from Separation Rapids to Boulder Canyon, 1978-1979. Final report, Contract No. 8-07-30-X0025. U.S. Bureau of Reclamation, 197 pp.

13. Jacobson, B. 1979. The Status of the Black Bass fishery in Lake Mead and a Program toward Restoration and Enhancement, annual report, U.S. Bureau of Reclamation, Contract No. 7-07-30-X0028, 52 pp.

14. Brooks, J. E., and S. A. Morgensen. 1981. The Status of the Black Bass Fishery in Lake Mead and a Program toward Restoration and Enhancement, annual report, U.S. Bureau of Reclamation, contract No. 7-07-30-X0028, 67 pp.

15. Morgensen, S. A., and C. O. Padilla. 1982. The Status of the Black Bass Fishery in Lake Mead and a Program toward Restoration and Enhancement, annual report, U.S. Bureau of Reclamation, Contract No. 7-07-30-X0028, 84 pp.

16. USDI. 1978-1981. Available Reservoir Elevation and Content, Lower Colorado Region. U.S. Bureau of Reclamation, Boulder City, Nevada.

17. Kramer, R. H., and L. L. Smith, Jr. 1960. First-year growth of the largemouth bass, *Micropterus salmoides* (Lacepede), and some related ecological factors. Trans. Am. Fish Soc. 89 (2): 222-233.

18. Kramer, R. H., and L. L. Smith, Jr. 1962. The formation of year classes in largemouth bass. Trans. Am. Fish. Soc. 91 (1): 29-41.

19. Laurence, G. C. 1969. The energy expenditure of largemouth bass larvae, *Micropterus salmoides,* during yolk absorption. Trans. Am. Fish. Soc. 98 (3): 398-405.

20. Kramer, R. H. 1961. The early life history of the largemouth bass, *Micropterus salmoides* (Lacepede), with special reference to factors influencing year class strength. PhD Thesis, University of Minnesota, Minneapolis, 122 pp.

21. Miller, K. D., and R. H. Kramer. 1971. Spawning and early life history of largemouth bass *(Micropterus salmoides)* in Lake Powell, pages 73-84 in G. H. Hall

(ed.) Reservoir Fisheries and Limnology. Am. Fish.
Soc. Spec. Publ. No. 8., Washington, D.C.

22. Von Geldern, C. E., Jr. 1971. Abundance and distri-
     bution of fingerling largemouth bass, *Micropterus
     salmoides,* as determined by electrofishing, at Lake
     Nacimiento, California, Cal. F. and G 57 (4): 228-
     245.

23. Pettengill, T. D. Oct. 1981. Personal communication.
     Lake Powell Fishery Investigations. Utah Division of
     Wildlife Resources, Page, Arizona.

24. Hulsey, A. H. 1957. Effects of a fall and winter
     drawdown on a flood control lake. Proc. 10th Ann.
     Conf. S.E. Assoc. G and F Comm. (1957): 285-289.

25. Heman, M. L., R. S. Campbell, and L. C. Redmond. 1969.
     Manipulation of fish populations through reservoir
     drawdown, Trans. Am. Fish. Soc. 98 (2): 293-304.

26. Bennett, G. W., H. W. Adkins, and W. F. Childers.
     1969. Largemouth bass and other fishes in Ridge Lake,
     Illinois 1941-1963. Ill. Nat. Hist. Surv. Bull. 30
     (1): 1-67.

27. Jester, D. B. 1971. Effects of commercial fishing,
     species introductions and drawdown control on fish
     populations in Elephant Butte Reservoir, New Mexico,
     pages 265-285 in G. E. Hall (ed.) Reservoir Fisheries
     and Limnology. Am. Fish. Soc. Spec. Publ. No. 8,
     Washington, D.C.

28. Aggus, L. R., and G. V. Elliot. 1975. Effects of
     cover and food on year-class strength of largemouth
     bass, pages 317-322 in R. H. Stroud and H. Clepper,
     (eds.) Black Bass Biology and Management Sport
     Fishing Institute, Washington D.C.

29. Zweiacker, P. N., R. C. Summerfelt, and J. N. Johnson.
     1973. Largemouth bass growth in relationship to
     annual variations in mean pool elevations in Lake Carl
     Blackwell, Okla. Proc. 26th Ann. S. E. Assoc. G and F.
     Comm. (1973): 530-540.

CHAPTER 34

ARTIFICIAL WIND POWERED CIRCULATION OF
HIGH MOUNTAIN LAKES OF THE UINTAS

Mark A. Shaw
   Wasatch-Cache National Forest
   Salt Lake City, Utah

INTRODUCTION

Artificial destratification of lakes has often been utilized as a fisheries management tool. Most often it has been used with varying success to control summer algae blooms and to prevent summer kill of cold water sport fish. The destratification has been accomplished most often by hypolimnetic aeration. A number of aeration systems have evolved over the years, usually involving air compressors powered by electricity.

Mechanical circulation, by directly pumping the lake water from the bottom to the surface has also been used successfully to destratify lakes. Vesuvius Lake in Ohio, and Boltz Lake in Kentucky were successfully destratified in 1972 by using a twelve inch, mixed flow pump driven by a gasoline engine. [1]

The purpose of the installation of the wind powered circulators was to break down the normal summer thermal stratification, and decrease the length of time that the lower lake depths were devoid of dissolved oxygen. These bottom depths of the lake are normally devoid of oxygen for ten to eleven months out of the year. Only during spring and fall turnover does the entire lake volume become oxygenated, exposing the bottom sediments to a brief period of breakdown in an oxygenated environment.

As soon as summer thermal stratification or winter ice cover develops, the oxygen is quickly depleted in the hypolimnion or lower lake depths. In this anaerobic environment, the organic compounds only partially breakdown, with such toxic biproducts as methane and hydrogen sulfide. During the period of six to seven months of ice cover, the dissolved oxygen can be completely utilized in the lake, leading to partial or complete winter fish kills.

If by circulation the hypolimnion could be oxygenated during the ice free months, the underlying sediments and their associated organic compounds would be decomposed in an aerobic environment for a period of six to seven months, compared to one to two months under natural conditions. This situation could then possibly reduce the dissolved oxygen demand during the ice covered months, and produce an environment which could overwinter game fish.

This has been the case in several North Dakota lakes. These lakes were not suited to overwintering fish because of high dissolved oxygen demands. The lakes were circulated by the use of the Lake-Aid wind powered circulator to produce an artificially destratified environment. Over a period of three to five years, this lengthened period of aerobic decomposition lead to favorable overwintering conditions in the lakes. [2]

Several high mountain lakes of the Uinta Mountains located in northeastern Utah are subject to partial or complete winter fish kills because of this depletion of the dissolved oxygen. This is due to the decomposition of the accumulated organic materials in the bottom sediments, and the previous summer's growth of aquatic vegetation. Both Marsh Lake and Teapot Lake at one time overwintered fish populations, but recently had developed partial or complete winter fish kill environments. No significant outside sources of increased organic enrichment had been introduced into these lakes, but Marsh Lake and Teapot Lake were natural lakes which were dammed for irrigation water storage. The gradual accumulation of organic sediments plus macrophyte establishment and subsequent decomposition may be responsible for the high dissolved oxygen demand during the ice covered months on these lakes.

To ascertain the effectiveness of wind powered destratification in the Uintas, the first wind powered circulator was installed on Marsh Lake in July of 1979. Subsequent circulators were installed in the summer of 1980, with one of these being installed in Teapot Lake on the Mirror Lake Highway, and the other in Sargent Lake.

The first signs that the wind powered circulator was able to artificially destratify Marsh Lake came in the late summer after initial installation. This pattern was not repeated in Teapot Lake or Sargent Lake during 1981. This

580

paper presents the results of the changes caused by the wind powered circulators on Marsh, Teapot and Sargent Lakes.

MATERIALS AND METHODS

A wind powered water circulator manufactured by Lake-Aid Inc. of Bismark, North Dakota was used in this study. The circulator (Figure 1), is a self contained wind powered water pumping device, which pumps the water from the bottom of the lake to the surface. The wind is captured in four stacked split savonious fan blades which are attached via a central axle to an impeller. The impeller is situated in the bottom of a suspended tub, to which is attached a telescoping pipe which is then lowered to within two or three feet of the lake bottom. At the water surface, the central axle runs through an oil filled cone which prevents the axle from being immobilized during periods of no wind, and complete ice freeze over. The circulator floats on four styrofoam filled pontoons.

The circulator is intended to be assembled on the lake shore. The parts can be transmitted in two long bed pickups to the site of installation, or in the case of a remote lake, by four to six sling loads via a helicopter. The circulator can be assembled in one day by two but preferably three people. No special tools are necessary for assembly of the circulator.

The circulator was towed to its installation point by the use of a six man rubber raft, powered by a 3.5 hp outboard motor. It was anchored into position by the use of four small (13 inch) car tires filled with rock and concrete. The anchors were attached to the circulator by means of one-quarter inch plastic coated steel cable. The actual installation of the circulator in the lake, and subsequent adjustment of the tub level took approximately one half to one full day. The circulator was then completely operational, with no further adjustments required. The only maintenance required was the yearly greasing of the main bearing on top of the central axle.

The temperature-oxygen profiles, and pH data were gathered by a variety of means depending upon the availability of equipment and personnel. The data up to March of 1980 were collected by the use of point water sampling devices such as a Forest Sampler, Hach Dissolved Oxygen Water Sampler, and a LaMotte Water Sampling Bottle. All of these devices are designed to take a water sample from a desired depth, and preserve the sample from contamination by water from other levels, or atmospheric aeration. A thermometer was placed inside the sampling device and allowed sufficient time to stabilize, to record the temperature at the desired depth.

The dissolved oxygen content was determined using the Winkler with Azide Modification process using wet chemicals, or dry chemicals from a Hach Water Analysis Kit. The pH was determined utilizing a Hach pH kit, which employs the colorimetric method, using a color disc. The above methods of determining dissolved oxygen content were both tedious and time consuming but resulted in accurate determination of the dissolved oxygen content. The pH determination was quick and simple, but was only accurate to the nearest one half measure of pH. From March of 1980, the temperature-oxygen profiles were determined by the use of a Hach Portable oxygen-temperature meter. The meter was calibrated using a nomogram for dissolved oxygen saturation at given temperatures and barometric pressures. The barometric pressure was determined by the use of a portable barometer. The dissolved oxygen and temperature were checked at the beginning and end of each sampling period by comparison with a water sample processed by the modified Winkler method for D.O., and a mercury thermometer for temperature validation. Lake depth profiles were obtained by direct measurement with a depth chord and sounding lead. The data were collected from a two or six man rubber raft in the summer months, and by drilling holes through the ice during the winter.

The wind data were collected by the use of a Meteorology Research, Inc. Mechanical Weather Station, Model 1071.

RESULTS

Marsh Lake

Temperature - Oxygen Profiles

Before the circulator was installed in Marsh Lake in 1979, the lake displayed the development of a strong thermal stratification (Figure 2). Approximately one month later the lake had developed a strong thermocline between four and five meters of depth. There had also been substantial warming in the thermocline to a minimum temperature of 9.5°C, and a continued depletion of oxygen below the point of the establishment of the thermocline (Figure 3).

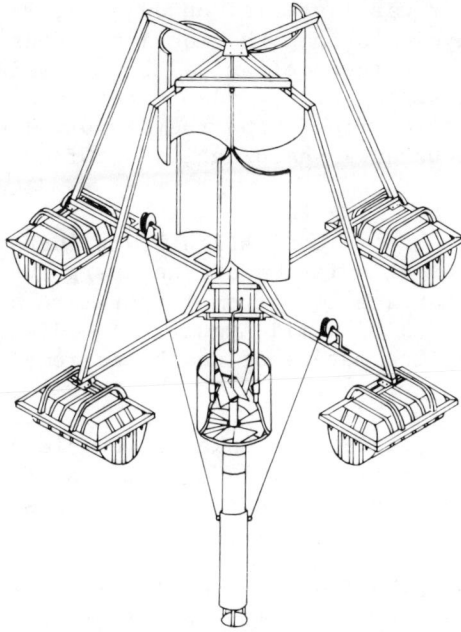

Figure 1. Wind powered water circulator

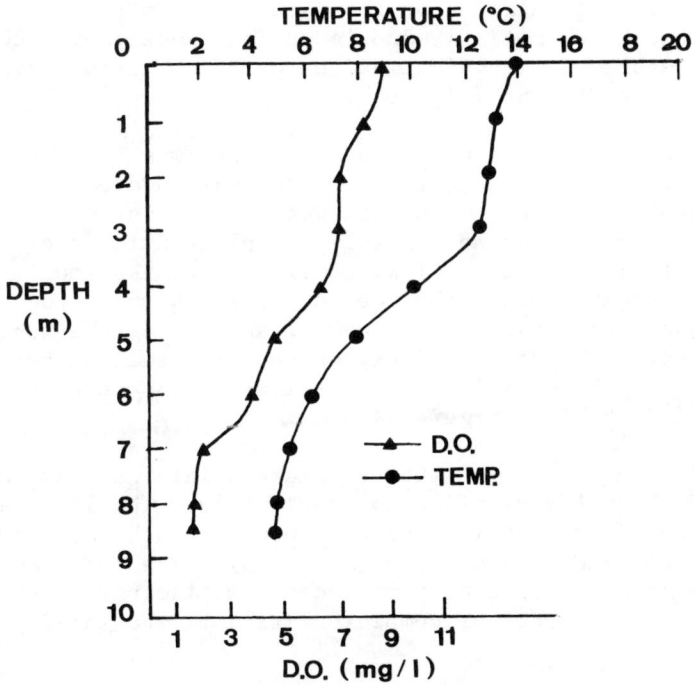

Figure 2. Marsh Lake 6/12/79

The circulator was installed on July 27, 1979, and the lake still displayed a strong thermocline. Four days later on July 31, 1979, the thermocline had been eliminated, but there was still a strong clinograde temperature profile. There had been little effect upon the dissolved oxygen profile, which still began a sharp decline at four meters of depth (Figure 4).

As the lake began to cool slightly in August there was a continued gradual alleviation of the sharp thermal stratification from a surface to bottom temperature difference of 7°C in July to 3.5°C in mid-August. The dissolved oxygen content below five meters of depth had remained essentially the same (Figure 5).

By September 1, 1979, the lake was completely mixed, with a temperature gradient of 12°C at the surface to 11°C at the bottom. The oxygen profile was also essentially orthograde (Figure 6).

The dissolved oxygen content during the first winter of circulation was decidedly different from previous years without the circulator. Before the circulator was installed the lake did not become anoxic until late into the ice covered season during the month of April. Because of the artificial circulation of the water beneath the ice, the lake became essentially anoxic during January of 1980. This is compared to a typical oxygen gradient taken during the same month of 1968 (Figure 7).

The small irrigation dam on Marsh Lake was breeched during July of 1980, and the lake level was lowered by approximately one meter. The dam was reconstructed at the end of the summer, but no filling took place until the spring of 1981. The lower lake level reduced the amount on aquatic vegetation under the ice cover, which correspondingly reduced the dissolved oxygen demand during the winter of 1980-1981. The dissolved oxygen content remained between two and three milligrams per liter which prevented a winter fish kill. (Figure 8).

During the summer of 1981 the lake maintained the same degree of thermal stratification as in 1979. The lake was significantly warmer below three meters of depth when compared to the same time in 1970 (Figure 9). The rapid surface warming after this date produced a sufficiently strong stratification to prevent complete mixing of the lake.

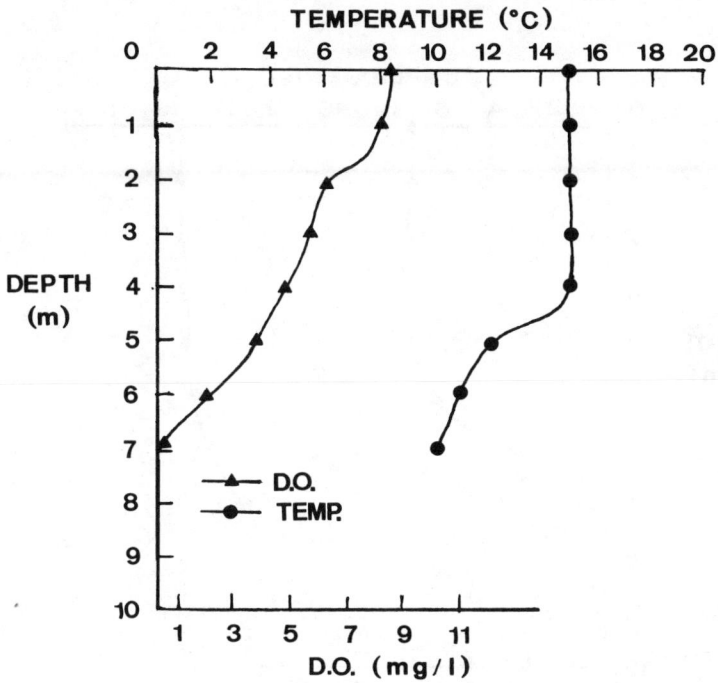

Figure 3. Marsh Lake 7/18/79

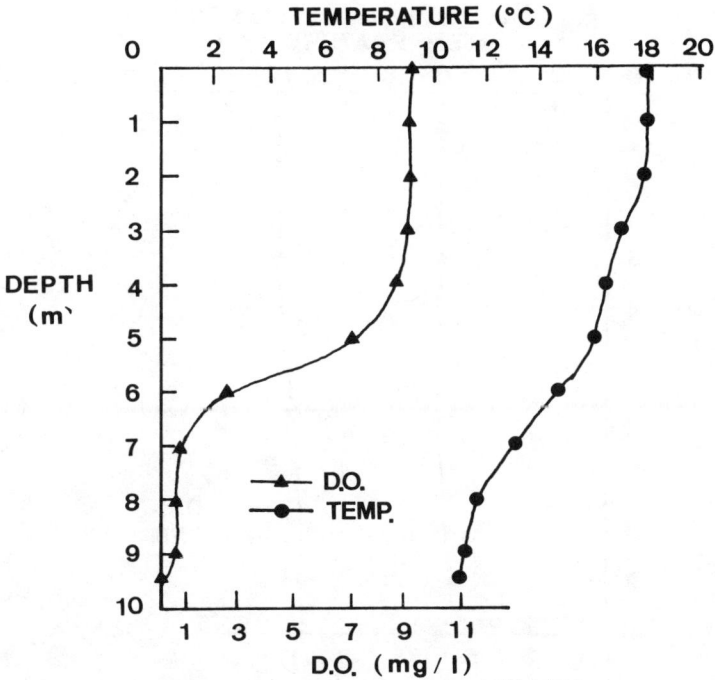

Figure 4. Marsh Lake 7/31/79

585

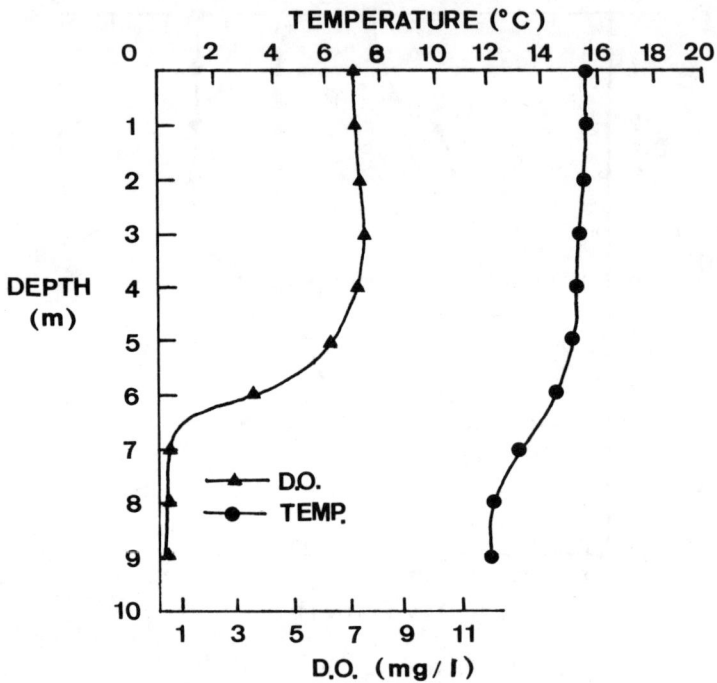

Figure 5.   Marsh Lake   8/15/79

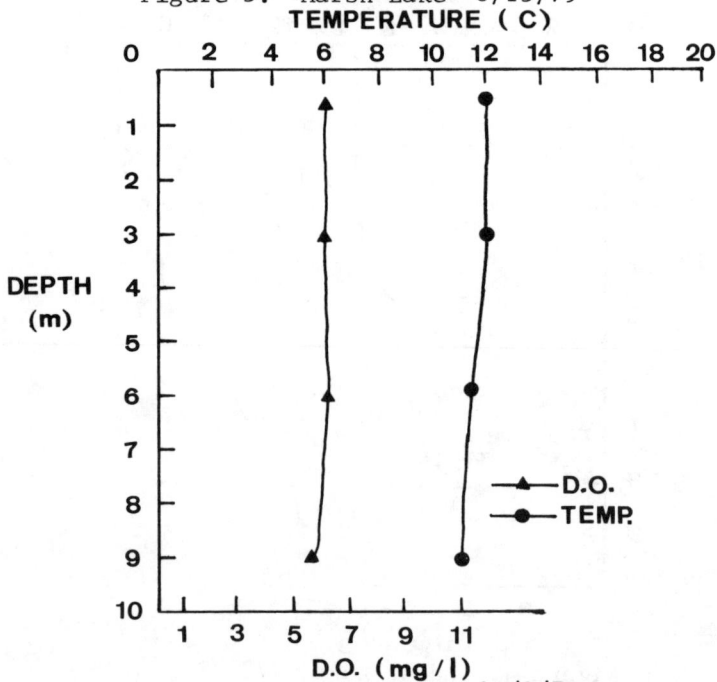

Figure 6.   Marsh Lake   10/1/79

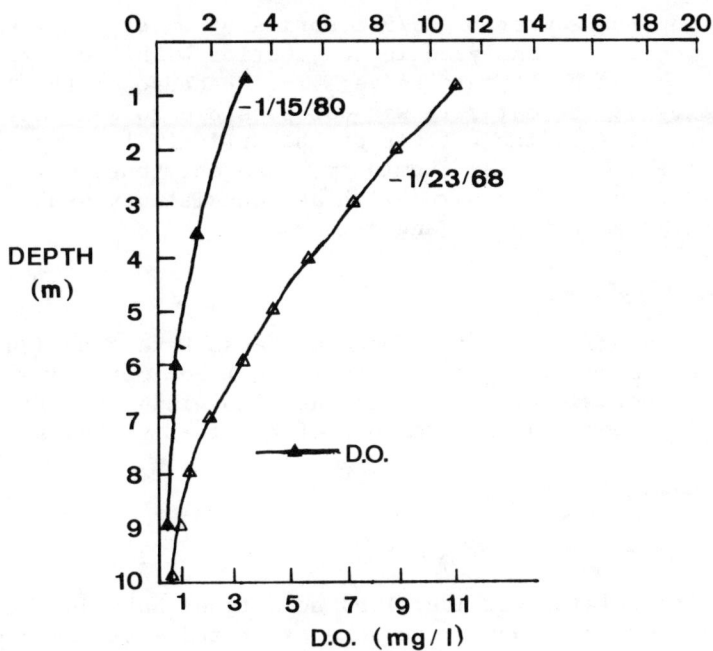

Figure 7.   Marsh Lake   1/23/68 & 1/15/80

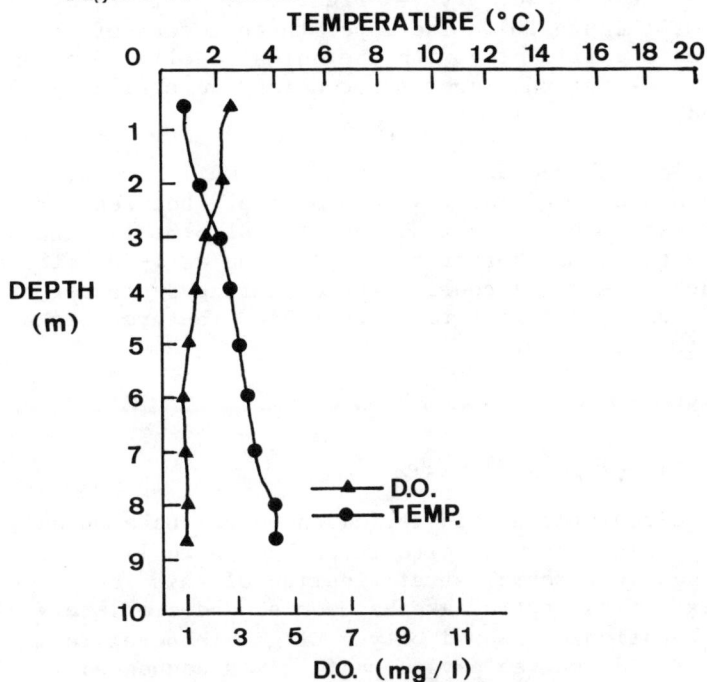

Figure 8.   Marsh Lake   4/10/81

Water Chemistry Measurements

During the summer of 1979 a series of water samples were collected and analyzed in conjunction with the Water Research Lab at Utah State University. Because of the variability of the data, it was not possible to make any interpretations of the effects of the artificial circulation on the lake's water chemistry. The only chemical parameter which was measured with any consistency was pH which varied between 7 and 8.

Wind Speed Data

The average daily wind speed on Marsh Lake from June through August of 1979 was two to three miles per hour. There were occasional storm patterns with winds from five to ten miles per hour for periods of two to five hours.

Sargent Lake

Temperature - Oxygen Profiles

The circulator was installed on Sargent Lake in the first week of July 1980. The lake exhibited a strong orthograde temperature profile but after a period of over one month, the lake showed little effect of any mixing by the circulator (Figure 10). The upper three meters of the lake had cooled significantly with the normal cooling process of mid summer but the lower three meters were relatively unchanged.

The lake showed an increase of 1.2 mg/l D.O. at 1/2 m of depth and 0.5 mg/l D.O. at 1 m of depth between April of 1980 and 1981. During the summer of 1981, the lake had essentially the same thermal stratification as 1980. There was a much more rapid cooling period during September of 1981, advancing the fall turnover to two to three weeks earlier than in 1980.

Teapot Lake

Temperature - Oxygen Profiles

The circulator was installed on Teapot Lake on July 22, 1980. By August 28, the circulator had shown little or no effect upon the thermal stratification of this lake. The upper six meters of the lake had cooled and completely mixed, but the hypolimnion showed only a slight temperature increase. The dissolved oxygen profile was unchanged from

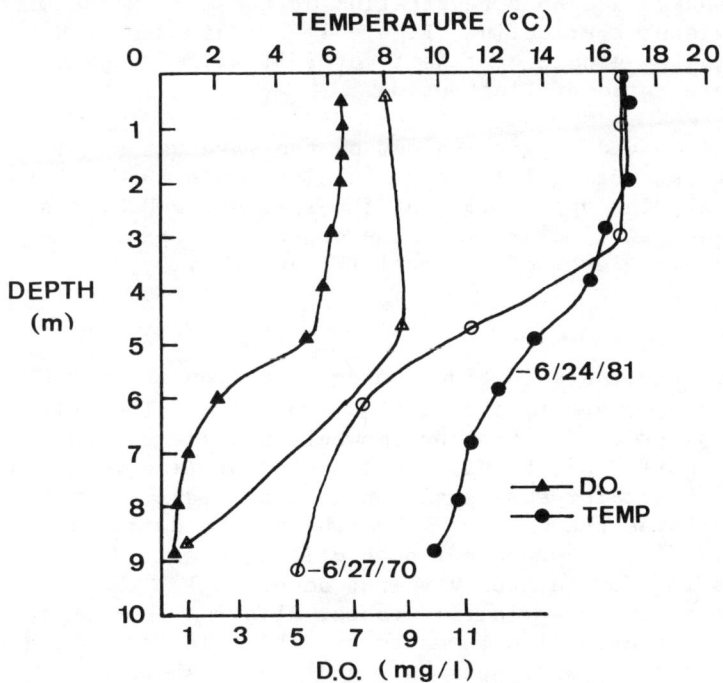

Figure 9. Marsh Lake 6/27/70 & 6/24/81

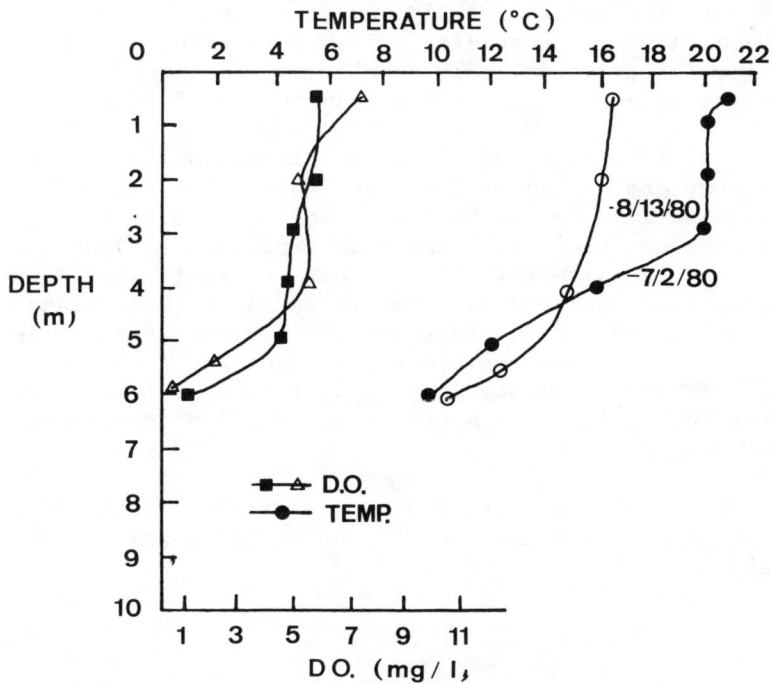

Figure 10. Sargent Lake 7/2/80 & 8/13/80

the time of installation. During the late fall of 1981 the lake showed a deeper penetration of the epilimnion with a less dominant thermocline (Figure 11). This may have been due to the greater intensity of fall storm activity and the more rapid cooling period.

The dissolved oxygen content of the lake was significantly greater during later April of 1981 than 1980. During 1980 only the upper 1.5 m of the lake had a D.O. content of 3-4 mg/l. During 1981 the upper 4 meters of the lake had a D.O. level of 3-4 mg/l (Figure 12).

DISCUSSION

All of the lakes have one thing in common in that they do not have a constant water inflow - outflow. The only water inflow comes from spring snowmelt and the subsequent overland runoff. After this short period of time there is no inflow except for some small springs or seeps. Marsh and Teapot Lakes are natural lakes which were dammed in the late 1920's, but have not been utilized as irrigation reservoirs for ten to twenty years; accordingly, their water levels have been fairly stable, allowing macrophyte communities to establish themselves. This has led to the accumulation of organic sediments and the establishment of the deoxygenated winter kill environments. Sargent Lake has been stocked experimentally several times, but has never been known to overwinter a fish population. From shore line observations, this lake's water level has remained relatively stable for a long period of time.

The ability of the wind powered circulators to bring about a complete mixing of the lakes' otherwise thermally stratified waters has not been realized. The effect on Sargent Lake has been much less than expected, even though it is relatively shallow. This contrasts with Marsh Lake which is deeper and has over three times the surface acreage of Sargent Lake. Here the circulator has significantly altered the temperature stratification of the lake. Although there are not any wind data to compare between the two lakes, Marsh Lake has a better orientation and location for wind mixing.

Table I presents a brief synopsis of the effects of the circulators after a period of one or two years of circulation.

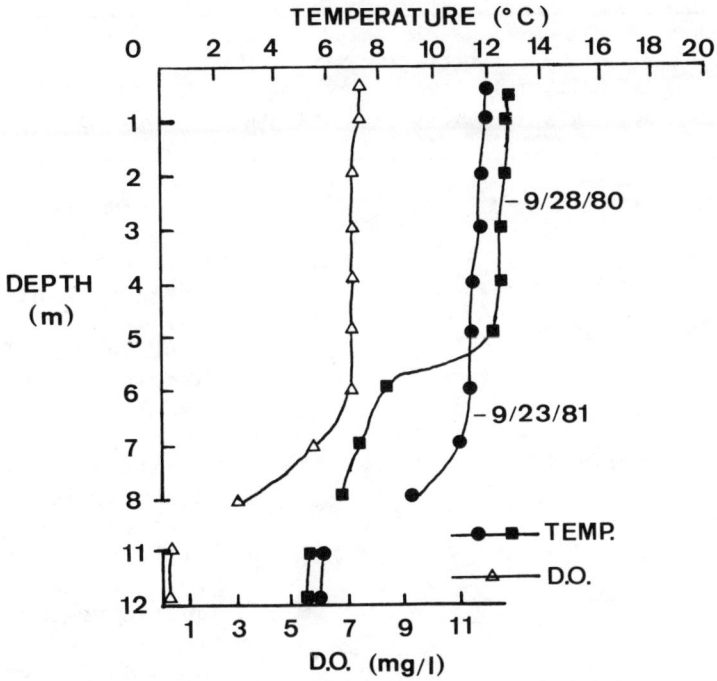

Figure 11.   Teapot Lake   9/28/80 & 9/23/81

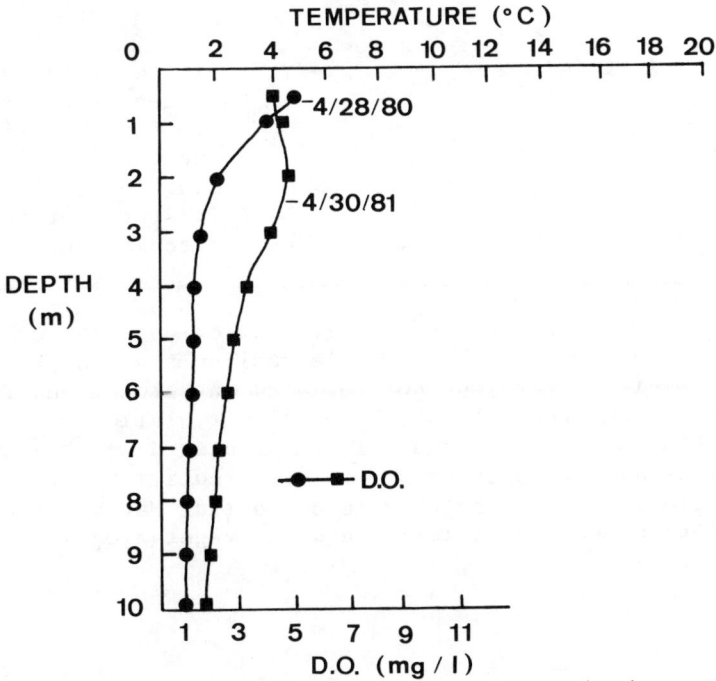

Figure 12.   Teapot Lake   4/28/80 & 4/30/81

Table I. Lake Characteristics and Circulator Effects

Lake	Natural or Artificial	Surface Area (Acres)	Maximum/ Average Depth(m)	Circulator Effects
Sargent	Natural	9	7/4.3	Little effect on oxygen-temperature profile, but increased winter D.O. level after first summer of circulation. No fish life.
Marsh	Artificial	33	9.5/5.1	Significant change in summer temperature profile, but little change in D.O. profile. Aquatic veg. reduction in 1980 resulted in over wintered fish population.
Teapot	Artificial	14	14/7.3	Little effect on oxygen-temperature profile. Significant increase in winter D.O. level after first summer of circulation.

During the first winter of circulation on Marsh Lake, the lake became anoxic three months earlier than in previous winters. This pattern was not repeated on Sargent and Teapot Lakes during their first winter of circulation. Both of these lakes showed an increase in the winter dissolved oxygen content after the first summer of circulation. The major difference between these lakes is that Marsh Lake had extensive beds of submerged aquatic vegetation which

had been undisturbed for ten to fifteen years, whereas
Sargent and Teapot Lakes do not have extensive areas of
aquatic plant growth.

During the winter of 1981, Marsh Lake had sufficiently
high levels of dissolved oxygen to overwinter a fish population for the first time in fifteen or twenty years. When
the small irrigation dam was breached in 1980, the amount
of submerged aquatic vegetation was reduced approximately
fifty percent, or roughly four acres. This reduction in
the amount of organic material captured under the ice,
plus the circulator, significantly lowered the dissolved
oxygen demand.

The lakes' small irrigation dam, which will once again
inundate a large shallow area of water, was reconstructed
in 1980. If macrophyte production again returns to Marsh
Lake, it may be necessary to periodically drain the lake
to reduce the dissolved oxygen demand.

Extensive amounts of toxic gases were vented from the
lakes by the circulators. During both summer and winter,
the distinct odor of hydrogen sulfide was evident at the
circulators. This was especially the case during the winter on all of the circulators, and on Teapot Lake during
the summer.

The circulators were able to maintain an ice free area
at least as large as the outside pontoons during the winter
months. There was an occasional thin, solid ice freeze
during persistent calm, cold weather; but the ice layer
was melted as soon as any wind began the pumping of the
warmer 4°C water from the bottom of the lake. This small
amount of ice free water was not large enough to bring
about any significant oxygenation of the lakes; and did
not inhibit the depletion of the dissolved oxygen.

The study of the effectiveness of the wind powered circulator on high mountain lakes has produced mixed results.
The circulator more effectively destratified a moderately
deep lake (Marsh Lake) than a shallow lake (Sargent Lake).
This was probably due to better wind conditions on Marsh
Lake. The deep lake's (Teapot) thermal stratification was
not altered significantly by the circulator.

Sargent and Teapot Lakes showed significant improvement in their winter dissolved oxygen content after only one summer of circulation. This indicates that in lakes without extensive beds of submerged aquatic vegetation, complete summer circulation is not necessary to lower the dissolved oxygen demand.

The circulator on Marsh Lake showed that the summer stratification can be altered, but in this case the long term effect of artificial circulation without alteration of the aquatic vegetation community will never be known. The unexpected lowering of the lake water level and subsequent macrophyte reduction proved to be an effective management technique, but changed the lake's original biological nature.

The wind powered water circulators have not been entirely successful in breaking down the normal summer thermal stratification of the lakes in this study. A combination of such factors as wind speed, lake orientation to the wind direction, and lake depth, all play a part in determining the circulators effectiveness. There has also not been a consistent inhibition of the rate of depletion of the dissolved oxygen after the lakes' ice cover developed in the winter. The severity of the oxygen depletion caused by accumulated organic materials and/or aquatic plant growth before the circulators were installed seemed to be the major factors which determined the effectiveness of the wind powered circulators.

REFERENCES

1. Ridley, J. E. and J. M. Symons. 1972. New Approaches To Water Quality Control in Impoundments. Water Pollution Microbiology, Ralph Mitchell, ed. Wiley Interscience. New York. pp. 389-412.

2. Henegar, Dale. 1979. Personal Conversation. Chief of Fisheries Division. North Dakota Game and Fish Department. Bismark, North Dakota.

PART 9

WATER: ECONOMICS

CHAPTER 35

PROJECTED IMPACTS ON WATER
RIGHTS, WATER LAWS AND POLICY
STIMULATED BY ENERGY DEVELOP-
MENT AND OTHER PRESSURES

Alten B. Davis, Professor
    Political Science Department
    Weber State College
    Ogden, Utah

INTRODUCTION

Many historians have contended that the American revolution began when the first European set foot on American
soil. The basis for their thesis is that the environment
was so different from that which the Europeans had left that
it would gradually bring about an evolutionary change in the
attitude of colonists which would produce demands for institutional changes which the British as representatives of the
European environment and institutions would refuse to grant
because they could not understand the rationale behind the
demands. Thus the colonists were forced to resort to
revolution.

The environment and demands for water were so different
in the arid and semi-arid West from that of the humid East
that water resource development would bring about a change
in water rights and water law that could be accurately
labeled as revolutionary in nature. Eastern water law had
developed naturally around the European concept of riparian
rights because the adequate amounts of rain usually provided
sufficient moisture to grow crops to maturity if cultivated
correctly to hold the moisture. Hutchins (1) gives the
following definition for the doctrine:

> The riparian doctrine accords to the owner
> of the land contiguous to a water source a right
> to use water on such land for various beneficial
> purposes. (p 158)

The West's arid climate would require water for crop
irrigation and western mining would differ from eastern
manufacturing in that the hydraulic water power would have
to be diverted from the stream and taken to the ore source

to be utilized. Hutchins (1972) defines the appropriative doctrine evolved to meet these new conditions as:

> Contemplating the acquisition of rights to use of water by diverting water and applying it to reasonable beneficial use for a beneficial purpose in accordance with procedures and under limitations specified by constitutional and statutory law or acknowledged by the courts. (p 157)

The pragmatic development of this new appropriation doctrine by Mormon pioneers irrigating in Utah and the 49ers mining in California is colorfully and accurately chronicled by such pioneers in western law and history as Wells A. Hutchins and Frank J. Trelease. They note that the practical evolution of the doctrine by users of the water is eventually given a form of legality by state laws and by state courts, but since most westerners were squatters on the federal lands, the important validation would have to come from that level. Finally, beginning with the Act of 1866 (2) Congress, by a series of legislative acts, gave de jure recognition to the evolved de facto water customs and laws of the states that had developed under the appropriative doctrine. But it was not until 1935 in an interpretation of the Desert Land Act of 1877 (3) in the California-Oregon Power Company case (295 U.S. 142, 1935) that the United States Supreme Court upheld the severance of water from public lands under the Desert Land Act and under other federal land laws as well. Thus was completed the official validation of the water rights revolution of the West which evolved a new doctrine to deal with the existing situational realities.

Today, as we draw near the end of the twentieth century, the West is involved in the beginning of an extended energy development boom. A developing minerals boom will start in the 1990s. Meanwhile the impacts of mushrooming population and the secondary effect spinoffs of energy and mineral development will also be making their demands felt. All of this translates into future demand for scarce water resources.

Will the resulting new environment cause a modern revolution in water rights and water laws in the West? The purpose of this paper is to assess some of the major problems that new pressures are either generating or exacerbating in an attempt to anticipate any changes that will be produced. While this paper will deal with the West in general, it will attempt to focus more particularly on the Colorado River Basin.

THE PRESSURE OF THE ENERGY CRUNCH

The Washington Post of September 28, 1981 (4) carried an article that was based primarily on the remarks of the president of Exxon Corporation who was explaining what Exxon intended to do with its oil shale holdings in the Piceance Basin in Colorado. The president enthusiastically observed that his company hoped to be producing and refining 8 million barrels of oil per day by 2000 A.D. He also optimistically observed that such a level of production would necessitate importing water from the Missouri/Mississippi River systems. Senator Gary Hart of Colorado had nothing but sobering remarks to make about both the effects of that level of production for shale oil and the consumption of water that visualized inter-basin transfers.

The Colorado Department of Resources (5) confirmed in 1979 the necessity to import water for an 8 million barrel per day production process when it concluded that in the Upper Colorado River Basin the water demands for coal gasification plants and for oil shale plants producing a combined output of 1.5 million barrels per day, as well as water demands for associated growth, could be satisfied from surface supplies without having to significantly reduce other projected uses in the Upper Basin.

At a meeting of the western governors in Park City, Utah in 1980, Governor Richard D. Lamm of Colorado noted (6) that coal production in the West will rise from 157,272,727 metric tons (173 million tons) in 1979 to 392,727,273 metric tons (432 million tons) by 1990. There will be a marked increase in the role of coal, shale oil, and synthetic fuels in supplying the energy needs after the year 2000, and the major source of all these alternatives to oil in the West.

Perhaps the most rational way to comprehend how the energy developments translate into water consumption is to start with the microcosmic view at the barrel level. Utah experts at the University of Utah and Utah's Energy Office (1980) have projected the estimate shown in Table I. Is there any wonder that Exxon's president was talking about sources of water outside the Colorado River Basin to quench the thirst of an 8 million barrel per day oil shale production refining complex?

A quick assessment of the project proposed by Intermountain Power Associates will help to focus more precisely on the immediate impacts that energy development will have on water rights and water law. The I.P.A. brochure (7) explains that:

Intermountain Power Project was conceived in
mutual need and a spirit of determined cooperation.
Its mission is to provide vital electrical energy
to areas of five western states with a combined
population of more than 5.5 million. To accomplish
this task the Intermountain Power Agency will
construct and operate the Intermountain Power
Project (IPP), an $8.7 billion, 3000 megawatt
coal-fired generating station to be located near
Delta, Utah. (p 3)

The I.P.A. is a separate legal entity and political
subdivision of the state of Utah which was organized under
the Utah Interlocal Cooperation Act in June 1977 by 23 Utah
municipalities. The Agency now has 36 partners committed to
purchasing all the power that will be generated by the
project.

The Agency has purchased sufficient shares of the
Sevier River and underground water for $83 million from five
irrigation companies to ensure the project of 18.2 hectare-
feet (45 thousand acre-feet) even during dry years. There
has been a conscientious attempt to mitigate the overall
impact of water loss to agriculture. The company's brochure
(I.P.A. 1980) proudly notes that:

It is expected that a net loss of agricultural use of only 5.665 hectare-feet (14,000
acre-feet) of water will result from the Project's
purchase of water rights. Reduced transpiration
and evaporation from wet lands and other water
surfaces combined with better management practices
and water storage facilities will reduce current
losses. Accordingly, it is expected that approximately 2.833 hectares(7,000 acres) of agricultural land--much of it marginal--will be taken
from production. (p 13)

TABLE I
ENERGY NEEDS FOR WATER

Energy Source	Barrels of Water	Barrels of Fuel
Oil Shale	3 to 5	1
Tar Sands	2 to 4	1
Coal Gasification	1½	1
Coal Liquification	3 to 5	1
Petroleum Refining	1½	1

While the IPP will have a significant impact in one of the most arid areas in Utah, the project will be implemented without any significant modification in water law. This is simply an example of the free market operation which has accompanied the appropriation doctrine. Water is not appurtenant to the land and so can be sold separately.

There have been nine law suits brought in the U.S. District Court challenging the proposed use by the Agency of the water rights it has acquired, and asking the court to reverse on various grounds the Utah State Water Engineer's decision to conditionally approve the change in use of water rights from agriculture to project use. The court has ruled for the Agency in five of these suits, two have been settled by stipulation and two are pending. Water use can and will be changed without really violating existing laws.

Clyde (8) indicates that putting a water right together for a major project, almost everywhere in the West, involves different criteria than a similar size project for agriculture. The water supply is needed for year-round use and must be dependable, even during periods of drought. A completely adequate supply to meet needs during a dry cycle will result in surplus water yield during normal and wet years. Clyde points out that if normal and wet cycles ". . .extend for substantial periods of time, and surplus water isn't used there could be a problem of statutory forfeiture for nonuse." (pp 2-4) Some states have already evolved a change in established riparian policy by providing for the issuance of nonuse permits. This is just one example of the many anticipated impacts of development on water rights, but also helps raise the question of how many unanticipated impacts there may be.

The IPP and other similar projects merely represent the beginning of a far-reaching transfer of uses. Irrigated agriculture has generally been on water priced between $0.25 and 0.74 per hectare-meter ($2 and $6 per acre-foot) and therefore cannot possibly hope to compete with industrial or municipal customers who will pay over $12.00 per hectare meter($100 per acre-foot). A slurry pipeline from the West to the Midwest cannot find water at its source for the slurry, so its builders are negotiating for huge blocks of water with the state of South Dakota. One does not have to be clairvoyant to perceive that somewhere in the near future the forces represented by the nine suits against the IPP may very well result in demands for and passage of legislation that will try to slow this cooption of agricultural water by creating the equivalent of a "water greenbelt." Nevada's State Engineer already has statutory authority to approve or

reject an application to appropriate water to be used for
generating energy to be exported out of state. (9)

One should not assume that most of the future demand
for water and the economic pricing of water will result from
attempting to meet the needs of energy development. There
are other factors, such as population growth and inflation,
that are forces in shaping demand and price. By the year
2000 A.D. the United States may begin to feel the serious
effects of shifting agricultural land and water to other uses.

Perhaps one can complete this brief overview by paraphrasing Omar Khayyam to note that the moving finger of
development pressure having writ on the scroll of water
rights and water laws moves on, leaving developers, legal
experts, legislators, administrators and citizens the responsibility of translating the needed changes into a new,
workable fabric of the law.

PUBLIC INTEREST DOCTRINE

Through a series of acts, including the Desert Land Act
of 1877, the federal government deferred its control over the
western waters to the states. The states had exercised de
facto control before this federal action and all would eventually establish a permit system as an institutional means
of establishing a valid claim to water. Prior to the establishment of the permit system, an individual could establish
a legally-protected claim to water simply by taking dominion
over the unappropriated water and placing it to beneficial
use. The whole process of claims to water was based on the
priority of filing and the existince of unappropriated water.

The rapid depletion of the availability of unappropriated water has led most western states to adopt statutes
that contain the concept that an application should not be
approved if the approval would interfere with a more beneficial use of water, or would be contrary to the public
interest. Clyde (1978) lists ten western states that have
statutes specifically introducing the criteria of the "more
beneficial use" or "contrary to the public interest" in
granting permits for appropriation. (8) The general trend
that produced these statutes has led some students of water
law to conclude that prior appropriation is dying. A further
claim is that some states have moved away from priority in
filing as being the principal criteria in approving or re-
jecting applications and have adopted a public interest
standard as the governing criteria for water allocation.
While there can be no question about a trend away from the
priority in time concept, Clyde (1978) dispels some of the

604

simplistic thinking in this area by determining that the court cases interpreting the "public interest" concept have had a general tendency to construe them narrowly as protection for only the public economic interest. (8) The specific standard used in one case to decide between two competing project applications for water was which project would produce the most economic benefits and the least costs.

While one may take a skeptical view of how far water law has really moved from literally no social criteria in granting a water right toward using the concept of the public interest as a major criteria, one must be aware of the qualitative change that is represented by 1976 Utah statute which allows the State Engineer to approve applications to appropriate water for industrial, power, mining or manufacturing purposes for a specific period of time, after which the water will revert to the state for reallocation. (10) The State of Utah has assumed the responsibility of being a water broker and will have to flesh up its passive public interest standard.

Perhaps a further indication of the positive development in water development and water rights is Dallin Jensen's (11) (1976) cryptic statement that:

> It is interesting to note that while many
> western states have resisted efforts to adopt
> state land use planning, virtually every western
> state has some form of water use planning. (p 2)

RESERVED WATER RIGHTS--INDIAN WATER RIGHTS

The evolution of Indian water rights in the West has been slow, but essentially consistent in the substantive development since the landmark case of Winters v. United States (207 U.S. 564). The Supreme Court concluded that the United States had reserved for the Indians of the Fort Belnap Reservation sufficient water from the Milk River to irrigate the lands included by the treaty in the reservation. Furthermore, the appropriative rights obtained by the other water users subsequent to the time that the water was reserved but prior to the time that it was put to use on the reservation were subordinate to the reservation's rights. These reserved Indian rights were valid even though they had not been filed for under state law and even though they had not been put to beneficial use.

Rifkind (12) enunciated in his Special Masters' Report of findings for Arizona v. California (1963) some of the logic underlying the Winters Doctrine approach which

605

essentially turned traditional western water law and rights upside down. He noted that all reservations were created with the intent to change the nomadic life of the Indians to an agricultural economy. He further observed that all reservations created in the Colorado River mainstream area were arid and could not support the existence of an agricultural economy without water. Therefore, in creating the original Indian reservations, the United States implicitly reserved water for such development even though water was apparently never explicitly reserved for such a purpose in the creation of any Indian reservation (Rifkind, 1960). While other potential water users have to file with the state to develop a water claim, the Indian lands are presumed to have automatically established claims on waters available and appropriation of an amount sufficient to develop lands susceptible to agricultural development and related water needs. Such claims are assumed to be established at the time the reservation was created. The quantity of the water reserved could not be determined on the basis of the Indian population because the intent was to permanently establish the Indians and their future population growth could not be anticipated. This logic leads to the conclusion that the claims or rights to water must be somehow calculated in terms of numbers of acres and not numbers of Indians. Such a method of calculation would be essentially contrary to established policy because most water rights in the West are divorced from and not appurtenant to the land. Indian use of water could then be expanded when Indians saw fit to irrigate their under-developed lands or as Rifkind put it, "I have concluded that enough water was reserved to satisfy the future expanding agricultural and related water needs of each reservation." (p 260) Non-Indian claimants using water would have had to put their waters to beneficial use or lose the right to the unused water.

Most Indian lands, just like most lands in the West, were not susceptible to irrigation so the courts concluded that the implied intent to establish a water right for Indian reservations would in reality be

> . . .the amount of water necessary to irrigate all (Practibly) irrigable acreage on the reservation and to satisfy related needs (stock and domestic uses), subject only to the priority of appropriative rights established before a particular reservation was created and water reserved for its benefit. (Rifkind, p 262)

Quantification of Indian water rights then seems to hinge primarily on determining the number of acres susceptible to irrigation in any reservation.

While "susceptibility to irrigation" and related uses had been the criteria for determination of the quantity of Indian water resources, would the use of the water have to be restricted to agricultural development and related uses? Arizona v. California (373 U.S. 595, 1963) raised the question as to whether the water reserved for Indian reservations may not be used for purposes other than agriculture and related uses, but concluded that this question was not before them, exercised judicial restraint and did not answer the question relating to use. While the court had refused to address the question of possible use, it had established all the needed basis for assuming that Indian use of their rights would not be restricted to use for agriculture and related needs. One solicitor for the Department of the Interior (13) has thus concluded that where circumstances warrant the use of Indian lands for recreational, commercial, or industrial rather than for agricultural purposes, these reserved water rights remain available for those other purposes.

The major problem remaining to be solved relevant to Indian water rights is the specific quantification of the waters reserved for the Indians. President Jimmy Carter (14) placed a high priority on the quantification process and in his Water Policy Message of June 6, 1978 he instructed all federal agencies dealing with such rights to promptly inventory and quantify them. These unquantified Indian and other federal reserved water rights hang like an oppressive cloud over water resources planning, project planning, and water rights throughout the West. The General Accounting Office Report (15) indicated that there were 145,002 hectares (358.3 million acres) of federal land in 11 western states and estimated that perhaps 77,459 hectares (187 million acres) of that amount might carry reserved water rights. The same study determined that there were 13,677 hectares (33,796,000 acres) of Indian reservation lands in the seven states comprising the Colorado River Basin. Some estimates indicate that 5% of the 145,690 hectares (350.8 million acres) of private land in the 17 western states are irrigated. If we were to assume that Indian lands had an equal percentage of land susceptible to irrigation and that each hectare of land would require an average of .0012 hectare-feet (3 acre feet) of water per year to produce crops, we would reach the following gross approximations for the seven states of the Colorado River Basin: 683 hectares (1,689,000 acres) of land which would consume 2,052 hectare-feet (5,069,400 acre feet) of water. One must remember that not all Indian lands in these states have claims on Colorado River water and also readily conclude that Indian lands at best might be half as susceptible to irrigation as private lands, but these are sobering statistics for a river basin which is already suffering from water overdrafts.

The process of quantification will be slow because the private users who are junior to Indian water rights do not wish to lose those rights and some Indians hope that the future will lead to an expansion of their rights beyond the concept of "lands susceptible to irrigation and related purposes." Furthermore, the alternative methods for identifying and quantifying the rights are processes fraught with interminable delays. The usual method of determining water rights is by adjudication in the courts. Judicial adjudication suits are complex, protracted and costly litigation which could continue for decades. When one multiplies this incremental approach of a small segment of a basin by an area as vast as the upper and lower Colorado River Basins, the breadth of the problem in terms of complexity and time become obvious. This bleak picture is rendered even more nebulous by noting that under the McCarran Amendment (16) the state courts were given permission to sue the federal government in order to bring all of the parties to the bar whose water rights were being adjudicated. That should have helped clarify the fact as to whether the adjudication of water rights would take place in the state or federal courts, but the courts in the case Colorado River Water Conservation District v. United States (424 U.S. 800, 1976) determined that state courts have concurrent jurisdiction with federal courts to adjudicate Indian reserve water rights in a general stream adjudication. Confusion and delay continue to reign in the seeking of an adjudicated solution to the question of actual amounts of Indian water.

The Carter Administration recommended the use of the second means of quantifying rights whenever possible--a negotiated settlement-- an agreement negotiated and approved by a state and an Indian tribe specifically defining and quantifying the water rights of the tribe. This process requires an extended period of negotiation to reach a general consensus between state representatives, federal representatives and Indian representatives. The proposal must then be submitted to the state legislature for passage as a legislative act and to the tribe for referendum. Utah is the only state that has formally negotiated a proposed solution to water rights with the Indians. A brief review of the process as carried out in Utah may be instructive. Utah experienced several failures to reach even a tentative consensus with tribal negotiators and then, when that hurdle was cleared, suffered through several failures to get legislative acceptance. Senate Bill 64, Ute Indian Water Compact (17), was passed by the legislature and signed by the governor. However, when it was submitted to the tribe for the needed approval, the proposal was defeated in a referendum in which fewer than a majority of the tribal members chose to vote.

There are hopes that the Indians will decide to take a
second vote, with the needed members participating, and
ratify the agreement.

President Carter's message had directed that efforts be
made not only to quantify Indian rights, but to evaluate
projects for the development of Indian water resources. The
Interagency Task Force on Water Policy Implementation reported their progress in June 1979 by identifying 50 programs
in 17 agencies which had potential for making a significant
contriuution to Indian water development. While many Indians
are still hopeful of keeping their "trusteeship" water rights
open-ended and undefined, other Indians are realizing that
the special status enjoyed by Indian rights has been suffering from erosion lately and that now is a propitious time to
resolve those rights. (18)

RESERVED WATER RIGHTS--FEDERAL NON-INDIAN RIGHTS

We have seen that reserved water rights were created at
the time the reservation was established or withdrawn from
the public domain by the United States. The courts in
Winters (297 U.S. 564, 1908) and related cases have upheld
the contention that the United States also by implication
reserved sufficient water for the irrigation of lands susceptible to irrigation and for related uses.

Obviously the same logic can be held to apply for all
federal lands which are withdrawn from the public domain for
various purposes by the United States. This is exactly what
Rifkind (12) concluded in his report on Arizona v. California
that:

> If the U.S. can set aside public lands for
> an Indian reservation and at the same time reserve
> water for the future requirement of the land, I
> can see nc reason why the U.S. cannot equally
> reserve water for public land which it sets aside
> as a national recreation area. (pp 292-3)

Reserved water rights are water rights held by the U.S.
which were created at the time the United States reserved or
withdrew land from the public domain. The U.S. has the right
to use water for the purpose of land reservation and the
water right bears a priority date of the day of the land
reservation. If the logic of Winters applies to all federal
reservations of the public domain, are there any real differences between Indian and non-Indian rights? The U.S. Task
Force on Non-Indian Water Rights (19) answers this question
by noting (a) that Indian rights also involve the federal

government's trust responsibility for Indians, and (b) that
Indian reserved water rights typically relate to purposes
substantially different from those of non-Indian federal
agencies.

Clyde (8) concludes that the basis of the federal power
that underlies the reservation doctrine originates from two
sources: first, the power of the federal government as a
proprietor of the western public domain with the right to
control, reserve or dispose of federal property like any
other proprietor (see Article IV, Section 3, Clause 2, U.S.
Constitution); second, the government acts in the role of a
sovereign exercising the specific powers, primarily the
commerce and treaty making power, granted in the Constitution.

There is no longer any real legal dispute over the
existence of federal non-Indian reserved water rights. The
statements of Rifkind and the Supreme Court in Arizona v.
California were reinforced by the acceptance by two presti-
gious national commissions, Public Land Law Review Commission
(20) and the National Water Commission (21), of the existence
of the doctrine. Any remaining doubts were resolved in
Cappaert v. United States (422 U.S. 1041, 1975 and 426 U.S.
1281, 1976) by a unanimous decision of the Supreme Court and
the cursory way in which the court examined the Winters doctrine of implied reservation and concluded that it could be
applied to virtually all types of federal reservations
(enclaves) in the arid West.

What does this federal non-Indian reservation doctrine
translate into in terms of actual acreage? The decision does
not apply to all federal lands but specifically those lands
which have been reserved and placed in federal enclaves for
various purposes. How are we to distinguish between such
lands? The Supreme Court specifically defined the terms involved in Federal Power Commission v. Oregon (349 U.S. 435,
1955) as follows:

> FEDERAL PUBLIC DOMAIN - Land owned by the
> United States by virtue of sovereignty, that has
> never been in state or private ownership, and
> that is available for disposition or use under
> the general laws applicable to federally owned
> land.

> FEDERAL WITHDRAWN LAND - Land owned by the
> United States that has been formally designated
> for a particular purpose, therefore is withdrawn
> from disposition or use under the general laws
> applicable to federally owned land. (pp 443-4)

The G.A.O.Report (15) lists the total public lands in 11 western states at 145,002 hectares (358,3 million acres) and estimates the federal hectares which may carry reserved water rights at 77,459 hectares (187.2 million acres).

The purpose underlying water law is to insure that the water users will receive a water supply that will enable them to continue their uses, plan for the future, and realize their expectations. But the G.A.O. Report (1978) notes that:

> Undetermined federal and Indian reserved water rights is one of the difficult water resource problems in the eleven western states. This problem is causing controversies and litigation, can lead to economic and social disruptions, and inhibits the efficient use of scarce western water resources. The problem probably will get worse as increased demands are made on the available water supply. . ..

> The lack of information on the amount of reserved water rights makes it virtually impossible for potential water users and state administrators to determine what, if any, waters are available for appropriation under state law for new projects and uses, and what water uses may be superceded by **ex**ercixe of reserved rights. (p 1)

What are the major issues and controversies created by federal and Indian reserved water rights? We have just covered most of the first controversial question--the definition and scope of the federal reserved water rights. The second problem is one of quantification with all its accompanying problems of interpretation. The court has determined that there will be no federal reservation construed unless such water rights can be implied as necessary to accomplish the purpose of the reservation, and that such implied reservation was within the intent of the federal government. Furthermore, the amount of water which the federal government may claim pursuant to the Winters doctrine must reflect the nature of the federal enclave and is subject to quantification in an appropriate proceeding. This statement indicates the nature of the third controversial issue--what courts will have jurisdiction to adjudicate the rights and what governmental entities shall have the right to administer federal reserved water rights. Finally there is the question of possible compensation for private water users who have been operating for years on valid water permits but find themselves as junior rights holders for non-existent water when the federal reserved rights are finally adjudicated for an enclave established prior to the private filings.

The President's Water Policy Statement (14) encouraged the expeditious quantification of both Indian and non-Indian rights. The U.S. Task Force on Non-Indian Federal Water Rights (19) takes a brief look at the procrastination and causes thereof, and spills a little guilt on all parties. The Supreme Court has held that reserved water rights may be asserted to fulfill primary purposes for which the land was withdrawn and reserved from the public domain. (Cappaert v. United States, 426 U.S. 1281 and New Mexico v. United States, 438 U.S. 696.)

But the quantity is to be sufficient to encompass both existing and reasonable future water uses necessary to fulfill those primary reservation objectives (Arizona v. California, 373 U.S. 600, 1963). The U.S. Task Force on Non-Indian Federal Water Rights (1980) states that the methodology for quantifying actual current consumptive uses has been well developed, but methods for quantifying non-consumptive uses (instream flows) and reasonably foreseeable future uses are still in their infancy. Furthermore, the Cappaert decision raised the question of whether federal reserved rights also extended to groundwater and then, for most interpretors, avoided reaching a conclusion.

The question of who shall have jurisdiction in adjudicating such rights was seemingly skewered in favor of the state courts and state administrators by the McCarran Amendment (43 U.S. s666). Even so, confusion has developed in interpreting the act. In United States v. District Court of Eagle County (401 U.S. 520) the Supreme Court modified the previous interpretation of the McCarran Amendment to hold that the United States could be joined in a water adjudication suit if a substantial portion of a river was being adjudicated and not just if an entire stream or watershed was being dealt with. The court also concluded that the Act gave state courts jurisdiction over the reserved water rights of federal non-Indian reservations and enclaves instead of just those federal rights acquired pursuant to state law. Neither Eagle County nor its companion cases involved Indian water rights, so the issue is confused in that area and there has been a resultant "rush to the courts" to place Indian reserved rights conflicts before either the federal or state courts depending on the party's concept of court bias.

The U.S. Task Force on Non-Indian Federal Water Rights (19) 1979 Report begins and concludes optimistically. It assumes the problem is essentially identified, that the differences between the contesting parties are not that great, and that Congress could and probably should pass legislation compensating holders of state water rights taken

612

by the establishment of a federal reserved right. They
decided that the major energy of the Task Force should be
concentrated on an accelerated program to quantify the
federal water rights. They see a firm basis for optimism
and believe that a consensus can be reached because (a) quantification benefits all by establishing certainty for water
rights, (b) most federal reservations are located high on the
watershed and will not impair upstream private users because
they are limited in number, (c) most federal reserve rights,
stockwatering, recreation and human consumption, do not involve substantial consumptive uses, and the non-consumptive
uses primarily involve instream flows, which preserve water
for appropriation under state law at points downstream from
federal reservations. One need not be as optimistic as the
Task Force to conclude that the drift of change in non-
Indian federal water rights has become quite predictable but
the areas of non-reserved rights and instream flows remain
areas of uncertainty for the future.

NON-RESERVED FEDERAL WATER RIGHTS

A Solicitor's Opinion issued on June 25, 1979 (22)
(hereinafter the "Prior Opinion") added new dimension to the
legal uncertainties of water rights in the West. The opinion
first reviewed the general state of federal non-Indian
reserved water rights for the major governmental land holders:
National Park Service, Fish & Wildlife Service, Bureau of
Reclamation and the Bureau of Land Management. The Prior
Opinion then broke new legal ground by announcing the existence of "non-reserved federal water rights". The Solicitor
defined "non-reserved" water rights as rights that may
possibly be claimed by the United States for congressionally
authorized programs. These non-reserved rights were to be
automatically appropriated by a mere application of water to
a beneficial use and are acquired by the United States without regard or compliance with state substantive law.
Solicitor Coldiron (23) summarizes the general rationale of
the non-reserved rights in a brief paragraph:

> In brief, the proponents of the federal non-
> reserved rights theory assert that by enactment
> of various land use statutes "Congress authorized
> the United States to appropriate unappropriated
> water available on the public domain" implicitly,
> without regard to the substantive provisions of
> state water law, and that such "federal non-
> reserved water rights are not dependent upon the
> substantive contours of state water law." The
> Prior Opinion asserts that, since the federal
> government has never granted away its right to

make use of unappropriated water on federal lands,
". . .the United States has retained its power to
vest in itself water rights in unappropriated
waters and may exercise such power independent of
substantive state law." Such water rights were
asserted to be available to fulfill authorized
congressional purposes on the public domain, reserved and acquired lands, could be consumptive
and could be used for "fish and wildlife, scenic
values, and areas of critical environmental
concern." The priority date was said to be the
date of initial use, and the quantity of the right
determined by the requirements necessary to carry
out "congressionally authorized management objectives on federal lands." (p 2)

Coldiron, the new solicitor, reversed the Prior Opinion
in a new opinion issued September 11, 1981 (23). He cryp-
tically notes that, "To the extent that the Prior Opinion
and the Supplemental Opinion are inconsistent with the conclusion reached herein, they are recinded." (p 1) The
conclusions of Solicitor Coldiron are not surprising when
one considers the known opinions of the Secretary of the
Interior, James Watt, in whose jurisdiction Solicitor
Coldiron was exercising his authority to interpret the law.
The following two paragraphs are direct quotes from the
Coldiron Opinion (1981):

    A review of the applicable federal consti-
    tutional, legislative and judicial authorities
    demonstrate the power of Congress to control the
    usage of water appurtenant to the federal lands.
    The legislative and case law authorities also
    demonstrate congressional intent to defer control
    of water to the states, in all but the most
    limited circumstances. Congress has chosen to
    displace state control of water appurtenant to
    federal lands only when necessary to accomplish
    the original purpose of formal reservations.
    When not necessary to accomplish such original
    purpose, Congress has uniformly permitted and
    the Supreme Court has recognized state control.

    Within this framework, there is an insuf-
    ficient legal basis for the creation of what has
    been called federal "non-reserved" water rights,
    especially in the wake of the Supreme Court pro-
    nouncements in United States v. California and
    New Mexico v. United States. I must conclude
    therefore that there is no federal "non-reserved"

614

water right. Federal entities, including, without
limitation, the National Park Service, Fish and
Wildlife Service, Bureau of Reclamation and the
Bureau of Land Management, may not, without congressionally created reserved rights, circumvent
state substantive or procedural laws in appro-
priating water. Rather, consistent with the
express language in the New Mexico decision,
federal entities must acquire water as would any
other private claimant within the various states.
(p 12)

Uncertainty of course will continue. A Californian in
the Interior Department decided in the 1930s that the land
law provisions did not apply to Imperial Valley water users
using federal water. About forty years later the courts reversed that decision. Coldiron has overturned the opinion of
a previous solicitor and one can conclude that this opinion
can possibly be flipped again by another solicitor. Additional uncertainty is added by the footnote statement in the
Coldiron Opinion (1981) that:

This opinion is not intended to modify or
supercede any portion of the Prior Opinion dealing with the reserved water rights of the non-
Indian land management agencies in the Department.
I may further review those portions of the Prior
Opinion at a future date as specific circumstances
warrant. (p 1 footnote 2)

RIGHTS TO INSTREAM FLOWS

The discussion of issues in this paper has focused primarily on federal actions because all too often they have
represented the leading cutting edge of the pressures of
change. Here in the brief coverage to be given to the
important concept of rights to instream flows we will present
a cryptic summary of state actions in that field by Jensen in
1976 (11):

In recent years many states have taken steps
to preserve instream flow values. This has been
accomplished in a variety of ways and no attempt
will be made here to list or evaluate these programs, except to point out that the recognition
of instream values has had a significant impact
on some state water allocation programs. In some
instances, states have attached sufficient importance to instream values to withdraw certain streams
or portions of streams from appropriation under

615

scenic rivers acts. Other states have authorized
a state agency to establish minimum stream flows
and thus make future appropriations subject to
this action, or to allow a state agency to file
an application and secure a right to a minimum
stream flow. The impact of some of this legislation is to--in effect--give instream flow values
a priority in the allocation of a state's unappropriated water and, in other instances, to place
these uses on a par with other potential uses in
competition for the state's available water supply.
Thus, it is clear that in those states which
recognize instream flow needs the competition
for the unallocated water will be substantially
increased. (p 7)

One must pause to give credit to the foresight of
organizations who conceived, established and cooperated with
the Cooperative Instream Flow Service Group, a portion of
the Western Energy and Land Use Team, the United States Fish
and Wildlife Service, established in 1976 and based in Fort
Collins, Colorado, with primary funding by the United States
Environmental Protection Agency. The Cooperative Instream
Flow Service Group was instrumental in getting numerous
studies completed including one for each western state on the
current status of instream flow rights in their states and
projected strategies to improve the status of those rights.
The Group also sponsored numerous water law short courses
throughout the West in cooperation with the Natural Resources
Law Institute of the Lewis and Clark School of Law located in
Portland, Oregon. Such courses were attended by federal
administrators, Indians, private water users, attorneys,
academics, and others resulting in an excellent cross-
pollination of ideas. The general objective of the studies,
publications and short courses was probably to capitalize on
the ancient Greek concept--that to know is to do.

The claiming of instream flow rights in most cases represents recent ecological awareness and so in most cases may
not be incorporated into most federal land and water management agency goals as spelled out in legislation until the
1970s. That would make their requests for water, even though
part of a federal reserved water right, junior to most other
appropriations of water because New Mexico v. United States
made it clear that reserved water existed from the creation
of an organization managing land and water only for specific
goals. Many of the streams in the West were oversubscribed
long before 1970.

The Prior Opinion (1979) of the Solicitor, later rejected

616

by Solicitor Coldiron (1981), was obviously an attempt to give instream flow rights a legal basis that was equal to other federal non-Indian reserved water rights. Solicitor Coldiron was quite correct in blowing the whistle on this legal interpretive end run that sought to bypass both congressional intent and previous court decisions.

The U.S. Task Force on Non-Indian Federal Water Rights (19) did have some encouraging observations that applied to the instream flow area. The Report notes that such flow rights, by definition, involve no consumptive use and therefore preserves the water for appropriation under state systems at points lower in the watershed. The Report also noted that such uses had few upstream water users that might use sufficient water to impair the needs of instream flow. These optimistic observations must be matched against the mushrooming demands for water for energy and mineral development, some of which will come in the heart of the overthrust belt or in the instream flow area. The second bitter reality for hopes to improve instream flow rights in the near future is that the existing financial crunch on government at all levels will tend to leave this new demand for public monies essentially unfunded.

While the future of instream flows may continue to be a thorn-filled path, the trend to recognize the public interest concept in granting state water rights, the increasing use of our streams for both active and passive (aesthetic) recreation and the growing awareness of the ecological basis for instream flows will help to somewhat offset its junior position of water rights.

CONCLUSION

This paper started out by noting that a water rights revolution took place in the West when it was first settled because the environment and needs were hostile to the riparian doctrine and so spawned a new pragmatic riparian doctrine. Obviously the availability of water will be one of the limiting factors in the magnitude and number of energy related projects that can develop in the West. Additional pressures will be exerted in identifying and quantifying Indian and federal reserve water rights. The demand to reserve water for instream flows will continue to grow as will the effort to extend the use of the "public interest" doctrine. The combined push-pull-shove effect of the above forces will result in extensive changes in the use of water and in continuing changes in water rights and water policy, but the system appears to be so resiliant that it will be able to deal with these problems without capsizing.

LITERATURE CITED

1. Hutchins, W. A. 1971. Water right laws in 19 western states. Miscellaneous Publication #206, Volume I. U.S. Dept. of Agriculture. Washington, D.C. 650 pp.

2. (Mining Act of 1866) U.S. Congress. 14 Stat. 253 (1866) 30 U.S.C. Sec. 51 (1952).

3. (Desert Land Act of 1877) U.S. Congress. 19 Stat. 377 (1877); as amended 43 U.S.C. Secs. 321-323 (1952).

4. The Washington Post. September 28, 1981. p. A-16.

5. Colorado Dept. of Natural Resources. 1979. The availability of water for oil shale and coal gassification development in the upper Colorado River basin. Water Resources Council. Denver, Colorado. 5 pp.

6. Lamm, R. D. September 4, 1980. Energy activities in the west. Western Governors Policy Office. Park City, Utah. 10 pp.

7. Intermountain Power Agency. 1980. An illustrated prospectus on the intermountain power project. 23 pp.

8. Clyde, E. W. March 17, 1978. Current problems--legal overview. Rocky Mountain Mineral Law Foundation. Tucson, Arizona.

9. Nevada State Legislature. Nevada Revised Stat. ₴533.370.

10. Utah State Legislature. 1976 Budget Session. Chapter 32, Laws of Utah.

11. Jensen, D. W. 1976. State water law reforms in water allocation. Conference on Energy and the Public Lands. Park City, Utah. 23 pp.

12. Rifkind, S. H. December 5, 1960. Report of the special master on Arizona v. California (373 U.S. 546). Filed and accepted January 16, 1961 (394 U.S. 940).

13. Bell, D. C. 1979. Report on the reservation doctrine. Western States Water Council.

14. Carter, J. E. June 6, 1978. The president's water policy statement on federal and Indian reserved water rights. IN: Public papers of the president, vol. 1. pp. 1043-1224.

618

15. (G.A.O. Report) U.S. Congress, Office of the Comptroller General. November 16, 1978. Reserved water rights for federal and Indian reservations: a growing controversy in need of resolution. Washington, D.C.

16. (McCarran Amendment 1952) Dept. of Justice Appropriation Act of July 10, 1952. Ch. 51 Sec. 208, 66 Stat. 560 (1952); 43 U.S.C. Sec. 666 (1970).

17. (Ute Indian Water Compact) Utah State Legislature. April 3, 1980. Utah Code Annotated 1953, as amended, Section 73.

18. Foreman, R. L. 1981. Indian water rights: a public policy and administrative mess. Danville, Illinois. 233 pp.

19. U.S. Task Force on Non-Indian Federal Water Rights. 1980. President's Water Policy Implementation Task Force 5A. 121 pp. and appendices.

20. Public Land Law Review Commission. June 1970. One-third of the nation's land: a report to the president and to the congress by the public land law review commission. Washington, D.C. n.p.

21. National Water Commission. June 1971. Water policies for the future: final report to the president and to the congress of the United States. Washington, D.C. n.p.

22. (Prior Opinion) Opinion of the Solicitor, U.S. Dept. of Interior. June 25, 1979. 86 I.D. 533.

23. (Coldiron Opinion) Opinion of the Solicitor, U.S. Dept. of Interior. September 11, 1981. 88 I.D. 1055. 12 pp.

CHAPTER 36

AN ECONOMIC EVALUATION OF ALTERNATIVE STRATEGIES
FOR MAINTAINING INSTREAM FLOWS

Rangesan Narayanan
    Economics Department and Utah Water Research Laboratory,
    Logan, Uah

Dean Larson
    Utah Water Research Laboratory, Logan, Utah

A. Bruce Bishop
    Civil and Environmental Engineering Department and
    Utah Water Research Laboratory, Logan, Utah

Parvaneh Amirfathi
    Economics Department and Utah Water Research Laboratory,
    Logan, Utah

INTRODUCTION:

Many beneficial uses of streams and adjacent riverine lands depend on instream flows. Activities for which instream flows are valuable include outdoor recreation, hydropower, navigation, waste transport and assimilation, fish and wildlife maintenance and preservation of riverine ecosystems. The amount of streamflow that is necessary to maintain instream values is referred to as the instream flow requirement.

These instream uses, however, conflict with water diverted from streams for culinary uses, and for the production of food and fiber, energy, and industrial products. These offstream uses often directly compete for water for instream uses, particularly in the western states where water is relatively scarce. The demand for water use offstream has been increasing and is expected to increase substantially, intensifying the competition for water between instream and offstream purposes.

The legal framework which has evolved to govern the use of water in the arid west is the appropriation doctrine [1]. Basically, the appropriation doctrine provides that a person may acquire an exclusive right to divert a specific amount of a public water source and apply it to a specific beneficial use, in preference to subsequent appropriators but without impairing existing rights. Priorities for use, then, are on a "first-in-time is first-in-right" basis. Proponents contend that in the promotion of comprehensive development and efficient use of the resource, an appropriative system is superior to its main alternative, the riparian system [2].

The doctrine's evolution, however, has not been hospitable to instream values.* The traditional view has been that there can be no exclusive right to water freely flowing in the stream; the water right can only be perfected by establishing physical control over the quantity of water needed and actually applying it to the intended beneficial use. Historically, the lack of institutional provision of rights for instream uses might be explained by the relative abundance of instream flow compared to the demand for water for offstream activities. However, with the cumulative effects of over a hundred years of offstream development, continued flow availability for instream values cannot be taken for granted. Realizing that instream flows produce benefits worthy of protection, which can thus be regarded as a legitimate use of the resource, the question now is how to provide for a balanced allocation of water between the two balanced types of water use.

Advocates of special protection for instream flows find two main obstacles to integrating instream uses within the appropriative system. The first, alluded to above, is the difficulty of satisfying the requirements of appropriation. The elements of a valid appropriation have been assumed to consist of (1) a notice of intent to appropriate; (2) an actual diversion, reducing the water to possession; (3) an application to a beneficial use [3]. While the status of instream uses as beneficial has been controversial, it is the diversion requirement that has effectively barred instream appropriations. There is evidence that this obstacle can be overcome: Colorado and Montana, for example, have made statutory provisions for instream appropriation. In any case, other options for protecting flows are available [4].

The other obstacle to integrating instream uses into the appropriative system is the, apparently dominant, view that instream uses are different in kind from the traditional offstream uses. Instream values are alleged to be best protected as a public trust by the state, and it would be improper to subject them to exclusive or private use. As noted, many states have statutory provisions through which,

---

*The exception is hydroelectricity generation, which has been easily accommodated because of the need to control the water flow driving the turbines. That is, the actual diversion requirement is met.

explicitly or indirectly, instream flows can be protected, and exercise of these provisions is usually justified by some version of the public trust doctrine (even in Colorado and Montana). While this "special consideration" approach has achieved some desired results in protecting instream uses, the disadvantage is that it does not really provide a balanced view of the resource because it does not integrate instream with offstream uses.

Two problems have characterized the approach. One is its inflexibility. It is still difficult at best to secure instream flows on heavily appropriated streams, and administrative arrangements make it unlikely that instream reservations could be changed in response to changes in social priorities. A second problem is that the methods for determining instream flow needs have not been tied to the economic viewpoint that permeates the appropriative system, making more difficult the allocation of water between instream and offstream uses according to relative values. The typical approach now is to define stream flow regimes required to meet some qualitative use criteria, usually the flow required to support a population of some aquatic organism(s) at various life stages. These criteria are used to establish minimum levels of instream flows through strategies that preempt new appropriations, or changes in existing uses, that might deplete the stream.

In the long run, instream use allocations must be integrated with allocations to uses traditionally regarded as beneficial. Whether instream values are exclusively protected by the state, or state protection and private appropriation are combined, rational allocation decisions require information on the relative benefits of instream flows and the costs of various proposed methods for obtaining needed flows. Economic cost-benefit analysis is the major tool for generating this information.

ECONOMIC APPROACH TO INSTREAM FLOW MAINTENANCE

Instream flow is an essential input in the production of beneficial activities within the stream channel. This input has, in economic terms, 'public good' characteristics. In general, this means that for a given level of instream flow, many different instream uses can take place without any one use excluding one or more other uses. Whereas conflicts among instream uses may occur, competition for water (of a given quality) among the uses does not since there can be only one level of instream flow possible at any given time. The offstream uses, on the other hand, directly compete for the total water supply.

The demand for water for each instream use can be derived using market data or survey methods [5]. Household production theory [6] has been used to derive these demands

623

and questionnaires and interviews have been used to evaluate demands for fishing, white water boating and stream-side activities [7]. Other important instream flow benefit studies were done by Daubert et al. [8] and Walsh et al. [9]. Since only one level of instream flow can prevail at any given time, the aggregate demand can be derived by vertical summation of the derived demands. In Figure 1(a), the demands for two instream uses are represented by curves D_1 and D_2. The vertical sum D_i of these demands is the aggregate demand for instream flows. In Figure 1(b), D_o represents the demand for water for offstream uses. The horizontal sum of D_o and D_i is the aggregate demand D for water. A horizontal summation is used here because the same water cannot satisfy both uses.

The supply curve S represents the minimum cost of supplying various quantities of water. If the water supply is fixed, then S will be a vertical line. Otherwise, minimum cost combinations of such alternatives as water importation, reservoir construction or enlargements, or groundwater pumping for flow augmentation would be used to derive S. The intersection of S and D at E_1 represents the benefit maximizing allocation, with the optimal level of instream flow $W_1 = W_2$ as shown in Figure 1(a) and the optimal offstream water use W_o as shown in Figure 1(b).

Another way of looking at this allocation is that the optimal instream flows are determined by the intersection of demand for instream flows D_i and the marginal opportunity cost of water taken from offstream uses S'. This is seen as point E_2 in Figure 1(c). The marginal resource cost S' is obtained by plotting the residual supply for water represented by the horizontal difference between S and D_o. The benefit maximizing condition, therefore, is that the sum of marginal benefits for instream uses should be equal to the marginal benefit of each offstream use, which in turn should be equal to the marginal cost of water.

It is also useful to conceptualize the balances between instream and offstream use in the context of a stream channel that transports water to downstream users. This transport function provides a certain amount of instream flow in many reaches. The exact amount of flow depends on the distribution in time and space of diversion points and return flows along the stream. If, in any given stream reach, the normal flow is large enough to satisfy all instream uses, the marginal benefit of additional instream flow is zero. Otherwise, appropriate policies are needed to increase instream flows.

In Figure 2, the marginal benefits of downstream uses (on the left vertical axis) for various amounts of water (measured from 0 to the right) is shown by D_d. The total quantity of water available is fixed and represented by the length 00'. D_u indicates the marginal benefits of water for upstream uses (on the right vertical axis) for various

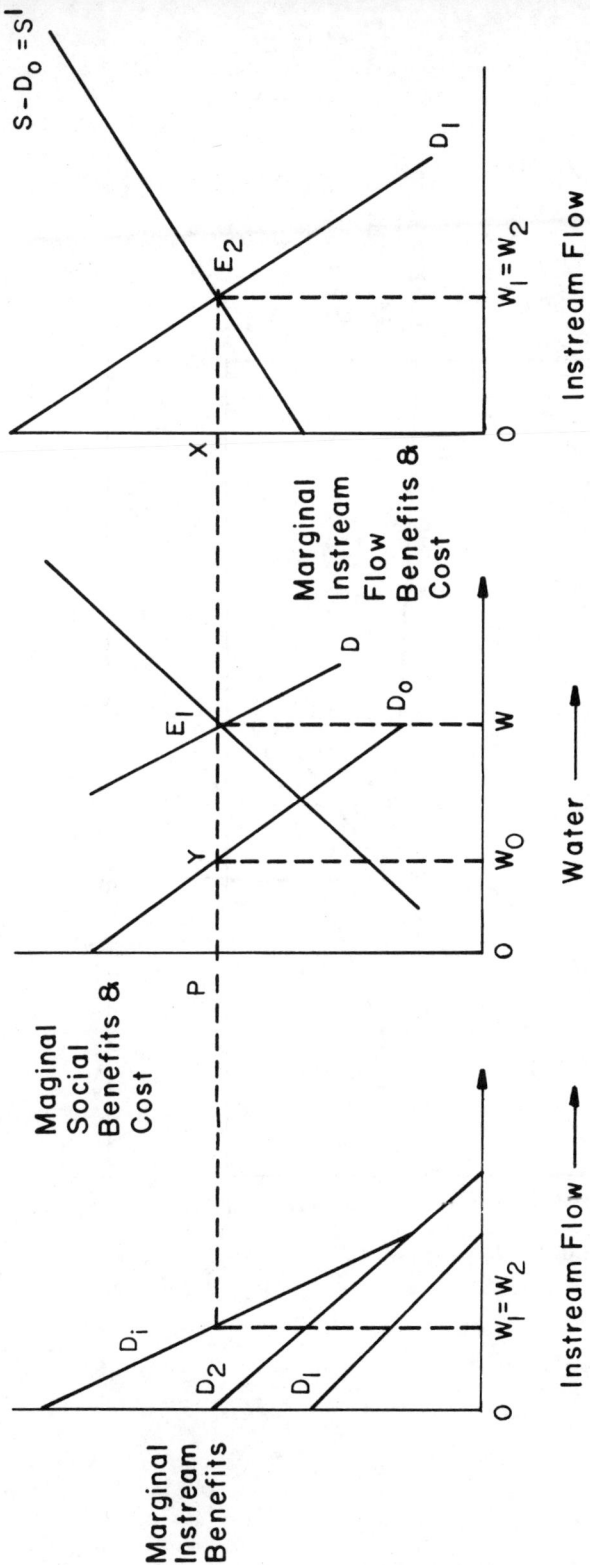

Figure 1. Optimal allocation of water.

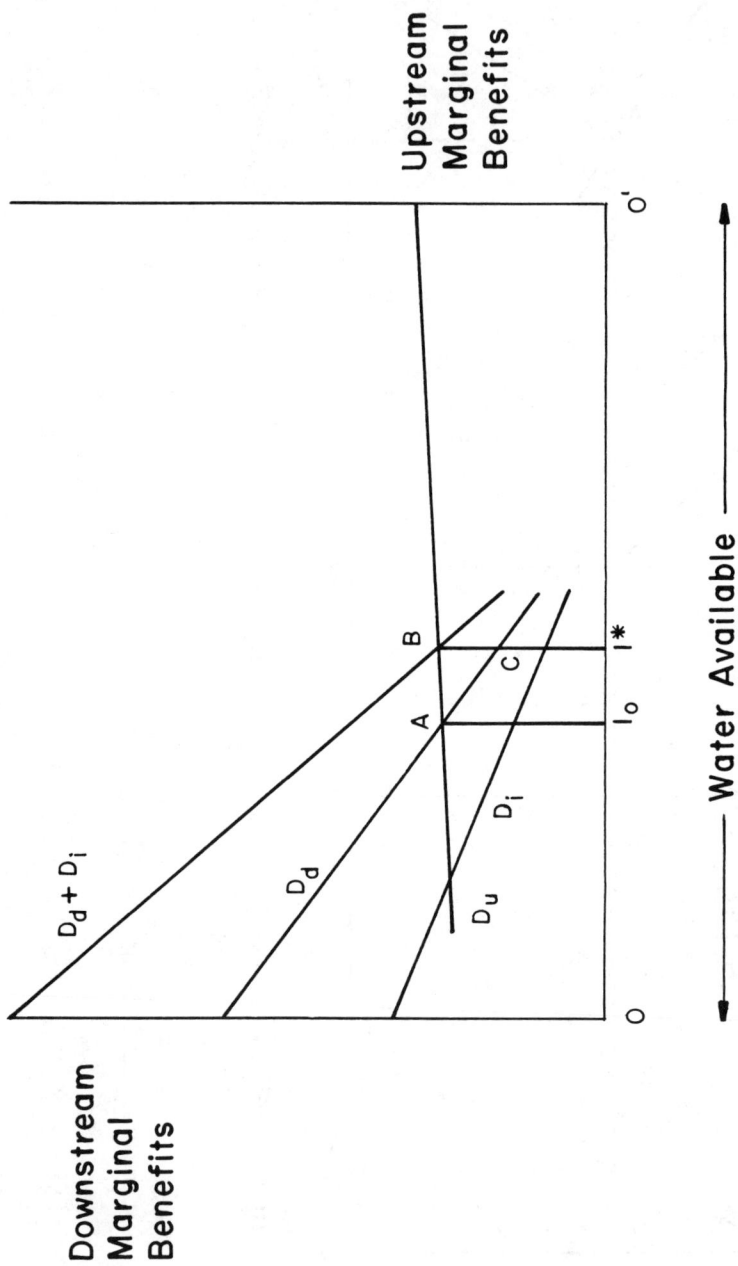

Figure 2. Optimal instream flow determination.

amounts of water use (measured from point 0' to the left). If water rights are freely transferable, OI_o and I_o0' will represent the water rights held by down stream and upstream users, respectively. Therefore, the resulting instream flow will be OI_o, the amount of water passed through the stream channel to meet downstream rights.

If the aggregate demand for instream flows is D_i, then the combined marginal benefits can be represented by $D_d + D_i$, the vertical sum D_d and D_i at each flow level. The intersection of $D_d + D_i$ and D_u at B represents the benefit maximizing point, and OI^* is the optimal level of instream flow. This implies that the instream flows should be increased by I_oI^* to maximize the benefits to society.

Determination of an optimal level of instream flow for any given stream reach is difficult in practice. First, variations in the quantity of water available from year to year is an important consideration. While the techniques of cost-benefit analysis under uncertainty developed by Hirshleifer [10] and Arrow and Lind [11] can be applied in this situation, these techniques require additional information on the attitudes toward risk of affected individuals. Furthermore, the computational effort is increased manyfold for practical applications. However, the probabilistic nature of stream flows should be recognized in making instream flow provisions due to the enormous differentials in the values of certain and uncertain water supplies. Second, estimates of instream flow demands are lacking, largely due to the unavailability of site specific data and the high cost of obtaining such data. Finally, there are theoretical controversies in estimating benefits and no generally accepted empirical framework to do so.

Without quantitative information on instream flow benefits and individual attitudes toward risk, randomness in water availability cannot be incorporated satisfactorily within the cost-benefit framework. Moreover, the deterministic approach is somewhat insensitive to the structure of appropriative water rights, which incorporates the condition of uncertain supply by ordering rights according to seniority. The deterministic approach to instream flow thus has a tendency either to displace senior rights by reserving a base flow to be met before diversions are allowed, or to provide less instream flow protection than intended because other claims will have priority in times of shortage.

However, the deterministic approach to instream flow requirements can be modified to account for uncertain water supply conditions. Under this approach, the deterministic "instream flow requirement" criterion is replaced by an "expected instream flow requirement".

To compare these approaches a general stochastic linear programming model was used to estimate the expected cost, in terms of the foregone value of agricultural products, of maintaining expected instream flows using each approach. A direct conflict between offstream agricultural use and the maintenance of instream flows is assumed. However, conflicts with other water uses as well as additional water management alternatives such as reservoir construction, modification of reservoir operating rules, groundwater pumping and interbasin transfers, can be accommodated within the model framework. A case study is used to illustrate the model application.

DESCRIPTION OF STUDY AREA

The Blacksmith Fork and Little Bear River drainages located in the southwest portion of Cache County in northern Utah were selected as a case study area. The Little Bear, draining an area of 339 square miles, flows roughly south to northwest to its confluence with the Bear River. The Blacksmith Fork, draining 267 square miles, flows roughly east to west to join the Logan River which later flows into the Bear River. The headwaters of both rivers originate in the Wasatch mountains, and annual discharge volumes are dependent on the winter snowpack.

About 15% of the Little Bear drainage and 63% of the Blacksmith Fork drainage lie in the Cache National Forest or state lands. In the downstream reaches, approximately 32,000 acres in the Little Bear drainage, and 2,000 acres in the Blacksmith Fork drainage, are irrigated. Irrigation constitutes by far the heaviest use made of the water. Other uses are made for municipal, culinary, and hydroelectricity generation purposes. Both rivers support excellent brown trout fisheries. The Blacksmith Fork in particular, however, is dewatered over part of its lower reaches during the middle and late summer of years with below normal flows. Such dewatering occurred in the summer of 1981, resulting in the loss of a large number of fish. A proposal by the City of Hyrum to rehabilitate its power plant on the Blacksmith could dewater another stretch above the canyon mouth by diverting the flow into a pipe for conveyance to the downstream generation site. This area, already experiencing conflicts between water use for irrigated-agriculture and instream flows for fish habitat, presents a good situation for demonstrating the model application.

DESCRIPTION OF THE MODEL

A stochastic linear programming model [12] is developed in this study to analyze various instream flow strategies. The annual water availability is assumed in the model to be a discrete random variable that can take on any one of eight flow levels. Each flow realization is an independent event with an associated probability. The monthly flows are calculated as a fixed proportion of the annual water availability (Table I). This array of possible flow events can be regarded as describing a hypothetical water rights structure, such that the senior rights are associated with the lowest flow.

In general, instream flow strategies are conceived as the manner in which the portions of the total streamflow are identified and reserved for instream use once the desired instream flow level is determined. In the context of random water availability, this desired level of instream flow should be defined in terms of expected instream flow requirement. Three basic strategies for meeting this requirement are examined under two conditions of water rights transferability. The expected instream flow strategy (EIF) determines the combination of rights needed to maintain the desired level of expected instream flow at least cost in terms of expected agricultural output foregone. The minimum flow strategy (IF) meets the desired level by reserving the required amount of the most senior rights. The critical flow strategy (CF) combines the previous two, using IF to obtain a base flow that minimizes the risk of irreversible damages, and EIF to obtain the remainder of the desired expected instream flow. Under water rights condition 1, transfers between agriculture and instream uses are restricted; under

Table I. Stream Flow at Different Probability in Acre/Feet

States	Probability	Months					
		May	June	July	Aug.	Sept.	Seasonal Total
1	0.029	9000	4200	2600	2200	2000	20000
2	0.1176	18000	8400	5200	4400	4000	40000
3	0.1764	27000	12600	7800	6600	6000	60000
4	0.3235	36000	16800	10400	8800	8000	80000
5	0.1470	45000	21000	13000	11000	10000	100000
6	0.1470	54000	25200	15600	13200	12000	120000
7	0.029	63000	29400	18200	15400	14000	140000
8	0.029	72000	33600	20800	17600	16000	160000

629

condition 2 short term transfers are permitted. These conditions correspond to a program of long term water rights acquisitions, and one of annual administrative allocations, respectively. Of the six strategies that could be thus conceived, results are presented for only five (EIF 1, EIF 2, IF, CF 1, CF 2) because transferability is not applicable to IF. An example of the schedule of rights acquired under the different strategies is given in Table II.

The agricultural sector is the only offstream user included in the model. It is assumed to be competitive in both input and output markets. Water rights are transferable freely within the agricultural sector, and farmers attempt to maximize expected profits. In addition, it is assumed that fairly accurate estimates of the quantity of water available each month of the irrigation season are provided to farmers so they can decide on cropping patterns and allocate land optimally among various crops.

In formulating the model P_{jr} represents the net revenue (value-added) per acre of jth crop produced on the rth class of land. Irrigated land is classified into three classes based on productivity levels. The values of P_{jr} for the six major crops are shown in Table III. Z_{jr}^k is the number of acres of rth class of land devoted to the production of jth crop when water availability event k occurs with an associated probability π^k. The expected returns to irrigated agriculture are given by:

$$\sum_k \sum_j \sum_r P_{jr} \, \pi^k \, Z_{jr}^k \qquad (1)$$

The problem is to maximize Equation 1 subject to the following constraints. The amount of irrigated land is restricted to be less than the available acres L_r^* for each of the land classes for every event k.

$$\sum_j Z_{jr}^k \le L_r^* \qquad r = 1,2,3; \ k = 1,2, \ \ldots \ 8 \qquad (2)$$

The amount of water A_t^k used for irrigation in month t and probability state k is defined by the following equation:

$$\sum_r \sum_j w_{jt} \, Z_{jr}^k - A_t^k = 0 \quad t = 1,2,\ldots5; \ k = 1,2,\ldots8 \qquad (3)$$

where w_{jt} represents the consumptive use requirement for crop j and month t. The values for w_{jt} used in the analysis are

630

Table II.   Distribution of Instream Flow Reservations by Strategy Over Flow Events in August (Expected Instream Flow = 70%)

Strategy	Flow Events							
	1	2	3	4	5	6	7	8
EIF1	2200	4214	6414	8614	10814	13014	15214	17414
EIF2	2200	4045	6463	8445	11000	13200	15400	17600
IF	2200	4400	6600	8800	11000	13200	13489	12669
CF1	2200	4214	6414	8614	10814	13014	15214	17414
CF2	2200	4045	6463	8445	11000	13200	15400	17600

Table III.   Net Revenues Per Acre For Different Crops and Land Classes (P_{jr})

Class	Crop					
	Alfalfa Full	Alfalfa Partial	Barley	Corn Grain	Beets	Nurse Crop
Class 1	107.49	82.13	106.68	156.63	72.44	64.21
Class 2	86.83	68.29	89.75	120.22	48.47	50.98
Class 3	67.81	62.38	74.96	77.32	43.85	39.98

computed using the Modified Blaney-Criddle equation and are shown in Table IV.   Six crop rotational constraints are used in the model [13].   A general representation of these equations is given by:

$$\sum_r \sum_j V_{jr}^i \, Z_{jr}^k \gtrless 0 \quad i = 1,2,\ldots5; \; k = 1,2,\ldots8 \qquad (4)$$

where V_{jr}^i represents the proportion of various crop acreages required for good crop rotation.
    The water supply constraint,

$$I_t^k + A_t^k = Q_t^k \quad t = 1,2,\ldots5; \; k = 1,2,\ldots8 \qquad (5)$$

631

Table IV. Water Requirement For Crops Per Acre
In Acre-Inches (w_{jt})

Month	Crop					
	Alfalfa Full	Alfalfa Partial	Barley	Corn Grain	Beets	Nurse Crop
May	3.828	3.190	1.772	1.311	1.240	1.772
June	5.727	4.713	7.805	3.801	3.345	7.805
July	7.597	6.228	7.665	7.392	7.528	7.665
August	6.416	5.508	1.513	6.235	7.566	1.513
September	3.644	3.197	0.930	2.417	4.239	0.930
Total	27.212	22.836	19.685	21.156	23.918	19.685

restricts the sum of the amounts of water used in irrigated agriculture A_t^k and the instream flows I_t^k to be equal to the water availability Q_t^k. The distribution of values of Q_t^k are shown above in Table I. The expected instream flow requirement is imposed by the constraint

$$\sum_k \pi^k I_t^k \geq \bar{I}_t^* \qquad t = 1, 2, \ldots 5 \qquad (6)$$

where $\bar{I}_t^*$ is the desired expected instream flow level. The values of $\bar{I}_t^*$ examined in this analysis correspond to 40%, 50%, 60%, and 70% of average flows and are shown in the first column of Table V. To restrict water right transfers between irrigation and instream flows, the following constraints are included.

$$A_t^{k+1} - A_t^k \geq 0 \quad k = 1, 2, \ldots 7; \quad t = 1, 2, \ldots 5$$

$$\qquad\qquad\qquad\qquad\qquad\qquad\qquad\qquad (7)$$

$$I_t^{k+1} - I_t^k \geq 0 \quad k = 1, 2, \ldots 7; \quad t = 1, 2, \ldots 5$$

A_t^{k+1} represents the irrigation water use corresponding to event Q_t^{k+1}. The difference between A_t^{k+1} and A_t^k, therefore, can be interpreted as the water right of k+1th seniority. For example, A_t^1 is the amount of most senior water rights, $A_t^2 - A_t^1$ represents the amount of water rights

Table V. Minimum Instream Flow Requirements In Acre-Feet (I_t^*)

Expected Instream Flow	Time				
	May	June	July	August	September
40% (4667)	4667	4681	4729	4790	4852
50% (5834)	5834	5882	6034	6158	6283
60% (7000)	7000	7084	7401	7772	8003
70% (8167)	8167	8285	9029	10127	12850

having the next lower priority and $A_t^8 - A_t^7$ represents the most junior water rights in the stream.

Strategy EIF 1 is given by the stipulation of constraint (7), in which the model does not allow transfer of water rights between irrigated agriculture and instream flows because the constraint fixes the allocation between them for any flow event. More technically, the variables in constraint (7) can be regarded as first stage decision variables in a two stage linear programming model with the cropping pattern regarded as the second stage decision variable. Strategy EIF 2 is obtained if constraint (7) is not imposed, in which the model implies that water rights can be transferred between agriculture and instream flows after observing the event Q_t^k. Without constraint (7), the model reduces to solving the independent linear programming problems for each flow event. All decision variables are therefore second stage decision variables.

The model is solved with and without constraint (7) for various levels of expected instream flow requirements. While the first case solution (EIF 1) may yield a lower value of the objective function (1), the second case solution (EIF 2) might involve large transaction costs due to the necessity of transferring water rights.

In addition, minimum flow requirements $\bar{I}_t^k$ are imposed by stipulating

$$I_t^k \geq \bar{I}_t^k \qquad t = 1, 2, \ldots 5; \ k = 1, 2, \ldots 8 \qquad (8)$$

633

These constraints are used in two different ways. First, by implicitly finding $\bar{I}_t{}^k$ such that the expected value of $\min(Q_t{}^k, \bar{I}_t{}^k) = \bar{I}_t{}^*$, the minimum instream flow reservation consistent with the expected instream flow requirements can be found. These minimum requirements by month are shown in Table V under columns 2-6. Then, constraint (8) is imposed so that $I_t{}^k \geq \min(Q_t{}^k, \bar{I}_t{}^k)$. By imposing (8), the decrease in the objective function (1) value corresponding to minimum instream flow strategy (IF) is determined.

The minimum flow constraints can also be used to set critical instantaneous flows that may be required to prevent irreversible damages, since simple EIF requirements could allow zero flows. Critical flows $I_t{}^c$ were set at 20% of average flows by stipulating in Equation 8, $I_t{}^k \geq \min(Q_t{}^k, I_t{}^c)$. The critical flow strategy is used with and without Equation (7) for conditions 1 and 2, to give results for strategies CF 1 and CF 2, respectively.

ANALYSIS OF RESULTS

Solutions for the five different strategies discussed earlier were obtained for expected instream flow levels of 40%, 50%, 60%, and 70% of average flows. For comparison purposes, a base solution with $\bar{I}_t = 0$ is obtained without Equations (7) and (8). The objective values for all solutions were subtracted from the base solution value to arrive at the cost of instream flow maintenance for the five alternative strategies. These costs are shown in Table VI.

It is interesting to note that while the differences in objective values with (EIF 1, CF 1) and without (EIF 2, CF 2) the transferability constraint (7) are negligible, the

Table VI. Minimum EXPECTED COSTS of Instream Flow Maintenance

Strategies	Expected Flow			
	40%	50%	60%	70%
EIF1	603653	764760	938858	1397799
EIF2	603648	764755	938856	1397796
IF	625171	829248	1103966	1424417
CF1	605351	766458	940556	1397799
CF2	605346	766454	940553	1397796

corresponding pattern of water allocation for the two conditions is different. Although an allocation decision rule without Equation (7) would be theoretically preferable, its implementation would probably incur high water right transaction costs.

The difference in objective function values between the simple EIF strategies and the critical flow strategies are not significant. This implies that critical instantaneous flows can be provided with minimal impacts on present agriculture use. However, the differences in costs between the minimum flow strategy and other strategies are substantial. As expected instream flow requirements are increased from 40% to 70%, the cost differences at first increase and then decrease both in absolute and in relative terms. This is because for lower flow requirements relatively more senior water rights are held by agriculture under the EIF and CF strategies. As flow requirements are increased, costs increase as water is witheld from irrigation use. At higher expected instream flow levels, more senior rights are held for instream flows. Therefore, the minimum flow strategy and the expected flow strategies tend to become similar at higher expected instream flow requirements. The maximum differences in costs, however, do not exceed 10% among various strategies.

Figure 3 shows irrigated land acreages to utilize various levels of water availability under alternative strategies for 40%, 50%, and 60% expected instream flow levels. The expected flow and the critical flow strategies have almost flat curves over a wide range of flow levels, indicating a stable situation for maintaining irrigated acreages. However, under the minimum flow strategy, more land is irrigated at higher flows. This is because at lower stream flows a relatively greater amount of "certain" water is reserved for instream purposes. The correspondingly more "uncertain" water is available at higher flows for irrigation purposes. In the expected and critical flow cases, the junior and senior water rights are more evenly distributed.

In Figure 4, hydrographs labeled "streamflow" for 80% and 30% of flows are shown for the irrigation season. The corresponding solution for instream flow values under expected flow (EIF 1), critical flow (CF 1) and minimum flow (IF) strategies are shown for 50% expected instream flow level. Compared to the other two strategies, the higher positioning of the curve for instream flows for the minimum flow strategy clearly indicates that this approach requires a larger base of water rights, thus allowing a much smaller amount for agriculture during critical periods of water demand.

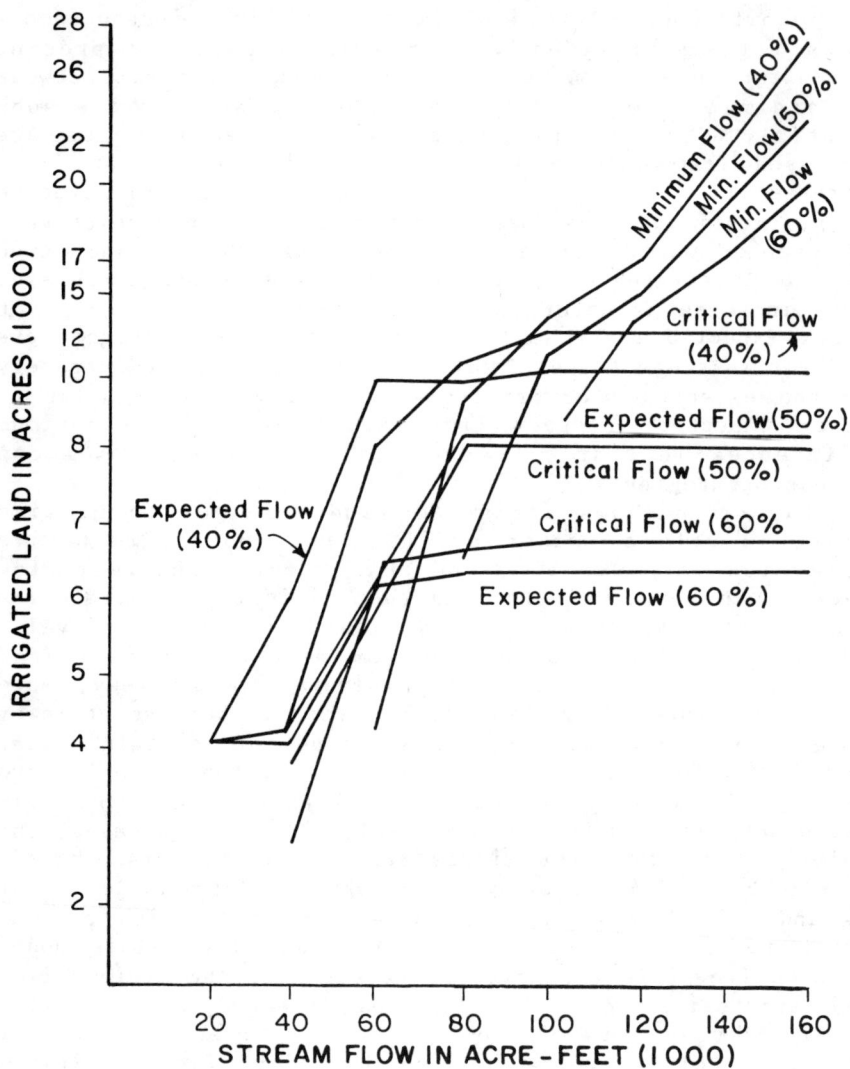

Figure 3. Irrigated land under various strategies.

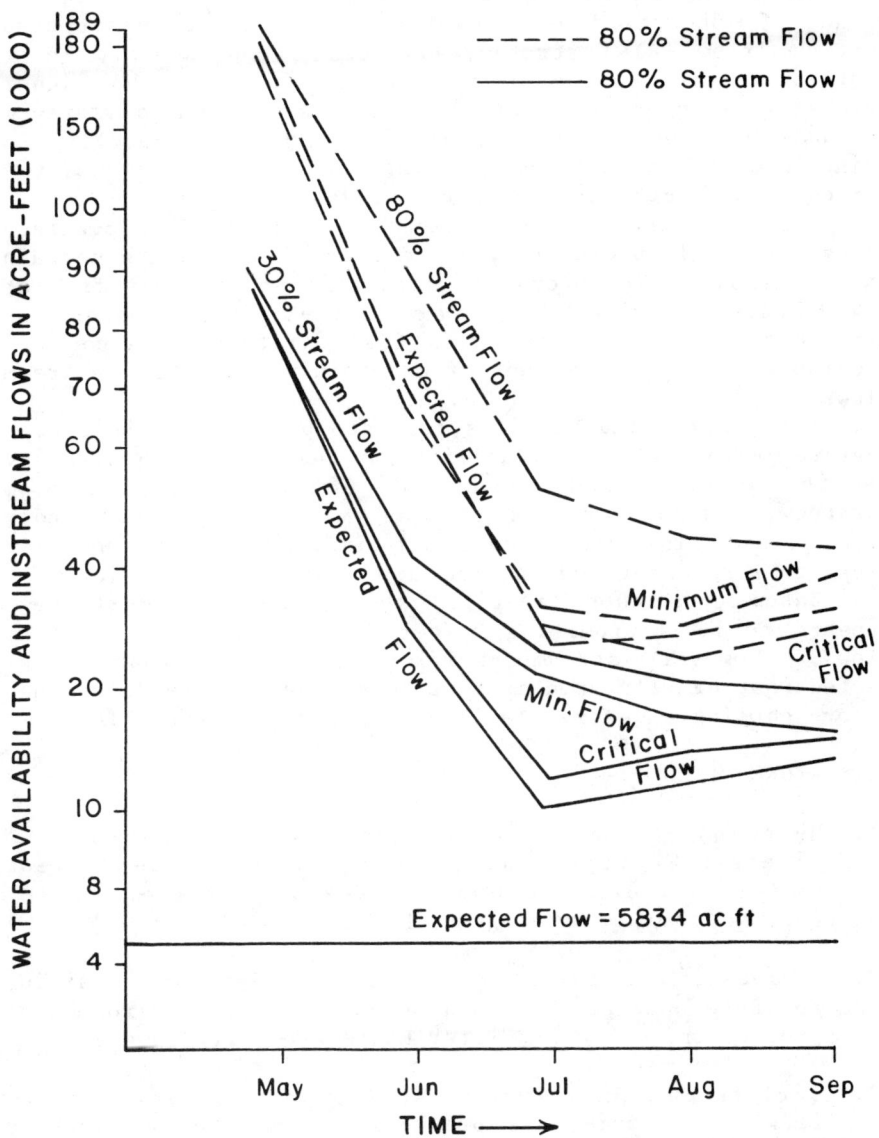

Figure 4. Instream flows under alternative strategies.

CONCLUSIONS

By using a stochastic linear programming model, the expected cost of maintaining various level of expected instream flows was determined. Due to the difference in the values of junior and senior water rights, using an expected instream flow (EIF) strategy produces a consistently lower cost as compared to a minimum flow (IF) strategy. At higher levels of expected instream flow requirements, the difference in costs between the two strategies narrowed. However, to maintain expected instream flows at 70% of average flows, the agricultural sector has to be virtually eliminated.

One disadvantage of the expected instream flow strategy is that it could prescribe zero instream flows during certain short periods. Therefore, a critical instantaneous flow of 20% of average flows to prevent irreversible damages was stipulated. This critial flow (CF) strategy does not appreciably increase the cost of maintaining expected instream flows.

The irrigated acreage is found to be fairly stable over most ranges of water availability for the simple EIF strategy and the critical flow strategy. Under the minimum flow strategy, larger acreages of land may be irrigated under high stream flow conditions, but irrigated land generally drops to zero under low and medium stream flow conditions.

Based upon these results, the critical flow strategy appears to be a promising criterion for providing instream flows. However, stream-specific costs of alternative expected instream flow requirement levels need to be determined before chosing a desired level of expected instream flow.

LITERATURE CITED

1. Hutchins, W. A. 1971. Water Rights Laws in the Nineteen Western States. Vol. I. U. S. Department of Agriculture, Economic Research Service. Misc. Pub. No. 1206. Washington, D. C.

2. Trelease, F. 1977. New water legislation: drafting for development, efficient allocation and environmental protection. Land and Water Law Review, 12:385-429.

3. Tarlock, A. D. 1978. Appropriation for instream flow maintenance: a progress report on "new" public western water rights. Utah Law Review, 1978:210-247.

4. Dewsnup, R., and D. Jensen. 1977. State laws and instream flows. Department of the Interior, Fish and Wildlife Service, Office of Biological Services. FWS/OBS-77/27. Washington, D. C.

5. Freeman, A. M, III. 1979. The Benefits of Environmental Improvements. Baltimore, MD: Johns Hopkins University Press.

6. Becker, G. S. 1965. A theory of allocation of time. Economic Journal, 75:493–517.

7. Daubert, J. T., and R. A. Young. 1981. Recreational demands for maintaining instream flows: a contingent valuation approach. American Journal of Agricultural Economics, 63:666–676.

8. Daubert, J. T., R. A. Young, and S. L. Gray. 1979. Economics benefits from instream flow in a Colorado mountain stream. Colorado Water Resources Research Institute, Colorado State University, Fort Collins, Colorado.

9. Walsh, R. G., R. K. Ericson, D. J. Arosteguy, and M. P. Hansen. 1980. An empirical application of a model for estimating the recreation value of instream flow. Colorado Water Resources Research Institute, Colorado State University, Fort Collins, Colorado.

10. Hirshleifer, J. 1966. Investment decisions under uncertainty: applications of the state preference approach. Quarterly Journal of Economics, 80:252–277.

11. Arrow, K. J., and R. C. Lind. 1970. Uncertainty and the evaluation of public investment decisions. American Economic Review, 60:364–378.

12. Wagner, H. M. 1975. Principles of Operations Research. Prentice-Hall, Englewood Cliffs, NJ, pp. 667–672.

13. Keith, J. E., K. S. Turna, S. Padungchai, and R. Narayanan. 1978. The impact of energy resource development on water resource allocation. UWRL/P-78/005. Utah Water Research Laboratory, Logan, Utah.

CHAPTER 37

ENERGY IMPACTS OF MAN-MADE
WATER RECREATION AREAS

David A. Bell
    Mechanical Engineering Department
    Utah State University
    Logan, Utah.

J. Clair Batty
    Mechanical Engineering Department
    Utah State University
    Logan, Utah.

E. Bruce Godfrey
    Economics Department
    Utah State University
    Logan, Utah.

J. Paul Riley
    Civil & Environmental Engineering Department
    Utah State University
    Logan, Utah.

Thomas C. Stoddard
    Mechanical Engineering Department
    Utah State University
    Logan, Utah.

INTRODUCTION

Energy has moved to the front of American consciousness during the past decade as never before. The efficiencies of energy conversion devices, from dishwashers to cars to jet aircraft, have become a principal selling point for a specific brand or product. Indeed, the notion that a more energy conservative product is superior to a less conservative one has ostensibly amended the American creed, with

great emphasis now being placed on thermodynamic efficiencies for a miriad of products and devices, thus presenting a complex empirical challenge to the energy analyst
[1].

Coincident with the popular emphasis on energy
efficient toasters and diesels, the concept of overall
energy effectiveness for large scale projects has attracted
much attention among those concerned with resource management. There appears to be an increasing demand for energy
impact analyses as an extension of current environmental
impact statements [2]. It is, therefore, not unreasonable
to assume that in the future, energy impact statements for
alternative projects will be weighed in the same balance as
the gasoline mileage ratings of new automobiles.

In anticipation of future requirements for energy
impact statements, and in an effort to satisfy the need for
a working methodology for their preparation, this paper
deals with a pilot study of an energy impact analysis for
recreation use of multi-purpose man-made reservoirs or
impoundments.

ENERGY FLOW ANALYSIS

Unlike establishing the thermal efficiency of a jet
engine, the identification of energy flows in the environment is generally difficult. The number of variables involved is large and requires a systematic sorting and classification methodology to maintain order and reliability -
hence the name, "energy accounting."

Identifying and organizing the energy inputs to such
activities as the processing of oil shale, the production of
fuel alcohol, or the construction and operation of a
hydropower project has become a topic for debate [3, 4].
Various forms of energy flow analyses have appeared in the
popular and technical literature, with identifying and
quantifying of energy flows becoming the principal activity
of many researchers. The complexity involved in dealing
with diverse inputs and processes has become one of the
major criticisms of energy accounting.

Net Energy

Energy accounting is based on the concept of net
energy. A pioneer in the energy accounting field, H. T.
Odum, defines net energy as the amount available for consumer use after the energy costs of finding, producing,
refining, transporting, and so on have been paid [5]. One

may think of this concept in terms of the axiom, "It takes energy to convert energy [6]."

Net energy = Quantity of energy represented by the quantity of reserve extracted − Energy spent in converting the energy resource to a form with social utility

Though net energy can be negative, it should normally be positive for a project to warrant further consideration.

Energy Accounting and Economics

One of the major motivations for energy accounting is a widespread feeling that the marketplace does not accurately reflect the true value of certain forms of energy, and that the cost of many energy forms will shift dramatically upwards in the future relative to other goods and services [7, 8]. Thus traditional economic benefit/cost analyses and life cycle costing procedures are suspect by some people as a sole predictive mechanism for the future viability of a particular process or product. For example, the official oil industry view in 1972 was that the relatively costly oil shale technology would be competitive when the price of oil reached $3.73/bbl [9]. In 1974 [10] that threshold price was pegged at $6.80/bbl, $15.00/bbl in 1975 [11], $21.00/bbl in 1976 [12] and revised upward to $25/bbl later that same year [13]. Today the price of OPEC oil is $38/bbl, and the oil shale boom is still only a whimper. Certain energy accounting advocates have claimed that only when energy inputs to alternate sources equal the energy inputs to recovery of oil from shale will oil shale development be economically viable [51].

Despite the fact that economic predictive mechanisms are not perfect, due to such factors as government intervention, inadequate value system definitions, lack of knowledge about future technology, foreign cartels, and vagaries of the marketplace, replacement of the benefit-cost analysis by an energy based theory of value is not warranted [14]. Rather, this paper supports the thesis that energy accounting may provide certain useful insights to policy makers to supplement traditional economic analysis. We attempt to demonstrate, using a sensible approach to energy accounting, that the energy implications of millions of people being attracted hundreds or thousands of miles for recreation at major man-made water areas should be considered in evaluating water management alternatives.

There seem to be as many methods of energy accounting as there are energy accountants. However, the common characteristic of all methods is that before one even approaches the determination of net energy, many obstacles must be overcome. The greatest challenge is to identify and then quantify the significant energy flows that cross the project control boundary as shown in Figure 1.

TYPICAL INPUTS

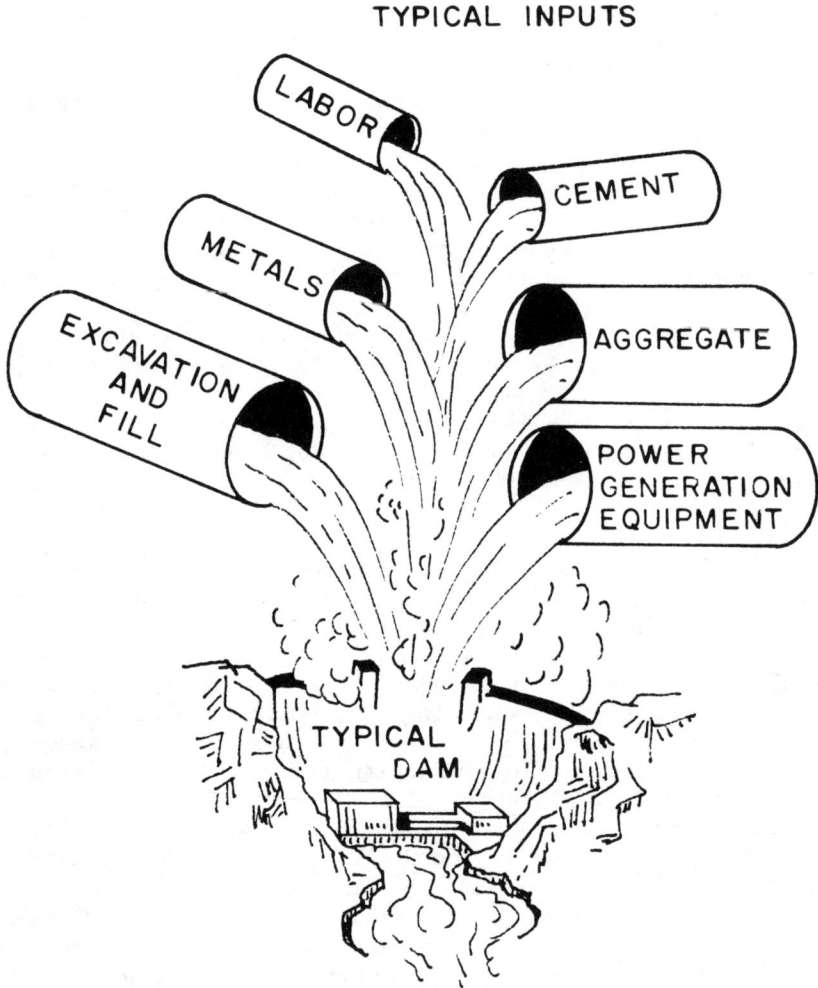

Figure 1. Typical energy resource inputs to a power project.

Types of Inputs

Energy flows may be classified as either Direct or Indirect. Direct inputs come from energy sources applied directly to the project such as concrete, steel, and gasoline. Direct social subsidies are those energy flows which represent a societal allocation of controlable energy resources such as coal, oil, natural gas, and electricity. Direct natural subsidies are those as yet unmanaged energy flows such as solar energy, potential and kinetic energy associated with river systems, and biological phenomena such as photosynthesis and chemical potentials.

Indirect energy inputs tend to be somewhat nebulous and constitute a "catch-all" category for anything that is not a material or process expenditure. Typical indirect inputs include labor, engineering, maintenance, and personnel services. Obviously, a certain number of indirect inputs are associated with each direct input.

The type and number of inputs considered in the analysis are dependent upon the energy accounting scheme. An all-inclusive approach to energy analysis of man made flat water recreation sites would show not only energy inputs for the items such as concrete and steel, but also would include the direct subsidy of solar energy falling on the site and would subtract the photo-synthetic energy of the foliage lost to the impounded water [15].

Quantification of the many indirect energy forms associated with the "holistic" type of analysis is difficult and highly subjective [16]. For reasons justified below, the holistic approach to energy accounting do not seem consistent with the objectives of this study and are rejected in favor of the approach based on the following three assumptions:

1. Oil is used as the standard fossil fuel equivalent. Energy inputs to water based recreation are expressed in terms of kilojoule equivalents of oil rather than of coal, electricity or other forms of energy. Oil is the form of energy currently attracting the greatest national concern.

2. Natural subsidies are ignored. Only those inputs which represent a societal allocation of energy resources are considered. Some holistic models claim that the loss of vegetation and its photosynthetic potential represents an energy cost which is chargeable against the project. For water projects in the west, vegetation losses are often in

the form of lost sagebrush and other plants adapted to
steep, rugged and arid land. Lake Powell is cited as an
example to illustrate the magnitude of this potential
loss. Taking the net vegetation support area to be equal to
the reservoir's total surface area, and using an average
value for typical biomass growth rates in the region, the
annual photosynthetic loss from construction of the Glen
Canyon dam is estimated at 86 TJ (86×10^{12} joules) [17].
This quantity is less than 2 percent of the total annual
estimated energy investment at the site of 4340 TJ. It thus
seems to us that these natural subsidies, while important
from certain perspectives, may reasonably be assumed
negligible from a societal cost point of view.

3. <u>Indirect inputs are not extensively considered in
the analysis.</u> This assumption does not preclude use of
indirect inputs that warrant consideration. Attempts to
itemize and prioritize the multitude of indirect inputs,
such as labor, quickly entangle the analyst in a web of
complexity and judgmental bias which is not unlike the now
defunct Labor Theory of Value [22].

To understand the effects of converting labor inputs
from dollars to energy units, based on a ratio relating net
energy expenditure to gross national product (GNP) or some
such relationship, a strong person can typically produce
0.0373 kW (0.05 hp) continuously. Thus, the total energy
produced in an 8-hour working day amounts to about 0.3 kw-hr
(0.4 hp-hr). If the same work could be done with
electricity costing about 5¢/kw-hr, the value of the day's
labor would be less than 2 cents. It is probable that this
conversion procedure would distort the energy accountng
analysis. Greater distortion exists if one contrasts manual
with professional labor as shown in Table I. Further, if
labor is included, why not the education process that
qualifies one for the position in question? The process of
including indirect inputs can go on and on.

Indirect energy inputs, for the most part, constitute a
type of social overhead or base load. This load exists and
is carried by society whether or not a specific project is
constructed. If a project is not constructed the load is
carried by some other form of societal activity. These
costs do not appear suddenly when a contract is let. They,
along with all the amenities to the worker's family, are
present regardless of job or location. As long as the human
resource is available to do the job, the baseload is present. Simple relocation of this resource and temporary
committment to a particular project does not change the
societal burden. For this reason, unless specific condi-

Table I. Relative value of various forms of energy [1].

Cost (1980 dollars) of 10^9 joules

Fossil Fuels		
Natural Gas	$	0.97
Coal		1.28
Diesel		2.39
Gasoline		4.40
Electricity		
Industrial		4.17
Residential		10.83
Animal Feed		
Alfalfa Hay		6.83
Grain Corn		10.59
Mechanical Energy		
Farm Tractor		29.59
Automobile		98.57
Foods		
Wheat		23.94
Rice		49.31
Bread		78.33
Potatoes		85.69
Turkey		126.18
Beef		246.06
Human Labor		
Manual Labor	9	944.26
Skilled Labor	22	729.75
Professional	119	331.17

tions make these inputs particularly relevant to a project, they are ignored in our approach.

STAFF EVALUATION

In this study the energy requirements of six selected water recreation areas located either entirely or principally in Utah are evaluated. These sites are listed in Table II. From a preliminary review of the selected sites, three facets of energy usage appeared to be significant; namely, construction of the facility, energy expended at the site in pursuit of recreation, and energy used in travel to and from the site.

Table II. Characteristic recreation data for selected sites
(obtained through government agencies - 1978).

Impoundment	Size (Hectares) Land	Water	No. of Visitors	Visitor Origin (Percentage) Local	Other	No. of Boat Days
Lake Powell	434 707	65 843	2 127 419	7	93	215 463
Flaming Gorge	38 866	17 011	680 870	20	80	110 800
Willard Bay	1 082	4 014	560 195	93	7	24 255
Rockport	312	437	293 804	95	5	9 460
East Canyon	112	277	141 695	95	5	6 663
Hyrum	118	192	164 195	95	5	7 827

Construction Energy

Estimates of the energy used in the construction of a
water recreation site were compiled through detailed review
of construction plans, specifications, and log books.
Energy input values for most major material inputs were
summarized from the literature or calculated from manufacturing data. Figure 2 illustrates for portland cement the
detail involved in this process and depicts the energy
inputs, consistent with the energy accounting approach
previously described.

Major input items and their respective energy values are
presented in Table III. Since certain components are manufactured from a percentage of recycled material and raw
material, the recycle energy savings associated with these
material inputs is accounted for in the table. Application
of the energy input values to the total estimated lifetime
material requirements of a project such as a hydropower dam
yields a net societal energy investment or project construction and maintenance energy input.

Table IV is a typical computer printout of estimates of
the construction energy inputs for the Glen Canyon and
Flaming Gorge Dams.

On-site and Travel Energy

From available sources of data, including existing
visitation records and the results of on-site surveys, an
attempt was made to determine distances and modes of travel
to each recreation site. A careful statistical analysis of
the data was carried out to adjust for multiple purpose
travel, most probable points of origin for out-of-state
visitors, and other factors [21].

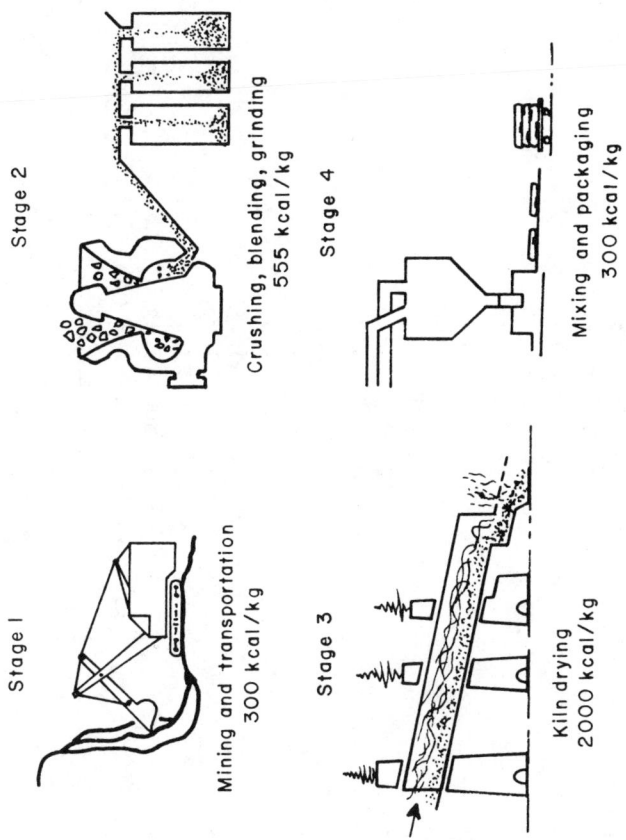

Stage 1

Mining and transportation
300 kcal/kg

Stage 2

Crushing, blending, grinding
555 kcal/kg

Stage 3

Kiln drying
2000 kcal/kg

Stage 4

Mixing and packaging
300 kcal/kg

Figure 2. Energy inputs to portland cement [1].

Table III. Energy inputs to major materials or processes on a raw material basis [1].

Material	Raw Material Input Value	Recycled Input Value
steel cast	33 520 kJ/kg	16 760 kJ/kg
steel rebar	41 900 kJ/kg	20 950 kJ/kg
carbon steel	58 600 kJ/kg	29 330 kJ/kg
stainless	67 040 kJ/kg	33 520 kJ/kg
aluminum	276 541 kJ/kg	108 940 kJ/kg
copper	142 461 kJ/kg	83 800 kJ/kg
cement	12 570 kJ/kg	not applicable
excavation and fill	29 kJ/kg	not applicable
aggregates	46 kJ/kg	not applicable
diesel fuel	155 031 kJ/kg	not applicable
gasoline	129 891 kJ/kg	not applicable
PVC	108 941 kJ/kg	not applicable
polethylene	125 700 kJ/kg	not applicable
glass plate	46 090 kJ/kg	not applicable
electric motors	83 800 kJ/kg – 419 002 kJ/kg	62 850 kJ/kg
generators	83 800 kJ/kg	62 850 kJ/kg
hydro-turbines	87 990 kJ/kg	41 900 kJ/kg
steam turbines	104 750 kJ/kg	62 850 kJ/kg
pumps	83 800 kJ/kg	188 550 kJ/kg
engines	83 800 kJ/kg – 14 146 068 kJ/kg	41 900 kJ/kg

PROJECT: Glen Canyon Dam (Lake Powell)

Installed Capacity: 900 000 .kW   Net energy investment in fossil fuel equiv.
Ave. Output for 1973: 525 000 .kW   Ave. output in electricity kW
Evaluation Period of 50 years

	Installed Mass or Quantity (Million kg.)	Mass Replacement Schedule	Energy Inv. Raw Mat.	Energy Inv. Recycled Mat.	Replacement Factor	Net Mat Energy Invest. TJ (10^{12} joules)
Steel Rebar	12.998	0.0	41 900	20 950	1.0	544.606
Steel Carbon	46.392	0.0	58 660	29 330	1.0	2 721.372
Aluminum	0.950	1.1	276 541	108 940	1.4	376.494
Copper	0.250	1.3	142 461	83 800	1.8	62.754
Cement	969.834	0.0	12 570	12 570	1.0	12 191.206
Aggregate	9 298.408	0.0	46	46	1.0	428.564
Excavation & Fill	11 997.946	0.0	29	29	1.0	351.526
Turbines Hydro.	4.426	1.2	87 990	87 990	1.6	612.028
Generators	5.625	1.3	83 800	83 800	2.0	931.018

Total Energy Investment   18 219.652

Hours to break even point at average output   9 631.
Hours to break even point at total capacity   5 618.

PROJECT: Flaming Gorge Dam

Installed Capacity: 10 500 .kW   Net energy investment in fossil fuel equiv.
Ave output for 1973: 2 100 .kW   Ave. output in electricity kW
Evaluation period of 50 years

	Installed Mass of Quantity (Million kg.)	Mass Replacement Schedule	Energy Inv. Raw Mat.	Energy Inv. Recycled Mat.	Replacement Factor	Net Mat. Energy Invest. TJ (10^{12} joules)
Steel Rebar	2.300	0.0	41 900	20 950	1.0	96.354
Steel Carbon	6.699	0.0	58 660	29 330	1.0	392.957
Aluminum	0.005	1.1	276 541	108 940	1.4	1.982
Copper	0.005	0.0	142 461	83 800	1.0	0.712
Cement	179.969	0.0	12 570	12 570	1.0	2 262.225
Aggregate	1 699.709	0.0	46	46	1.0	78.341
Excavation & Fill	1 399.760	0.0	29	29	1.0	41.054
Turbines Hydro.	0.550	1.2	87 990	87 990	1.6	76.036
Generators	0.680	1.3	83 800	83 800	2.0	112.523

Total Energy Investment   3 062.184

Hours to break even point at average output   10 351.
Hours to break even point at total capacity   7 869.

(The replacement factor is given as $\left[\frac{\text{Energy Inv. in Recycled Material}}{\text{Energy Inv. in Raw Material}}\right]$ x No. times replaced + 1.0] = RF
and the net investment is given by: Net Material Energy - RF (Energy Inv. in Raw Material) Installed Mass)

Following extensive on-site interviewing, estimates
were made of the recreation activities at each site.
Visitor information and sizes of boats were used to estimate
fuel consumption on the lake. Fuel consumption figures from
various manufacturers were obtained [18], and correlations
for fuel consumption rates as a function of engine size were
developed as shown in Figure 3. These estimates of fuel
consumption from visitation data were supplemented with
marina fuel sales records. The price of fuel seems to have
had an important influence on visits to water recreation
areas in recent years, as indicated in Figure 4.

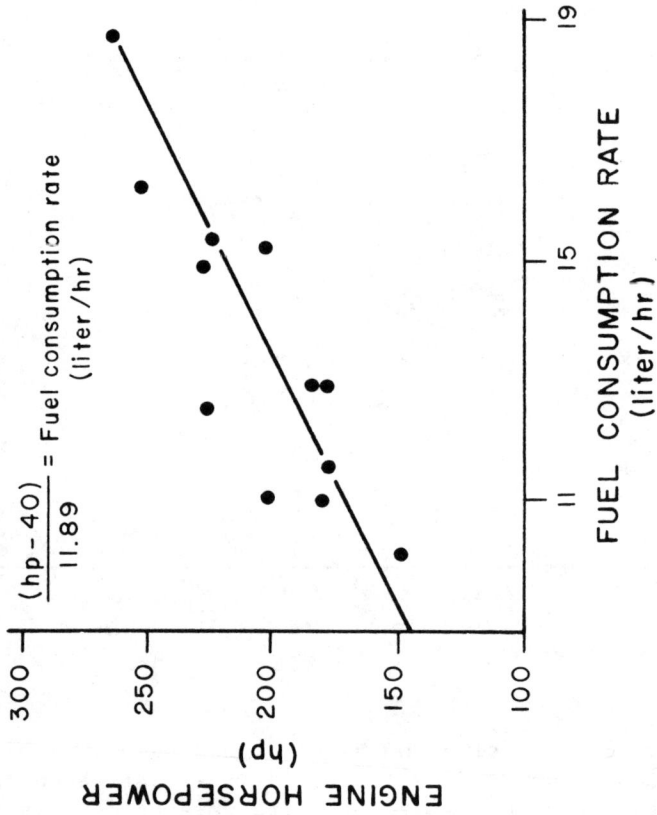

Figure 3. Average estimated fuel consumption rate for power boats as a function of engine size.

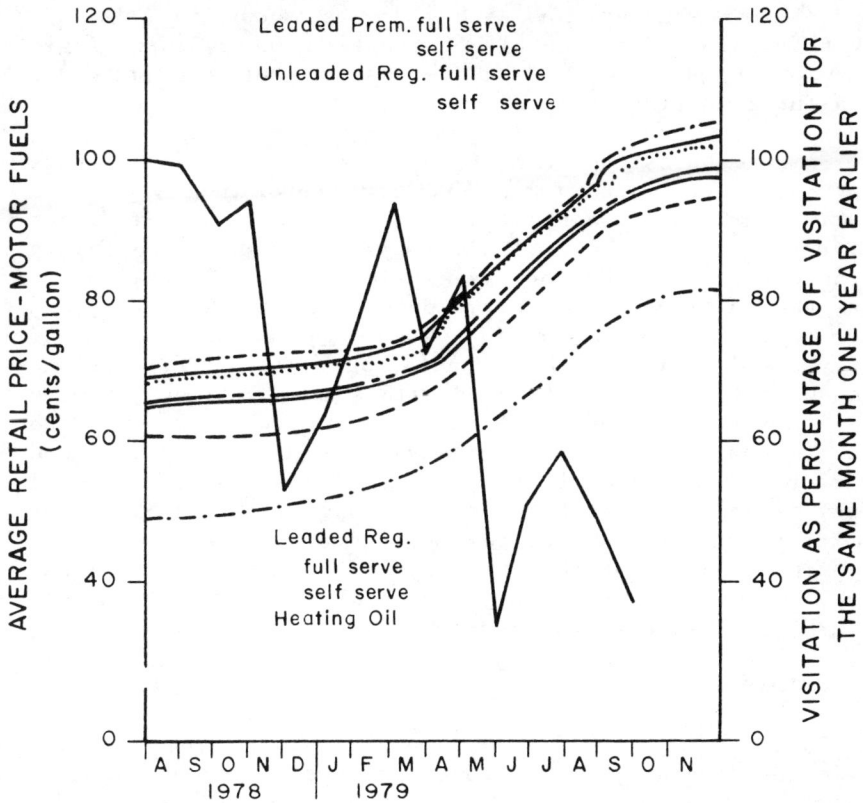

Figure 4.   1979 fuel prices and Lake Powell attendance as
percentage of the same month the previous year
[19].

        Personnel employed by the Utah Department of Parks and
Recreation record information concerning users of each of
the reservoirs selected for this study.   These data were
supplemented with on-site interviews that were used to
provide estimates of selected user characteristics such as
origin, destinctions, mode of travel, activity participation
as well as size and type of equipment used. These data were
used to estimate the energy used by recreationists at the
various sites.

TOTAL ENERGY BUDGET

        The estimated energy expenditures for construction were
distributed over the estimated useful life of the project,
and the resulting annual cost estimate was combined with
annual on-site and travel energy costs to produce a total

annual energy budget for each site as shown in Table V. Only a fraction of the total construction energy input logically could be allocated to recreation but the total is shown for comparison purposes.

Some interesting comparisons can be made at this point. In 1978, Utah farms required approximately 2000 TJ to accomplish their agricultural functions based on direct fuel usage reports [20]. This total is roughly two thirds of the energy expended for travel alone to Lake Powell. Thus, the total annual energy expenditure for recreation on this reservoir is apparently about twice the amount of energy used annually for Utah's agriculture production. This quantity may be contrasted with a much smaller but more local impoundment behind Hyrum dam in northeastern Utah which is characterized by 0.3 percent of the water area and 8 percent of the annual visitations of Lake Powell. Total energy expenditures at Hyrum dam are approximately 85 TJ, or 4 percent of the 1978 agricultural amount. Based on these data, it is estimated that the total flat water recreational energy expenditure in Utah is several times that of agriculture, the major industry of the state.

CONCLUSIONS

Table V depicts for various reservoirs the relative magnitudes of energy expenditures for recreation as compared to those for construction of the facilities. This comparison clearly shows that even though the energy devoted to construction is substantial, and must be of concern in

Table V. Comparison of construction, travel and on-site annual energy expenditures associated with recreational activities at selected man-made water recreation sites.

Site	Construction TJ	Travel TJ	On-Site TJ
Lake Powell	364	3000	976
Flaming Gorge	61	0[1]	0[1]
Willard Bay	12	55	55
Rockport	9	38	28
East Canyon	8	24	30
Hyrum	2	57	26

[1]The on-site survey yielded inconclusive data.

654

deploying resources, the amount of energy that continues to be expended in travel and on-site recreation should receive earnest attention. Thus, energy account analysis could be an important technique in the planning procedure for future water resource projects containing a recreation component. This paper proposes and demonstrates a viable procedure for conducting this form of analysis.

REFERENCES

1. D. A. Bell, The Development of a Thermodynamic Energy Accounting System and its Application to Power Project Evaluation, Master's Thesis, Utah State University, Logan, Utah, 1977.

2. Non-Nuclear Energy Research and Development Act of 1974 (PL 93-577), 93rd Congress, 1974.

3. D. Heuttner, Net Energy Analysis: An Economic Assessment, Science 4235, 1976, 101-105.

4. N. Georgescu-Roegen, Energy and Economic Myths, Southern Economic Journal, 41 (1975), 347-381.

5. H. T. Odum, Energy, Ecology, and Economics, Ambio, 2 (1973), 220-227.

6. W. Clark, It takes energy to get energy; the law of diminishing returns is in effect. Smithsonian 5:(No. 9), 84-91, 1974.

7. E. Cook, Saving and environment gout and economy cabbage. Exxon, USA 13, No. 2, 11-15, 1975.

8. J. C. Batty, D. A. Bell, and B. C. Jensen, Energy Accounting-Perspectives and Guidelines, Utah State University, 1976.

9. G. U. Dinnen, and L. Cook, "Oilshale and the Energy Crisis," Paper presented at the Winter Annual Meeting of American Society of Mechanical Engineers, New York, NY, 1972.

10. M. A. Adelman, et al., Energy Self-Sufficiency: An Economic Evaluation, Technology Review, 76 (1974), 45.

11. L. B. Rothfield, Waste and Energy Requirements for the Oilshale Industry, Proceedings at the Conference on Water Requirements for Lower Colorado River Basin Energy Needs. Tucson Arizona, 1975.

12. M. MacCormick, Proceedings of Technology subcommittee, Energy R & D and Demo., Speech. Washington, 1D. LC., 10-12 March 1976.

13. W. H. Wiser, College of Science Distinguished Guest Lecture Series, Address, Utah State University, June 1980.

14. H. T. Odum, and E. LC. Odum, Energy Basis for Man and Nature, McGraw-Hill, 1976.

15. S. Bayley, et al., Energetics and Systems Modeling: A Framework Study for Energy Evaluation of Alternative Transportation Modes, Engineering and Industrial Experiment Station, University of Florida, 1977.

16. C. D. Myers, Energetics: Systems Analysis with Applications to Water Resources Planning and Decision-Making, U. S. Army Engineer Institute for Water Resources, 1977.

17. J. C. Batty, D. A. Bell, and D. C. Jensen, Energy Accounting-Perspectives and Guidelines, Utah State University, 1976.

18. Mercury Outboard Motor Division Testing Center, "Fuel Conservation Tests," Pamphlet, 1979.

19. Unpublished data available from Utah Department of Parks and Recreation, Salt Lake City, Utah.

20. Utah State Tax Commission, Report of Gasoline Tax Refunds for Agricultural Off-Road Vehicles, Utah State Printing Office, 1978.

21. Stoddard, Thomas C., Energy Accounting as a Tool for Evaluating Man-made Water Recreation. Unpublished M.S. Thesis, Utah State University, Logan, Utah, 1981.

22. Baug, Mark, Economic Theory in Retrospect, Irwin, 1968.

PART 10

PANEL DISCUSSION

INTEGRATING AQUATIC RESOURCE VALUES INTO
COLORADO RIVER MANAGEMENT

L. Douglas James
        Director, Utah Water Research Laboratory,
        Utah State University, Logan, Utah

For generations, men have used the Colorado River as a source of water for an arid land. They developed its waters to make deserts productive agriculturally and to serve municipal and industrial uses that permitted millions of people to move into the Southwest. Water availability has been extended through low flow seasons and drought years by the construction of reservoirs that store runoff from melting snow and release the water at times and to places where people want it.

As the economy built by developing this water resource flourished, many river reaches lost much of their natural character. Particularly in the downstream reaches, the annual flow pattern changed from spring peaks followed by flow recession over a long hot summer to one artificially determined by return flows.

Furthermore, as consumptive uses have increased, total river flows have diminished. As the flow reduction has been relatively larger than any increased trapping of salinity and as irrigation water has been used to leach previously stored salts from agricultural soils, the total dissolved solids content of the remaining river flow has increased. Simultaneously, urban and industrial growth (including potential massive development of presently lightly explored fossil fuel resources) threaten to discharge new toxic wastes into the river environment.

Many people are becoming concerned as they observe deterioration in the Colorado's aquatic environment and extrapolate long term harm to the total ecosystem. The aquatic environment provides many important amenities, and its vitality is widely taken as an indicator of the health of life support systems vital to human well-being.

In short, river development for beneficial offstream use often conflicts with instream aquatic resource values. Full river development is unnecessarily harmful to the natural riverine ecosystem. Full protection of the natural system is unnecessarily restrictive to beneficial use of river flows.

Wise management practice goes to neither extreme. Rather, it searches out compromise, a water management policy that balances the benefits from water use against the aquatic values sacrificed or (when taken in the opposite direction) the aquatic gains by more natural river flows against beneficial offstream use denied. Such a policy is found by identifying and analyzing tradeoffs.

The present situation combines a specific combination of beneficial water uses with a specific state of the aquatic environments. Opportunities exist to develop additional water for beneficial use, but these are likely to inflict losses on aquatic values. Opportunities exist to enhance the aquatic environments, but these are likely to reduce values gained through the beneficial use of water.

In tradeoff analysis, we recognize that many sources and methods exist for developing additional water supply (reservoir or groundwater storage; reservoir location, materials of construction, and operating policy; extent of reuse within the system; etc.), and some of these would be much less detrimental than others to the aquatic environment. Other factors being equal, the preferred water development would supply the needed water with minimum adverse environmental impact. Conversely, the preferred method to enhance the aquatic environment is associated with minimum loss of water for beneficial use.

The potential contribution tradeoff analysis can make to water project design or reservoir operating policy formulation is obvious in principle. In making practical application of these principles, however, we are severely handicapped by a lack of understanding of how marginal changes in runoff volumes and patterns in the timing and quality of flow affect aquatic values. This symposium was organized to 1) inventory what we know, 2) enhance coordination among research groups in learning more, and 3) promote interaction between water resource planning and aquatic resource specialists to direct river management practice toward the state of the art.

The technical sessions of this 1981 Symposium on Aquatic Resource Management of the Colorado River Ecosystem contained over 40 papers. Various authors presented the politics of

660

the river, a conceptualization of economic tradeoffs between diversion and instream water use, effects on the river of fossil fuel development, the potential of cloud seeding for augmenting river flows, characterization of stream and reservoir ecosystems and the effects of river water uses on that characterization, habitat requirements of various fish species and how habitats are being altered by river development and water management, salinity and nutrient movement through streams and reservoirs, and chemical and biological changes occurring in stream and reservoir waters.

These papers present considerable valuable information but come far short of supplying quantitative methods for defining marginal tradeoffs between water development and aquatic resource values. The purpose of this panel discussion is to survey what remains to be done. Paraphrasing Robert Browning, we have come to the last of the symposium for which the first was made.

The technical papers reported specific studies. This final session has the harder job of synthesizing information from these papers into the state of the art for assessing 1) what aquatic resource needs are not well enough understood for defining marginal tradeoffs but have requirements that can be defined in a more general way for river planning, 2) what we now know about marginal tradeoffs that can be applied, 3) what research topics need to be pursued to develop better understanding for future applications − aquatic resource conflicts deserve high priority, and 4) what research approaches are most promising for pursuing the priority topics. We can use present knowledge to reduce current clashes between economic benefits and aquatic values and pursue research to improve our understanding of the tradeoffs in pending decision situations.

For this synthesis, five sets of questions will be probed:

1. What harms have been already caused to the river's aquatic resources and how do these rank in terms of the severity of their impact? Conversely, what opportunities are currently most promising for aquatic resource enhancement in terms of probability of success and value received through the effort?

2. What characteristics of existing flow and quality patterns have the greatest impacts on the aquatic resources in the Colorado River and its reservoirs? What changes in those characteristics and resulting impacts are most threatening? Stream characteristics can be defined in terms of the

temporal patterns in the runoff hydrograph, the geometry of the flow cross sections, and sediment and salinity content. Reservoir characteristics encompass spatial patterns of different quality parameters, time patterns of those parameters at a given site, patterns of changes in reservoir contents of water and sediment, and movement of density currents through the lake.

3. What changes in reservoir operating policy would be most effective in reducing present or projected harms? What changes would have the most positive effect in enhancing aquatic resource quality?

4. How do the changes in reservoir operating policy that would be best for the aquatic resources affect offstream beneficial uses?

5. What legal, political, and institutional forces constrain shifts in reservoir operating policy that would better balance beneficial use and aquatic resource values? Are current policies too set by law to be changed? Where the law would have to be changed, what are the political prospects for effecting the changes? Would desirable shifts in operating policy be accepted by consensus as needful? How is our freedom to change constrained by financial commitments to pay for past projects? Could we equitably shift to a management policy more favorable to aquatic resources without changing repayment requirements?

Of the above questions, the most critical require further research into the mechanisms through which a flow of given characteristics affects stream or reservoir aquatic environments. Once these mechanisms are known, we can estimate the aquatic impacts of managed changes in flow characteristics. We can go on to consider the processes that changes in flow characteristics that have already occurred may have set in motion that will continue to impact aquatic resources over time. What are the direct effects of changes in reservoir operation? What are the indirect effects of river development stimulating population growth, land use change, increased outdoor recreation, and changes in human values?

During this symposium, we have discussed answers to these questions. Still more could be learned by consulting experts who are not with us. From this total body of knowledge, much could be applied to guide reservoir operation and river management policy, we could learn much more and do much better through further research.

The magnitude of the need and the administrative decentralization in river management place a burden of cooperation on the research community. We need to seek consensus in research prioritization (in cooperation with river system managers), in formulating complementary study designs with information exchange on research in progress, and in coordinating the collection and exchange of data. We can no longer afford the luxury of independent studies and independent data collection. It is much more economical to gather different samples for simultaneous studies from the same boat than it is to send separate boats. It is more meaningful to be able to compare studies on related topics if their data are taken at common times and places.

The Colorado River has been described as the most managed major river in the world. Large uncontrolled flows have not reached the sea in 50 years. Yet, do we understand what we are doing to the river? Could we do better in meeting present needs? Is our management system flexible enough to adjust to future needs? Much that we have heard in this symposium raises doubts in our ability to answer these questions. Such doubts should motivate us to learn more.

I have come to this conference while returning home from India and being involved in planning for the development of the Narmada River, a river that carries four times the annual runoff of the Colorado and is one of the largest free flowing rivers left in the world. It has gone undeveloped for so long because of a water rights dispute between upstream and downstream states; and, recently, an Indian Supreme Court Tribunal allocated the water between a wetter upstream and a dry downstream state. With the way to river development now open, both states see water development as a source of income for poor farmers, a source of food for a hungry nation, and a source of hydroelectric power for cities which experience rotating blackouts during peak demand periods. India plans to build 30 dams to control their river as they have seen us do on the Colorado, and no thought is being given to aquatic resources.

The world is following our example. Are we satisfied with what we have done? It is one thing to answer negatively and another to convincingly set forth principles that will help the third world avoid our mistakes.

We have four panelists who have been listening to the symposium papers with instructions to abstract information that will help us. Five individuals were asked the following sets of questions:

663

1.   How is the present course of land and water use
change in the Colorado River Basin expected to alter river
flows and water quality? What opportunities exist to alter
water management practices to improve (protect) the aquatic
environment?   In short, considering the economic demands
for river use and the economic limitations on the degree to
which river flows and water quality can be controlled, what
sorts of water management and land use policies are feasible.
What actions seem most reasonable?

2.   The system for Colorado River management as currently institutionalized has evolved through a long history of
compromise among competing river users and uses (Upper Basin
v. Lower Basin, California v. Arizona, agriculture v. urban,
etc.).   Should revisions in river management practice
be beneficial, how severely would their implementation be
constrained by existing institutions?   Are current water
users or other water right holders going to forego water use
to protect aquatic resources?   Should they be compensated?
How much?

3.   The need to give special attention to aquatic
resource management depends on just how severe a threat to
river aquatic resources would otherwise develop over time.
What will predicted flow and water quality changes do to the
aquatic environment?   How severe is the threat to aquatic
resources by such dangers as containment pond spills?   In
short, how severe a problem is expected? What specific
flow and water quality situations are most threatening?
What actions will really contribute to correcting problem
situations?

4.   Reservoir aquatic environments are determined by
inflow amounts and quality as well as by reservoir bed
materials.   They develop continually while the reservoir
fills and ages, are affected as new reservoirs are built
upstream, and tend to stabilize over time. What environments
will develop in the long run in the Colorado River reservoirs?   How would the reservoir environments be affected by
such dangers as containment pond spills? What problem
situations threaten damage to them? What precautions would
be effective in protecting the desirable features in reservoir environment?

5.   The most sure response to the above question is that
we really do not know all the answers.   Further research
is needed. Without more answers, major aquatic damage
can still occur as good intentions prove to be bad errors.
What are the prospects for research funding for improved
aquatic resource management? What sorts of funding will have
highest priority?

We were not successful in finding someone in the federal research establishment who was willing during this transition period in government to discuss the prospects of research funding with us. This lack of research leadership is unfortunate, but it does provide an opportunity and increase the burden on the research community to formulate and orchestrate a research agenda.

The four panelists will discuss the above questions. Afterwards, we will assess our ability to provide information in marginal tradeoffs between water use benefits and aquatic resource values for scientific river management.

INSTITUTIONAL PERSPECTIVES ON
COLORADO RIVER MANAGEMENT

Vernon E. Valantine
        Assistant Chief Engineer
        Colorado River Board of California

Any discussion of possible revisions in the management practices of the Colorado River should be based on an understanding of the institutional bases for current management practices. The institutional bases for the management of Colorado River water have evolved over many years and incorporate many documents that spell out the compromises reached among the many competing river users in seven states and the Republic of Mexico. Collectively, these compromises are known as "The Law of the River." Each of the Basin states has its own framework of water rights laws and regulations, and they are all based on the Appropriation Doctrine. Those documents in the Law of the River that are most applicable to the question of river management requirements, and the fundamentals of the Appropriation Doctrine, are briefly reviewed, and their significance is discussed in view of possible changes in management practices.

COLORADO RIVER COMPACT

The key element of the Law of the River is the Colorado River Compact, signed by representatives from the seven Basin states in 1922. The Colorado River Compact divides the water between the Upper and Lower Basins with a division point at Lee Ferry, 17 miles below Glen Canyon Dam. Each Basin was apportioned the right to beneficial consumptive use of 7.5 million acre-feet per year (maf/yr) from the Colorado River system. In addition, the Lower Basin was given the right to increase its use by 1 maf/yr. The Compact states that any required delivery of water to Mexico shall be supplied first from water surplus to the foregoing apportionments (a total of 16.0 maf/yr) and that if the surplus is insufficient, the burden of the deficiency shall be borne equally by the Upper and Lower Basins. It also provides that the Upper Basin will not cause the flow at Lee Ferry to be depleted below 75 maf for any period of 10 consecutive years.

667

BOULDER CANYON PROJECT ACT

The Boulder Canyon Project Act, which became effective in June 1929, provided federal approval for the Colorado River Compact and authorized construction of the Hoover Dam and Powerplant and the All-American Canal. One of the requirements of the Act was that it would not take effect until California adopted legislation setting a limit on its use of Colorado River water, as was done by the California Legislature when it passed the California Limitation Act in March 1929.

WATER DELIVERY CONTRACTS

Water delivery contracts for water service from the Colorado River at Lake Mead and below were executed between the Secretary of the Interior and agencies of the Lower Basin states beginning in 1930. The California water delivery contracts established priorities among the various agencies within California.

THE MEXICAN WATER TREATY

Agreed to in 1944, the U.S.-Mexico Treaty divided the waters of the Colorado River, as well as the Rio Grande and Tijuana Rivers, between the United States and Mexico. This Treaty was ratified by the Senate in 1945, and gave Mexico a guarantee of 1.5 maf/yr from the Colorado River. Davis Dam was completed in 1950 as required by the Treaty to reregulate the flows released from Hoover Dam.

UPPER COLORADO RIVER BASIN COMPACT

Entered into in 1948, the Upper Colorado River Basin Compact apportioned water within the Upper Colorado River system. The compact gave 50,000 af/yr to Arizona, which has a small area in the Upper Basin, and divided the remainder of the Upper Basin's water among the states as follows: Colorado, 51.75 percent; New Mexico, 11.25 percent; Utah, 23 percent; and Wyoming, 14 percent.

COLORADO RIVER STORAGE PROJECT ACT

The Colorado River Storage Project Act, passed in 1956, authorized major developments in the Upper Basin consisting initially of four large storage units and eleven participating water projects. Nine additional participating projects have been subsequently authorized. The storage units are Glen Canyon Dam; Flaming Gorge Dam; the Curecanti Unit on the Gunnison River consisting of Blue Mesa, Morrow Point, and

Crystal Dams; and Navajo Dam. The storage units provide about 33.6 maf of storage capacity.

The participating water projects use revenues generated from the hydroelectric plants at the storage units to help repay the cost of irrigation features which are beyond the ability of the water users to repay. Several projects have been completed at this date, several more are under construction or are nearly completed, and a few projects are still at the planning stage or have been indefinitely postponed.

ARIZONA V. CALIFORNIA

The March 1964 U.S. Supreme Court Decree in Arizona v. California established several additional dimensions to the apportionment of Colorado River water. It provided that if sufficient mainstream water is available to satisfy 7.5 maf/yr of consumptive use for the States of Arizona, California, and Nevada, then Arizona is apportioned 2.8 maf/yr; California, 4.4 maf/yr; and Nevada 0.3 maf/yr. If more than 7.5 maf is available, then California is apportioned 50 percent of such surplus, and Arizona, 50 percent, with the United States having the right to contract with Nevada for 4 percent, to come out of Arizona's share. During shortage conditions, the Secretary of the Interior is directed to first satisfy present perfected rights and then to apportion the amount remaining to the states.

The five Indian reservations located along the mainstream in Arizona, California, and Nevada were allocated present perfected rights for annual quantities not to exceed 1) diversions of 905,496 acre-feet, or 2) the quantity of water necessary to supply the consumptive use required for irrigation of 136,636 acres and related uses, whichever of the two is less.

COLORADO RIVER BASIN PROJECT ACT

Enacted in 1968, this law culminated several years of negotiation and compromise among the Colorado River Basin states, the Columbia River Basin states, the federal government, conservation groups, and others. Major features of the Act are 1) the Central Arizona Project and five Upper Basin water projects were authorized; 2) in the event of a water shortage, California's basic apportionment of 4.4 maf/yr, and similar uses of like character in Arizona and Nevada, have priority over the Central Arizona Project; 3) the United States assumed responsibility for meeting the entire Mexican Water Treaty obligation when the river is augmented by 2.5 maf/yr; and 4) the Secretary of the Interior was directed to

669

establish coordinated long-range operating criteria for major
Colorado River reservoirs based upon priorities listed in the
Act. This last assignment is particularly relevant to the
purposes of this symposium because it designates a target
for our results.

COORDINATED LONG RANGE OPERATING CRITERIA
FOR COLORADO RIVER RESERVOIRS

As required by the Colorado River Basin Project Act, the
Operating Criteria were issued in June 1970 by the Secretary
of the Interior. They have as an objective the release of a
minimum of 8.25 maf/yr at Lee's Ferry and provide that each
year, in the context of its flow situation, that a reservoir
operating plan be developed by the Secretary after consultation with the seven Basin states. The criteria provide for a
determination by the Secretary of the amount of water to be
retained in Upper Basin reservoirs in order to meet obligations to the Lower Basin without impairment of the Upper
Basin's consumptive uses. When the Upper Basin's storage is
greater than the amount determined above, the reservoir
contents above the minimum will be released in order to
maintain, as nearly as practicable, active storage in Lake
Mead equal to active storage in Lake Powell.

SIGNIFICANCE OF LAW OF THE RIVER

The Congressional Acts, interstate compacts, court
decrees, international treaty, contracts, and administrative
regulations, which comprise the Law of the River, establish
constraints or limits on the water available within each
state in the Colorado River Basin. It should be noted that
they establish an apportionment for each state, which is not
a fixed quantity but rather is a share of a total supply,
which is dependent upon hydrologic conditions of annual
runoff quantities and amount of water in storage.

It should also be noted that these are annual apportionments. Since stream aquatic resources need, in many cases,
flows to be kept at certain levels at certain times during
the year or certain minimum flow rates to be maintained,
these management criteria can frequently be met within the
overall apportionments set by the Law of the River. For
example, the releases of water from Glen Canyon Dam that are
required pursuant to the Law of the River are on an annual
basis. Within the limits set by that annual release requirement, the releases can be varied to different rates of flow
during the different months or even on a day-to-day basis.
With that flexibility, flows in the river below Glen Canyon
Dam can be, and are being, varied in the patterns needed to

serve many societal purposes: hydroelectric generation, recreational use of the river through the Grand Canyon, sustaining the trout fishery below Glen Canyon Dam, and enhancing the bass fishery in Lake Mead. Thus the tradeoffs involving aquatic resources on the main stream are more among these instream uses than between instream uses and diversions.

While the apportionments of consumptive use made by the Law of the River cover all uses within the Colorado River Basin, including uses on tributaries, the principal constraints are in regard to mainstream flows, not to tributary flows. Thus, the flow regimes of the various tributaries in the Upper and Lower Basins are only indirectly circumscribed by the Law of the River. If, for example, a state wishes to keep one of its streams in a free-flowing condition, and use its apportioned share of Colorado River System water by depleting other streams within that state, it may do so.

In addition to the water quantity restraints set by the Law of the River, there are water quality restraints on Colorado River water uses. Maximum allowable salinity concentrations are prescribed at Imperial Dam. The states in the Colorado River Basin have a mutual interest in each state minimizing any deleterious impacts on the river's salinity that may result from the diversion and use of water.

STATE WATER LAWS

Within each state, water allocations are made in accordance with that state's water laws and regulations. The states of the Colorado River Basin all employ the appropriation doctrine as the basis for their water laws (California has a dual appropriation-riparian doctrine). While the states vary in application of this doctrine, certain basic principles apply to all. These principles are 1) the priority of appropriation is set by the earliest date for filing and/or for placing the water to beneficial consumptive use, 2) the right is based upon beneficial consumptive use, and, if a water right holder ceases to make beneficial use of a portion or of all of his water rights, he may be forced to surrender the unused portion, and 3) the rights may be sold or leased to others, provided that other holders of appropriative rights are not injured by the transfer.

An example of what can be done within the framework of the Law of the River and of existing state water laws to enhance aquatic resources was presented at this Symposium, in connection with the Central Utah Project. It was reported that the Central Utah Water Conservancy District, which will be paying for and distributing the water conveyed into the Bonneville Basin, has agreed to surrender a portion of its water developed by the Central Utah Project for use in maintaining live streams in the Uinta Mountains for fishery habitat.

SUMMARY

An initial reaction to the many documents that make up the Law of the River may be that the apportionment of Colorado River water is tightly constrained and there is little leeway to revise river management practices. However, closer analysis shows that the major concerns of aquatic managers for the mainstem stream environment can be accommodated within the framework of the Law of the River. This is because many of the problems can be solved and benefits realized through establishing minimum flow rates (a reordering of the flow pattern within the year) or manipulating the flow on streams that are fully within the control of one state. These situations are only generally constrained by the Law of the River, and a great deal of flexibility remains to improve aquatic environments within the limits of the existing constraints. In other words, the issues are not resolved in the legal domain but through examining such issues, continuing with the above example, as the hydroelectric power benefits lost versus the aquatic gains from altering the pattern of reservoir releases.

The flexibility to reorder tributary streamflow to meet aquatic resource needs is more tightly constrained by state water laws. State water rights deal in much smaller numbers and with many more right holders than does the Law of the River, and these fine divisions work against flexibility. Second, requests for water for aquatic resource use must be processed through state water rights procedures, and transfers of water for aquatic resource use require funds to purchase the rights and proof of minimal adverse impacts to other holders of diversion rights. Both factors work against change.

Finally, the Department of the Interior has been assigned responsibility for reservoir operation, and its operating policies have evolved over a period which the three major reservoirs have been filling. They are now full, and the basin is entering a new operating situation for which we know very little about the effects of reservoir operation on lake aquatic resources, the changes in operating policy that may be beneficial, or whether there will be significant constraints on those changes. No specific legal problems, however, are yet known.

RIVER AQUATIC SYSTEMS

J. A. Stanford
   University of Montana,
      Biological Station

The vast amounts of data presented on the effects of
water development on river aquatic systems found in the
present symposium, past symposia, various environmental
impact statements, agency reports and informative journal
publications basically boil down to assessments of how stream
regulation has apparently altered biogeochemical processes to
favor salmonid production and concomitantly eliminate habitat
of endemic species. Conditions in riverine segments that
presently contain extremely productive trout fisheries (e.g.,
in the Black Canyon of the Gunnison and below Lake Powell)
result more by accident than by planned management. Likewise,
the remaining viable populations of endemic fish species
(e.g., squawfish, bony-tailed chub, hump-backed chub, razorback sucker) are protected not by a planned design but because the deleterious effects of regulation in Upper Basin
segments are ameliorated by natural physico-chemical processes and/or by infusions of water from unregulated side flows.
In spite of countless site-specific impact assessments, the
art of predictive management, based on scientifically-derived
facts about life histories as functions of riverine processes, has not progressed beyond infancy in the Colorado
River Basin.

Some of the difficulty in synthesizing management plans
for riverine species is technical. It results from incomplete understanding of the life histories and behavior of the
endemic fishes, an inability to predict the timing of water
delivery and uncertainty over the possible effects of potentially toxic wastes (e.g., oil shale leachates) in various
river segments. Also, the myriad of water rights in the Basin
place complicated constraints on implementing any management
option. However, the biggest difficulty in formulating a
management plan, based on scientific interpretations of
time-series data, is a lack of interagency coordination.

Too many factions are trying to manage the Colorado River system solely to meet their individual objectives. At a time when the gap between science and management is very wide, dozens of state and federal agencies are individually interpreting relatively meager scientific data from which they fervently try to justify favored management policy or additional funding for "monitoring" programs. One result is an indecisive message to the political machinery that eventually assigns water use in the River system.

Given that dozens of agencies are involved in management of aquatic resources in the Colorado River Basin, some mechanism of coordination has to be developed that matches water use allocations in both the Upper and Lower Basins with water availability and the maintenance of water quality and wildlife values in, at least, selected riverine segments. This idea was heralded to the Congress of the United States by the Comptroller General (1979) on May 4, 1979. The report states that unless interagency management plans coalesce, many water users in the Colorado Basin will be out of water by the year 2000. One of those users would undoubtedly be the endemic riverine fishes.

Symposia, such as the present and ones previous (see especially Spofford et al. 1980), promote fact sharing but often raise more questions than they answer. Authors at this symposium presented various river segments as critical to either native or introduced fishes (Table I). This list is only partial as many Arizona and other tributaries were not covered and data on the razorback sucker, another important endemic species, were not summarized.

Table I. Segments of the Colorado River System Which are Inhabited by Important Populations of Trout or Native Cyprinids

Area Name	Segment	Fishery
Hot Sulphur Springs	Upper Colorado River	Trout
Dinosaur National Monument	Yampa River	Cyprinidae
Black Canyon	Gunnison River	Trout
Flaming Gorge	Green River	Trout
Grey Rocks	Green River	Cyprinidae
Westwater	Upper Colorado River	Cyprinidae
Lee Ferry	Lower Colorado River	Trout
Grand Canyon	Lower Colorado River	Cyprinidae
Black Canyon	Lower Colorado River	Trout

It is desirable and apparently possible to maintain both trout and other introduced species as well as endemic fishes within the Colorado River system. But, they cannot co-exist in the same river segment because they have very different habitat requirements. Therefore, various river segments must be individually managed within a basin-wide plan which sets both reach water quality and quantity.

I strongly feel that enough information presently exists to set temperature and discharge criteria for priority river segments. Temperature and discharge regima have been measured over long time periods, are fairly predictable, and are primary factors controlling energetics of riverine organisms (Stanford and Ward 1979). Firmly substantiated criteria, emphasizing these variables and specifying ranges in their values required for optimization of the various fisheries must be established now.

The state-of-the-art in the ecology of regulated streams has reached the point that many processes can be empirically predicted (see Ward and Stanford 1979). When processes maintaining the fisheries are better understood (see for example Caster, this volume), river management criteria should be based on an ecosystem-level approach. Intense process-oriented research is needed to understand stream ecosystems well enough to make this possible. Once this research has built a usable foundation, each priority segment can be evaluated to refine long-term management goals in the face of changing water demands.

In segments presently influenced or potentially impacted by oil shale leachates or salt inflows, it may be necessary to monitor various chemical parameters to prevent excessive point-source loading. But, at present it appears that the salt problem is over-emphasized and is best understood and managed by critical evaluations of the sink-effect of specific ions in the mainstream reservoirs (see Paulson's comments, this volume).

However, synthesis and implementation of management plans will continue to be elusive unless basin-wide leadership develops. I suggest that a technical river management advisory commission be established to evaluate existing scientific data and apply a systems approach to derive flow and quality criteria. Most of the scientific understanding upon which meaningful aquatic resource management is based is being generated by university departments under contract with various state and federal agencies. Yet, the scientists doing the work have very little input into implementation of the management process, thus enhancing the possibility of weak or even erroneous interpretations being implemented

675

as policy. In order to help correct this situation, the proposed commission should include university scientists, representatives of state and federal agencies with management authority within the Colorado River Basin, and informed public participants.

It would be useful to separate the commission into four working groups: Upper Basin rivers, Upper Basin reservoirs, Lower Basin rivers, and Lower Basin reservoirs. The commission should derive a management plan for maintenance of specific wildlife and other aquatic and riparian resource values, with emphasis on specific, priority locations (perhaps at the sacrifice of the same values at other locations within the basin).

From a riverine systems standpoint, the important objectives are to:

1. define priority river segments for intense management (e.g., Table I);

2. establish environmental criteria (e.g., temperature and discharge regima) that will provide successful reproduction of the target species selected for the reach;

3. pursue the research necessary to formulate ecosystem-level management plans in cooperation with other working groups;

4. integrate segment aquatic resource water needs into a systems model that also considers water availability as affected by upstream dams, diversions, and return flows (the Colorado River System Simulation, Cowan, this volume, is a good start but does not incorporate enough of the scientific data base from an ecosystem perspective); and

5. communicate management options in meaningful ways to agencies and the general public.

This plan obviously needs refinement and probably requires legislative action. Failure to act, however, will undoubtedly generate continued interagency squabbling to the continued consternation of the public which wants aquatic resources protected and those researchers who know how to do it.

LITERATURE CITED

Comptroller General. 1979. Colorado River Basin water problems: How to reduce the impact. U. S. General Accounting Office, CED-79-11.

Spofford, W. O. Jr., A. L. Parker and A. V. Kneese (Eds.). 1980. Energy Development in the Southwest. Problems of Water, Fish, and Wildlife in the Upper Colorado River Basin. Two Volumes. Resources for the Future, Washington, D. C.

Ward, J. V. and J. A. Stanford (Eds.). 1979. The Ecology of Regulated Streams. Plenum Press, New York.

Jerry Miller,
        U.S. Bureau of Reclamation
        Salt Lake City, Utah

This symposium has brought together a number of scientists who are working independently on related aspects of Colorado River potamology and limnology. As we have addressed common problems, we have identified some differences of opinion but also much larger areas of agreement. All of us have been reinforced by sharing common conclusions and grown intellectually by discussing our differences. Continuing this exchange is very important to the future management of the Colorado River.

The most important concept concerning the understanding and management of river water quality is that we are dealing with a dynamic system. Conditions and responses will not be the same each year. Specific responses of aquatic populations to the hydrologic environment will be based on existing conditions and threshold levels.

One example of what needs to be done through this exchange is accomplishment of a better understanding of the role of individual ions in river management. We need to define how the individual ions comprising total dissolved solids (salinity) individually and collectively impact various water uses. Only then can water users (including aquatic resource needs) better define and prioritize their needs with clearer reference to the individual ions.

Several papers discussed ways in which reservoirs affect the routing or change the ion constituency of salinity, such as calcium carbonate precipitation. We also heard several descriptions of potential salinity reductions due to the influences of reservoir storage. We heard discussions on how various ions impact various users from the effects of calcium and hardness on municipal and industrial users to sodium impacts on agriculture and public health. Since an important river management objective is to reduce the overall impacts of salinity, it is apparent that we need to understand how the various ions contribute to the total salinity impact and then consider the results in salinity management planning.

I recall Jay Messer's introduction to his paper titled Natural Salinity Removal in Mainstem Reservoir Mechanism, Occurrence and Water Resources Impacts, when he said essentially "calcium carbonate precipitates in reservoirs, so what? Calcium carbonate (hardness) is one of the most easily treated components of salinity by lime softening. However, calcium precipitation increases the sodium absorption ratio and thereby potentially increases the agricultural impact."

Not only do we need to know what is happening to salinity in the reservoirs, but we should also understand the significance of these changes in terms of economics and public health. Biogeochemical salinity controls in the Colorado River Basin present a major research gap, which needs to be given serious attention in formulating a preventative salinity control program. Sodium, for example, is not restricted by these biogeochemical controls; and the oil shale development can potentially expose a tremendous sodium supply to hydrologic forces. A preventative program is essential.

Another topic of interest is the movement of sediments and nutrients through the Colorado River Basin. The trapping of natural phosphorus loads in the Upper Basin reservoirs encourages blue-green algae blooms and eutrophication. Furthermore, the increasing population and development of natural resources in the Upper Basin is adding significant nutrient loads to the system and increasing both cultural and natural eutrophication problems. The significance of this eutrophication is becoming more apparent as the Upper Basin reservoirs are increasingly utilized for municipal water use. It is my opinion that the blue-green algae and eutrophication problem in the Upper Colorado River reservoirs could develop into a water quality problem approaching the magnitude of the salinity problem of the Lower Basin.

Downstream, Lake Powell is such an efficient sediment and nutrient trap that the biological productivity of Lake Mead has significantly declined, adversely impacting its sports fishery. It is also most interesting that we have not yet documented any significant nutrient related blue-green algae blooms in Lake Powell.

Hydrologically, Glen Canyon Dam separates two Colorado River flow regimes. Upstream, the river and its tributaries

experience seasonal high and low flows and periods of
drought. Downstream, flow variations in the river are
largely responses to changing power, irrigation, and municipal demands.

Salinity and nutrient movements have been particularly
hard to measure in the upper basin. Salts and nutrients
accumulate in channels and floodplains until picked up by
flood hydrographs. The irregularity of these sudden loadings
makes them very difficult to measure. The isolation of many
of these streams adds to the problem. The Colorado River
inflow to Lake Powell in Cataract Canyon, for example, is
approximately 100 miles downstream from the nearest U.S.
Geological Survey gauges. Between the key reservoir in the
Colorado River and the upstream monitoring sites, lies miles
of hydrologically unstudied wilderness. In the symposium
sessions and individual discussions, I have heard some good
recommendations on how to change the hydrologic monitoring
system to better fit the management needs we have come to
recognize in the last few years.

Lake Powell has essentially doubled the water storage in
the Colorado River Basin. The initial filling of the major
Colorado River reservoirs, underway during the 1970s,
concluded with the initial filling of Lake Powell in 1980.
While Lake Mead has been in operation for over 40 years, it
has developed new chemical and biological equilibria in
the past 7 years in response to changes in the sediment,
nutrient, and temperature loadings caused by the establishment of Lake Powell.

Chemically and biologically, Lake Mead has changed from
a eutrophic system to a low productivity oligotrophic system.
The change has affected the reservoir fishery. Past studies
have identified a eutrophic condition as undesirable; however, controlled eutrophic conditions benefit the aquatic
system. I believe that blue-green vs nonblue-green algae
dominance is a more significant index of water quality than
eutrophic vs oligotrophic.

We need to combine our various skills in a common effort
aimed at improved overall management of the Colorado River
Basin. Many operating possibilities for improvement exist.
For example, we might change the withdrawal elevations in
Lake Mead from the hypolimnion to the epilimnion, seeking to
trap nutrients and increase productivity. At Flaming Gorge
Reservoir, we made this change in 1978; therefore, it is
important that a limnological survey be made at Flaming
Gorge to determine the actual changes in reservoir dynamics
that occurred. Examining an actual change in operation at

Flaming Gorge can be done to help decide if a similar change will produce desired results in Lake Mead. Properly designed and funded reservoir limnology studies provide the needed management tools for reservoirs in the future.

Flaming Gorge Reservoir has several fascinating characteristics. Releases are by a selective epilimnion withdrawal for downstream temperature reduction in the summer but by a much deeper release in the winter: the trophic status varies from hypereutrophic in the inflow area to oligotrophic near the dam. A chemocline is present in the lower portion of the reservoir near the dam. The chemocline and thermodynamics of Flaming Gorge have apparently been shifting since operation of the selective withdrawal in 1978, and we suspect that that complete turnover might occur during the winter of 1981-82.

Information gained by carefully studying Flaming Gorge Reservoir would be used to determine if an evaporation reduction can actually be realized by substituting epilimnion for hypolimnion withdrawal. This is vital information to consider before changing the withdrawal zone on Lake Mead as has been proposed in this symposium.

This example of how research at one site can be used in making management decisions elsewhere in the basin illustrates how essential it is that we as scientists are aware of each other's work. As we make the opportunities to share our experiences, we help each other make better management decisions and improve overall utilization of the Colorado River.

If we are going to model the salt movement through the Colorado River system, it is important that we understand the routing, bank storage, precipitation, and leaching now occurring in the reservoirs. An unpredicted decline of salinity occurred at Imperial Dam from 1970 to 1980. Further work is needed to determine the significance of this decline. Some feel that it is only temporary, and others wonder about ion constituent shifts and their economic impact. By working together, we can resolve these issues over the next few years.

Finally, according to the title, this is a "management symposium" for the Colorado River ecosystem. We have talked alot about specialized and technical research items, but somehow synthesizing all this into the subject of management has not yet occurred. As scientists and water resource managers, we definitely need to combine our efforts and start working for the overall best management of the Colorado River resource. Without doubt, there are conflicts between

many management objectives; however, I believe that as we
continue to work together, we will find the most practical
and overall beneficial management of the Colorado River
aquatic resources.

The focal point to Colorado River ecosystem management
is the reservoirs. Understanding the physical, chemical, and
biological balances developing in the Colorado River reservoirs is the key to improved management for multiple use of
the water resources of the total system. Since the mainstem
reservoir system has now completed initial filling and is
approaching a dynamic equilibrium, we must get on with the
business of understanding what we have created so that we can
manage it in the best interest of us all.

SCIENTIFIC PERSPECTIVES ON INTEGRATED AQUATIC
RESOURCES MANAGEMENT OF THE COLORADO RIVER

Larry J. Paulson,
 Lake Mead Limnological Research Center
 University of Nevada, Las Vegas

It is the opinion of Mann [1] that "management of the Colorado River should reflect a comprehensive assessment of alternatives and explicit recognition of tradeoffs in uses to which the river's water is put." This view is perhaps shared by most of us but, as Mann [1] further points out, "the existing political arrangements and practices – based on complex constitutional, legal and financial arrangements – make such analysis difficult." It appears that it was these "arrangements" that led to Congressional authorization of four, multi-million dollar salinity control projects, under Title II of the Colorado River Salinity Control Act of 1974, before salinity standards were even adopted on the river. Similar "arrangements" apparently led to construction of a 53 million dollar advanced wastewater treatment plant in Las Vegas, when there was no consensus among scientists that extreme phosphorus removal was required to protect water quality in Lake Mead [2] and despite mounting evidence that the operation of the plant would further reduce food resources available for the fish populations [3,4]. These "arrangements" have implemented several, costly management practices in the Colorado River Basin, before adequate studies have been conducted to evaluate alternatives, or despite scientific results that contradict the management decision.

Science has unfortunately played a minor role in de-velopment of these recent management practices. The water quality standards, which the salinity control projects and the advanced wastewater treatment plant are being used to pursue, were based on very limited technical knowledge. The Colorado River Basin states were hesitant to establish standards for this reason, and because they felt poorly grounded standards would not be equitable and enforceable [5]. Nonetheless, in order to meet requirements of the

Federal Water Pollution Control Act of 1965 (P.L. 89-234) and the Amendments of 1972 (P.L. 92-500), the Basin states agreed to a non-degradation policy after lengthy negotiations with the U.S. Environmental Protection Agency. The Basin states then established standards on the basis of flow-weighted average nutrient and total dissolved solids (TDS) concentrations for years when data were available and could be agreed upon as reflecting baseline conditions in the river. Scientific studies have still not been conducted to determine if water quality standards on the main stem are appropriate for protecting or enhancing beneficial uses.

It is equally discouraging to observe the manner in which most government research programs have been conducted in the past decade. Scientific research has become subservient to environmental impact assessment. Studies are something we must do to satisfy the requirements of the National Environmental Policy Act of 1969 or the Endangered Species Act of 1973. Study objectives, research methodology, and even the format and length of reports are governed by law. It is this replacement of science that has resulted in neglect of the extensive literature and data base available on the river. Predictions that evaporation in Lake Mead would decrease with cold-water inflows from Glen Canyon Dam, or that salt dissolution would decline as the reservoir aged have, until recently [6,7], laid idle in government files without follow-up studies. Ion constituent data, which date back to 1926 in Grand Canyon and 1935 below Hoover Dam, have been overlooked in government research programs dealing with salinity [5]. Recent reexaminations of these data have provided clues why TDS concentrations in the Lower Basin may have decreased since 1972 [7] and uncovered possible alternative methods of salinity control [7,8].

The traditional scientific approach - the one where research is funded because it promises new discovery and application by building on previous findings - has been largely abandoned in aquatic studies in our efforts to comply with the flow of environmental legislation enacted in the past two decades. We have allowed management of the Colorado River to be controlled by legislation designed to cure problems perceived by non-scientists as generally occurring throughout the nation.

Our problems, however, are not commonly shared by the rest of the country. The Colorado River is the principal water source serving 19 million people in the arid southwest. We are all concerned about protecting the environment so that the river does not become polluted, native fishes have habitat, and people have high quality recreation. We also

realize that we must have power, irrigation and municipal
water supplies and a place to dispose of wastes. The Colorado
River has many uses, each of which is aesthetically or
economically important to some, or all of us. It must be
managed accordingly, but it is the users, the ones most
knowledgeable about the problems, that should decide how to
manage the river, what legislation we require and how we
should implement the programs.

When management requires water quality standards, we
should establish standards in a scientific and logical manner
to serve the long-term needs of the users. It was this very
process that led to enactment of the Reclamation Act of 1902
and the Boulder Canyon Project Act of 1928 - legislation that
still serves the needs of users in the Colorado River Basin.
This process can work again if we can make the "tradeoffs"
that are so vital in management of a multi-purpose river.

In order to do this, however, we must first be able to
identify "tradeoffs." This requires good scientific research
and a thorough understanding of the Colorado River system.
The papers presented at this symposium demonstrate that
considerable progress is being made in this direction.
Several areas were also identified where research would be
beneficial in the future. Nonetheless, research will not
improve management of the Colorado River unless the results
can be better integrated into the decision making process.
Perhaps a regional management board of the kind proposed by
Stanford, comprised of knowledgeable users, managers and
scientists, could be established in the basin to help identify problems, assess research findings, develop management
alternatives, and advise decision makers of the tradeoffs
associated with implementation of alternative programs. This
might be a necessary first step in our efforts to develop
integrated resource management on the Colorado River.

REFERENCES

1. Mann, D. E. 1975. Politics in the United States and
    the salinity problem of the Colorado River. Pages
    113-129. In International Symposium on Salinity of the
    Colorado River. Nat. Resour. Jour. Vol. 15.

2. Goldman, C. R. 1975. A review of the limnology and
    water quality standards for Lake Mead. Ecol. Research
    Assoc. Davis, Calif. 101 pp.

3. Paulson, L. J., J. R. Baker and J. E. Deacon. 1980. The limnological status of Lake Mead and Lake Mohave under present and future power plant operations of Hoover Dam. Lake Mead Limnological Res. Ctr. Tech. Rept. No. 1. Univ. Nev. Las Vegas. 229 pp.

4. Baker, J. R. and L. J. Paulson. 1981. The effects of limited food availability on the striped bass fishery in Lake Mead. In Symposium on the Aquatic Resources Management of the Colorado River Ecosystem. Nov. 16-19, 1981. Las Vegas, NV (this volume).

5. U.S. Dept. of Interior (USDI). 1981. Quality of water. Colorado River Basin Progress Report No. 10. 190 pp.

6. Paulson, L. J. 1981. Use of hydroelectric dams to control evaporation and salinity in the Colorado River system. In Symposium on the Aquatic Resources Management of the Colorado River Ecosystem. Nov. 16-19, 1981. Las Vegas, NV (this volume).

7. Paulson, L. J. and J. R. Baker. 1981. The effects of impoundments on salinity in the Colorado River. In Symposium on the Aquatic Resources Management of the Colorado River Ecosystem. Nov. 16-19, 1981. Las Vegas, NV (this volume).

8. Messer, J. J., E. K. Israelsen and V. D. Adams. 1981. Natural salinity removal processes in reservoirs. Utah Water Res. Lab., Water Quality Series UWRL/Q-81/03. 83 pp.

CONCLUDING REMARKS

L. Douglas James
        Director, Utah Water Research Laboratory,
            Utah State University, Logan, Utah

This panel discussion, on integrating aquatic resource values with beneficial economic uses for scientific Colorado River management, began by describing the need for information on how the several uses are affected by various streamflow and reservoir content characteristics. Qualitative understanding of these effects can be used in formulating general management guidelines for interim use. Quantitative understanding provides the firmer basis required for estimating tradeoffs and negotiating choices.

Throughout the symposium and particularly in the panel discussion, I have sensed a feeling of gratification that our efforts here are moving toward meeting this need. However, we are still a long way from our goal. We have advanced toward the needed technical relationships, but we could still greatly profit from systematic classification of what we know and its exposure of critical gaps of what we need to know. We have heard descriptions of the riverine and lacustrine Colorado Basin resources, the river management system for benefical use, and the institutions responsible for that management, but experts in all three areas could still greatly profit from systematic exchange of information on studies underway and data collected. This symposium provides needed information exchange on studies underway and results achieved. It has also hopefully helped with the interagency contacts needed to begin coordinated data collection and common data banking.

Realistically, however, periodic symposia do not do the job. The suggestion made by Stanford for establishing a technical river management advisory commission combining aquatic resource scientists, water user groups, responsible agency management, and representatives of the public interest is a good one. Its success, however, will require strong leadership, a well-structured relevant agenda, and close

689

interaction with those currently setting river management policy within the Department of the Interior.

While this symposium, as was expected, did not reach explicit tradeoff definitions, we did make real progress. Specific conclusions include:

1. The responsibility assigned by the Colorado River Basin Project Act of 1968 to the Department of the Interior and largely administered by the Bureau of Reclamation makes that agency the focus of any effort to combine management for beneficial use with management for aquatic resource values on a scientific basis. The Bureau's participation in the symposium and on the panel is thus particularly encouraging.

2. The requirements the "Law of the River" puts on water management do not appear to constrain shifts in reservoir operation that would be more favorable to aquatic resources. This situation is favorable to aquatic interests in that it means that little if any change in legislation at the federal level will be needed to enhance aquatic resource protection. Needed changes can be accomplished administratively through existing agencies.

3. The programs in which changes may currently be needed are principally in Bureau reservoir operation policy and in water rights administration within the individual states. These changes should be implemented with relative minor loss to competing beneficial uses. The greatest institutional resistance is likely to be associated with prior water rights and instream flow uses at the state level. The current water allocations and methods for reallocation are largely biased toward the interests of those diverting water.

4. Needs for legislative and administrative change are less clear with respect to water quality. Salinity control standards and "zero discharge" requirements seem favorable to aquatic resources in principle, but current efforts to achieve these goals are not based on sufficient scientific understanding of the total aquatic system to give assurance that we will avoid major ecological disasters as side effects. At the state level, water quality control tends to be biased toward requirements arbitrarily set from general perceptions of aquatic needs with little analysis of economics related to waste disposal.

5. Efforts to adjust reservoir operation, state water rights, basin salinity control, and other water quality control to promote healthy aquatic environments are severely

constrained by insufficient understanding of the relevant ecological processes. We need to be able to articulate aquatic requirements scientifically and quantitatively. Consequences of the present situation include management guesses that harmfully impact aquatic resources and management overlooking aquatic needs because they are only voiced in environmental principles.

6. Research priorities and designs need to be driven by deficiencies in scientific understanding for articulating the needs of aquatic environments and not be generalistic environmental perceptions. Recent emphases on the latter have deflected research talent from real needs and produced products called research reports which are essentially useless and have contributed to reductions in research funding.

7. Presently, we are in a much stronger position for articulating aquatic resource needs with respect to flow quantities and perhaps temperature than we are with respect to water quality. These flow needs should be articulated quantitatively, with needed caveats explicitly stated, and provided to river management decision makers.

8. River management with respect to water quality has been handicapped by institutionalization of a perception of water quality entirely captured in the concept of total dissolved solids. In fact, water quality has multiple dimensions encompassing many ions, organic chemicals, colloidal particles, and biological organisms which interact with one another dynamically as flows rise and fall, hydrographs move downstream, and water is held in reservoirs. These complex notions must be sorted into indices that can be used to relate stream and reservoir characteristics to the health of aquatic communities and the benefits from economic uses. Total dissolved solids is too general an expression of quality to be used as a single parameter as a basis for scientific river management.

9. Current priority research needs for aquatic resource management applications include a) interactions among water quality parameters and resulting stream and reservoir quality characteristics by location and over time, b) effects of stream and reservoir, quantity and quality characteristics on ecological processes in the short and long runs (dynamic integration of the effects of patterns in characteristics occurring over time), c) determination of what can and cannot be achieved through reservoir operation and other implementable management options in altering stream and reservoir characteristics, recognizing interdependencies among those

characteristics. Information in these three areas will greatly contribute to formulating the initial guideline and later quantitative tradeoffs needed for scientific river management.

10. The Upper and Lower Colorado Basins have different flow and quality regime situations that require different management guidelines and tradeoff quantifications. Because of differences between stream and reservoir environments, separate relationships must be used for scientific management for a) Upper Basin reservoirs, b) Upper Basin streams, c) Lower Basin reservoirs, and d) mainstem Colorado below Glen Canyon Dam. Lower Basin tributary streams can be, at least initially, grouped with the Upper Basin streams.

11. The Upper Basin reservoirs are most threatened by blue-green algae and eutrophication; and for many smaller ones, the problem is being accentuated by tributary economic development. Management alternatives include varying withdrawal elevations, artificial mixing, and inflow treatment. Problems are being caused as organic chemicals from these reservoirs flow into municipal water supply systems. Specific guidelines and tradeoffs are needed for the managers of the individual reservoirs.

12. The Upper Basin streams (and Lower Basin tributaries) are largely managed indirectly as they experience the consequences of water allocations and reallocations among competing off-stream uses. In this situation, the guidelines and tradeoffs need to be expressed in forms that can be used by the state water rights agencies in assessing the effects on aquatic resources of the water rights changes they review. For some Upper Basin streams below reservoirs, more intensive management to provide special environments for targeted fish species through flow and temperature control may be selected and can be implemented by following guidelines that we are now able to determine. The major missing link in both water rights and targeted stream management applications is a lack of understanding of water quality impacts. These need to be defined in a way that can be used by both water rights administrators and reservoir managers.

13. The Lower Basin reservoirs experienced major water quality changes during the filling of Lake Powell. Nutrient loading and productivity declined markedly in Lake Mead. Now that Lake Powell has been initially filled, the Lower Basin reservoirs are entering a new era for which we have only limited understanding of how operating decisions will affect reservoir quality characteristics, aquatic productivity, and salinity routing. Guidelines are needed for lake management

and the regulation of waste discharges flowing into the reservoirs (locally or from energy development further upstream).

14. The flows in the lower mainstem Colorado are controlled by reservoir releases to the point where natural events have only minor effects. Releases are set by the water volumes needed to satisfy downstream diversion requirements and follow time patterns set by hydroelectric power generation and recreation needs. In this situation, the tradeoffs between the economic benefits resulting from flow releases timed to meet these uses and having whatever quality characteristics are associated with these patterns and the effects on aquatic resources need to be defined and evaluated so that the balance can be adjusted as appropriate.

15. A systems approach needs to be taken to Colorado River flow and quality management. Flow and quality need to be considered in their interactive effects on aquatic resources. The affects need to be evaluated in terms of their dynamic properties through determination of reaction rates, impact rates, and durations provided in the system for various reaction processes.

At this symposium we have made important progress toward more scientific river management. We have through this panel enriched our understanding of the research and institutional adjustments needed for scientific management to become a reality. We must all work for hopes to bear fruit.